Methods in Enzymology

Volume 236

BACTERIAL PATHOGENESIS

Part B

Interaction of Pathogenic Bacteria with Host Cells

METHODS IN ENZYMOLOGY

EDITORS-IN-CHIEF

John N. Abelson   Melvin I. Simon

DIVISION OF BIOLOGY
CALIFORNIA INSTITUTE OF TECHNOLOGY
PASADENA, CALIFORNIA

FOUNDING EDITORS

Sidney P. Colowick and Nathan O. Kaplan

Methods in Enzymology

Volume 236

Bacterial Pathogenesis

Part B

Interaction of Pathogenic Bacteria with Host Cells

EDITED BY

Virginia L. Clark

Patrik M. Bavoil

DEPARTMENT OF MICROBIOLOGY AND IMMUNOLOGY
SCHOOL OF MEDICINE AND DENTISTRY
UNIVERSITY OF ROCHESTER
ROCHESTER, NEW YORK

ACADEMIC PRESS

San Diego New York Boston London Sydney Tokyo Toronto

This book is printed on acid-free paper. ∞

Academic Press, Inc.
525 B Street, Suite 1900, San Diego, California 92101-4495

United Kingdom Edition published by
Academic Press Limited
24–28 Oval Road, London NW1 7DX

International Standard Serial Number: 0076-6879

International Standard Book Number: 0-12-182137-4

PRINTED IN THE UNITED STATES OF AMERICA
94 95 96 97 98 99 MM 9 8 7 6 5 4 3 2 1

Table of Contents

Section I. Bacterial Effects on Immune Cells

Section II. Phagocytosis and Killing Assays

Section III. Adherence

Section IV. Invasion and Intracellular Survival in Eukaryotic Cells

Section V. Identification of Genes Involved in Invasion

Contributors to Volume 236

Article numbers are in parentheses following the names of contributors.
Affiliations listed are current.

JOHN F. ALDERETE (22), *Department of Microbiology, The University of Texas Health Sciences Center, San Antonio, Texas 78284*

MICHAEL A. APICELLA (17), *Department of Microbiology, University of Iowa College of Medicine, Iowa City, Iowa 52242*

ROSSANA ARROYO (22), *Department of Microbiology, The University of Texas Health Sciences Center, San Antonio, Texas 78284*

PATRICK BERCHE (41), *Laboratoire de Microbiologie, Faculté de Médecine Necker-Enfants Malades, 75015 Paris, France*

THOMAS BORÉN (25), *Department of Molecular Microbiology, Washington University School of Medicine, St. Louis, Missouri 63110*

FRANCES BOWE (37), *Department of Microbiology and Immunology, Oregon Health Sciences University, Portland, Oregon 97201*

HELMUT BRADE (1), *Forschungsinstitut Borstel, Institut für Experimentelle Biologie und Medizin, 23845 Borstel, Germany*

CHRISTOPHER M. CLEMENS (3), *Research and Development Division, Ophidian Pharmaceuticals, Inc., Madison, Wisconsin 53711*

PASCALE COSSART (41), *Laboratoire de Génétique Moléculaire des Listeria, Institut Pasteur, 75724 Paris, France*

DAVID L. COX (27), *Treponema Immunobiology Section, National Center for Infectious Diseases, Centers for Disease Control, Atlanta, Georgia 30333*

CAROLYN D. DEAL (24), *Department of Bacterial Diseases, Walter Reed Army Institute of Research, Washington, D.C. 20307*

MELINDA S. DETRICK (2), *Department of Microbiology, University of Tennessee, Knoxville, Tennessee 37996*

KAREN DODSON (20), *Department of Molecular Microbiology, Washington University School of Medicine, St. Louis, Missouri 63110*

SHAYNOOR DRAMSI (41), *Laboratoire de Génétique Moléculaire des Listeria, Institut Pasteur, 75724 Paris, France*

PETER ELSBACH (14), *Departments of Microbiology and Medicine, New York University School of Medicine, New York, New York 10016*

ERIC A. ELSINGHORST (28), *Department of Bacterial Diseases, Walter Reed Army Institute of Research, Washington, D.C. 20307*

PER FALK (25), *Department of Molecular Biology and Pharmacology, Washington University School of Medicine, St. Louis, Missouri 63110*

STANLEY FALKOW (39), *Department of Microbiology and Immunology, Stanford University School of Medicine, Stanford, California 94305*

B. BRETT FINLAY (30, 33), *Biotechnology Laboratory, Departments of Biochemistry and Microbiology, and Canadian Bacterial Diseases Network, University of British Columbia, Vancouver, British Columbia, Canda V6T 1Z3*

HANS-DIETER FLAD (1), *Forschungsinstitut Borstel, Institut für Experimentelle Biologie und Medizin, 23845 Borstel, Germany*

JEAN-LOUIS GAILLARD (41), *Laboratoire de Microbiologie, Faculté de Médecine Necker-Enfants Malades, 75015 Paris, France*

JORGE E. GALÁN (35), *Department of Microbiology, School of Medicine, State University of New York at Stony Brook, Stony Brook, New York 11794*

TOMAS GANZ (13), *Department of Medicine, University of California, Los Angeles, Center for the Health Sciences, Los Angeles, California 90024*

JANINA GOLDHAR (16), *Department of Human Microbiology, Sackler School of Medicine, Tel Aviv University, 69978 Tel Aviv, Israel*

HEIKE U. C. GRASSMÉ (29), *Max-Planck-Institut für Biologie, Infectionsbiologie, D-72076 Tübingen, Germany*

SUNITA GULATI (11), *The Maxwell Finland Laboratory for Infectious Diseases, Boston City Hospital, Boston University School of Medicine, Boston, Massachusetts 02118*

DAVID A. HAAKE (31), *Department of Medicine, Division of Infectious Diseases, West Los Angeles Veterans Administration Medical Center, Los Angeles, California 90073*

SYLVIA S. L. HARWIG (13), *Department of Medicine, University of California, Los Angeles, Center for the Health Sciences, Los Angeles, California 90024*

FRED HEFFRON (37), *Department of Microbiology and Immunology, Oregon Health Sciences University, Portland, Oregon 97201*

DAVID G. HICKS (6), *Departments of Orthopaedics, and Pathology and Laboratory Medicine, University of Rochester School of Medicine, Rochester, New York 14642*

STANLEY C. HOLT (18), *Department of Periodontics and Microbiology, The University of Texas Health Sciences Center at San Antonio, San Antonio, Texas 78284*

SCOTT J. HULTGREN (20), *Department of Molecular Microbiology, Washington University School of Medicine, St. Louis, Missouri 63110*

RALPH R. ISBERG (21, 42), *Howard Hughes Research Institute, and Department of Molecular Biology and Microbiology, Tufts University School of Medicine, Boston, Massachusetts 02111*

FRANÇOISE JACOB-DUBUISSON (20), *Department of Molecular Microbiology, Washington University School of Medicine, St. Louis, Missouri 63110*

SIAN JONES (32, 38), *Department of Microbiology, University of Pennsylvania School of Medicine, Philadelphia, Pennsylvania 19104*

JOHN R. KALMAR (7, 9), *Department of Periodontology, Eastman Dental Center, Rochester, New York 14620*

SUSAN A. KINDER (18), *Section of Periodontics, University of California, Los Angeles, School of Dentistry, Los Angeles, California 90024*

ROBERT KREISBERG (2), *Department of Microbiology, University of Tennessee, Knoxville, Tennessee 37996*

HOWARD C. KRIVAN (24), *American University of the Carribean, School of Medicine, Plymouth, Montserrat, West Indies*

META J. KUEHN (20), *Department of Molecular Microbiology, Washington University School of Medicine, St. Louis, Missouri 63110*

CATHERINE A. LEE (39), *Department of Microbiology and Molecular Genetics, Harvard Medical School, Boston, Massachusetts 02115*

HACKEN LEFFLER (17), *Department of Psychiatry, Langley Porter Psychiatry Institute, University of California, San Francisco, San Francisco, California 94143*

MICHAEL W. LEHKER (22), *Department of Biological Sciences, University of Texas, El Paso, El Paso, Texas 79902*

ROBERT I. LEHRER (13), *Department of Medicine, University of California, Los Angeles, Center for the Health Sciences, Los Angeles, California 90024*

R. LINDSTEDT (17), *Department of Psychiatry, Langley Porter Psychiatry Institute, University of California, San Francisco, San Francisco, California 94143*

HARALD LOPPNOW (1), *Forschungsinstitut Borstel, Institut für Experimentelle Biologie und Medizin, 23845 Borstel, Germany*

MICHAEL A. LOVETT (31), *Departments of Medicine, and Microbiology and Immunology, Division of Infectious Diseases, University of California, Los Angeles, School of Medicine, Los Angeles, California 90024*

ROBERT E. MANDRELL (17), *Oak Research Institute, Oakland Childrens Hospital, Oakland, California 94609*

ZELL A. MCGEE (3), *Center for Infectious Diseases, Diagnostic Microbiology, and Immunology, University of Utah School of Medicine, Salt Lake City, Utah 84132*

DANIEL P. MCQUILLEN (11), *The Maxwell Finland Laboratory for Infectious Diseases, Boston City Hospital, Boston University School of Medicine, Boston, Massachusetts 02118*

ROBERT MÉNARD (36), *Unité de Pathogénie Microbienne Moléculaire, Institut National de la Sante et de la Recherche Medicale U 389, Institut Pasteur, 75724 Paris, France*

VIRGINIA L. MILLER (40), *Department of Microbiology and Molecular Genetics, and Molecular Biology Institute, University of California, Los Angeles, Los Angeles, California 90024*

ROBERT N. MOORE (2), *Department of Microbiology, University of Tennessee, Knoxville, Tennessee 37996*

STAFFAN NORMARK (25), *Department of Molecular Microbiology, Washington University School of Medicine, St. Louis, Missouri 63110*

REGIS J. O'KEEFE (6), *Department of Orthopaedics, University of Rochester School of Medicine, Rochester, New York 14642*

JOHN L. PACE (35), *Department of Microbiology, School of Medicine, State University of New York at Stony Brook, Stony Brook, New York 11794*

DANIEL A. PORTNOY (32, 38), *Department of Microbiology, University of Pennsylvania School of Medicine, Philadelphhia, Pennsylvania 19104*

M. GRACIELA PUCCIARELLI (30), *Biotechnology Laboratory, Departments of Biochemistry and Microbiology, and Canadian Bacterial Diseases Network, University of British Columbia, Vancouver, British Columbia, Canada V6T 1Z3*

J. EDWARD PUZAS (6), *Departments of Orthopaedics, and Pathology and Laboratory Medicine, University of Rochester School of Medicine, Rochester, New York 14642*

SUSANNAH RANKIN (42), *Department of Molecular Biology and Microbiology, Tufts University School of Medicine, Boston, Massachusetts 02111*

RICHARD F. REST (8, 10), *Department of Microbiology and Immunology, Hahnemann University School of Medicine, Philadelphia, Pennsylvania 19102*

SUSAN D. REYNOLDS (6), *Department of Orthopaedics, University of Rochester School of Medicine, Rochester, New York 14642*

PETER A. RICE (11), *The Maxwell Finland Laboratory for Infectious Diseases, Boston City Hospital, Boston University School of Medicine, Boston, Massachusetts 02118*

ERNST THEODOR RIETSCHEL (1), *Forschungsinstitut Borstel, Institut für Experimentelle Biologie und Medizin, 23845 Borstel, Germany*

ILAN ROSENSHINE (33), *Biotechnology Laboratory, Departments of Biochemistry and Microbiology, and Canadian Bacterial Diseases Network, University of British Columbia, Vancouver, Canada V6T 1Z3*

EVA ROZDZINSKI (23), *Laboratory of Molecular Infectious Diseases, The Rockefeller University, New York, New York 10021*

SHARON RUSCHKOWSKI (33), *Biotechnology Laboratory, Departments of Biochemistry and Microbiology, and Canadian Bacterial Diseases Network, University of British Columbia, Vancouver, Canada V6T 1Z3*

PHILIPPE J. SANSONETTI (36), *Unité de Pathogénie Microbienne Moléculaire, Institut National de la Sante et de la Recherche Medicale U 389, Institut Pasteur, 75724 Paris, France*

JULIUS SCHACHTER (26), *Chlamydia Research Laboratory, Department of Laboratory Medicine, University of California, San Francisco, San Francisco General Hospital, San Francisco, California 94110*

GARY K. SCHOOLNIK (19), *Departments of Medicine, Microbiology, and Immunology, Division of Infectious Diseases and Geographic Medicine, and Howard Hughes Medical Institute, Stanford University Medical School, Stanford, California 94305*

LYNN SLONIM (20), *Department of Molecular Microbiology, Washington University School of Medicine, St. Louis, Missouri 63110*

GERALD SONNENFELD (5), *Department of General Surgery Research, Carolina's Medical Center, Charlotte, North Carolina 28232*

DAVID P. SPEERT (8), *Departments of Pediatrics and Microbiology, and The Canadian Bacterial Diseases Network, Research Centre, Vancouver, British Columbia, Canada V5Z 4H4*

THOMAS H. STEINBERG (12), *Department of Medicine, Washington University School of Medicine, St. Louis, Missouri 63110*

BARBARA J. STONE (40), *Department of Microbiology and Molecular Genetics, University of California, Los Angeles, Los Angeles, California 90024*

GISELA STORZ (15), *Cell Biology and Metabolism Branch, National Institute of Child Health and Human Development, National Institutes of Health, Bethesda, Maryland 20892*

ROBERT STRIKER (20), *Department of Molecular Microbiology, Washington University School of Medicine, St. Louis, Missouri 63110*

JOEL A. SWANSON (12), *Department of Cell Biology, Harvard Medical School, Boston, Massachusetts 02115*

KEN-ICHI TANAMOTO (4), *Division of Microbiology, National Institute of Health Sciences, Tokyo 158, Japan*

LEWIS G. TILNEY (34), *Department of Biology, University of Pennsylvania, Philadelphia, Pennsylvania 19104*

MARY S. TILNEY (34), *Department of Biology, University of Pennsylvania, Philadelphia, Pennsylvania 19104*

MICHEL B. TOLEDANO (15), *Cell Biology and Metabolism Branch, National Institute of Child Health and Human Development, National Institutes of Health, Bethesda, Maryland 20892*

GUY TRAN VAN NHIEU (21, 42), *Department of Molecular Biology and Microbiology, Tufts University School of Medicine, Boston, Massachusetts 02111*

ELAINE TUOMANEN (23), *Laboratory of Molecular Infectious Diseases, The Rockefeller University, New York, New York 10021*

THOMAS E. VAN DYKE (7), *Department of Periodontology, Eastman Dental Center, Rochester, New York 14620*

JOSEPH P. M. VAN PUTTEN (29), *Max-Planck-Institut für Biologie, Infektionsbiologie, D-72076 Tübingen, Germany*

JAN F. L. WEEL (29), *Max-Planck-Institut für Biologie, Infektionsbiologie, D-72076 Tübingen, Germany*

JERROLD WEISS (14), *Departments of Microbiology and Medicine, New York University School of Medicine, New York, New York 10016*

PRISCILLA B. WYRICK (26), *Department of Microbiology and Immunology, University of North Carolina School of Medicine, Chapel Hill, North Carolina 27599*

Preface

This volume contains contributions covering the wide spectrum of interactions between bacterial pathogens and their eukaryotic hosts. As such, it complements Volume 235 of *Methods in Enzymology* in which methods for the isolation and identification of bacterial pathogens and associated virulence determinants are described. Since the study of bacterial pathogenesis borrows from a great variety of technologies and disciplines, our aim has been to provide the reader with a representative sample of the most advanced pathogenesis research. Methods and approaches included in this volume should allow those interested in studying pathogenesis to derive inspiration, if not direct application, for the design of their own experimental approaches.

The volume is divided into five sections. First, methods to evaluate the various effects of pathogenic bacteria and bacterial products on immune cells are presented. This is followed by a section in which techniques to characterize and measure the "response" of the eukaryotic cell to the bacterial invader are described. The last three sections contain chapters focusing on the analysis of the actual interaction between bacterial pathogens and susceptible host eukaryotic cells. A variety of cellular, molecular, and genetic methods to study the bacterial determinants of adherence to, entry into, survival within, and growth inside eukaryotic cells are presented. Complementary to these articles are contributions relating to the characterization of the complex array of eukaryotic molecules and signal transduction reactions which are intimately involved with the parasitic process.

We are greatly indebted to all the contributors for sharing their research expertise and making this volume possible.

VIRGINIA L. CLARK
PATRIK M. BAVOIL

METHODS IN ENZYMOLOGY

VOLUME 75. Cumulative Subject Index Volumes XXXI, XXXII, XXXIV–LX
Edited by EDWARD A. DENNIS AND MARTHA G. DENNIS

VOLUME 76. Hemoglobins
Edited by ERALDO ANTONINI, LUIGI ROSSI-BERNARDI, AND EMILIA CHIANCONE

VOLUME 77. Detoxication and Drug Metabolism
Edited by WILLIAM B. JAKOBY

VOLUME 78. Interferons (Part A)
Edited by SIDNEY PESTKA

VOLUME 79. Interferons (Part B)
Edited by SIDNEY PESTKA

VOLUME 80. Proteolytic Enzymes (Part C)
Edited by LASZLO LORAND

VOLUME 81. Biomembranes (Part H: Visual Pigments and Purple Membranes, I)
Edited by LESTER PACKER

VOLUME 82. Structural and Contractile Proteins (Part A: Extracellular Matrix)
Edited by LEON W. CUNNINGHAM AND DIXIE W. FREDERIKSEN

VOLUME 83. Complex Carbohydrates (Part D)
Edited by VICTOR GINSBURG

VOLUME 84. Immunochemical Techniques (Part D: Selected Immunoassays)
Edited by JOHN J. LANGONE AND HELEN VAN VUNAKIS

VOLUME 85. Structural and Contractile Proteins (Part B: The Contractile Apparatus and the Cytoskeleton)
Edited by DIXIE W. FREDERIKSEN AND LEON W. CUNNINGHAM

VOLUME 86. Prostaglandins and Arachidonate Metabolites
Edited by WILLIAM E. M. LANDS AND WILLIAM L. SMITH

VOLUME 87. Enzyme Kinetics and Mechanism (Part C: Intermediates, Stereochemistry, and Rate Studies)
Edited by DANIEL L. PURICH

VOLUME 88. Biomembranes (Part I: Visual Pigments and Purple Membranes, II)
Edited by LESTER PACKER

VOLUME 89. Carbohydrate Metabolism (Part D)
Edited by WILLIS A. WOOD

VOLUME 90. Carbohydrate Metabolism (Part E)
Edited by WILLIS A. WOOD

VOLUME 91. Enzyme Structure (Part I)
Edited by C. H. W. HIRS AND SERGE N. TIMASHEFF

Edited by Giovanni Di Sabato, John J. Langone, and Helen Van Vunakis

Volume 109. Hormone Action (Part I: Peptide Hormones)
Edited by Lutz Birnbaumer and Bert W. O'Malley

Volume 110. Steroids and Isoprenoids (Part A)
Edited by John H. Law and Hans C. Rilling

Volume 111. Steroids and Isoprenoids (Part B)
Edited by John H. Law and Hans C. Rilling

Volume 112. Drug and Enzyme Targeting (Part A)
Edited by Kenneth J. Widder and Ralph Green

Volume 113. Glutamate, Glutamine, Glutathione, and Related Compounds
Edited by Alton Meister

Volume 114. Diffraction Methods for Biological Macromolecules (Part A)
Edited by Harold W. Wyckoff, C. H. W. Hirs, and Serge N. Timasheff

Volume 115. Diffraction Methods for Biological Macromolecules (Part B)
Edited by Harold W. Wyckoff, C. H. W. Hirs, and Serge N. Timasheff

Volume 116. Immunochemical Techniques (Part H: Effectors and Mediators of Lymphoid Cell Functions)
Edited by Giovanni Di Sabato, John J. Langone, and Helen Van Vunakis

Volume 117. Enzyme Structure (Part J)
Edited by C. H. W. Hirs and Serge N. Timasheff

Volume 118. Plant Molecular Biology
Edited by Arthur Weissbach and Herbert Weissbach

Volume 119. Interferons (Part C)
Edited by Sidney Pestka

Volume 120. Cumulative Subject Index Volumes 81–94, 96–101

Volume 121. Immunochemical Techniques (Part I: Hybridoma Technology and Monoclonal Antibodies)
Edited by John J. Langone and Helen Van Vunakis

Volume 122. Vitamins and Coenzymes (Part G)
Edited by Frank Chytil and Donald B. McCormick

Volume 123. Vitamins and Coenzymes (Part H)
Edited by Frank Chytil and Donald B. McCormick

Volume 124. Hormone Action (Part J: Neuroendocrine Peptides)
Edited by P. Michael Conn

Volume 125. Biomembranes (Part M: Transport in Bacteria, Mitochondria, and Chloroplasts: General Approaches and Transport Systems)
Edited by Sidney Fleischer and Becca Fleischer

VOLUME 143. Sulfur and Sulfur Amino Acids
Edited by WILLIAM B. JAKOBY AND OWEN GRIFFITH

VOLUME 144. Structural and Contractile Proteins (Part D: Extracellular Matrix)
Edited by LEON W. CUNNINGHAM

VOLUME 145. Structural and Contractile Proteins (Part E: Extracellular Matrix)
Edited by LEON W. CUNNINGHAM

VOLUME 146. Peptide Growth Factors (Part A)
Edited by DAVID BARNES AND DAVID A. SIRBASKU

VOLUME 147. Peptide Growth Factors (Part B)
Edited by DAVID BARNES AND DAVID A. SIRBASKU

VOLUME 148. Plant Cell Membranes
Edited by LESTER PACKER AND ROLAND DOUCE

VOLUME 149. Drug and Enzyme Targeting (Part B)
Edited by RALPH GREEN AND KENNETH J. WIDDER

VOLUME 150. Immunochemical Techniques (Part K: *In Vitro* Models of B and T Cell Functions and Lymphoid Cell Receptors)
Edited by GIOVANNI DI SABATO

VOLUME 151. Molecular Genetics of Mammalian Cells
Edited by MICHAEL M. GOTTESMAN

VOLUME 152. Guide to Molecular Cloning Techniques
Edited by SHELBY L. BERGER AND ALAN R. KIMMEL

VOLUME 153. Recombinant DNA (Part D)
Edited by RAY WU AND LAWRENCE GROSSMAN

VOLUME 154. Recombinant DNA (Part E)
Edited by RAY WU AND LAWRENCE GROSSMAN

VOLUME 155. Recombinant DNA (Part F)
Edited by RAY WU

VOLUME 156. Biomembranes (Part P: ATP-Driven Pumps and Related Transport: The Na,K-Pump)
Edited by SIDNEY FLEISCHER AND BECCA FLEISCHER

VOLUME 157. Biomembranes (Part Q: ATP-Driven Pumps and Related Transport: Calcium, Proton, and Potassium Pumps)
Edited by SIDNEY FLEISCHER AND BECCA FLEISCHER

VOLUME 158. Metalloproteins (Part A)
Edited by JAMES F. RIORDAN AND BERT L. VALLEE

VOLUME 159. Initiation and Termination of Cyclic Nucleotide Action
Edited by JACKIE D. CORBIN AND ROGER A. JOHNSON

VOLUME 160. Biomass (Part A: Cellulose and Hemicellulose)
Edited by WILLIS A. WOOD AND SCOTT T. KELLOGG

VOLUME 161. Biomass (Part B: Lignin, Pectin, and Chitin)
Edited by WILLIS A. WOOD AND SCOTT T. KELLOGG

VOLUME 162. Immunochemical Techniques (Part L: Chemotaxis and Inflammation)
Edited by GIOVANNI DI SABATO

VOLUME 163. Immunochemical Techniques (Part M: Chemotaxis and Inflammation)
Edited by GIOVANNI DI SABATO

VOLUME 164. Ribosomes
Edited by HARRY F. NOLLER, JR., AND KIVIE MOLDAVE

VOLUME 165. Microbial Toxins: Tools for Enzymology
Edited by SIDNEY HARSHMAN

VOLUME 166. Branched-Chain Amino Acids
Edited by ROBERT HARRIS AND JOHN R. SOKATCH

VOLUME 167. Cyanobacteria
Edited by LESTER PACKER AND ALEXANDER N. GLAZER

VOLUME 168. Hormone Action (Part K: Neuroendocrine Peptides)
Edited by P. MICHAEL CONN

VOLUME 169. Platelets: Receptors, Adhesion, Secretion (Part A)
Edited by JACEK HAWIGER

VOLUME 170. Nucleosomes
Edited by PAUL M. WASSARMAN AND ROGER D. KORNBERG

VOLUME 171. Biomembranes (Part R: Transport Theory: Cells and Model Membranes)
Edited by SIDNEY FLEISCHER AND BECCA FLEISCHER

VOLUME 172. Biomembranes (Part S: Transport: Membrane Isolation and Characterization)
Edited by SIDNEY FLEISCHER AND BECCA FLEISCHER

VOLUME 173. Biomembranes [Part T: Cellular and Subcellular Transport: Eukaryotic (Nonepithelial) Cells]
Edited by SIDNEY FLEISCHER AND BECCA FLEISCHER

VOLUME 174. Biomembranes [Part U: Cellular and Subcellular Transport: Eukaryotic (Nonepithelial) Cells]
Edited by SIDNEY FLEISCHER AND BECCA FLEISCHER

VOLUME 175. Cumulative Subject Index Volumes 135–139, 141–167

VOLUME 176. Nuclear Magnetic Resonance (Part A: Spectral Techniques and Dynamics)
Edited by NORMAN J. OPPENHEIMER AND THOMAS L. JAMES

VOLUME 177. Nuclear Magnetic Resonance (Part B: Structure and Mechanism)
Edited by NORMAN J. OPPENHEIMER AND THOMAS L. JAMES

VOLUME 178. Antibodies, Antigens, and Molecular Mimicry
Edited by JOHN J. LANGONE

VOLUME 179. Complex Carbohydrates (Part F)
Edited by VICTOR GINSBURG

VOLUME 180. RNA Processing (Part A: General Methods)
Edited by JAMES E. DAHLBERG AND JOHN N. ABELSON

VOLUME 181. RNA Processing (Part B: Specific Methods)
Edited by JAMES E. DAHLBERG AND JOHN N. ABELSON

VOLUME 182. Guide to Protein Purification
Edited by MURRAY P. DEUTSCHER

VOLUME 183. Molecular Evolution: Computer Analysis of Protein and Nucleic Acid Sequences
Edited by RUSSELL F. DOOLITTLE

VOLUME 184. Avidin–Biotin Technology
Edited by MEIR WILCHEK AND EDWARD A. BAYER

VOLUME 185. Gene Expression Technology
Edited by DAVID V. GOEDDEL

VOLUME 186. Oxygen Radicals in Biological Systems (Part B: Oxygen Radicals and Antioxidants)
Edited by LESTER PACKER AND ALEXANDER N. GLAZER

VOLUME 187. Arachidonate Related Lipid Mediators
Edited by ROBERT C. MURPHY AND FRANK A. FITZPATRICK

VOLUME 188. Hydrocarbons and Methylotrophy
Edited by MARY E. LIDSTROM

VOLUME 189. Retinoids (Part A: Molecular and Metabolic Aspects)
Edited by LESTER PACKER

VOLUME 190. Retinoids (Part B: Cell Differentiation and Clinical Applications)
Edited by LESTER PACKER

VOLUME 191. Biomembranes (Part V: Cellular and Subcellular Transport: Epithelial Cells)
Edited by SIDNEY FLEISCHER AND BECCA FLEISCHER

VOLUME 192. Biomembranes (Part W: Cellular and Subcellular Transport: Epithelial Cells)
Edited by SIDNEY FLEISCHER AND BECCA FLEISCHER

VOLUME 193. Mass Spectrometry
Edited by JAMES A. MCCLOSKEY

VOLUME 194. Guide to Yeast Genetics and Molecular Biology
Edited by CHRISTINE GUTHRIE AND GERALD R. FINK

VOLUME 195. Adenylyl Cyclase, G Proteins, and Guanylyl Cyclase
Edited by ROGER A. JOHNSON AND JACKIE D. CORBIN

VOLUME 213. Carotenoids (Part A: Chemistry, Separation, Quantitation, and Antioxidation)
Edited by LESTER PACKER

VOLUME 214. Carotenoids (Part B: Metabolism, Genetics, and Biosynthesis)
Edited by LESTER PACKER

VOLUME 215. Platelets: Receptors, Adhesion, Secretion (Part B)
Edited by JACEK J. HAWIGER

VOLUME 216. Recombinant DNA (Part G)
Edited by RAY WU

VOLUME 217. Recombinant DNA (Part H)
Edited by RAY WU

VOLUME 218. Recombinant DNA (Part I)
Edited by RAY WU

VOLUME 219. Reconstitution of Intracellular Transport
Edited by JAMES E. ROTHMAN

VOLUME 220. Membrane Fusion Techniques (Part A)
Edited by NEJAT DÜZGÜNEŞ

VOLUME 221. Membrane Fusion Techniques (Part B)
Edited by NEJAT DÜZGÜNEŞ

VOLUME 222. Proteolytic Enzymes in Coagulation, Fibrinolysis, and Complement Activation (Part A: Mammalian Blood Coagulation Factors and Inhibitors)
Edited by LASZLO LORAND AND KENNETH G. MANN

VOLUME 223. Proteolytic Enzymes in Coagulation, Fibrinolysis, and Complement Activation (Part B: Complement Activation, Fibrinolysis, and Nonmammalian Blood Coagulation Factors)
Edited by LASZLO LORAND AND KENNETH G. MANN

VOLUME 224. Molecular Evolution: Producing the Biochemical Data
Edited by ELIZABETH ANNE ZIMMER, THOMAS J. WHITE, REBECCA L. CANN, AND ALLAN C. WILSON

VOLUME 225. Guide to Techniques in Mouse Development
Edited by PAUL M. WASSARMAN AND MELVIN L. DEPAMPHILIS

VOLUME 226. Metallobiochemistry (Part C: Spectroscopic and Physical Methods for Probing Metal Ion Environments in Metalloenzymes and Metalloproteins)
Edited by JAMES F. RIORDAN AND BERT L. VALLEE

VOLUME 227. Metallobiochemistry (Part D: Physical and Spectroscopic Methods for Probing Metal Ion Environments in Metalloproteins)
Edited by JAMES F. RIORDAN AND BERT L. VALLEE

VOLUME 228. Aqueous Two-Phase Systems
Edited by HARRY WALTER AND GÖTE JOHANSSON

Section I

Bacterial Effects on Immune Cells

[1] Induction of Cytokines in Mononuclear and Vascular Cells by Endotoxin and Other Bacterial Products

By HARALD LOPPNOW, HELMUT BRADE, ERNST THEODOR RIETSCHEL, and HANS-DIETER FLAD

Introduction

Bacteria or bacterial products activate host responses during infectious and inflammatory processes. These substances cause some of their various effects in the host by induction of regulatory mediators, such as prostaglandins or cytokines. Evaluation of cytokine production following stimulation with bacteria or bacterial structures *in vitro* might be helpful in determining the capacity of these compounds to induce host responses. In this article we describe the quantification of the capacity of bacterial stimuli to induce cytokines, such as interleukin 1 (IL-1) and interleukin 6 (IL-6), by *in vitro* assay systems. We focus on the lipopolysaccharides (LPS) of gram-negative bacteria[1,2] and LPS partial structures.[3]

Assay Methods to Determine Cytokine Production in Human Cell Cultures *in Vitro*

Cells of the monocyte/macrophage lineage and vascular cells are among the main target cells of bacteria or lipopolysaccharides during bacterial infections. The isolation, culture, stimulation, and assessment of cytokine production of these cells in response to bacterial stimuli are discussed below.

Isolation and Stimulation of Human Mononuclear Cells

For the isolation of mononuclear cells (MNCs), blood is taken from healthy donors.[3] The heparinized (20 IE/ml) blood is mixed with the same volume of Hanks' balanced salt solution and 18 ml of this solution is layered on 6 ml of Ficoll–Hypaque[4] or Percoll. The interphase layer of mononuclear cells is isolated after 40 min of centrifugation (21°C, 400 *g*), including 5 min of acceleration. The mononuclear cells are centrifuged

[1] E. T. Rietschel and H. Brade, *Sci. Am.* **267,** 54 (1992).
[2] C. R. H. Raetz, *Annu. Rev. Biochem.* **59,** 129 (1990).
[3] H. Loppnow, H. Brade, I. Dürrbaum, C. A. Dinarello, S. Kusumoto, E. T. Rietschel, and H.-D. Flad, *J. Immunol.* **142,** 3229 (1989).
[4] A. Böyum, *Scand. J. Clin. Lab. Invest.* **21,** 77 (1968).

(10 min, 21°C, 300 *g*) three times in serum-free RPMI 1640 containing 2 m*M* L-glutamine, 100 U/ml penicillin, and 100 μg/ml streptomycin (Biochrom KG, Berlin, Germany). The cells are resuspended in serum-free medium, and the cell number and viability are calculated following trypan and Türk's stain.

After the final centrifugation the mononuclear cells are resuspended at 10×10^6/ml, and 50 μl of this suspension is added to each well of a 96-well culture plate (Greiner und Söhne, Nürtingen, Germany). Serial 2- to 10-fold dilutions of the stimuli are prepared in polypropylene tubes (1.5 ml), and 50 μl thereof is immediately added to the mononuclear cells. The cultures are incubated for 24 hr at 37°C and 5% CO_2. Subsequently the supernatants are carefully removed without disturbing the cell pellet and stored in the presence of carrier protein (2% fetal calf serum) at $-20°$. Finally the cytokine content of these samples is determined in the respective assay.

Isolation and Stimulation of Human Endothelial and Smooth Muscle Cells

Human vascular endothelial cells (ECs) or smooth muscle cells (SMCs) are isolated from unused portions of saphenous veins obtained following bypass surgery. These usually discarded specimens are transported in heparinized blood at 4°C. The endothelial cells are isolated by treatment with collagenase[5] and cultured in medium 199 with 5% fetal calf serum, endothelial cell growth factor (ECGF; 50 μg/ml), heparin (25 μg/ml), antibiotics, and L-glutamine. Subsequently, smooth muscle cells are isolated from medial explants using the outgrowth technique established by Ross and Kariya.[6] These cells are cultured in Dulbecco's modified Eagle's medium (DMEM) containing 10% fetal calf serum, antibiotics, and L-glutamine.

The vascular cells are seeded into 96-well culture dishes at 10,000 cells/cm^2 and incubated for 3 days at 37°C, 7.5% CO_2. The medium of these confluent cultures is replaced by fresh medium. Endothelial cells are stimulated in medium 199 containing 5% fetal calf serum, antibiotics, and L-glutamine. Smooth muscle cells are stimulated in DMEM containing 10% fetal calf serum, antibiotics, and L-glutamine. The cultures are incubated for 24 hr; the supernatants are harvested and assayed as described above.

[5] E. A. Jaffe, R. L. Nachmann, C. G. Becker, and C. R. Minick, *J. Clin. Invest.* **52,** 2745 (1973).

[6] R. Ross and B. Kariya, *in* "Handbook of Physiology" (D. F. Bohr, A. P. Somlyo, and H. Y. Sparks, Jr., eds.), Sect. 2, Vol. II, p. 66. Am. Physiol. Soc., Bethesda, MD, 1980.

Competition Assay

In some experiments the antagonistic activity of inactive LPS or partial structures thereof is analyzed by coincubation with active LPS. In these competition assays, the antagonists are prepared (4-fold concentrated) and applied to the cells in one-fourth of the final volume. The agonists are prepared accordingly and added to the cultures 10 to 30 min later. The cultures are then assayed as described above.

Detection of Cytokines in Bioassays

The cytokines are determined in biological assays or enzyme-linked immunosorbent assays (ELISA). The biological assays are described in more detail elsewhere. IL-1 is detected with murine thymocytes,[3] human fibroblasts,[7] or the murine T-cell line D10S[8] with comparable results. Interleukin 6 is measured using murine B9[9] or 7TD1 cells.[10]

Interleukin 1β ELISA

IL-1 is also detected in an enzyme-linked immunosorbent assay. ELISA kits were kindly donated by Dr. H. Galati (Hoffmann–LaRoche, Basel, Switzerland). ELISA is performed as suggested by the manufacturer. Briefly, microtiter plates are coated with goat anti-human IL-1α or IL-1β antibody (24 hr) and washed. Samples or controls and the peroxidase-linked goat anti-human IL-1β are added simultaneously. After 24 hr the plates are washed again. Substrate (tetramethylbenzidine–hydrogen peroxide) is added, followed by incubation for 10 min, and absorption is measured at 450 nm. The IL-1 content of the samples is determined in reference to an internal standard.

Determination of Minimal Concentration Necessary for Cytokine Induction

To determine the capacity of a given substance to induce cytokines the minimal concentration still active for cytokine induction is estimated. For this purpose serial dilutions of the compounds are added to the MNC

[7] H. Loppnow, H.-D. Flad, I. Dürrbaum, J. Musehold, R. Fetting, A. J. Ulmer, H. Herzbeck, and E. Brandt, *Immunobiology* **179,** 283 (1989).

[8] S. F. Orencole and C. A. Dinarello, *Cytokine* **1,** 14 (1989).

[9] L. A. Aarden, E. R. De Groot, O. L. Schaap, and P. M. Lansdorp, *Eur. J. Immunol.* **17,** 1411 (1987).

[10] J. van Snick, S. Cayphas, A. Vink, C. Uyttenhove, P. G. Coulie, M. R. Rubira, and R. J. Simpson, *Proc. Natl. Acad. Sci. U.S.A.* **83,** 9679 (1986).

and the cytokine activity in the supernatants is determined. The data are then plotted, and the last concentration of stimulus still inducing cytokine production (minimal concentration) is defined. An example showing the determination of the minimal concentrations of two compounds (LPS-A and LPS-B) is given in Fig. 1.

Lipopolysaccharide A induces maximal production of IL-6 at concentrations of 200 ng LPS/ml to 64 pg LPS/ml. Further dilution of LPS-A results in a dose-dependent decrease in cytokine production. The minimal concentration of this LPS still active in cytokine production in this particular experiment is 0.5 pg LPS/ml. The second test compound (LPS-B; Lipid A) stimulated maximal cytokine production at a range of 200 ng LPS/ml to 1600 pg LPS/ml, the minimal concentration being 320 pg LPS/ml.

Comparable results can be obtained in other test systems, i.e., IL-1 production detected in murine thymocytes or human fibroblasts (data not shown). As an example the results obtained by IL-1 measurement in an IL-1β ELISA of the same samples used in Fig. 1 are shown in Fig. 2. The minimal concentrations necessary for cytokine induction are 320

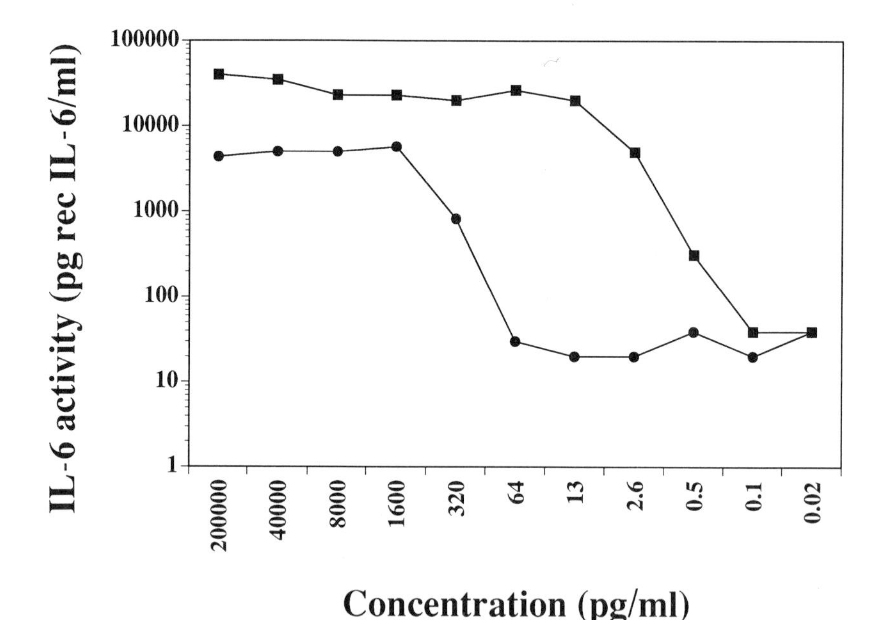

FIG. 1. Determination of the minimal concentration necessary for cytokine induction using biological cytokine assay. Mononuclear cells were stimulated with lipid A (●; LPS-B) or a 640-fold more potent lipopolysaccharide (■; LPS-A) as described in the text. IL-6 activity is measured in 7TD1 assay.

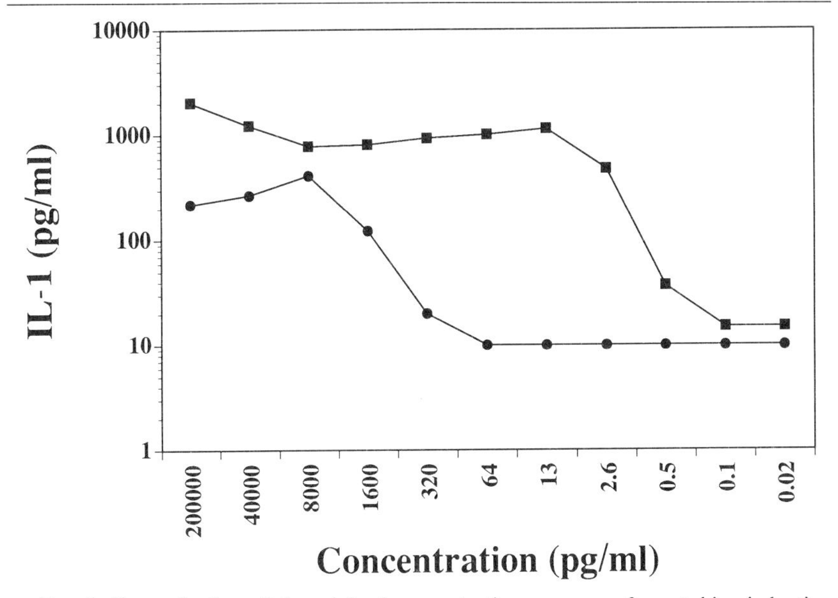

FIG. 2. Determination of the minimal concentration necessary for cytokine induction using ELISA. Mononuclear cells were stimulated with lipid A (●) or a 640-fold more potent lipopolysaccharide (■) as described in the text. IL-1β activity in the same supernatants as analyzed in Fig. 1 was measured in IL-1β ELISA.

(LPS-B) and 0.5 pg/ml (LPS-A). Different experiments usually provided the same results, irrespective of whether IL-1 or IL-6 is measured and the assay that is used.

These experiments provide information regarding the capacity of test compounds for cytokine induction. The minimal concentrations of a variety of different materials are summarized[3,11–13] in Table I. The LPSs of gram-negative bacteria are among the most potent stimuli tested in these assay systems. Heat-killed gram-positive or gram-negative bacteria at concentrations of 10^4 cells/ml or greater can induce higher levels of cytokine activity in the culture supernatants; however, no comparison in terms of minimal concentrations is possible because of the particulate character of these stimuli. Furthermore, an estimation of the amount of active

[11] H. Loppnow, L. Brade, H. Brade, E. T. Rietschel, S. Kusumoto, T. Shiba, and H.-D. Flad, *Eur. J. Immunol.* **16,** 1263 (1986).

[12] N. L. Kovach, E. Yee, R. S. Munford, C. R. H. Raetz, and J. M. Harlan, *J. Exp. Med.* **172,** 77 (1990).

[13] H. Loppnow, P. Libby, M. A. Freudenberg, J. H. Krauss, J. Weckesser, and H. Mayer, *Infect. Immun.* **58,** 3743 (1990).

TABLE I
MINIMAL CONCENTRATION FOR CYTOKINE INDUCTION OF BACTERIAL STRUCTURES

Compound	Minimal concentration (pg/ml) necessary to induce cytokine production
S-form LPS	1–100
R-form LPS	1–100
Free lipid A	10–10,000
Oligoacyl lipid A	100–100,000
Deacylated LPS	4000
Monophosphoryl lipid A	500,000
Muramyl dipeptide	≥1,000,000
Lipoteichoic acid	≥1,000,000
Core oligosaccharides	≥1,000,000
Lipid A precursor Ia	≥10,000,000
Lipid X	≥10,000,000

compound (LPS) in terms of pg LPS/ml is possible only in the gram-negative bacteria.

As pointed out in previous publications,[3,11–13] the *S*- and *R*-form LPSs induce cytokines if applied at doses of 1 to 100 pg/ml, depending on the donor of mononuclear cells and reflecting possible differences in the state of preactivation or responsiveness of the different mononuclear cell isolates. LPS partial structures, such as lipid A, are somewhat less active, i.e., they stimulate cytokine production, if 10 to 10,000 pg/ml is applied. In the case of other partial structures, such as pentaacylated lipid A or core oligosaccharides, microgram amounts are needed to induce cytokine production. Some compounds, like the lipid A precursor Ia, also termed lipid IVa, compound 406, or LA-14-PP, do not induce cytokine production at all[3] under the conditions used.

Vascular cells, such as saphenous vein endothelial or smooth muscle cells, also produce IL-6 when stimulated with LPS. In contrast to mononuclear cells, these cells release more IL-6 following IL-1 stimulation.[14,15] Moreover, these cells are not as responsive to LPS as mononuclear cells. Here, 10- to 100-fold higher concentrations of LPS are required for the induction of IL-6[16] (data not shown).

[14] H. Loppnow and P. Libby, *J. Clin. Invest.* **85,** 731 (1990).

[15] H. Loppnow and P. Libby, *Cell. Immunol.* **122,** 493 (1989).

[16] H. Loppnow, H.-D. Flad, E. T. Rietschel, and H. Brade, *in* "Pathophysiology of Shock, Sepsis and Organ Failure" (G. Schlag and H. Redl, eds.), p. 403. Springer-Verlag, Berlin/New York, 1993.

Determination of Antagonistic Properties of Lipopolysaccharides or Lipopolysaccharide Partial Structures

In the above-mentioned cytokine induction system the inactive lipopolysaccharides or partial structures thereof can inhibit the capacity of active LPSs to induce cytokine production. This phenomenon is dose dependent and specific for LPS. The activation of mononuclear cells, endothelial cells, as well as smooth muscle cells by endotoxin is reduced by substances such as synthetic lipid A precursor Ia, *Rhodobacter capsulatus* LPS, or *Rhodopseudomonas sphaeroides* LPS. We and others[17–19] also showed that LPS-induced adhesion of leukocytes to ECs is reduced by inactive LPS structures. To inhibit the activity of bioactive compounds, the antagonists have to be added to the test cells prior to or simultaneously with the agonists.

The antagonistic capacity of the inactive lipopolysaccharides or LPS partial structures are specific, as cytokine production induced by protein mitogens, cytokines, gram-positive bacteria, or phorbol myristate acetate is not affected by antagonists.[18,20] The competition experiments indicate that the induction of cytokines by LPS is mediated by specific mechanisms.

Discussion

Lipopolysaccharides of gram-negative bacteria induce cytokine production or other host responses during infection and inflammation. As structure–function investigations have shown[21] it is the lipid A moiety that constitutes the structure of the LPS molecule responsible for the induction of these responses. The induction by LPS is probably mediated through specific receptors[22] or other specific mechanisms not determined to date. LPS is a very potent stimulus of mononuclear cells. It induces

[17] T. H. Pohlmann, R. S. Munford, and J. M. Harlan, *J. Exp. Med.* **165,** 1393 (1987).

[18] H. Loppnow, E. T. Rietschel, H. Brade, U. Schönbeck, P. Libby, M.-H. Wang, H. Heine, W. Feist, I. Dürrbaum-Landmann, M. Ernst, E. Brandt, E. Grage-Griebenow, A. J. Ulmer, S. Campos-Portugez, U. Schade, T. Kirikae, S. Kusumoto, J. Krauss, H. Mayer, and H. D. Flad, *in* "Bacterial Endotoxin: Recognition and Receptor Mechanisms" (J. Levin, C. R. Alving, R. S. Munford, and P. L. Stütz, eds.) Vol. 2, p. 337. Elsevier, Excerpta Medica, Amsterdam/New York/Oxford, 1993.

[19] U. Schönbeck, H.-D. Flad, E. T. Rietschel, E. Brandt, and H. Loppnow, *J. End. Res.* **1,** 4 (1994).

[20] D. T. Golenbock, R. Y. Hampton, N. Qureshi, K. Takayama, and C. R. H. Raetz, *J. Biol. Chem.* **266,** 19490 (1991).

[21] E. T. Rietschel, L. Brade, K. Brandenburg, H.-D. Flad, J. De Jong-Leuveninck, K. Kawahara, B. Lindner, H. Loppnow, T. Lüderitz, U. Schade, U. Seydel, Z. Sidorczyk, A. Tacken, U. Zähringer, and H. Brade, *Rev. Infect. Dis.* **9,** 527 (1987).

[22] S. D. Wright, *Curr. Opin. Cell Biol.* **3,** 83 (1991).

cytokines if picogram amounts are applied, whereas other bacterial stimuli appear to be less potent inducers of cytokines.[23] These results are obtained *in vitro*,[3] but have been reproduced by others in human whole blood *ex vivo*.[12] The cellular response to LPS appears to be specific, as inactive LPS[13] or synthetic lipid A partial structures[3] inhibit the stimulation of cells by active LPS. The capacity of the various inactive compounds to reduce activities appears to be different in various test systems. Although deacylated LPSs[24] reduced adhesion of polymorphonuclear cells to umbilical vein endothelial cells,[17] these lipopolysaccharides did not reduce production of tumor necrosis factor α- induced by LPS.[12]

The potencies of various bacterial compounds to induce cytokine production can be determined if serial dilutions of the test compounds are assayed in the same experiment. Determination of cytokine activity in single dilutions of stimuli, using high concentrations, might show identical cytokine activity, although the potencies of the tested compounds necessary to induce cytokine production are a thousand- or a millionfold different. By way of cytokine measurements in multiple dilutions, the capacity of a compound can be determined by calculation of the ED_{50} or the minimal concentration. Both methods yield essentially the same results and information.

A limitation of these systems is the dependency of cytokine production on the donor of cells. In numerous experiments, 100-fold different minimal concentrations of a given substance are measured. However, the differences in minimal concentrations of various compounds are reproducible in separate experiments. Thus, compounds to be compared have to be tested in the same assay and a number of assays have to be performed to determine the minimal concentration of a substance necessary to induce cytokines.

Acknowledgments

This work was partially supported by Grants Lo 385/1-1, Lo 385/4-1 (H. Loppnow), and SFB 367 (B2; E. T. Rietschel) from the Deutsche Forschungsgemeinschaft and by the Fonds der Chemischen Industrie (H.-D. Flad, and E. T. Rietschel).

[23] P. Hoffmann, S. Heinle, U. Schade, H. Loppnow, A. J. Ulmer, H.-D. Flad, G. Jung, and W. Bessler, *Immunobiology* **177,** 158 (1988).

[24] R. S. Munford and C. L. Hall, *Science* **234,** 203 (1986).

[2] Effect of Bacterial Products on Colony-Stimulating Factor Production

By MELINDA S. DETRICK, ROBERT KREISBERG, and ROBERT N. MOORE

Introduction

Bacterial products such as lipopolysaccharide (LPS) and lipoarabinomannan (mycobacterial cell wall-associated glycolipid, LAM) stimulate the production of cytokines that can modulate the host response to infection. Colony-stimulating factors (CSFs) are a class of such cytokines.[1–3] Not only are the CSFs produced in direct response to bacterial products or in a cytokine cascade triggered by bacterial products, but also the CSFs are lymphokines that stimulate the production and differentiation of myeloid cells from hematopoietic precursors.[3–5] To date, four major CSFs are known, granulocyte–macrophage CSF (GM-CSF), granulocyte CSF (G-CSF), macrophage CSF (M-CSF), and interleukin 3 (IL-3). G-CSF induces the production and differentiation of granulocytes and M-CSF functions to stimulate macrophage production and maturation, whereas GM-CSF produces both granulocytes and macrophages. IL-3 is responsible for the production of several cell types: granulocytes, macrophages, eosinophils, megakaryocytes, and mast cells. Interleukin 5 (IL-5) has been implicated as a fifth CSF specific for eosinophils.[6]

In addition to the role CSFs play in producing mature cells from hematopoietic precursors, they also have the dual function of activating and prolonging the survival of mature cells.[3,4] For instance, GM-CSF enhances inhibition of neutrophil migration and primes polymorphonuclear cells for increased superoxide anion production.[3,7] G-CSF not only induces granulopoiesis, but also can activate mature neutrophils by enhancing

[1] R. N. Apte and D. H. Pluznik, *J. Cell. Physiol.* **89,** 313 (1976).

[2] D. C. Dale, S. Lau, R. Nash, T. Boone, and W. Osborne, *J. Infect. Dis.* **165,** 689 (1992).

[3] R. N. Moore, *in* "Advances in Inflammation Research: New Perspectives in Anti-inflammatory Therapies" (A. Lewis, N. Ackerman, and I. Otterness, eds.), p. 33. Raven Press, New York, 1988.

[4] R. N. Moore, J. T. Hoffeld, J. J. Farrar, S. E. Mergenhagen, J. J. Oppenheim, and R. K. Shadduck, *in* "Lymphokines" (E. Pick, ed.), Vol. 3, p. 119. Academic Press, New York, 1981.

[5] E. R. Stanley, *in* "The Lymphokines" (J. W. Hadden and W. E. Stewart II, eds.), p. 101. Humana Press, Clifton, NJ, 1981.

[6] H. E. Broxmeyer, *Am. J. Pediatr. Hematol. Oncol.* **14,** 22 (1992).

[7] J. Fleischmann, D. W. Golde, R. H. Weisbart, and J. C. Gasson, *Blood* **68,** 708 (1986).

antibody-dependent cellular cytotoxicity (ADCC), phagocytosis, and survival and by potentiating the production of toxic oxygen metabolites.[3,8] Macrophages respond to M-CSF by producing prostaglandin E_2 (PGE_2), IL-1, α/β-interferon (α/β-IFN), and plasminogen activator.[3] In addition, the CSFs are chemoattractants for their specific leukocyte target cells.[3]

Because of their ability to augment the production of mature cells from hematopoietic precursors and to modulate the secretory capacity, chemotaxis, and killing mechanisms of phagocytes, the CSFs function as important regulators of inflammatory and immune system responses. Evidence in support of this concept has accumulated over the years. For example, GM-CSF enhanced the phagocytosis of opsonized *Staphylococcus aureus,* the anticandidal activity of human monocytes, and the uptake and killing of the intracellular parasite *Leishmania tropica* by macrophages.[7,9,10] In addition, the production of GM-CSF by murine spleen cells after intravenous injection of heat-killed *Corynebacterium liquefaciens* and by human monocytes and large granular lymphocytes after stimulation with *Mycobacterium avium–M. intracellulare* has been documented.[11,12] G-CSF is released by alveolar macrophages during bacterial pneumonia, elevated in the serum of dogs after intravenous injection of endotoxin, and elevated in mice infected with *Listeria monocytogenes*.[2,8,13,14] Increases in the level of M-CSF in the serum of mice infected with *L. monocytogenes* have also been reported.[13–15] M-CSF has been implicated in the enhancement of murine peritoneal macrophage killing of *Candida albicans,* as well as the enhancement of human monocyte-mediated anticandidal activity.[10,16] IL-3 also enhances similar anticandidal activity in human monocytes.[10] Salmonellae increase general colony-stimulating activity (CSA) in the serum of infected animals.[17–20] Additionally, increased

[8] A. Tazi, S. Nioche, J. Chastre, J.-M. Smiéjan, and A. J. Hance, *Am. J. Respir. Cell Mol. Biol.* **4,** 140 (1991).

[9] E. Handman and A. W. Burgess, *J. Immunol.* **122,** 1134 (1979).

[10] M. Wang, H. Friedman, and J. Y. Djeu, *J. Immunol.* **143,** 671 (1989).

[11] D. K. Blanchard, M. B. Michelini-Norris, C. A. Pearson, S. McMillen, and J. Y. Djeu, *Infect. Immun.* **59,** 2396 (1991).

[12] T. Yoshida, T. Hotta, K. Shimokata, M. Ichihara, K.-I. Isobe, and I. Nakashima, *Infect. Immun.* **59,** 1032 (1991).

[13] P. Egan and C. Cheers, *Immunology* **70,** 191 (1990).

[14] C. Cheers, A. M. Haigh, A. Kelso, D. Metcalf, E. R. Stanley, and A. M. Young, *Infect. Immun.* **56,** 247 (1988).

[15] C. Cheers and E. R. Stanley, *Infect. Immun.* **56,** 2972 (1988).

[16] A. Karbassi, J. M. Becker, J. S. Foster, and R. N. Moore, *J. Immunol.* **139,** 417 (1987).

[17] P. A. Chervenick, *J. Lab. Clin. Med.* **79,** 1014 (1972).

[18] T. J. MacVittie and R. I. Walker, *Exp. Hematol.* **6,** 613 (1978).

[19] D. Metcalf, *Immunology* **21,** 427 (1971).

[20] A. Trudgett, T. A. McNeill, and M. Killen, *Infect. Immun.* **8,** 450 (1973).

CSA was found in mice with pneumonia caused by *Chlamydia trachomatis*.[21]

The importance of CSFs during infection by bacteria as producers of mature, infection-fighting cells and as activators of those mature cells is an active area of research. The rest of this article is therefore dedicated to the description of assays for the detection and identification of CSA in the serum of infected animals, in tissue homogenates, and in tissue culture supernatants.

Soft Agar Colony Formation Assay

The soft agar colony formation assay is based on the premise that bone marrow cells suspended in soft agar undergo clonal proliferation in the presence of CSF. Clonal proliferation results in the formation of discrete proliferative centers visible as colonies on magnification. This assay was developed approximately 27 years ago and has been used to study hematopoietic progenitor cells and the CSFs.[22,23] The assay allows for quantification of the CSF present and with minor adaptations can determine the identity of the CSF(s) present in the test samples.

Preparation of Soft Agar Medium

1. Complete RPMI 1640 (RPMI 1640 supplemented with 15% fetal bovine serum, 100 U/ml penicillin, 100 μg/ml streptomycin sulfate, and 2 m*M* L-glutamine) is prewarmed to 42° in a water bath.
2. Bacteriological-grade agar is prepared at 4% (w/v) in sterile water by boiling for 3 min in a water bath. The molten agar is diluted to 0.4% (v/v) in the prewarmed (42°) complete RPMI.
3. The soft agar medium is tempered to 42° in a water bath for the duration of experimental preparations (at least 30 min before addition of cells).

Preparation of Assay Plates

1. Serial dilutions of filter-sterilized test samples (e.g., serum, tissue culture supernatants) in complete RPMI are added to sterile 35 × 10-mm plastic tissue culture dishes. The volume added to each dish should not exceed 0.2 ml.
2. Prepared plates are stored at 4° until addition of soft agar medium.

[21] D. M. Magee, D. M. Williams, E. J. Wing, C. A. Bleicker, and J. Schachter, *Infect. Immun.* **59,** 2370 (1991).
[22] T. R. Bradley and D. Metcalf, *Aust. J. Exp. Biol. Med. Sci.* **44,** 287 (1966).
[23] D. H. Pluznik and L. Sachs, *J. Cell. Comp. Physiol.* **66,** 319 (1965).

Preparation of Bone Marrow Cells

1. Femurs are removed aseptically from 4- to 7-week-old mice (euthanized by cervical dislocation). Approximately 1 to 2 $\times$ 10^7 marrow cells can be recovered per two femurs.

2. Bone marrow cells are flushed from the femurs with sterile Ca^{2+},Mg^{2+}-free Hanks' balanced salt solution (HBSS) by aspiration using a syringe fitted with a 23-gauge needle or by grinding the bones in a sterile mortar. Approximately 10- to 12 ml HBSS should be sufficient to recover marrow cells.

3. The cells are collected in a sterile 15-ml tissue culture tube and vortexed briefly to ensure complete cell dispersal.

4. The cell solution is set aside for 5 min. During this period, bone fragments settle from the marrow cell suspension.

5. The marrow cells can be carefully removed from the bone fragments with a pipet.

6. The marrow cells are counted and added to the tempered soft agar medium at a final concentration of 5 $\times$ 10^4 nucleated cells/ml of medium.

7. One milliliter of soft agar/cell suspension is added to each of the assay plates. Each plate is gently swirled immediately after the addition of medium.

8. The plates are left at room temperature for 15 min or until agar mixture has gelled.

9. The assay is incubated for 5 to 7 days at 37° in a humidified, 7% CO_2 incubator.

10. Colony formation is scored by observing the plates under 35$\times$ magnification with an inverted microscope. Colonies, i.e., discrete proliferative centers consisting of more than 25 cells, are counted.

11. If plates cannot be scored immediately, they may be fixed with 3 to 4 drops of sterile 1% (v/v) acetic acid in HBSS and stored at 4° for 7 days.

Comments

1. The colony formation assay as designed measures only the total CSA in a test sample; however, CSF-specific, neutralizing antibodies (available commercially) may be added to the assay as appropriate to determine CSF identity. Complete or partial abrogation of colony response by a CSF-specific, neutralizing antibody is indicative of CSF identity.

2. This assay is used with consistent results in our laboratory. Scoring plates consistently for colony formation, however, takes practice and may be an area of concern for the inexperienced technician.

3. Dulbecco's modified Eagle's medium (DMEM), α minimum essential medium (αMEM) or other media may be substituted for RPMI 1640 in the procedure described above.

4. The amount of CSA in a sample can be determined from the linear portion of a dose–response curve using the following formula: number of colonies × 1/dilution × 1/volume of sample added to plate = units of CSA/milliliter of sample.

Cautions

1. All equipment and solutions used in preparing the soft agar assay must be sterile.

2. The soft agar medium must be maintained at 42° before and after the addition of molten agar. Temperatures below 42° result in premature gelling of the agar mixture. Temperatures greater than 42° "fry" the marrow cells.

3. Fetal bovine serum (FBS) should be selected carefully. Most FBS contains endotoxin. FBS with endotoxin levels greater than 10 ng/ml can suppress colony formation. FBS with endotoxin levels less than 0.01 ng/ml is not optimally supportive of colony formation. If necessary, LPS at a final concentration of 0.1 ng/ml can be added to each assay plate as a supplement.[24]

4. Water and other medium supplements should also be monitored for endotoxin contamination.

5. Indomethacin at a final concentration of 0.1 μg/ml may be added to reduce endogenous production and potential interference by PGE_2.[25]

Uptake of [^{3}H]Thymidine by Bone Marrow Cells

Bone marrow cells in the presence of CSA are stimulated to proliferate. In the presence of [^{3}H]thymidine, the actively growing cells incorporate the labeled thymidine into newly synthesized DNA. Cells cultured in the presence of increasing amounts of CSA proliferate proportionately and incorporate increasing amounts of [^{3}H]thymidine. Measurement of the amount of [^{3}H]thymidine incorporated into the cells during culture can be used as an indicator of CSA in test samples. The method described is based on that of Horak *et al.*[26]

[24] R. N. Moore, J. G. Joshi, D. G. Deana, F. J. Pitruzello, D. W. Horohov, and B. T. Rouse, *J. Immunol.* **136,** 1605 (1986).

[25] R. N. Moore, F. J. Pitruzello, H. S. Larsen, and B. T. Rouse, *J. Immunol.* **133,** 1 (1984).

[26] H. Horak, A. R. Turner, A. R. E. Shaw, and O.-W. Yau, *J. Immunol. Methods* **56,** 253 (1983).

Preparation of Bone Marrow Cells

1. Bone marrow cells are recovered from murine femurs as described in the Soft Agar Colony Formation Assay (this section).
2. After the cells are counted, the cell solution is gently centrifuged at 100 *g* for 10 min at room temperature.
3. Bone marrow cells are resuspended in complete RPMI 1640 (described in the Soft Agar Colony Formation Assay, this section) at a final concentration equal to 7.5×10^5 nucleated cells/ml.

Preparation of Bone Marrow Cell Culture

1. Serial dilutions of filter-sterilized test samples (serum, tissue culture supernatants) are made in 0.1 ml complete RPMI 1640 in sterile 96-well flat-bottomed microtiter plates. Each dilution should be made in triplicate or quadruplicate. Control wells with 0.1-ml complete RPMI 1640 should also be included to assess background radioactivity.
2. One-tenth-milliliter of the cell suspension is added to each well. The final volume of each well is adjusted (if necessary) to 0.2 ml.
3. Plates are incubated for 4 to 5 days in a 37°, 7% CO_2 humidified incubator.
4. For the final 6 hr of incubation, the cells are pulsed with 1 μCi [^{3}H]thymidine/well.
5. The cells are harvested on glass fiber filters with an automatic cell harvester (following manufacturer's instructions) and dried overnight.
6. Radioactivity of the dried filters is assessed by liquid scintillation counting.
7. The relative activity (Δcpm) of each sample in stimulating proliferation of bone marrow cells can be determined from the average counts per minute of a dilution set minus the average counts per minute of the control wells.

Comments

1. DMEM, Iscove's modified Dulbecco's medium (IMDM), αMEM, and McCoy's 5A media are assay-compatible alternatives to RPMI 1640.
2. The values for cell number and [^{3}H]thymidine concentration given in the above procedure are those used successfully in our laboratory. This assay in other hands may require optimization of these parameters before adequate results can be attained. As a guide, cell numbers ranging from 4×10^4/well to 1×10^5/well[2,13,14,26] are used and the concentration of [^{3}H]thymidine used in uptake assays ranges from 0.5 to 2 μCi/well.[13,21,26]

3. Any scintillation fluid formulated for high-efficiency tritium counting and quench resistance may be used in this assay; however, once an acceptable scintillation fluid has been chosen, it should be used consistently to minimize counting variability between different scintillation fluid formulations.

4. CSF-specific, neutralizing antibodies (available commercially) can be used in this assay to determine identity of CSA samples. The antibodies are serially diluted and combined with optimally stimulating dilutions of CSA samples. The total volume in each well should not exceed 0.2 ml. Abrogation of the proliferative response indicates CSF identity.

Cautions

1. Radioactive waste must be handled with care and disposed of according to local, state, and federal guidelines.

2. All equipment and solutions prior to addition of [^{3}H]thymidine must be sterile to minimize risk of contamination.

Proliferation of Factor-Dependent Cell Lines

Cell lines dependent on the presence of a specific growth factor(s) can be used to detect CSA. The following procedure is a general outline for the detection of CSA using a factor-dependent cell line. This procedure can be adapted as necessary to accommodate cell lines with different growth requirements.

Maintenance of Factor-Dependent Cell Lines

Before acquiring factor-dependent cell lines, a source(s) of growth factor must be obtained. Cell lines producing particular growth factors and recombinant growth factors (available commercially) are acceptable sources. The growth factor is used as a supplement to medium used for the maintenance of the cell line. Table I is a partial list of factor-dependent cell lines and includes the growth factor requirement(s) of each cell line and references on the use of each line. A review of the references will provide more detailed information on the cell line in question.

Detection of Colony-Stimulating Activity by Factor-Dependent Cell Line Proliferation

1. Filter-sterilized test samples (serum, tissue culture supernatant) are serially diluted in triplicate in a sterile 96-well flat-bottomed microtiter plate. Dilutions are made in 0.1 ml culture medium minus growth factor.

TABLE I
FACTOR-DEPENDENT CELL LINES AND THEIR COLONY-STIMULATING FACTOR REQUIREMENTS

Cell line	Source	CSF requirement	Reference(s)
FDC.P1	murine	IL-3, GM-CSF	*a–c*
NFS.60	murine	IL-3, G-CSF[d]	*a, e*
32D Cl3	murine	IL-3, G-CSF[d]	*a*
NFS/N1.M6	murine	IL-3	*a*
32D Cl23	murine	IL-3, GM-CSF	*f, g*
32D GM1	murine[h]	IL-3, GM-CSF	*i*
32D GM2	murine[h]	IL-3, GM-CSF	*i*
32D G1	murine[h,j]	G-CSF	*i*
32D G2	murine[h,j]	G-CSF	*i*
Mo7e	human[k]	IL-3, GM-CSF	*e, l*
TALL-101	human[k]	GM-CSF	*e*

[a] H. G. Derigs, G. S. Burgess, D. Klingberg, T. S. Nahreini, D. Y. Moehizyki, D. E. Williams, and H. S. Boswell, *Leukemia* **4,** 471 (1990).
[b] T. M. Dexter, J. Garland, D. Scott, E. Scolnick, and D. Metcalf, *J. Exp. Med.* **152,** 1036 (1980).
[c] D. M. Magee, D. M. Williams, E. J. Wing, C. A. Bleicker, and J. Schachter, *Infect. Immun.* **59,** 2370 (1991).
[d] Also responds to human G-CSF.
[e] D. C. Dale, S. Lau, R. Nash, T. Boone, and W. Osborne, *J. Infect. Dis.* **165,** 689 (1992).
[f] J. S. Greenberger, M. A. Sakakeeny, R. K. Humphries, C. J. Eaves, and R. J. Eckner, *Proc. Natl. Acad. Sci. U.S.A.* **80,** 2931 (1983).
[g] T. Yoshida, T. Hotta, K. Shimokata, M. Ichihara, K.-I. Isobe, and I. Nakashima, *Infect. Immun.* **59,** 1032 (1991).
[h] Derived from cell line 32D Cl3.
[i] G. Migliaccio, A. R. Migliaccio, B. L. Kreider, G. Rovera, and J. W. Adamson, *J. Cell Biol.* **109,** 833 (1989).
[j] Grows in response to G-CSF in medium without FBS.
[k] Grows in response to human cytokines.
[l] D. K. Blanchard, M. B. Michelini-Norris, C. A. Pearson, S. McMillen, and J. Y. Djeu, *Infect. Immun.* **59,** 2396 (1991).

2. The factor-dependent cell line is harvested, centrifuged at 100 *g* for 10 min at room temperature, and washed with HBSS to remove residual growth factor. The cells are counted, centrifuged at 100 *g* for 10 min, and diluted in fresh culture medium minus growth factor to a concentration of 7.5×10^5 cells/ml.

3. One-tenth milliliter of cells is added to each well of the microtiter plate. The total volume in each well should not exceed 0.2 ml.

4. Plates are incubated for 24 hr at 37° in an humidified, 7% CO_2 incubator. For the last 6 hr of incubation, the wells are pulsed with 1 μCi [^{3}H]thymidine.

5. The cells are harvested onto glass filters with an automatic cell harvester according to the manufacturer's directions.

6. The filters are dried overnight and radioactivity is assessed by liquid scintillation counting.

Comments

1. Several parameters of the assay may need to be optimized for different cell lines and media. The length of incubation prior to [^{3}H]thymidine addition may be increased or decreased to suit the assay results. The number of cells added to each well may need to be altered as may the concentration of [^{3}H]thymidine used for pulsing. Some previously reported ranges that may be useful in establishing acceptable parameters to assess proliferation are cell ranges from 5×10^3 to 1×10^5 cells/ml, 14 to 48 hr incubation prior to [^{3}H]thymidine pulsing, and 0.5 to 2 μCi [^{3}H]thymidine/well.[2,8,11,12,21]

2. The medium used in the assay is dependent on the cultivation medium of the cell line used. The only variation of the assay medium from the growth medium is that no growth factor is added.

3. The choice of scintillation fluid is left to the discretion of the investigator involved (refer to Comment under 3 Uptake of [^{3}H]Thymidine by Bone Marrow Cells).

4. All equipment and solutions must be sterile until [^{3}H]thymidine addition to avoid contamination.

5. Standard curves for individual growth factors may be prepared by using serial dilutions of recombinant CSFs and assessing proliferation in the above assay. These curves may be used to determine the quantity of CSA in a sample; however, the identity of the CSF in the test sample must be established prior to quantification.

6. CSF-specific, neutralizing antibodies (available commercially) may be added to the assay to determine the identity of the CSF(s) present. Dilutions of CSF-specific, neutralizing antibody are added to wells containing optimal stimulatory activity and abrogation of the proliferative response is measured (decrease in uptake of [^{3}H]thymidine).

Cautions

1. All radioactive waste must be properly disposed according to federal, state, and local guidelines.

2. Many of the factor-dependent cell lines are able to proliferate in response to more than one CSF. Controls should be performed for each cell line to verify CSF responsiveness.

Differentiation of WEHI-3B Cells and Radioimmunoassay for M-CSF

The WEHI-3B cell line, while not dependent on cytokines for growth, differentiates in response to G-CSF. This unique characteristic of the WEHI-3B cell line can be used to detect the presence of G-CSF in samples of unknown composition.[14,27] This procedure is detailed by Nicola[27] and will not be described in this chapter.

A radioimmunoassay (RIA) allows for the detection and quantification of an antigen in solution. An RIA is a competition assay between radiolabeled and unlabeled antigen to antigen-specific antibody. The greater the concentration of unlabeled antigen in a sample, the less radiolabeled antigen bound. An RIA procedure for M-CSF is described by Stanley[28] and will not be described in this chapter. His procedure details the purification of M-CSF, iodination of M-CSF, preparation of antibody against M-CSF, and procedure for the RIA.

Enzyme-Linked Immunosorbent Assay

An enzyme-linked immunosorbent assay (ELISA) is a straightforward technique that can be used to detect the presence of specific antigen in a sample. A capture ELISA is commonly used to detect the presence of CSF in a sample. The capture ELISA uses two antibodies, a trapping antibody (1°Ab) and a developing Ab (2°Ab). Both antibodies are specific for different epitopes of the same antigen. The 1°Ab is adsorbed to an ELISA microtiter 96-well plate. This antibody pulls the antigen out of the sample. After the antigen is trapped by the 1°Ab, the 2°Ab, i.e., an antigen-specific antibody conjugated with horseradish peroxidase (HRP) or alkaline phosphatase (AP), is added. This antibody binds to the antigen trapped by the 1°Ab. The assay is developed colorimetrically via a specific enzyme reaction for either HRP or AP. The amount of color generated in each well corresponds directly to the amount of antigen present in the sample. The procedure described below is a general protocol for a capture ELISA.[29]

[27] N. A. Nicola, this series, Vol. 116, p. 600.

[28] E. R. Stanley, this series, Vol. 116, p. 564.

[29] E. Harlow and D. Lane, "Antibodies: A Laboratory Manual." Cold Spring Harbor Lab., Cold Spring Harbor, NY, 1988.

Capture ELISA Protocol

1. An antigen-specific antibody(1°Ab) is adsorbed to each well of a 96-well microtiter plate at a concentration of 1 μg protein/well in 50 μl carbonate buffer, pH 9.6 (1.59 g Na_2CO_3, 2.93 g $NaHCO_3$ in 1 liter distilled H_2O, store at 4°C) by incubation overnight at 5° or for 4 hr at 37°.

2. The plate is washed three times with phosphate-buffered saline (PBS)–0.05% Tween 20, pH 7.4 (8.0 g NaCl, 0.2 g KH_2PO_4, 1.2 g NaH_2PO_4, 0.2 g KCl, 0.5 ml Tween 20 in 1 liter distilled H_2O, store at 4°).

3. Unoccupied protein binding sites are blocked by incubating 100 μl blocking buffer, i.e., 3% bovine serum albumin in PBS, pH 7.2 (8.2 g $NaH_2PO_4 \cdot H_2O$, 72 g $Na_2HPO_4 \cdot 7H_2O$, 280.52 g NaCl, final volume is 1.6 liters of 20× PBS), in each well for 60 min at 37°.

4. The plate is washed with PBS–0.05% Tween 20 three times.

5. Serial dilutions of samples (serum, tissue culture supernatants, tissue homogenate preparations) and standards (purified or recombinant CSF, commercially available) are made in triplicate in PBS–0.05% Tween 20 (final volume/dilution is 50 μl).

6. The plate is incubated with shaking for either 1 hr at 37° or for 4 hr at room temperature.

7. The plate is washed three times with PBS–0.05% Tween 20.

8. The developing antibody (2°Ab) conjugated with HRP or AP is added to each well following the manufacturer's guidelines for dilution.

9. The plate is incubated with shaking for either 1 hr at 37°C or 4 hr at room temperature.

10. The plate is washed three times with PBS–0.05% Tween 20.

11. The plate is developed with a chromagen substrate for HRP or AP, appropriately, according to the manufacturer's guidelines.

12. The absorbance of each well is measured spectrophotometrically with a microplate reader.

13. The amount of CSF in a sample can be determined by comparing the absorbance values for the unknown with absorbance values generated with known concentrations of specific CSF and plotted as a standard curve (absorbance vs CSF concentration).

Comments

1. Various sources of CSF can be tested in a capture ELISA. Samples of serum, tissue culture supernatants, and preparations from tissue homogenates can be quantified. Preparations from tissue homogenates are difficult to quantify by bioassay as they may contain by-products that are toxic to cells. For the testing of such samples, the capture ELISA and RIA procedures are recommended.

2. The lengths of incubation listed in the above protocol are used with success in our laboratory. The time of incubation may, however, need to be altered dependent on the assay in question. For example, a longer incubation may be necessitated if an antibody with a lower antigen binding affinity is used.

3. Different types of plates are available for use in an ELISA and may vary in the initial adsorption procedure. The adsorption procedure used commonly in our laboratory is achieved with carbonate buffer, pH 9.6, and is listed in the above protocol. Some adsorption protocols use a neutral PBS buffer in place of carbonate buffer.

4. Washing of the plate is performed by filling each well with washing buffer and flicking the plate over a sink or pile of paper towels to remove buffer. This procedure is repeated three times for each washing step. Excess buffer on the edges of the plate may interfere with dilutions and can be wiped off with absorbent paper.

Cautions

1. The ELISA plate should never be allowed to dry out during the assay. If a solution for a step is not ready, the plates can be washed and filled with the washing buffer until the next step is prepared.

2. During incubation periods, the plate should be covered to minimize evaporation.

Conclusions

The assays covered in this section provide a basis for examining the effects of bacterial products on CSF production. These assays when used alone or preferably in combination can document the presence of CSFs in a various samples. In choosing which assays are appropriate for a given situation, the specificity, number of samples, cost, and time involved in each assay must be considered. Each assay has its own advantages and disadvantages; the most notable are cost and sample range. RIAs and ELISAs are the assays of choice when test samples include preparations from tissue homogenates, as these preparations may contain by-products that are toxic to cells used in the bioassays. RIAs and ELISAs are, however, expensive if the components must be purchased; therefore, in cases where a large number of samples must be assayed, they are generally not feasible. A bioassay is extremely sensitive and cost effective in most applications. Bioassays with slight modifications can be extremely specific for individual CSFs; however, as with RIAs and ELISAs, a bioassay is not practical for all experiments. Bioassays, as mentioned previously, can be negatively affected by toxic by-products in certain types of sample preparations. For use in limited testing situations in which a laboratory is not actively using bioassays, this procedure may not be as cost effective as a one-time purchase of an immunoassay kit. From the descriptions and comments on each assay, an investigator can choose assays that will be effective in a given situation.

[3] Effect of Bacterial Products on Tumor Necrosis Factor Production: Quantitation in Biological Fluids or Tissues

By ZELL A. McGEE and CHRISTOPHER M. CLEMENS

Introduction

An increasing body of evidence indicates that the septic shock and tissue damage that characterize many infectious diseases are not the direct result of the action of microbial toxic moieties such as lipopolysaccharide/endotoxin (LPS) released by gram-negative bacteria, but rather, that the shock component,[1,2] coagulation component,[3] and fever component[4] of septic shock are mediated by hormones of the immune system called lymphokines or cytokines. Two of the most potent of these cytokines are tumor necrosis factor α (TNF-α) and interleukin 1β (IL-1),[2,4-7] which display some synergistic and overlapping activities.[8] TNF-α was first described as a tumor-necrotizing factor found in the serum of Bacillus Calmette–Guérin (BCG)-primed, endotoxin-treated mice.[9] Later, after Beutler *et al.* had shown that endotoxin-induced shock was mediated by

[1] B. Beutler and A. Cerami, *Clin. Res.* **35,** 192 (1987).

[2] S. Okusawa, J. A. Gelfand, T. Ikejima, R. J. Connolly, and C. A. Dinarello, *J. Clin. Invest.* **81,** 1162 (1988).

[3] T. Van Der Poll, H. R. Büller, H. Ten Cate, C. H. Wortel, K. A. Bauer, S. J. H. Van Deventer, C. E. Hack, H. P. Sauerwein, R. D. Rosenberg, and J. W. Ten Cate, *N. Engl. J. Med.* **322,** 1622 (1990).

[4] M. J. Kluger, *Physiol. Rev.* **71,** 93 (1991).

[5] B. Beutler and A. Cerami, *N. Engl. J. Med.* **316,** 379 (1987).

[6] C. A. Dinarello, J. G. Cannon, N. S. Wolff, H. A. Bernheim, B. Beutler, A. Cerami, I. S. Figari, M. A. Palladino, Jr., and J. V. O'Connor, *J. Exp. Med.* **163,** 1433 (1986).

[7] J. Le and J. Vilcek, *Lab. Invest.* **56,** 234 (1987).

[8] K. Last-Barney, C. A. Homon, R. B. Faanes, and V. J. Merluzzi, *J. Immunol.* **141,** 527 (1988).

[9] E. A. Carswell, L. J. Old, R. L. Kassel, S. Green, N. Fiore, and B. Williamson, *Proc. Natl. Acad. Sci. U.S.A.* **72,** 3666 (1975).

TNF-α,[10] the apparent incongruity that surfaced was the clinical observation that septic shock indistinguishable from endotoxin shock of gram-negative etiology could be observed in septicemia caused by gram-positive bacteria,[11] which lack endotoxin. Soon, however, others showed that TNF-α could be induced by peptidoglycan[12] or lipoteichoic acid,[11] one or the other of which is found in all gram-positive bacteria. TNF-α can also be induced by toxic shock syndrome toxin 1[13] and staphylococcal enterotoxin A.[14] But lest we think we understand the microbial products that induce TNF-α production, incongruity again prods us to note that fungemia produces septic shock and TNF-α is induced by a variety of parasites such as schistosomes,[15] leishmania,[16] and plasmodia (malaria).[17,18] Because various strategies for blocking the production or the action of TNF-α in naturally occurring[19] or experimental[10,20] diseases decrease morbidity and mortality,[21] it is important to expeditiously identify those diseases in which TNF-α plays a deleterious role.

Whereas much of the research on the role of TNF-α in infectious diseases has been performed in the context of septic shock[5,22] and bacteremia,[23] evidence indicates that mucosal pathogens induce mucosal tissues to produce TNF-α and the TNF-α mediates mucosal damage.[24] Therefore, with the increasing focus on studies of the role of TNF-α as part of the molecular mechanisms of pathogenicity of a variety of infectious disease

[10] B. Beutler, I. W. Milsark, and A. C. Cerami, *Science* **229,** 869 (1985).

[11] G. Wakabayashi, J. A. Gelfand, W. K. Jung, R. J. Connolly, J. F. Burke, and C. A. Dinarello, *J. Clin. Invest.* **87,** 1925 (1991).

[12] M. Parant, *J. Leukocyte Biol.* **42,** 576 (1987).

[13] C. Jupin, S. Anderson, C. Damais, J. E. Alouf, and M. Parant, *J. Exp. Med.* **167,** 752 (1988).

[14] H. Fischer, M. Dohlsten, U. Andersson, G. Hedlund, P. Ericsson, J. Hansson, and H. O. Sjören, *J. Immunol.* **144,** 4663 (1990).

[15] K. Zwingenberger, E. Irschick, J. G. Vergetti Siqueira, A. R. Correia Dacal, and H. Feldmeier, *Scand. J. Immunol.* **31,** 205 (1990).

[16] M. S. Chiofalo, D. Delfino, G. Mancuso, E. La Tassa, P. Mastoeni, and D. Iannello, *Microb. Pathog.* **12,** 9 (1992).

[17] I. A. Clark, K. A. Rockett, and W. B. Cowden, *Lancet* **337,** 302 (1991).

[18] D. Kwiatkowski, A. V. S. Hill, I. Sambou, P. Twumasi, J. Castracane, K. R. Manogue, A. Cerami, D. R. Brewster, and B. M. Greenwood, *Lancet* **336,** 1201 (1990).

[19] M. H. Lebel, B. J. Freij, G. A. Syrogiannopoulos, D. F. Chrane, M. J. Hoyt, S. M. Stewart, B. D. Kennard, K. D. Olsen, and G. H. J. McCracken, *N. Engl. J. Med.* **319,** 964 (1988).

[20] F. B. Taylor, Jr., A. Chang, W. Ruf, J. H. Morrissey, L. Hinshaw, R. Catlett, K. Blick, and T. S. Edgington, *Circ. Shock* **33,** 127 (1991).

[21] R. C. Bone, *JAMA, J. Am. Med. Assoc.* **266,** 1686 (1991).

[22] B. Beutler, *Annu. Rev. Biochem.* **57,** 505 (1988).

[23] A. Waage, A. Halstensen, and T. Espevik, *Lancet* **1,** 355 (1987).

[24] Z. A. McGee, C. M. Clemens, R. L. Jensen, J. J. Klein, L. R. Barley, and G. L. Gorby, *Microb. Pathog.* **12,** 333 (1992).

processes, it is timely to examine methods for quantitating TNF in serum and other body fluids as well as in tissues that are the target of infectious disease processes.

Choice of Assays

The major choices for assaying TNF are (1) the enzyme-linked immunosorbent assay (ELISA) method, using self-generated plates or commercially prepared kits from any one of a dozen manufacturers (e.g., Endogen, Boston, MA; Genzyme, Cambridge, MA) and (2) the bioassay described by Zacharchuk *et al.*[25]

If the purpose of the study is to show that tissues are producing biologically active TNF, the bioassay method is preferable, because immunologically reactive TNF, which is detected by the ELISA method, is not necessarily bioactive. Møller *et al.*[26] found that ELISA tests measured biologically inactive as well as biologically active TNF in cerebrospinal fluid.

Because commercial ELISA kits provide the instructions for the methods they employ, and methodology for self-generated ELISA plates is described in detail elsewhere,[27] this chapter focuses on the bioassay for TNF.

Bioassay

Background

Cells from the murine fibroblast line L-929 (American Type Culture Collection, Rockville, MD, No. CCL-1) are particularly sensitive to the cytotoxic effects of TNF. This cytotoxic effect is further enhanced in the presence of actinomycin D. Because of the direct dose–response relationship between TNF concentration and L-929 cell cytotoxicity, this cell line is well suited for the detection and quantitation of TNF by the bioassay method. The TNF bioassay provides a very sensitive method by which native human, recombinant human, and murine TNF can be accurately quantitated.

[25] C. M. Zacharchuk, B.-E. Drysdale, M. M. Mayer, and H. S. Shin, *Proc. Natl. Acad. Sci. U.S.A.* **80,** 6341 (1983).

[26] B. Møller, S. C. Mogensen, P. Wendelboe, K. Bendtzen, and C. M. Petersen, *J. Infect. Dis.* **163,** 886 (1991).

[27] P. E. Hurtubise, S. Bassion, J. Gauldie, and P. Horsewood, *in* "Clinical Chemistry: Theory, Analysis, and Correlation" (L. A. Kaplan and A. J. Pesce, eds.), p. 205. Mosby, St. Louis, MO, 1984.

The following bioassay procedure was adapted from the method of Zacharchuk *et al.*[25] The bioassay can be used for blood or other biological fluids or for specimens generated *in vitro,* such as organ or tissue culture supernatant fluid and filtrates from homogenized organ cultures, other cells, or tissues.

Preparation of Samples for Bioassay

It is important to note that blood specimens for TNF assay should not be collected in commercially available heparinized glass tubes, because these tubes contain varying amounts of endotoxin, which can be responsible for the rapid and massive production of TNF *in vitro*.[28,29] Serum is adequate for TNF assays.

The authors have measured TNF in human tissues (fallopian tube mucosal organ cultures[24]) by homogenizing the tissues in at least 2 ml of tissue culture medium, using a Tissuemizer (Tekmar, Cincinnati, OH) fitted with a 10 N head. The Tissuemizer head is cleaned between specimens by running it full speed sequentially in three to four tubes of sterile tissue culture medium and then running it in 70% ethanol in water. The homogenate can be passed through a sterile, 0.45-μm pore-size membrane filter to sterilize the sample and remove cellular debris. Such tissue homogenates have maintained good TNF activity for up to 6 months when frozen and maintained at $-70°$.

The experience of some investigators has suggested the possibility that serum and other biological fluids may contain inhibitors of TNF or of the L-929 cells. Such effects can sometimes be abrogated by diluting the specimens up to 1 : 8 or greater prior to assay.

Because the assay may reflect the cytotoxic effects of TNF-α or TNF-β, assurance that the cytotoxic effect of a sample is caused by TNF-α should include demonstration that the cytotoxic effect is neutralized by treating the sample with specific anti-TNF-α antibody.

L-929 Cell Culture and Propagation

L-929 cells can be obtained from the American Type Culture Collection.

The culture medium recommended for propagation of the L-929 cell line is α-MEM (GIBCO, Grand Island, NY), supplemented with 10% (v/v) fetal calf serum (FCS) (GIBCO) and penicillin–streptomycin (GIBCO) at a final concentration of 100 U/ml penicillin and 100 μg/ml streptomycin.

[28] G. Leroux-Roels, J. Philippé, F. Offner, and A. Vermeulen, *Lancet* **336,** 1197 (1990).
[29] G. Leroux-Roels, F. Offner, J. Philippé, and A. Vermeulen, *Clin. Chem.* (*Winston-Salem, N.C.*) **34,** 2373 (1988).

If necessary, the pH of the medium should be adjusted to 7.2–7.4 by bubbling CO_2 gas through the medium and filter-sterilizing the pH-adjusted medium. After the cells have been propagated, they can be stored as seed stock by freezing them in αMEM–FCS in liquid nitrogen. Viability of the stock is best maintained if the cells to be frozen are suspended in αMEM–FCS containing 10% (v/v) Cryoserv [sterile, pyrogen-free 99% dimethyl sulfoxide (DMSO) available in 10-ml ampules (Research Industries Corp., Pharmaceutical Division, Midvale, UT). L-929 cell preservation is enhanced by adjusting the cell density to 1 to 3 $\times$ 10^6/ml and by using a controlled freezing rate of $-1°$/min down to $-30°$ and then transferring the cryovials to liquid nitrogen for long-term storage.

The L-929 cells can be grown in T-25 tissue culture flasks (Corning 25100 Baxter Scientific Products, McGaw Park, IL) using 5 ml of medium per flask or in T-75 flasks (Falcon 3024 Baxter Scientific Products, McGaw Park, IL) using 15 ml of medium per flask. One or two of the T-25 flasks provide sufficient cells for at least one assay plate, whereas one or two T-75 flasks provide sufficient cells for multiple assay plates. Optimal growth conditions are at 37° in a humidified atmosphere of 5% CO_2 in air. The cells do best if the culture medium is changed every 3 days and if the L-929 cells are subcultured at the point where they have grown to form a confluent monolayer on the bottom of the flask (confluence usually occurs 5 to 6 days following inoculation of flasks with L-929 cells). Cell suspensions for subculture are made using an inoculum which contains 1 $\times$ 10^5 cells/ml (as assayed by a hemocytometer chamber). A cell scraper (S/P T4206-1 Baxter Scientific Products, McGaw Park, IL) may be used to remove the adherent cells from the inner surface of the tissue culture flasks.

Bioassay Procedure

Preparation of L-929 Cell Suspension. Remove the adherent cells from the inside of the tissue culture flask(s) using a cell scraper. Pool the resulting cell suspensions and centrifuge for 10 min at 4° at 750 *g*. Resuspend the pellet in αMEM–FCS and, with a hemacytometer, perform a viable cell count using the trypan blue dye exclusion method.[30] Adjust the cell suspension to contain 4 to 5 $\times$ 10^5 viable cells/ml using αMEM–FCS as the diluent. The optimal inoculating cell suspension density is one that yields, after solubilization of the cells near the end of the assay, an optical density (OD) of 0 to 0.1 with the highest concentration of TNF used and an OD of 1.0 or higher with the zero TNF concentration.

[30] R. I. Freshney, "Culture of Animal Cells: A Manual of Basic Technique," 2nd ed., p. 245. Alan R. Liss, New York, 1987.

Setup of Microtiter Plates. Map out the sample arrangement on a plate layout diagram such as that shown in Fig. 1, being sure to include TNF standards, TNF dilutions for LD_{50} determination (if necessary), TNF-containing unknown samples, and control samples. To each appropriate well of a 96-well, flat-bottom microtiter plate(s) (Falcon, Microtest III, No. 3072, polystyrene, Baxter Scientific Products, McGaw Park, IL), add 0.1 ml of the L-929 cell suspension prepared as described above. Incubate the plate(s) at 37° in a humidified incubator with an atmosphere of 5% CO_2 in air for 18 hr. Following the initial incubation, decant the supernatants from each well by shaking and pounding the inverted plate(s) on a sterile cloth or a thin stack of autoclaved paper towels. Into each cell-containing well of the plate(s), add 0.1 ml of medium that was supplemented to contain 2.0 μg/ml of actinomycin D (Sigma, St. Louis, MO) and filter-sterilized. Immediately following the addition of the actinomycin

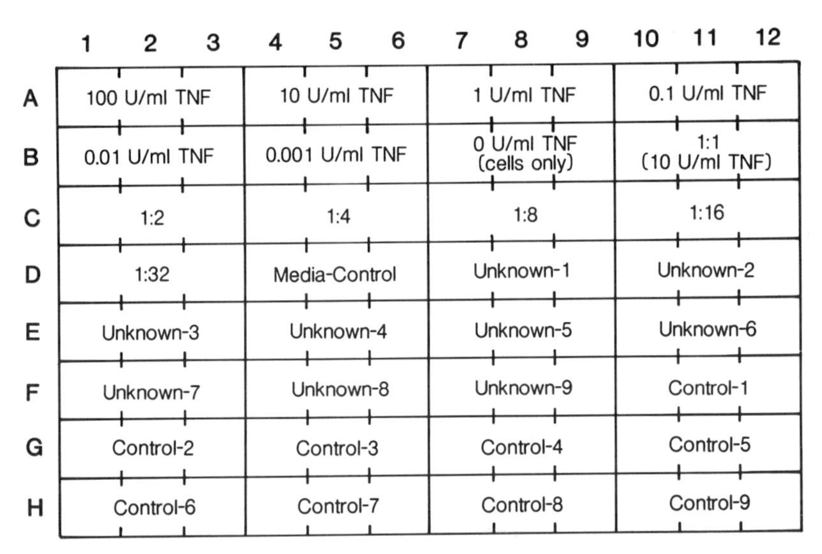

FIG. 1. Example of an assay plate layout. The plate layout should accommodate the experimental design. In the experiment for which this plate layout design was used, a series of infected experimental specimens (unknown, test samples) were matched with their paired, uninfected control specimens, and TNF assays were performed to determine the effect of infection on TNF production. Note that each standard, unknown, and control is performed in triplicate. Wells B-10 through D-3 are used to determine the specific activity of the TNF being used as standards, and the values for the standards are appropriately adjusted prior to determining the standard curve and values for the unknowns and controls. The specific activity of the TNF preparation being used is determined by making serial twofold dilutions of the 10 U/ml standard and determining the titer that lysed 50% of the L-929 cells (one unit of TNF is defined as the amount necessary to effect 50% cytotoxicity of L-929 cells[25]).

D-containing medium, add 0.1-ml volumes of one of the following: TNF standards, TNF LD_{50} dilutions, TNF-containing unknown samples to be assayed, or control samples (see Fig. 1). All standards and samples should be run in triplicate. Note that the standards and dilutions of TNF-α for specific activity are made up to the concentrations shown in Fig. 1 and are added to the appropriate wells in that concentration. Because all standards and samples are diluted in a parallel manner, correction for dilution that occurs in the course of the assay is not necessary.

Again, incubate the plate(s) at 37° in a humidified, 5% CO_2 incubator for 18 hr.

Spectrophotometric Measurement of Results. Following the above final 18-hr incubation, decant the supernatant fluid from each well by inverting and shaking the plate(s) as described above. Stain the remaining adherent (viable) cells by immersing the plate(s) in 0.2% (w/v) crystal violet in 2% (v/v) ethanol in water for 10 min. Rinse the plate(s) of excess stain using tap water. Solubilize the stained cells by adding to each well 0.1 ml of 1% (w/v) sodium dodecyl sulfate (SDS) in water. Place the plate(s) on a rotary shaker for 1 hr at 180 rpm. Measure the absorbance of each well at 570 nm using an ELISA plate reader that has been blanked to an OD of 0 at 570 nm using a separate plate with a well containing 0.1 ml of the 1% SDS solution.

Interpretation of Results. First, calculate the mean absorbance for each triplicate set of standards and other samples assayed. Using the mean absorbance data and the following formula, calculate the percentage cytotoxicity for each standard and sample:

$$\%\ \text{cytotoxicity} = [1 - (\text{absorbance of sample or standard}/\text{absorbance of 0 U/ml TNF})] \times 100$$

A TNF standard curve may be constructed on 5-cycle semilog paper by plotting the percentage cytotoxicity values for the TNF standards on the vertical, linear axis and the TNF standard concentrations (adjusted for specific activity as described below) on the horizontal, log axis. As an alternative, the TNF standard data may be analyzed by performing a linear regression analysis by the least squares method, and deriving a quadratic equation to describe the standard curve.

The TNF concentrations of the test samples can be determined using the TNF standard curve by finding the TNF concentration on the horizontal axis which corresponds to the percentage cytotoxicity value calculated for each sample. If a linear regression analysis is performed, the TNF concentration can be calculated by using the percentage cytotoxicity value of each test sample to solve the regression equation using the quadratic formula. The sensitivity of the assay is generally in the range 0.1 to 0.01 U/ml.

Tumor Necrosis Factor Standards. Generally, it is best to use TNF standard concentrations estimated to be 100, 10, 1, 0.1, 0.01, and 0.001 U/ml. However, because the specific biological activity of TNF-α varies from lot to lot and tends to degrade over time with storage, the activity of the standards is only an estimate, and it is preferable to determine the actual specific activity of the TNF standard preparation by the LD_{50} method described below, and to adjust the estimated TNF standard concentrations to the true values when constructing the TNF standard curve for each assay. The specific activity of the TNF standards is determined with each assay in the laboratory of McGee.

The activity of TNF-α is best preserved if the TNF-α to be used for standards in the assay is divided, immediately upon arrival in the laboratory, into aliquots of appropriate volume and concentration to be used in one assay and is kept frozen at -70 to $-100°$. Repeated freezing and thawing of TNF-α results in a substantial loss of activity.

TNF Specific Activity Determination by the LD_{50} Method. By definition, 1.0 U/ml of TNF activity is equivalent to the TNF concentration which results in a percent cytotoxicity value of 50% (the LD_{50}) in the L-929 cell bioassay system.[25] Therefore, using this definition, it is relatively simple to determine the true specific activity of any TNF standard preparation by performing an LD_{50} determination on it. This is best accomplished by serially diluting the TNF standard preparation of interest in two-fold steps and assaying the different dilutions for percent cytotoxicity as described above. The resulting percent cytotoxicity values should then be plotted against the corresponding *final* TNF dilution factor on linear graph paper, and determining the TNF dilution factor that corresponds to the 50 percent cytotoxicity value; this dilution is the LD_{50} titer. The LD_{50} titer can also be determined by performing a linear regression analysis by the least squares method and solving the regression equation using a percent cytotoxicity value of 50%. For the purpose of calculating the specific activity, the final 1 : 2 dilution of the TNF sample inherent in the cytotoxicity assay (when the 100 μl of the TNF standard dilution is added to the 100 μl of actinomycin-D-containing medium in the assay plate well) is compensated for by mathematically doubling the initial two-fold dilution factor prior to the construction of the LD_{50} plot (e.g., the dilution, "1 : 2," is actually 1 : 4 in the assay plate, therefore is designated 1 : 4 when plotting the LD_{50} cytotoxicity data. In other words, the dilution factor used in the plot of the LD_{50} titer determination should represent the *final* dilution factor of the original TNF-α stock in the assay plates. The reciprocal of the resulting LD_{50} titer represents the true specific activity in U/ml, of the TNF stock sampled (e.g., for an LD_{50} titer of 1 : 8, the specific activity of the undiluted preparation is 8 U/ml).

The authors have found the estimated 10 U/ml standard convenient to use for the specific activity determination, because the five serial two-fold dilutions used (Fig. 1) usually yield the 50% cytotoxicity endpoint. If the 10 U/ml standard is used to derive the LD_{50} titer, the reciprocal of that dilution divided by 10 yields the correction factor by which the concentration of the standards should be multiplied to get the actual TNF activity of the standards (e.g., for an LD_{50} titer of 1 : 8, the correction factor is 0.8, and the estimated 10 U/ml actually contains 8 U/ml, as indicated above.

If the investigator prefers, the LD_{50} titer can be calculated mathematically rather than being derived from a graph. If a percent cytotoxicity value of 50% is used in the regression equation, the LD_{50} titer can be solved for using the quadratic formula and used to correct the estimated standards as indicated above. The estimated concentration values for the TNF-α standards in the assay (100, 10, . . . , etc.) are corrected using the result of the LD_{50} determination before the standard curve and values of the unknown and control samples are determined.

[4] Induction of Prostaglandin Release from Macrophages by Bacterial Endotoxin

By KEN-ICHI TANAMOTO

Introduction

Endotoxin from gram-negative bacteria is known to exert numerous activities both *in vitro* and *in vivo*.[1] Endotoxin is chemically a lipopolysaccharide (LPS), consisting of a hydrophilic polysaccharide part and a hydrophobic lipid part, named lipid A, which has been proven to be the active center of almost all the biological activities of endotoxin. The chemical structure of typical lipid A derived from *Escherichia coli* was determined with the help of chemically synthesized lipid A; i.e., the pure synthetic lipid

[1] O. Lüderitz, M. Freudenberg, C. Galanos, V. Lefmann, E. T. Rietschel, and D. H. Show, *Curr. Top. Membr. Transp.* **17,** 79 (1982).

METHODS IN ENZYMOLOGY, VOL. 236

A structure exhibited activity qualitatively and quantitatively identical to that of natural free *E. coli* lipid A.[2,3]

Endotoxin, on injection into a host body, acts on various targets and induces a complicated reaction. Host cells exposed to LPS release distinct endogenous mediators including prostaglandins, which possess endotoxic activity.[3] The action of endotoxin is replaced by that of the mediators. The primary target of endotoxin is thought to be the reticuloendothelial system. Cells of the mononuclear phagocytic system, especially macrophages, produce, via the cyclooxygenase pathway, arachidonic acid metabolites which play an essential role in several endotoxic activities.[4] The concept of involvement of prostaglandins in endotoxic action is supported partly by the observation that certain endotoxic activities such as the induction of fever, abortion, and the early (but not late) phase of endotoxic shock are suppressed by an inhibitor of cyclooxygenase.[5–7] It has also been suggested that endotoxin shock is closely related to increased concentrations of several groups of prostanoids in blood and lymph.[8]

Treatment of macrophages from nonstimulated mouse peritoneum or rabbit lung with LPS induces a dose-dependent accumulation of prostaglandin (PG) E_2 and $PGF_{2\alpha}$ in the culture medium after 16 to 24 hr as described below. This induction of prostaglandins by LPS is completely inhibited by indomethacin. A similar dose-dependent release of PGE_2 and $PGF_{2\alpha}$ by LPS has been observed in thioglycolate-stimulated mouse peritoneal macrophages,[9] human blood monocytes,[10] and rat peritoneal macrophages,[11,12] the last of which produces thromboxane B_2 (TXB_2) and

[2] J. Y. Homma, M. Matsuura, S. Kanegasaki, Y. Kawakubo, Y. Kojima, N. Shibukawa, Y. Kumazawa, A. Yamamoto, K. Tanamoto, T. Yasuda, M. Imoto, H. Yoshimura, S. Kusumoto, and T. Shiba, *J. Biochem. (Tokyo)* **98,** 395 (1985).

[3] C. Galanos, O. Lüderitz, E. T. Rietschel, O. Westphal, H. Brade, L. Brade, M. Freudenberg, U. Shade, M. Imoto, H. Yoshimura, S. Kusumoto, and T. Shiba, *Eur. J. Biochem.* **148,** 1 (1985).

[4] L. T. Old, *Nature (London)* **326,** 330 (1987).

[5] R. Siegert, W. K. Philipp-Dormston, K. Radsak, and H. Menzel, *Infect. Immun.* **14,** 1130 (1976).

[6] J. T. Flynn, *J. Pharmacol. Exp. Ther.* **206,** 555 (1978).

[7] R. C. Skarnes and M. J. K. Harper, *Prostaglandins* **1,** 191 (1972).

[8] J. T. Flynn, *in* "Handbook of Endotoxin" (L. B. Hinshow, ed.), Vol. 2, p. 237. Elsevier, Amsterdam, 1984.

[9] L. M. Wahl, D. L. Rosenstreich, L. M. Glode, A. L. Sandberg, and S. E. Mergenhagen, *Infect. Immun.* **23,** 8 (1979).

[10] J. I. Kurland and R. Bockman, *J. Exp. Med.* **147,** 952 (1978).

[11] J. A. Cook, W. C. Wise, and P. V. Halushka, *J. Reticuloendothel. Soc.* **30,** 445 (1981).

[12] N. Feuerstein, J. H. Bash, J. N. Woody, and P. W. Ramwell, *J. Pharm. Pharmcol.* **33,** 401 (1981).

6-keto-$PGF_{1\alpha}$ (a stable form of PGI_2) additionally. Peritoneal macrophages from the LPS nonresponder mouse strain C3H/HeJ do not produce PGE_2 and $PGF_{2\alpha}$ on stimulation with LPS.[9] In contrast to macrophages, mouse and bovine granulocytes do not produce PGE_2 and $PGF_{2\alpha}$ and B and T lymphocytes do not release PGE_2 on LPS stimulation.[10]

LPS of wild-type bacteria and R mutants, as well as lipid A, induced prostaglandin release to a comparable degree, indicating that lipid A is the active principle of LPS in this reaction. Surprisingly, induction of prostaglandin release also occurs in both mouse peritoneal and rabbit alveolar macrophages stimulated with the incomplete form of lipid A analogs,[13] which are structurally distinct from the complete *E. coli* type lipid A (principally regarding the position of fatty acids and phosphate substitution) and have been shown to be endotoxically inactive in many biological tests. This phenomenon was not observed with the macrophages from LPS nonresponder mouse strain C3H/HeJ, indicating that the induction of prostaglandins by incomplete lipid A analogs is due to an endotoxin-specific mechanism. These facts suggest that the structural requirements of lipid A for the induction of prostaglandins are not very strict.

The methods related to this review are divided into three sections: (1) preparation of bacterial endotoxin, (2) cell (macrophage) preparation, and (3) measurement of prostaglandins. Regarding cell preparation, macrophages from rabbit lung and mouse peritoneum are described as typical cellular sources for the induction of prostaglandins by endotoxin. Prostaglandins and their metabolites are generally measured by such techniques as radioimmunoassay (RIA), gas chromatography (GC), GC–mass spectrometry, and high-performance liquid chromatography (HPLC). All these techniques were described previously in this series.[14] Here, RIA and HPLC are described mainly on the basis of our experiments.[15]

Isolation of Lipopolysaccharide

It is important to use pure LPS to study the physical, chemical and also biological properties of LPS as contamination with bacterial components such as protein, nucleic acid, and peptidoglycan may alter these properties.

[13] K. Tanamoto, U. Shade, E. T. Rietschel, S. Kusumoto, and T. Shiba, *Infect. Immun.* **58,** 217 (1990).

[14] A. R. Whorton, this series, Vol. 141, p. 341.

[15] K. Tanamoto, U. Shade, and E. T. Rietschel, *Biochem. Biophys. Res. Commun.* **165,** 526 (1989).

The most widely accepted methods for the isolation of LPS are phenol–water extraction originally developed by Westphal *et al.*[16] and phenol–chloroform–petroleum ether (PCP) extraction developed by Galanos *et al.*[17] The former method is used for the extraction of *S*-form LPS and the latter for the isolation of *R*-form LPS. The PCP procedure is also used for the further purification of *S*-form LPS after isolation by the phenol–water method in which the LPS still contains contaminating protein (1–3%). The combination of these two procedures yields LPS completely free of other components of bacteria (protein content is less than 0.1%).[18]

Procedure

Gram-negative bacteria are cultured in a suitable medium until the late log phase and centrifuged, and the sediment is washed with distilled water twice. The cells are then treated successively with ethanol and acetone and twice with ether. They are dried *in vacuo* over $CaCl_2$ to constant weight.

Phenol–Water Extraction

The dried bacteria (20 g) are suspended in 350 ml of water at 68° in a water bath. The same volume of 90% phenol preheated at 68° is added and the mixture is stirred vigorously for 10 to 15 min. It is then cooled and centrifuged at 15,000 g for 15 min, which results in the formation of three layers, a water phase, (sometimes a water–phenol interface), a phenol phase, and the insoluble material, in that order from the top. The water phase is aspirated off, and another 350 ml of preheated water is added to the rest of the mixture. The same extraction process is repeated. The water phases are combined and dialyzed for 2 to 3 days against distilled water. The removal of phenol is confirmed by the loss of color induced by the addition of $FeCl_3$ to the solution. The solution is then centrifuged to remove the insoluble material and lyophilized. The extract is dissolved in water again (3% solution) and is centrifuged at 105,000 *g* for 3 hr to remove RNA. The ultracentrifugation is repeated twice. The pellet is dissolved in 25 m*M* Tris–HCl buffer, pH 7.4, containing 20 μg of RNase/ml and incubated at 37° for 3 hr. Purified LPS is finally obtained by ultracentrifugation of the mixture. As an alternative, Cetavlon (cetyltri-

[16] O. Westphal, O. Lüderitz, and F. Bister, *Z. Naturforsch. B: Anorg. Chem., Org. Chem., Biochem., Biophys. Biol.* **7B,** 148 (1952).

[17] C. Galanos, O. Lüderitz, and O. Westphal, *Eur. J. Biochem.* **9,** 245 (1969).

[18] C. Galanos, O. Lüderitz, and O. Westphal, *Zentralbl. Bakteriol., Parasitenkd., Infectionskr. Hyg., Abt. 1: Orig., Reihe A* **243,** 226 (1979).

methylammonium bromide) is also used to remove the residual RNA completely. The material is checked for RNA by measuring the absorbance of the solution at 260 nm. The final sediment is freeze-dried. The yield of LPS is 3 to 5% of the dry bacteria.

Petroleum Ether–Chloroform–Phenol Extraction

The dried bacteria (25 g) are suspended in 100 ml of extraction mixture [90% phenol : chloroform : petroleum ether = 2 : 5 : 8 (v/v)]. The suspension is sonicated well to obtain a fine suspension. The mixture is stirred for 10 min on a magnetic stirring plate at 4°. The bacteria are centrifuged at 5000 rpm for 15 min at 4°, and the residual bacteria are removed by filtration. The same extraction from the bacteria is repeated twice and the supernatants are combined. Petroleum ether and chloroform are evaporated and water is added dropwise to the remaining phenol phase until the LPS begins to precipitate. The solution is centrifuged at 5000 rpm for 10 min. The supernatant is entirely decanted carefully and the inside of the tube is wiped well. The precipitate is washed three times with 2.5 ml of 80% phenol–water, and the inside of the tube is wiped with filter paper after each decantation of the supernatant. The precipitate is washed three times with ether to remove the phenol and dried *in vacuo* over $CaCl_2$. The LPS is suspended in 25 ml of water, warmed to 45°, and immediately placed in a desiccator, and the vacuum is very carefully turned on to remove the air. It is then sonicated for 5 min and centrifuged at 100,000 *g* for 4 hr at 4°. The supernatant is discarded and the pellet is dissolved in 20 ml of water. Undissolved material is removed by centrifugation at 3000 rpm for 10 min and the supernatant is lyophilized.

Preparation of Lipid A

As the active center of LPS resides in lipid A, this lipid is used as the endotoxin itself for the biological assays in many test systems. Lipid A is easily cleaved from LPS by mild acid hydrolysis as follows.[19]

One hundred milligrams of LPS is dissolved in 5 ml of water in a sealed tube. The same volume of 2% acetic acid is added to the solution with vigorous stirring. The solution is hydrolyzed at 100° for 1.5 hr, after which insoluble floating lipid A appears. Cleaved lipid A is sedimented by centrifugation at 15,000 g for 20 min at 4°. After the supernatant is decanted, the pellet is washed three times with distilled water and once with acetone and dried *in vacuo* over $CaCl_2$.

[19] C. Galanos, E. T. Rietschel, O. Lüderitz, and O. Westphal, *Eur. J. Biochem.* **19,** 143 (1971).

Note that the lipid A preparation obtained from bacteria is very heterogeneous, as determined mainly from the lack of such components as fatty acids and phosphate. A chemically synthesized *E. coli* type lipid A preparation is now commercially available. As mentioned above, because the structural requirements of lipid A for induction of prostaglandins are not very strict, natural lipid A is adequate for the activation of macrophages. If, however, the strict chemical structure is essential to the experiment, the synthetic pure material is recommended.

Solubilization of Lipopolysaccharide or Lipid A

The solubility of LPS and lipid A depends mainly on their salt form. To reproduce the constant solubility, they should be converted to a uniform salt form. Various cations attached to the negatively charged groups of LPS or lipid A are deionized by electrodialysis.[20] They are then converted to the required salt form by neutralization with the corresponding base. Most soluble preparations are obtained by adding the triethylammonium salt. The salt is obtained by dissolving the LPS or lipid A in water containing 0.02% triethylamine. After conversion to the salt, *S*-form LPS is very soluble in both distilled water and any usual physiological buffer. Lipid A is less soluble than LPS. Saline or any other salt solution should be strictly avoided for dissolving lipid A. In both cases a fine suspension is obtained more effectively by sonication.

Preparation of Rabbit Alveolar Macrophages

Normal rabbit lung contains resident alveolar macrophages. They adhere gently to the lung tissue and can be easily harvested by bronchoalveolar lavage through the trachea with saline (or Hanks' balanced salt solution). The original method was described by Myrvik.[21]

The rabbit is sacrificed by intravenous injection of 10 ml of air into the marginal ear vein. After the skin is flushed with 70% (v/v) ethanol, the thoracic cavity is opened along the median line. The serous membrane is stripped off the lung, freeing the lung. Then the jugular is opened and the trachea is dissected from the connective tissue. Attention must be paid not to injure the lung during the procedure. Blood is removed by cutting the descending vena cava. The upper part of the trachea is dissected from the connective tissue and is clamped shut with a hemostat to prevent the influx of blood. The trachea is cut above the point where it is clamped.

[20] C. Galanos and O. Lüderitz, *Eur. J. Biochem.* **54,** 603 (1975).
[21] Q. N. Myrvik, E. S. Leake, and B. Fariss, *J. Immunol.* **86,** 128 (1961).

The lung, heart, and trachea are taken out as a block. They are washed with phosphate-buffered saline (PBS) and the heart is taken out carefully, avoiding any injury to the lung. The lung and the trachea are washed well with saline to remove the adhering blood. An oral cannula is inserted into the trachea and fixed to it by binding with a string. About 30 to 40 ml of sterile saline is injected into the lung with a syringe connected to the oral cannula and the lung is massaged gently. The fluid is then aspirated with the syringe. The procedure is repeated four or five times, injecting 20 ml of saline after the second time. A total volume of about 70 to 80 ml of saline is obtained finally. The cell suspension is then centrifuged for 5 min at 1500 rpm, and the cells are counted with a hemocytometer. About 1 to 2×10^8 cells are obtained from each rabbit. The cells obtained by this method contain more than 95% alveolar macrophages.

Alveolar macrophages can also be harvested without removing the lung. An oral cannula is inserted and fixed as described above, saline is instilled, the lung is massaged well, and the fluid is aspirated with a syringe. No significant differences in cell yield and contamination by erythrocytes are observed.

Preparation of Mouse Resident Peritoneal Macrophages

Mice of either sex weighing 25 to 30 g are usually used. The mice are sacrificed and pinned on a board. The abdominal skin is flushed with 70% ethanol. Mice are flayed carefully, avoiding injury to the peritoneum. Iscove medium (5.5 ml) containing L-glutamine and 25 m*M* *N*-2-hydroxyethylpiperazine-*N*′-2-ethanesulfonic acid (HEPES) buffer (GIBCO, Grand Island, N.Y.), penicillin (100 U/ml) and streptomycin (100 μg/ml) are injected forcefully by an injcctor with a 19-gauge needle. The peritoneal cavity is lavaged well by the solution to obtain adequate mixing. The peritoneum is then massaged well with the injector still sticking into the cavity. Care is needed not to break through the peritoneum with the syringe during the procedure. The fluid is aspirated with the same syringe. The cell suspension is stored in a plastic tube in ice. The volume of the solution obtained is about 5 ml. The yield is about 6 to 7×10^6 cells per mouse, of which approximately 40% are nonadherent lymphocytes. After purification by elimination of nonadherent cells as described below, the content of mononuclear cells is greater than 95% estimated by Giemsa and nonspecific esterase staining.

Mouse peritoneal macrophages elicited by thioglycolate are also stimulated by endotoxin to induce prostaglandin release. Thioglycolate-elicited macrophages are obtained from mice that are injected intraperitoneally 4 days before with 2 ml of thioglycolate medium, which has the following components:

Tryptone	17 g
Soy peptone	3 g
Glucose	6 g
Sodium chloride	2.5 g
Sodium thioglycolate	0.5 g
L-Cystine	0.25 g
Na_2SO_3	0.1 g
Agar	0.7 g
Distilled water	To final 1 liter
pH	7.0.

The peritoneal cells are harvested by the same procedure as described above. The yield is 2 to 3 × 10^7 cells, 80 to 90% of which are macrophages.

Cell Culture

The cell number of each preparation is adjusted to 5 × 10^5 cells/ml of Iscove medium in the case of mouse peritoneal macrophages and 1 to 5 × 10^5/ml for rabbit alveolar macrophages. One-milliliter aliquots of a cell suspension in Iscove medium are cultured in 24-well Costar plates (Costar, Cambridge, MA) at 37° with 5% CO_2 for 3 hr and the macrophages are allowed to adhere to the plate. After the cells are washed three times with PBS (37°) to remove nonadherent cells, 1 ml of Iscove medium is added to each well. Ten-microliter quantities of various concentrations of an endotoxin (either LPS or lipid A) solution as a stimulant are added to the culture which is incubated for an additional 24 hr at 37° with 5% CO_2 in a humid atmosphere. The supernatant of each culture is transferred to a plastic tube, the cells are centrifuged at 1500 rpm, and the supernatant is used for the determination of prostaglandin by RIA. Samples not assayed immediately are stored at −20°.

Quantitative Determination of Prostaglandin by Radioimmunoassay

Among the methods for measuring prostaglandins, RIA has the advantages of high sensitivity, specificity, and simplicity. Because the method for RIA of prostaglandin, including the method for preparation of antisera, is summarized in this series (Vol. 86), the method originally developed

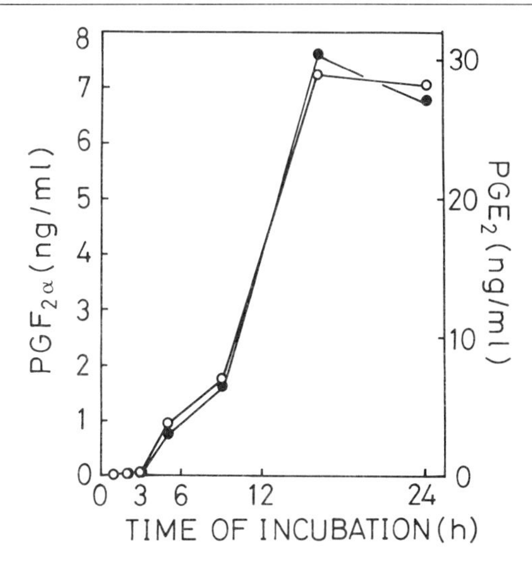

FIG. 1. Dose dependency of lipid A-induced prostaglandin release from rabbit alveolar macrophages. Cells (5×10^5) were cultured with different amounts of lipid A for 24 hr. Prostaglandins released were measured by radioimmunoassay. ●, PGE_2; ○, $PGF_{2\alpha}$.

by Pesker *et al.*[22–24] is described briefly. For more detailed information on the RIA of prostaglandins, there are several excellent reviews.[25–28]

Rabbit antibodies are obtained by using the prostaglandin–bovine serum albumin (BSA) complex as an immunogen. The concentration of antibody used for each metabolite is sufficient to bind 30 to 40% of its specific ligand. Fourteen concentrations of unlabeled prostaglandin standard ranging from 10 to 20,000 pg/ml are used for the standard curve. All reagents are adjusted to the appropriate concentration with PBS containing 0.1% gelatin. Each glass tube contains [^{3}H]prostaglandin (3000–3500 cpm), 0.1 ml of antibody, and 0.5 ml of either the prostaglandin standard or the test sample. The final volume of 1 ml is mixed by vortexing and the suspension is incubated at 4° for 18 hr. The unbound prostaglandin is removed by mixing it with 0.2 ml of charcoal (20 mg/ml) and centrifuging

[22] A. Jobke, B. A. Pesker, and B. M. Pesker, *FEBS Lett.* **37,** 192 (1973).

[23] B. A. Pesker, H. Anhut, E. E. Kroner, and B. M. Pesker, *in* "Advances in Pharmacology and Therapeutics" (J. P. Tillement, ed.), p. 341. Pergamon, Oxford and New York, 1979.

[24] J. Maclouf, this series, Vol. 86, p. 273.

[25] F. A. Fitzpatrick, this series, Vol. 86, p. 286.

[26] E. Granström, *Prostaglandins* **15,** 3 (1978).

[27] W. B. Campbell and S. R. Ojeda, this series, Vol. 141, p. 323.

[28] W. Van Rollins, S. H. K. Ho, J. E. Greenwald, M. Alexander, N. J. Dorman, L. K. Wong, and L. A. Horrocks, *Prostaglandins* **20,** 571 (1980).

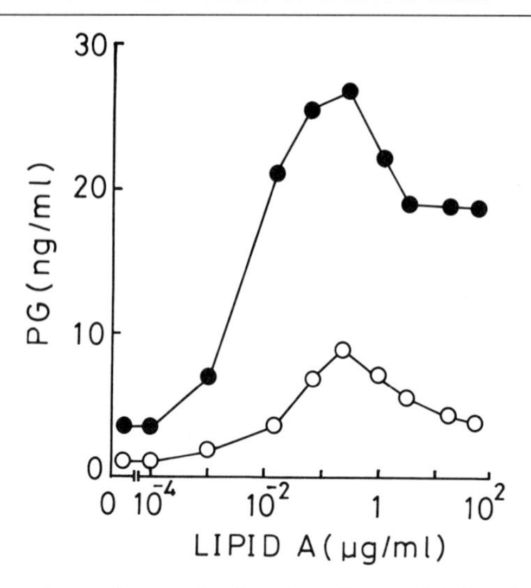

FIG. 2. Time dependency of prostaglandin release from rabbit alveolar macrophages with the stimulation of lipid A. Cells (5×10^5) were incubated with 1 μg of lipid A. Cultivation was stopped at the indicated times and the prostaglandins released were measured by radioimmunoassay. ●, PGE_2; ○, $PGF_{2\alpha}$.

the mixture at 2500 rpm for 10 min at 4°. The supernatant containing the antibody–prostaglandin complex is decanted into a 5-ml scintillation vial containing 3.5 ml of scintillation cocktail and counted in a scintillation counter. The prostaglandin content of the sample is calculated from the standard curve.

Separation and Purification of Prostaglandin by High-Performance Liquid Chromatography

Prostaglandins released in the supernatant (1 ml of Iscove medium) are extracted with 1 ml of ethyl acetate twice after the pH is adjusted to 3.5 with 1 *N* HCl. The ethyl acetate is evaporated by blowing with nitrogen gas and the remaining prostaglandins are dissolved in 20 μl of methanol. Samples are then analyzed by HPLC (Gilson, Model 802, Pump Model 302, Middleton, WI) using a column of ODS reversed phase (C_{18}) (5 μm, 4.6×250 mm) containing Lichrosorb. Elution is accomplished with a mixture of acetonitrile and water (32 : 68) containing 0.05% acetic acid at a flow rate of 1.5 ml/min. Labeled standards are used to establish retention times. The fractions corresponding to $PGF_{2\alpha}$, PGE_2, and 6-keto-$PGF_{1\alpha}$ are collected and evaporated. The remaining prostaglandins are dissolved in 1 ml of Iscove medium and saved for RIA.

The details of the methods for purification of the samples using HPLC are also described in other reviews.[25-29]

Dose and Time Dependency of Endotoxin-Induced Prostaglandin Release from Rabbit Alveolar Macrophages

Prostaglandins released into the medium are estimated in rabbit alveolar macrophages (5×10^5 cells/ml/well) treated with different concentrations of lipid A derived from *Salmonella minnesota* R 595 LPS. As shown in Fig. 1, the cells secrete PGE_2 and $PGF_{2\alpha}$ even at a concentration of 1 ng of lipid A/ml. Maximal prostaglandin production, 2.7×10^5 pg/ml (PGE_2) and 7.5×10^4 pg/ml ($PGF_{2\alpha}$), is observed at a concentration of 0.3 μg of lipid A/ml. At higher doses the production of prostaglandins decreases, probably because of the cell toxicity of lipid A. With stimulation by *S*-form LPS derived from *Citrobacter* the cells also produce prostaglandins. In this case no apparent cell toxicity is observed and maximal prostaglandin production (PGE_2, 7.2×10^4 pg/ml, and $PGF_{2\alpha}$, 4.5×10^4 pg/ml) occurs at the highest dose of LPS tested (100 μg/ml).

Time dependency of prostaglandin release from rabbit alveolar macrophages is tested with 1 μg of LPS/ml as a stimulant. After different culture times, prostaglandins released in the supernatant are measured by RIA. The time course of PGE_2 and $PGF_{2\alpha}$ production is shown in Fig. 2. Macrophages, when subjected to phagocytic stimuli such as zymosan, produce maximal amounts of prostaglandins within 1 to 2 hr of incubation. On the other hand, prostaglandin production is not apparent in such a short time when macrophages are stimulated by LPS; maximum prostaglandin release is observed after 16 to 24 hr.

Acknowledgments

I am grateful to Prof. Dr. Ernst Th. Rietschel and Dr. U. Shade, Borstel Institute, Germany, for their support throughout the course of this work.

[29] D. Venton, G. L. Breton, and E. Hall, this series, Vol. 187, p. 245.

[5] Effect of Bacterial Products on Interferon Production

By Gerald Sonnenfeld

Introduction

Interferons were described originally as agents produced by mammalian cells in response to viral infections.[1,2] Over the years, it has become apparent that interferons can be induced by several other stimulators, such as double-stranded RNAs and a wide variety of microorganisms, including bacteria.[3]

At first, it appeared unusual that interferons, known as antiviral agents, could be induced by bacteria; however, as knowledge of the interferon system and its interaction with the immune system grew, it became clear that induction of interferon by bacteria should have been expected.[4] It has become apparent that interferons can be produced as part of an immune response to foreign pathogens, and play a major role in regulating the response to those pathogens.[4,5] Therefore, the production of interferons by cells that play major roles in regulating immune responses, T lymphocytes, macrophages/monocytes, and natural killer cells, in response to bacterial infection is part of the host response to bacterial infection.[4,5]

There are three major classes of interferons: interferons-α, -β, -γ. Interferons-α and -β can often not be separated easily in the murine system and are referred to commonly as interferons-α/β when production *in vivo* in rodents is described.[6] Interferon-α is produced primarily by leukocytes and interferon-β is produced primarily by fibroblasts in response to viruses, double-stranded RNAs, and bacterial products such as lipopolysaccharide.[6] Interferon-γ is produced primarily by T lymphocytes in response to mitogenic challenge or by sensitized T lymphocytes in response to challenge with the sensitizing antigen.[6] In addition, interferon-γ can also be produced by natural killer cells in response to various microbial chal-

[1] A. Isaacs and J. Lindemann, *Proc. R. Soc. London, Ser. B* **147,** 258 (1957).

[2] I. Gresser, *Cell. Immunol.* **34,** 406 (1977).

[3] W. E. Stewart, II, "The Interferon System." Springer-Verlag, Wien and New York, 1979.

[4] G. I. Byrne and J. Turco, "Interferon and Nonviral Pathogens." Dekker, New York, 1988.

[5] G. Sonnenfeld, C. Czarniecki, C. Nacy, G. I. Byrne, and M. Degré, "Cytokines and Resistance to Nonviral Pathogenic Infections." Biomedical Press, Augusta, GA, 1992.

[6] A. Billiau, "Interferon-1: General and Applied Aspects." Elsevier Biomedical Press, Amsterdam, 1984.

lenges, including exposure to mycoplasma.[7] It is therefore not surprising that the exposure of the host to bacterial pathogens could yield great production of interferon.

Interferons have also been shown to play a role in host regulation of bacterial infection. Use of antibodies against interferon-γ in murine hosts yielding decreased resistance to infection with *Listeria monocytogenes* has suggested that the interferon could play a role in natural regulation of resistance to infection with bacterial pathogens.[8] In addition, exogenous interferons have been used to treat animal models of bacterial infection with some success.[9–15] Included among the bacterial infections treated either prophylactically or therapeutically with interferons are infections of rodents with *Salmonella typhimurium, Klebsiella pneumoniae, Staphylococcus aureus,* and *Escherichia coli.*[9–15] Clinical trials with interferon-γ have been carried out in patients suffering from bacterial infection after trauma surgery and chronic granulomatous disease.[16,17] Beneficial results regarding infection observed in the chronic granulomatous disease trial have led to U.S. licensure for interferon-γ. These results are additional evidence of a fundamental role for interferons in immune regulation of bacterial infections.

The list of microorganisms that can induce interferons is long. It is not limited to bacteria, as many fungi and protozoa can also induce interferons.[3] This is probably due to stimulation of similar cell types, T cells, macrophages/monocytes, and natural killer cells, by all of the microbes, again reinforcing the fundamental role of interferons in regulating resistance to microbial infections. The list of bacteria that can induce interferons includes both gram-negative and gram-positive bacteria such as *S. typhimurium, S. aureus,* and *Mycobacterium bovis* strain bacillus

[7] V. Kumar, J. Lust, A. Gifaldi, M. Bennett, and G. Sonnenfeld, *Immunobiology* **165,** 445 (1983).

[8] N. Buchmeier and R. D. Schreiber, *Proc. Natl. Acad. Sci. U.S.A.* **82,** 7407 (1985).

[9] Z. Izadkhah, A. D. Mandel, and G. Sonnenfeld, *J. Interferon Res.* **1,** 137 (1980).

[10] C. L. Gould and G. Sonnenfeld, *J. Interferon Res.* **7,** 255 (1987).

[11] M. J. Hershman, H. C. Polk, Jr., J. D. Pietsch, D. Kuftinec, and G. Sonnenfeld, *Clin. Exp. Immunol.* **73,** 406 (1988).

[12] A. F. Kiderlen, S. H. Kaufman, and M. L. Lohmann-Matthes, *Eur. J. Immunol.* **14,** 964 (1984).

[13] R. S. Kurtz, K. M. Young, and C. J. Czuprynski, *Infect. Immun.* **57,** 554 (1989).

[14] D. H. Livingston and M. A. Malangoni, *J. Surg. Res.* **45,** 37 (1988).

[15] H. Rollag, M. Degré, and G. Sonnenfeld, *Scand. J. Immunol.* **20,** 149 (1984).

[16] H. C. Polk, Jr., W. G. Cheadle, D. H. Livingston, J. L. Rodriguez, K. M. Starko, A. E. Izu, H. S. Jaffe, and G. Sonnenfeld, *Am. J. Surg.* **164,** 919 (1992).

[17] The International Chronic Granulomatous Diseases Cooperative Study Group, *N. Engl. J. Med.* **324,** 509 (1991).

Calmette–Guérin (BCG).[3] In addition, many bacterial products such as lipopolysaccharide from *S. typhimurium* and *E. coli,* enterotoxin A from *S. aureus,* and aqueous ether extract of *Brucella abortus* (BRU-PEL) have also been shown to be efficient inducers of various kinds of interferons.[18–23] The list of bacteria and bacterial products that induce interferons is too long to include here and is summarized in a monograph by Stewart.[3] In the remainder of this article, one technique for inducing interferon-α/β and one technique for inducing interferon-γ using bacteria and/or bacterial products that can be used readily in animal models systems are described. The interferons are not found normally in the circulation of the animals after infection, and these techniques for induction using bacteria or bacterial products require massive stimulation of the host to be able to detect circulating interferons. It should be noted that because these are infections to which the host is mounting an immune response, many additional cytokines besides the interferons are likely to be induced by bacteria.

Induction of Interferon-α/β by Bacterial Lipopolysaccharide

The induction of interferon-α/β by bacterial lipopolysaccharide could occur as a result of infection of the host with gram-negative bacteria that have lipopolysaccharide (endotoxin) as an integral component of the bacterial cell wall. Interferon-α/β induction has been achieved using several types of lipopolysaccharide from various gram-negative organisms. The model system now to be described uses lipopolysaccharide extracted and purified from *S. typhimurium* strain LT2 by means of the methods of Westphal and Jann.[24,25] Commercially available lipopolysaccharide from the same bacteria or other gram-negative bacteria could be substituted, but might not yield the same titer of interferon.

In this model, circulating titers of interferon-α/β can be induced with injection of lipopolysaccharide directly into the mice, but levels of circulating interferon were increased greatly if the mice were pretreated with *M. bovis* strain BCG or a cell wall extract from that organism.[24] It appears

[18] W. R. Stinebring and J. S. Youngner, *Nature* (*London*) **204,** 712 (1964).
[19] M. Ho, *Science* **146,** 1472 (1964).
[20] Y. Nagano, *Tex. Rep. Biol. Med.* **35,** 105 (1977).
[21] H. M. Johnson, G. J. Stanton, and S. Baron, *Proc. Soc. Exp. Biol. Med.* **154,** 138 (1977).
[22] J. S. Youngner, *J. Gen. Physiol.* **56,** 25s (1970).
[23] J. S. Youngner, G. Keleti, and D. S. Feingold, *Infect. Immun.* **10,** 1202 (1974).
[24] G. Sonnenfeld, S. B. Salvin, and J. S. Youngner, *Infect. Immun.* **18,** 283 (1977).
[25] O. Westphal and K. Jann, *Methods Carbohydr. Chem.* **5,** 83 (1965).

that the general immune stimulation induced by the treatment with *M. bovis* strain BCG can enhance the production of interferon.

To carry out the induction of interferon-α/β, adult (25–30 g) mice are used.[24] Outbred Swiss/Webster mice can be used, but inbred strains of mice may provide higher titers of interferon because of the genetic control of interferon production.[26] The sterile lipopolysaccharide preparation should be injected intravenously into the mice. One microgram of lipopolysaccharide per mouse is sufficient to produce maximum levels of interferon-α/β. The mice should be bled 2 hr after lipopolysaccharide injection to optimize production of the interferon. The interferon-α/β activity is found in the serum; however, it is not pure. Although the major species is interferon-α/β, some interferon-γ may be present, as may other cytokines. Purification of the interferon-α/β may be desirable.[27]

If enhancement of interferon-α/β production is desired by sensitization with *M. bovis* strain BCG, the following protocol should be followed.[28] *Mycobacterium bovis* strain BCG is grown for 10 to 12 days in Dubos medium. The bacteria are harvested and resuspended in phosphate-buffered saline. A reading is taken on a Klett colorimeter, and the bacteria are diluted with phosphate-buffered saline to a reading that had been predetermined to be equivalent to a concentration between 1×10^6 and 1×10^7 bacteria/ml. This should be confirmed by plating each batch of bacteria on Lowenstein–Jensen medium for a viable count. One-tenth milliliter of the bacterial suspension is injected intravenously into the mice.

Alternatively, the mice can be sensitized with a BCG cell wall preparation.[24,29] In this case, 24.75 mg of cells walls is placed in a 50-ml tissue grinder with a Teflon pestle, and 0.12 ml of light mineral oil is added. The mixture is ground to a paste, and then 16.7 ml of 0.85% NaCl containing 0.2% (v/v) Tween 80 is added. Grinding should be continued until a satisfactory emulsion is obtained. Three hundred micrograms of the cell wall is injected intravenously into the mice.

Four weeks after sensitization, 1 μg of lipopolysaccharide is injected intravenously into the mice. Two hours later, the mice are bled. The interferon-α/β activity is in the serum, but it should be noted that with the overall stimulation of immunity induced by BCG, it is more likely to be contaminated with other cytokines than the interferon-α/β induced by lipopolysaccharide without presensitization with BCG.

[26] E. DeMaeyer, J. DeMaeyer-Guignard, and D. W. Bailey, *Immunogenetics* **1,** 438 (1975).

[27] Y. Kawade, J. Fujisawa, S. Yonehara, Y. Iwakura, and Y. Yamamoto, this series, Vol. 78, p. 522.

[28] J. S. Youngner and S. B. Salvin, *J. Immunol.* **111,** 1914 (1973).

[29] S. B. Salvin, J. S. Youngner, and W. H. Lederer, *Infect. Immun.* **7,** 68 (1973).

Induction of Interferon-γ by *Mycobacterium bovis* Strain Bacillus Calmette–Guérin

The induction of interferon-γ in a host could occur as a result of stimulation of T cells or natural killer cells by bacteria that induce cell-mediated immune responses. The classic model system for induction of interferon-γ in mice involves the sensitization of mice with *M. bovis* strain BCG followed by challenge with tuberculin.[24,28,29] This technique results in the production of high titers of circulating interferon-γ, along with many other contaminating cytokines. Purification of the interferon-γ may be desirable after production.[30]

To induce interferon-γ, adult (25–30 g) mice are used.[24] Outbred Swiss/Webster mice can be used, but inbred strains of mice may provide higher titers of interferon because of the genetic control of interferon production.[26] *Mycobacterium bovis* strain BCG is grown for 10 to 12 days in Dubos medium.[28] The bacteria are harvested and resuspended in phosphate-buffered saline. A reading is taken on a Klett colorimeter, and the bacteria are diluted with phosphate-buffered saline to a reading that had been predetermined to be equivalent to a concentration between 1×10^6 and 1×10^7 bacteria/ml. This should be confirmed by plating each batch of bacteria on Lowenstein–Jensen medium for a viable count. One-tenth milliliter of the bacterial suspension is injected intravenously into the mice.

Alternatively, the mice can be sensitized with a BCG cell wall preparation.[24,29] In this case, 24.75 mg of cells walls is placed in a 50-ml tissue grinder with a Teflon pestle, and 0.12 ml of light mineral oil is added. The mixture is ground to a paste, and then 16.7 ml of 0.85% NaCl containing 0.2% Tween 80 is added. Grinding is continued until a satisfactory emulsion is obtained. Three hundred micrograms of the cell wall is injected intravenously into the mice.

Four weeks after sensitization with BCG, the mice are injected intravenously with 50 mg (0.2 ml) of old tuberculin. The mice are bled 3 to 4 hr later. The interferon-γ activity is in the serum. Alternatively, the spleens can be removed from BCG-sensitized mice 45 min after the old tuberculin challenge.[31] The spleens are dissociated into individual cells and placed into culture at a concentration of 10^7 cells/ml in Roswell Park Memorial Institute (RPMI) 1640 medium with 5% fetal bovine serum, 5×10^{-5} *M* 2-mercaptoethanol, and appropriate concentrations of L-glutamine, antibiotics, and sodium bicarbonate for tissue culture. The cells are incubated

[30] J. Wietzerbin and E. Falcoff, this series, Vol. 78, p. 552.

[31] G. Sonnenfeld, A. D. Mandel, and T. C. Merigan, *Cell. Immunol.* **34,** 193 (1977).

24 hr at 37° in an atmosphere of 7% O_2, 10% CO_2, and 83% N_2. After incubation, the interferon-γ activity is found in the culture supernatant fluid. This provides a interferon-γ preparation purer than that found in the serum, but it still contains other cytokine activities.

[6] Regulation of Osteoclastic Activity in Infection

By J. Edward Puzas, David G. Hicks, Susan D. Reynolds, and Regis J. O'Keefe

Introduction

The structural integrity of the skeleton is vital for normal human locomotion and function. The maintenance of the skeleton depends on a subtle balance between bone resorption and bone formation. In normal circumstances this balance is controlled by both systemic and locally acting signaling mechanisms. In disease states, however, the rates of bone resorption may exceed the rates of bone formation and lead to a net loss of bone mass. In some cases this process is generalized and affects the entire skeleton, as in osteoporosis and osteopenic bone diseases. In other cases, bone metabolism may be altered at specific sites. Focal lesions relating to infection, inflammation, tumors, etc., are situations that can result in severe localized osteolysis. The factors that are responsible for the net negative bone balance in infection and inflammation are the topic of this discussion.

Skeletal Homeostasis

In an adult human free from disease, the skeleton remains at a relatively stable total mass for more than two decades. This is due to an exact matching of the rates of bone formation with the rates of bone resorption. Such a balance can occur over a wide range of skeletal activity. That is, if resorbing cells (osteoclasts) are stimulated to resorb bone, a compensatory process stimulates the recruitment of forming cells (osteoblasts) to restore the resorbed bone after the resorption stimulus is removed. In fact, during this time of skeletal stability, it is actually quite difficult to alter bone mass at any site in the body. It then follows that in focal disease states such as infection and inflammation, where bone mass is permanently lost, exogenous agents must be acting to circumvent the normal regulatory pathways.

TABLE I
PRIMARY STIMULATORS OF BONE RESORPTION THAT REQUIRE A "HELPER" CELL TO ACTIVATE OSTEOCLASTS

Parathyroid hormone
1,25-Dihydroxyvitamin D_3
Interleukin-1 (α and β)
Lymphotoxin
Tumor necrosis factor α
Prostaglandins
Transforming growth factor β_1

A better understanding of the agents that stimulate bone resorption in these states should help to elucidate the mechanisms behind the permanent bone loss.

Activation and Recruitment of Osteoclasts

It is generally accepted that the origin of osteoclasts is the hematopoietic mononuclear cells that can be found in marrow and in the general circulation. The earliest identifiable stem cell in the osteoclast pathway is the granulocyte–macrophage colony-forming unit. This precursor cell is under the influence of a number of circulating cytokines that control its differentiation. One such critical molecule is colony-stimulating factor 1 (CSF-1). In animal mutations such as the *op/op* mouse, where there is impaired production of CSF-1, osteoclasts fail to develop.[1] Some of the early experiments that formed the direction for studying osteoclast ontogeny were performed by Walker.[2] His data showed that it was possible to cure both animals and humans that were deficient in osteoclast production by either transplanting marrow stem cells or creating a parabiotic union between the animals.

Once a CFU-gm stem cell becomes committed to the osteoclast lineage, a number of modulating factors and processes control its activity. From studies performed with isolated and purified preparations of osteoclasts and osteoclast precursors, it has been shown that a number of agents once believed to be primary stimulators of bone resorption do not directly affect osteoclastic activity.[3] A list of these molecules is presented in Table I. That is, most of the stimuli that are recognized as inducers of

[1] W. Wiktor-Jedrzejczak, E. Urbanowska, S. L. Aukerman, J. W. Pollard, E. R. Stanley, P. Ralph, A. A. Ansari, K. W. Sell, and M. Szperl, *Exp. Hematol.* **19,** 1049 (1991).
[2] D. G. Walker, *Clin. Orthop. Relat. Res.* **97,** 158 (1973).
[3] P. M. J. McSheehy and T. J. Chambers, *Endocrinology* **118,** 824 (1986).

bone resorption act through a secondary cell type. It appears that the secondary cell type is a mesenchymal lining cell or osteoblast. At present, there is an active search for the true osteoclast-inducing agent(s) produced by these cells.

In addition to a soluble mediator for osteoclast activation, the lining cells of bone may also direct the site at which bone will be resorbed. Although these mechanisms are not as well characterized as those for systemic and local control of osteolysis, they are a critical component of bone resorption. The current thinking in this regard is that lining cells or osteoblasts, on activation by a systemic agent such as parathyroid hormone, retract their membranous extensions and expose bone surfaces. Once the bone surface is exposed, the thin layer of osteoid overlying the surface may be removed by secretion of a collagenase from the lining cells.[4-6] Thus, by virtue of both a mechanical and enzymatic process a bony surface may be prepared for osteoclastic activity. The macroscopic regulation that determines which sites in the skeleton undergo bone resorption at any given time are not yet understood; however, it is possible that biomechanical factors play a role in this higher-order regulation.

Periodontal Disease and Osteomyelitis

Two processes that ultimately result in localized bone loss are periodontal disease and osteomyelitis. These afflictions may be manifestations of similar pathological processes, but research mechanisms could provide a more complete picture of both processes.

Periodontal Disease

End-stage periodontal disease manifests itself as a permanent loss of alveolar bone, resulting in tooth loosening and ultimately tooth loss. The bone loss is due to an imbalance in the normally coupled activities of osteoclasts and osteoblasts. A substantial amount of information is available regarding the activation of osteoclasts in this disease. It is now known that bacterial stimulation of bone resorption is a multistep process. Recent reports have documented that supernatants of peripheral blood mononuclear cell cultures in the presence of *Bacteroides gingivalis* or *Escherichia coli* show strong osteoclast-stimulating activity[7]; how-

[4] S. J. Jones, A. Boyde, N. N. Ali, and E. Maconnachie, *Scanning* **7,** 5 (1985).

[5] J. E. Puzas and J. S. Brand, *Biochim. Biophys. Acta* **429,** 964 (1975).

[6] J. E. Puzas and J. S. Brand, *Endocrinology* **104,** 559 (1979).

[7] A. A. Bom van Noorloos, J. W. van der Meer, J. S. van de Gevel, W. Schepens, T. J. van Steenbergen, and W. H. Burger, *J. Clin. Periodontol.* **17,** 409 (1990).

ever, the molecules stimulating bone resorption are not derived from the bacteria but rather appear to be cytokines such as interleukin-1 (IL-1) released from monocytes. This process does require viable bacteria as heating of the microorganisms eliminates the effect. The major bacterial agent responsible for initiating the effects is probably not lipopolysaccharide (LPS) as was once believed, because new data indicate that purified LPS from *B. gingivalis* and *E. coli* is not particularly effective in releasing bone-resorbing activity. Furthermore, cells other than the monocyte/macrophage can participate in the release of bone-resorbing cytokines. These include B cells, plasma cells, and spleen cells.[8,9] Thus, bacterial components that include LPS as well as other (possibly more potent) molecules activate host cells to begin the process of bone resorption.

Further experiments using bacterial and immune cell extracts in combination with isolated osteoclasts have documented that neither the bacterial components (i.e., LPS) nor the cytokines (i.e., IL-1) directly stimulate the activity of osteoclasts. That is, when isolated osteoclasts are exposed to bacterial stimulators of bone resorption there is no increase in cell activity. The same is true for IL-1. When these agents are exposed to osteoclasts in the presence of osteoblasts, however, a potent stimulation of bone resorption occurs.[10] These observations reinforce the thesis that the osteoblast or mesenchymal lining cell may be the pivotal cell with respect to the regulation of bone resorption.

The agent(s) released by osteoblasts (and other mesenchymal lining cells) that directly stimulates osteoclasts is, at this time, not known. A likely candidate that is receiving intense investigation as a possible mediator is interleukin-6 (IL-6) (see below for a more complete discussion). With the substantial amount of research being performed in this area the soluble mediator(s) that stimulates osteoclasts should be identified within a relatively short time.[11–16]

[8] M. Tew, D. Engel, and D. Mangan, *J. Periodontal Res.* **24,** 225 (1989).

[9] A. A. Bom van Noorloos, T. J. van Steenbergen, and E. H. Burger, *J. Clin. Periodontol.* **16,** 412 (1989).

[10] H. J. Sismey-Durrant and R. M. Hopps, *Arch. Oral Biol.* **32,** 911 (1987).

[11] T. Suda, N. Takahashi, and T. J. Martin, *Endocr. Rev.* **13,** 66 (1992).

[12] H. M. Perry, W. Skogen, J. Chappel, A. J. Kahn, G. Wilner, and S. L. Teitelbaum, *Endocrinology* **125,** 2075 (1989).

[13] M. J. Oursler and P. Osdoby, *Dev. Biol.* **127,** 170 (1988).

[14] J. D. Malone, S. L. Teitelbaum, G. L. Griffin, R. M. Senior, and A. J. Kahn, *J. Cell Biol.* **92,** 227 (1982).

[15] M. de Vernejoul, M. Horowitz, J. Demingnon, L. Neff, and R. Baron, *J. Bone Min. Res.* **3,** 69 (1988).

[16] H. Takahashi, T. Akatsu, H. Udagawa, T. Sasaki, A. Yamaguchi, J. Moseley, T. J. Martin, and T. Suda, *Endocrinology* **123,** 2680 (1988).

Although many of the mechanisms that stimulate osteoclastic bone resorption in periodontal disease are known, the underlying reason for an imbalance between the extent of resorption and the compensatory formation remains undiscovered. That is, if accelerated bone resorption were to occur in an environment that did not include bacterial and inflammatory cell activity, virtually all of the resorbed bone would be replaced and there would be very little change in net bone volume; however, because this does not occur in alveolar bone during a periodontal episode, other mechanisms must be interfering with a normal remodeling process. The results of histological studies in nonhuman primates suggest that bone formation at prior resorption sites in periodontal pockets is compromised. This is best appreciated when viewing the newly forming osseous tissue under polarized light (Fig. 1A). Such an examination demonstrates the presence of haphazard osteoblastic activity that leads to the formation of woven bone. This contrasts markedly with the lamellar bone that is found around unaffected teeth. Moreover, the reversal lines that signify the margin between the ending of bone resorption and the beginning of bone formation are discontinuous in periodontal remodeling sites as compared with normal sites (Fig. 1B). Thus, one potential mechanism for an inadequate degree of bone formation in infected and inflamed alveolar tissue may relate to disrupted orientational signals for osteoblastic activity. This disruption may be mediated by the multitude of regulatory agents and enzymes elaborated by bacteria or immune cells.

Osteomyelitis

Osteomyelitis (or bone infection) evokes an exuberant inflammatory response culminating in focal osteolysis. The affliction is characterized by necrosis and replacement of normal marrow (Fig. 2). In this disease, the affected bone is exposed to high concentrations of bacterial endotoxins, proteolytic enzymes, cytokines, and activated inflammatory cells. The bacterial products are usually derived from pathogenic strains of microorganisms. The predominant bacterium associated with osteomyelitis is *Staphylococcus*, with at least 95% of all pyogenic infections due to this microorganism. The other 5% of bone infections are due primarily to *Streptococcus* (3%), with strains of *Pneumococcus, Brucella,* and bacilli of the *Salmonella* group accounting for the remaining 2%.

Osteomyelitis usually occurs when bacteria from some other area of infection lodge in a bony site. The inflammatory cascade that then ensues is thought to be the origin of the signals that lead to rapid bone resorption and osteolysis. Although the details of bone resorption in osteomyelitis have not been elucidated, it is well recognized that factors mediating the acute inflammatory response are involved.

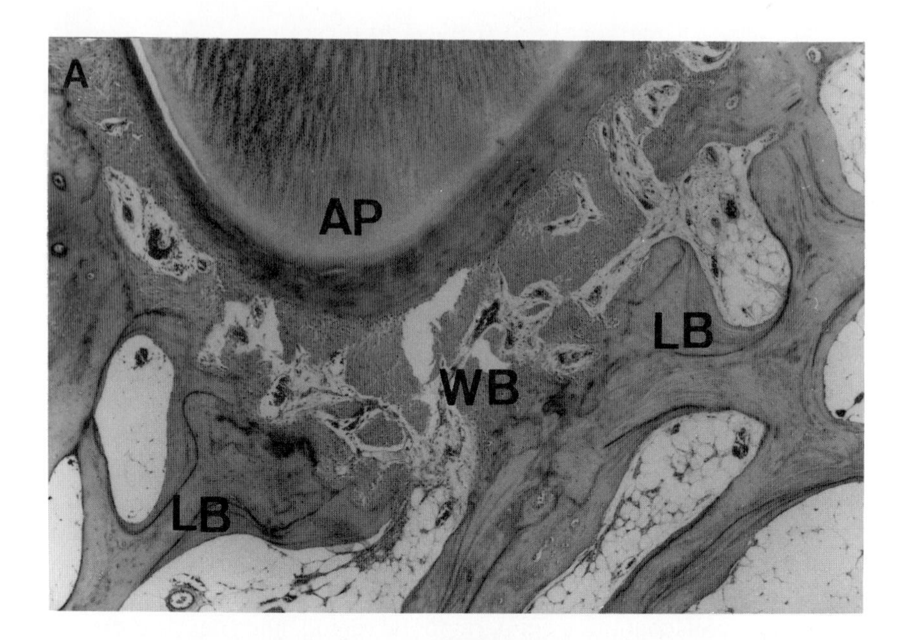

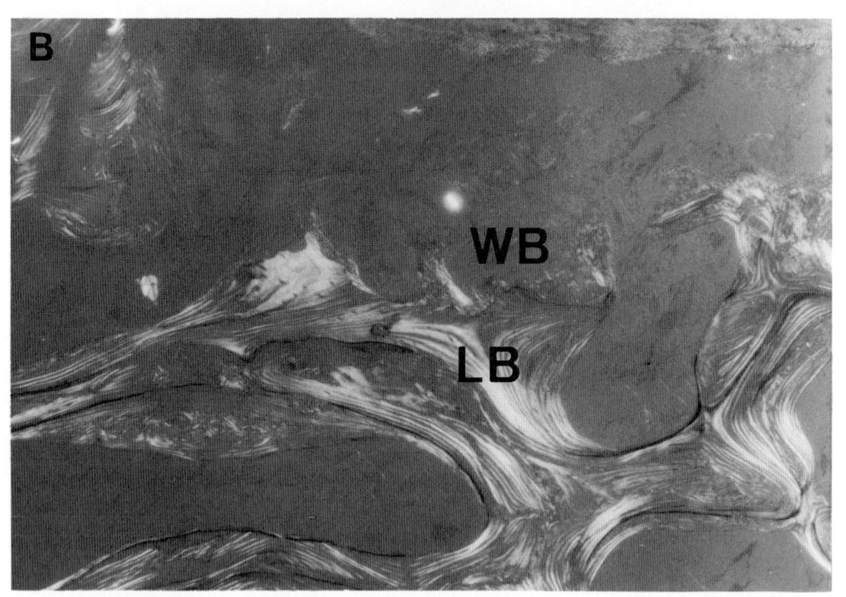

FIG. 1. Alveolar bone of nonhuman primate periodontal disease. (A) Regions of abnormal woven bone (WB) surrounded by normal lamellar bone (LB). The tooth apex is indicated by "AP". (B) Shows a similar section viewed under polarized light. The normal lamellar bone (LB) appears striated due to the laminar patterns of normal bone formation whereas woven bone (WB) shows none of this structure. The woven bone is characteristic of a structurally unsound tissue and contributes to tooth loosening.

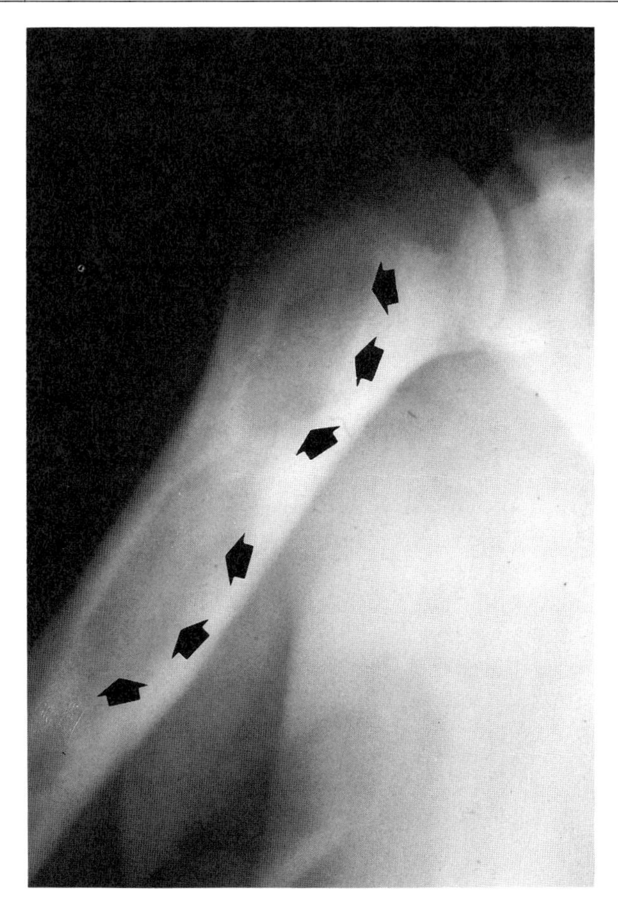

FIG. 2. Radiograph demonstrating the lytic nature of osteomyelitis. This photograph shows radiolucent areas (arrows) caused by the permanent bone loss mediated by infection. The focal osteopenia decreases the structural integrity and mechanical strength of the bone.

Factors important in the formation and stimulation of early osteoclastic cells in osteomyelitis are normally produced in the local marrow environment and include interleukins-1, -3, and -6, tumor necrosis factor, and the colony-stimulating factors gm-CSF and m-CSF. A recent study examining the role of the colony-stimulating factors in osteoclast differentiation showed that the effects of these cytokines were targeted to a population of marrow cells expressing the CD34 antigen. Although gm-CSF, m-CSF, and interleukin 3 all stimulated osteoclastic marker expression in the colonies, expression of calcitonin receptors (a more definitive marker for osteoclasts) was specific for cells treated with m-CSF and interleukin-3.

These latter factors were synergistic in combination and apparently are specific among the CSFs for stimulation of an osteoclastic phenotype.[17] These results are pertinent to *in vivo* pathways as injections of recombinant CSFs in animals results in a marked increase in osteoclast numbers and bone resorption.[18]

Similar to the colony-stimulating factors, the role of interleukin-6 in osteoclastic bone resorption is specific for effects on progenitor cells.[19–21] In bone-resorbing systems that contain only mature osteoclasts, IL-6 has had minimal effects on bone resorption, whereas large effects are seen in systems that contain osteoclast precursor cells.[20,22–24] Furthermore, in a cultured calvarial system, IL-6 stimulates a threefold increase in osteoclast number and a large increase in bone resorption over a 3-day period.[25] In animal models and in tumor patients in which IL-6 is produced in excess, dramatic increases in bone resorption are often associated with hypercalcemia.[26] Thus, the role of this molecule may be key in the progression of osteolysis in osteomyelitis. IL-6 is produced by a number of cell types, including monocytes/macrophages, lymphocytes, bone cells, and a number of neoplastic cell types.[27] IL-6 is typically not produced constitutively by normal cells, but is induced in cells of marrow origin by factors inciting an inflammatory response, including bacterial LPS, phytohemagglutinin, and products of the staphylococcal strains.[27,28] IL-6 is also produced in monocytes, macrophages, and osteoblasts in response to treatment with IL-1, parathyroid hormone, and tumor necrosis factor α (TNF-α).[25,28] As estrogen has been shown to inhibit IL-6 production strongly and basal synthesis of this factor is increased in marrow cell populations in the

[17] B. T. Povolony and M. Y. Lee, *Exp. Hematol.* **21,** 532 (1993).

[18] M. Y. Lee, R. Fukunaga, T. J. Lee, J. L. Lottsfeldt, and S. Nagata, *Blood* **77**(10), 2135 (1991).

[19] G. R. Mundy, *Int. J. Cell Cloning* **10,** 215 (1992).

[20] C. W. G. M. Lowik, G. van der Pluijm, H. Bloys, K. Hoekman, O. L. M. Bijvoet, L. A. Aarden, and S. E. Papapoulos, *Biochem. Biophys. Res. Commun.* **162,** 1546 (1989).

[21] N. Kurihara, C. Civin, and G. D. Roodman, *J. Bone Min. Res.* **6,** 257 (1991).

[22] G. G. Wong and S. Clark, *Immunol. Today* **9,** 137 (1988).

[23] R. Jilka, G. Hangoc, G. Girasole, G. Passeri, D. Williams, J. Abrams, B. Boyce, H. Broxmeyer, and S. Manolagas, *Science* **257,** 88 (1992).

[24] D. G. Roodman, N. Kurihara, Y. U. Ohsaki, A. Kukita, D. Hosking, A. Demulder, and F. Singer, *J. Clin. Invest.* **89,** 46 (1992).

[25] Y. Ishimi, C. Miyaura, C. H. Jin, T. Akatsu, E. Abe, Y. Nakamura, A. Yamaguchi, S. Yoshiki, T. Matsuda, T. Hirano, T. Kishimoto, and T. Suda, *J. Immunol.* **145,** 3297 (1990).

[26] K. Black, I. R. Garrett, and G. R. Mundy, *Endocrinology* **128,** 2657 (1991).

[27] J. Van Snick, *Annu. Rev. Immunol.* **8,** 253 (1990).

[28] R. Schindler, J. Mancilla, S. Endres, R. Ghordini, S. C. Clark, and C. A. Dinarello, *Blood* **75,** 40 (1990).

absence of estrogen, a role for IL-6 in the development of postmenopausal osteoporosis has recently been postulated.[23,29]

Two other key cytokines that are clearly involved in osteoclastic bone resorption in osteomyelitis are IL-1 and TNF-α. Both IL-1 and TNF-α stimulate bone resorption in mixed osteoclast/osteoblast cell cultures and stimulate up to a 60-fold increase in osteoclast formation in long-term marrow cultures. These effects are synergistic and are potentiated by 1,25-dihydroxyvitamin D which stimulates fusion of mononuclear cells.[30] Unlike IL-6, IL-1 and TNF-α both stimulate activation of mature osteoclasts. Although IL-1 has been shown to increase intracellular free calcium[31] in mature osteoclasts its effects on bone resorption are most likely mediated through osteoblasts. TNF-α effects on mature osteoclasts also are mediated through interactions with osteoblasts.[32] Similar to IL-6, IL-1 and TNF-α are both mediators of the acute inflammatory response and are produced by activated lymphocytes, macrophages, and polymorphonuclear leukocytes.[33,34] Under physiological conditions of bone resorption, TNF-α and IL-1 are produced by marrow progenitor cells, although the regulation of these factors under basal conditions is poorly understood. One of the major roles of these agents in bone resorption is the potent stimulation of IL-6 production, an effect that is inhibited in the presence of estrogen.[23] Although TNF-α and IL-1 stimulate IL-6 synthesis, they are in turn inhibited by this factor, demonstrating a feedback inhibition loop.

Figures 3A and 3B demonstrate the juxtaposition of all of the cells involved in osteomyelitis. These photographs document the presence of monocytes, macrophages, lymphocytes, and osteoclasts in focal resorption due to infection. Such histological evidence underscores the local nature of infection-mediated osteolysis.

Summary

Figure 4 is a diagrammatic representation of five pathways involved in the activation of osteoclastic and osteoblastic cell activity during an infectious process. Pathways 1 and 2 are involved in the recruitment and activation of osteoclasts. These pathways are controlled by systemic hormones and cytokines of the infection/immune axis. As described

[29] S. C. Manolagas and R. L. Jilka, *Calcif. Tissue Int.* **50,** 199 (1992).

[30] J. Pfeilschifter, C. Chenu, A. Bird, G. R. Mundy, and G. D. Roodman, *J. Bone Min. Res.* **4,** 113 (1989).

[31] H. Yu and J. Ferrier, *Biochem. Biophys. Res. Commun.* **191,** 343 (1993).

[32] B. M. Thomson, G. Mundy, and T. J. Chambers, *J. Immunol.* **138,** 775 (1987).

[33] K. J. Tracey, H. Vlassara, and A. Cerami, *Lancet,* May 20, **1 (8647),** 1122–1126 (1989).

[34] K. Tiku, M. L. Tiku, and J. L. Skosey, *J. Immunol.* **136,** 3677 (1986).

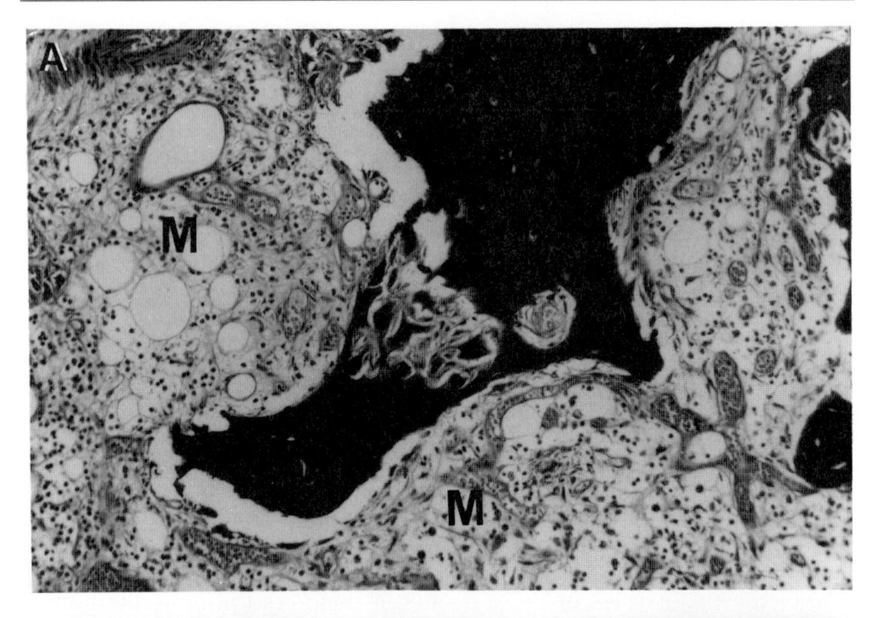

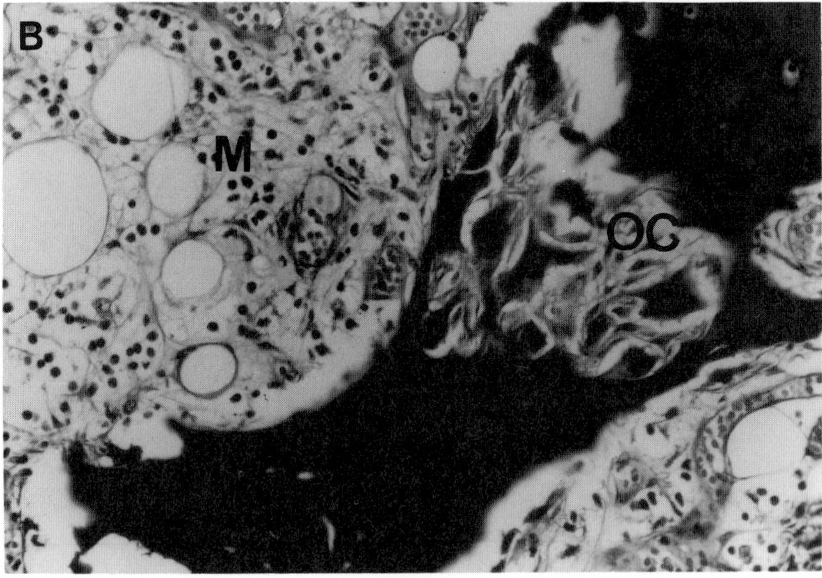

FIG. 3. Cellular activity in osteomyelitis. (A) Low power view of trabecular bone and marrow elements (M). There is a large infiltration of cells from the immune system in this region. (B) Higher-power view demonstrating intense focal osteoclastic activity (OC). Note the absence of osteoblastic cells in and around the resorbed surfaces.

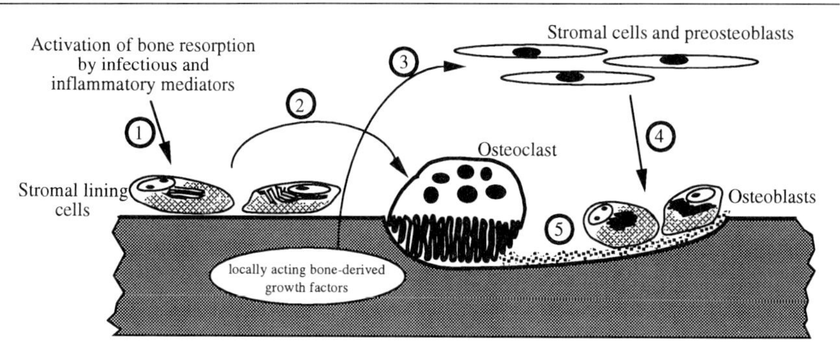

FIG. 4. Regulatory pathways for osteoclastic and osteoblastic activity. Pathway 1 represents the activation and recruitment of osteoclasts by cytokines and systemic hormones. Agents such as IL-1, IL-3, TNF-α, prostaglandins, CSFs, and parathyroid hormone activate stromal lining cells or authentic osteoblasts to release a soluble mediator that directly controls osteoclast function. Pathway 2 indicates that an unknown factor is responsible for controlling osteoclastic function. One of the molecules currently under investigation for this unknown factor is IL-6. Pathway 3 represents the release of bone-derived growth factors embedded in bone matrix. This pathway initiates the process for re-forming bone in the resorption lacunae. The factors present in bone matrix include transforming growth factor β insulin-like growth factor I, insulin-like growth factor II, basic fibroblast growth factor, platelet-derived growth factor, and bone morphogenetic proteins 1–7. Pathway 4 depicts the maturation of preosteoblasts to osteoblasts. Hormones and vitamins such as 1,25-dihydroxyvitamin D_3 and retinoic acid may play important roles in this pathway. Pathway 5 is intended to represent a spatial signal to orient osteoblastic bone formation. Current work in this area suggests that residual molecules remaining on resorption surfaces can direct the site of new bone formation. Factors that are elaborated during infection-mediated inflammation can stimulate osteoclastic activity via pathway 1. For this stimulation to lead to a net decrease in bone volume there must be a disruption somewhere in pathways 3–5. One likely site that may be disrupted by bacterial and immune cell products and lead to haphazard and woven bone formation in periodontal disease is pathway 5.

above, many of the cytokines are synergistic and can evoke very strong stimuli for bone resorption; however, under normal conditions for any given number of bone-resorbing sites, there is always an equivalent compensatory stimulus to enhance bone formation. Pathways 3 to 5 govern the formation stimuli. Thus, for bone to be permanently lost there must be a disruption in the cellular communication that exists between pathways 1 and 2 and pathways 3 to 5. Such a disruption occurs in periodontal disease and osteomyelitis. At present, the molecular mechanisms that create the disruption in cell communication are not known. They may be complex and involve as yet unidentified cell biological principles, or they may be relatively simple reactions involving known factors and enzymes.

Acknowledgments

Part of the work presented in this discussion was supported by NIH RO1 AR-28420 (J.E.P.), the Orthopaedic Research and Education Foundation (R.J.O.), and NIH T32 DK-07092 (S.D.R.)

[7] Effect of Bacterial Products on Neutrophil Chemotaxis

By JOHN R. KALMAR and THOMAS E. VAN DYKE

Introduction

The migration of leukocytes to sites of infection and trauma is a critical, highly conserved mechanism of innate host defense that occurs early in the inflammatory response to localize and limit damage to host tissues. In humans, the cellular infiltrate is initially characterized by a predominance of polymorphonuclear neutrophils (PMNs), followed by migrating monocytes or activated macrophages in more chronic inflammatory sites. Ultimately, this promotes the development of secondary or adaptive immune responses. The migration of these leukocytes is guided by gradients of chemical stimuli from the site of infection or tissue injury that induce morphological orientation and locomotion toward their source(s). Such movement, termed *chemotaxis,* occurs following the interaction of specific receptors on the leukocyte membrane with appropriate stimulant ligands (termed *chemoattractants*). Ligand binding results in the activation of membrane signal transduction pathways that stimulate intracellular cytoskeletal elements, promoting both changes in cell shape and increased cell movement.

Although chemotaxis implies directed cell movement, isotropic concentrations of chemotactic factors can increase nondirectional or random cell movement as well. Orthokinesis is defined as a change in the rate of locomotion along random paths and klinokinesis is a change in the rate at which the cell turns.[1] In the absence of stimulation, purified populations of peripheral blood PMNs suspended in buffer remain spherical and immotile. Addition of an isotropic concentration of chemoattractant to these suspensions is followed by morphological polarization of the cells with random axial orientation and an increase in cellular movement (kinesis). Introduction of a chemoattractant in an anisotropic or gradient format results in alignment of the cell axis, with a ruffled, lamellipodium visible

[1] P. C. Wilkinson, *Immunology* **6,** 273 (1985).

at the anterior surface and development of a narrow posterior tail. Chemotactic movement among PMNs is highly efficient with nearly straight-line paths observed toward the gradient source.

Putative chemotaxins or chemoattractants have largely been identified *in vitro* by using purified populations of PMNs or monocytes obtained from peripheral blood. A number of techniques, described below, have been used by various investigators to examine the response of these cell populations. Yet, although each has its merits and limitations, they have largely corroborated the chemotactic activity of several molecular species. The majority are host-derived (endogenous) and include C5a (a complement cleavage product), leukotriene B_4 (LTB_4), interleukin-8 (IL-8), platelet-activating factor (PAF), and platelet factor 4 (PF-4). Well-characterized exceptions are analogs of the formylated tripeptide, *N*-formylmethionyl-leucyl-phenylalanine, which are produced during protein synthesis by bacteria and have been proposed as major contributors to the recruitment of PMNs to sites of bacterial invasion.

Currently, the chemotactic activity of other substances produced by bacteria has not been well documented. Unless the bacterium is a successful intracellular parasite, however, bacterial products that could interfere with the recruitment of PMNs or monocytes would seem to be advantageous for the bacteria in the host–parasite relationship. This article addresses the complex molecular, biochemical, and cellular events through which bacterial substances influence the chemotactic responsiveness of professional phagocytes. The techniques used to study chemotaxis are briefly described followed by a summary of several known chemotactic agents. This is followed by data that examine the direct influence of various bacteria and bacterial products on the response of host leukocytes to these chemoattractants.

Methods of Study of Chemotaxis

A wide variety of techniques have been published that provide qualitative and quantitative information about chemotaxis and chemotactic agents. As mentioned, each has strengths and weaknesses and the reader is referred to several texts and reviews that provide methodologic detail beyond the scope of this article.[2,3] As most investigations to date have employed variants of two methods, micropore filter and under agarose, these are described here.

[2] G. DiSabato (ed.), this series, Vol. 162, [1]–[11].

[3] J. A. Metcalf, J. I. Gallin, W. M. Nauseef, and R. K. Root, *in* "Laboratory Manual of Neutrophil Function," Raven Press, New York, 1986.

Micropore Filter Method

Introduced by Boyden in 1962, the filter assay is a popular method for the study of leukocyte chemotaxis.[2-4] Simple in concept and material requirements, the assay uses a filter with pores or holes to divide a chamber into two compartments. The pore size is selected to permit actively motile cells to crawl through, while preventing cells from being drawn through passively by gravity. A chemoattractant in an appropriate buffer solution is placed in the lower compartment, and cell suspended in the same buffer without the chemotactic agent are placed in the upper chamber. Diffusion of the chemoattractant creates a gradient across the filter and responder cells migrate to and through the pores of the filter. After selected incubation times, the filters and associated cells are fixed and migration is assessed by microscopic examination.

Under Agarose Method

First described in 1974 by Cutler[4a] for guinea pig peritoneal neutrophils and later adapted by several authors for use with human leukocytes,[2,3,5,6] the under agarose measurement of chemotaxis has been used successfully with a variety of cell types. Conceptually, the method uses agarose as a semisolid matrix to control the diffusion of chemoattractant molecules. This creates the chemotactic gradient necessary to guide the movement of responsive cells between the agarose and the underlying surface. In its most basic configuration, a line of three wells is cut into an agarose gel. A cell suspension is placed into the central well, which is equidistant from the other two. Chemoattractant is added to one of the remaining wells and control solution or buffer in the other. Responsive cells migrate radially from the central well, eventually producing a flame or egg-shaped pattern that points toward the chemoattractant. The cells are then fixed and migration distances measured by direct observation.

Chemotactic Factors

Chemotactic factors are defined as those molecules that induce directional movement of cells exposed to a gradient of the agent. As discussed in a previous volume,[2] most chemoattractants activate numerous other cellular functions or pathways, including glycolysis, oxygen consumption, exocytosis, and NADPH oxidase activation and superoxide production.

[4] S. Boyden, *J. Exp. Med.* **45,** 453 (1962).
[4a] J. Cutler, *Proc. Soc. Exp. Biol. Med.* **147,** 471 (1974).
[5] R. D. Nelson, P. G. Quie, and R. L. Simmons, *J. Immunol.* **115,** 1650 (1975).
[6] D. E. Chenoweth, J. G. Rowe, and T. E. Hugli, *J. Immunol. Methods* **25,** 337 (1979).

These various pathways exhibit differential responsiveness that depends on not only the nature of the chemotactic factor but its concentration. It is important to keep in mind that the chemotactic response likely does not occur in isolation but is part of a complex stimulus–response coupling system. It is also suggested that the primary role of chemoattractants *in vivo* is not to effect directed movement but rather to induce increased locomotion (chemokinesis).[1,2] Moreover, the directional response is only seen when the concentration of the chemotactic gradient is low enough that the responder cell can discriminate the level of ligand from front to back of its cytoplasmic membrane. Such a condition leads to cell polarization, receptor redistribution, and movement toward the area of higher chemoattractant concentration. Movement continues until the ligand concentration is high enough to effect receptor saturation, at which point polarization and migration diminish rapidly. This implies the existence of an optimal concentration range for chemotactic responsiveness to a given factor, a feature that should be routinely examined when chemotaxis is being measured. Such a range, which depends on receptor number and recycling pathways, appears to be unique to individual cell types and for each chemoattractant.

Endogenous Chemoattractants

C5a

C5a is a small (11-kDa), glycosylated, cationic peptide derived from the amino-terminal end of the C5 alpha chain.[7,8] Produced through activation of both the classic and alternative complement pathways, it is an extremely potent chemoattractant for neutrophils. C5a also exhibits anaphylatoxic activities, increasing vascular permeability and inducing smooth muscle contraction.[9] C5a is susceptible to cleavage by carboxypeptidase N present in human serum to produce the less active C5a des-Arg. Two other serum proteins also play important modulatory roles for C5a. Gc-globulin (vitamin D-binding protein) binds to and increases the chemotactic activity of both C5a and C5a des-Arg.[10] Chemotactic factor inactivator decreases C5a potency, apparently by disturbing the Gc-globulin interaction with C5a.[11,12] Human PMNs express C5a receptors on the plasma

[7] T. E. Hugli and H. J. Muller-Eberhard, *Adv. Immunol.* **26,** 1 (1978).
[8] T. E. Hugli, *Complement* **3,** 111 (1986).
[9] T. E. Hugli, *Immunopathology* **7,** 193 (1984).
[10] H. D. Perez, E. Kelly, D. Chenoweth, and F. Elfman, *J. Clin. Invest.* **82,** 360 (1988).
[11] P. A. Ward and J. Ozols, *J. Clin. Invest.* **58,** 123 (1973).
[12] R. A. Robbins, J. K. Rasmussen, M. E. Clayton, G. L. Gossman, T. J. Kendall, and S. I. Rennard, *J. Lab. Clin. Med.* **110,** 292 (1987).

membrane that consist of a single glycoprotein (50 kDa) coupled to a guanine nucleotide-binding protein.[13–15]

Leukotriene B_4

Arachidonic acid released from cell membranes by the action of phospholipases is susceptible to oxygenation by two distinct enzymatic pathways: 5-lipoxygenase and cyclooxygenase. While the latter results in the production of various prostaglandin molecules, the former generates a family of reactive hydroperoxyeicosatetraenoic acid intermediates that are in turn converted into leukotrienes. Of the leukotrienes, identified by the presence of three conjugated double bonds, leukotriene B_4 (LTB_4) has exhibited activity as a chemotactic and chemokinetic factor.[16,17] Its importance as an inflammatory mediator has been demonstrated *in vivo*.[18] Substantial neutrophil accumulation occurs at sites of intradermal injection of LTB_4, whereas intravenous injection induces rapid, reversible neutropenia.[19,20] LTB_4 levels are elevated in gingival and periodontal inflammation, in psoriatic lesions, and in rheumatoid arthritis synovial fluid.[21,22] Moreover, 5-lipoxygenase inhibitors have been shown to decrease significantly the amount of inflammation associated with these disease sites.[23] Specific binding sites for LTB_4 have been identified on human PMNs and are apparently contained within a 60-kDa protein found in the plasma membrane.[24] Like the C5a receptor, the LTB_4 binding site is apparently associated with a guanine nucleotide-binding protein.[25]

Neutrophil-Activating Proteins

Several structurally related proteins that exhibit a variety of cytokine activities have been collectively termed *neutrophil-activating proteins*

[13] D. E. Chenoweth and T. E. Hugli, *Proc. Natl. Acad. Sci. USA* **75,** 3943 (1978).
[14] R. J. Johnson and D. E. Chenoweth, *J. Biol. Chem.* **260,** 7161 (1985).
[15] S. J. Siciliano, T. E. Rollins, and M. S. Springer, *J. Biol. Chem.* **265,** 19568 (1990).
[16] A. W. Ford-Hutchinson, M. A. Bray, M. V. Doig, M. E. Shipley, and M. J. H. Smith, *Nature* **286,** 264 (1980).
[17] C. L. Malmsten, J. Palmblad, A.-M. Uden, O. Radmark, L. Engstedt, and B. Samuelsson, *Acta Physiol. Scand.* **110,** 449 (1980).
[18] M. A. Bray, *Agents Actions* **19,** 87 (1986).
[19] M. A. Bray, A. W. Ford-Hutchinson, and M. J. H. Smith, *Prostaglandins* **22,** 213 (1981).
[20] H. Z. Movat, C. Rettl, C. E. Burrowes, and M. G. Johnston, *Am. J. Pathol.* **115,** 233 (1984).
[21] L. Harvath, *Experientia, Supplementum* **59,** 35 (1991).
[22] P. A. Heasman, J. G. Collins, and S. Offenbacher, *J. Periodontal Res.* **28,** 241 (1993).
[23] V. Kassis, *IRCS Med. Sci.* **13,** 182 (1985).
[24] D. W. Goldman, L. A. Gifford, R. N. Young, and E. J. Goetzl, *Fed. Proc.* **44,** 781 (1985).
[25] D. W. Goldman, L. A. Gifford, T. Marotti, C. H. Koo, and E. J. Goetzl, *Fed. Proc.* **46,** 200 (1987).

(NAPs). The NAPs are small (8–10 kDa), basic, single-chain proteins that commonly contain four internal cysteine residues forming two disulfide bridges, a feature shared by other related chemotactically active proteins, including PF4 and the GRO protein family.[21,26]

NAP-1 was initially identified as a product of lipopolysaccharide (LPS)-stimulated monocytes and termed interleukin-8.[27] Since then, its production has been associated with a number of cell types, including fibroblasts, endothelial cells, keratinocytes, hepatocytes, and even neutrophils.[28,29] Interleukin-1 (IL-1) and tumor necrosis factor (TNF) apparently serve as primary stimuli for NAP-1 expression among all cells studied to date, while induction by LPS is restricted to phagocytic and endothelial cells.[26] NAP-1 is secreted as a 79-residue protein that yields several biologically active variants following extracellular proteolytic processing. All forms are chemoattractive for PMNs *in vitro,* and injection of NAP-1 intradermally into rabbits, rats, and humans induces perivascular neutrophil infiltration, suggesting broadly conserved inflammatory activity.[21,28,30] Human PMNs express plasma membrane receptors for NAP-1 that also bind NAP-2 and GRO-α but not PF4.[26,31] NAP-2 is a 70-amino acid protein formed by cleavage of a 15-residue fragment from the amino terminus of connective tissue-activating peptide III (CTAP-III), a chemotactically inactive molecule found in the platelet α-granule.[32]

NAP-3 is a 73-residue protein derived from LPS-stimulated monocytes that is structurally identical to GRO-α (also known as melanoma growth-stimulatory activity protein, MGSA) and structurally similar to β-thromboglobulin.[33,34] NAP-4, an 8-kDa protein isolated from human platelet lysates, is chemotactic for human PMNs although it exhibits less potency than NAP-1. Among the NAP family members, it uniquely contains only two cysteine residues.[21] The NAP family members apparently

[26] M. Baggiolini and I. Clark-Lewis, *FEBS Lett.* **307,** 97 (1992).

[27] T. Yoshimura, K. Matsushima, J. J. Oppenheim, and E. J. Leonard, *J. Immunol.* **139,** 788 (1987).

[28] E. J. Leonard and T. Yoshimura, *Am. J. Respir. Cell Mol. Biol.* **2,** 479 (1990).

[29] F. Bazzoni, M. Cassatella, F. Rossi, M. Ceska, B. Dewald, and M. Baggiolini, *J. Exp. Med.* **173,** 771 (1991).

[30] I. Lindley, H. Aschauer, J.-M. Seifert, C. Lam, W. Brunowsky, E. Kownatzki, M. Thelen, P. Peveri, B. Dewald, V. von Tscharner, A. Walz, and M. Baggiolini, *Proc. Natl. Acad. Sci. USA* **85,** 9199 (1988).

[31] E. J. Leonard, T. Yoshimura, A. Rot, K. Noer, A. Walz, M. Baggiolini, D. A. Walz, E. J. Goetzl, and C. W. Castor, *J. Leukocyte Biol.* **49,** 258 (1991).

[32] A. Walz and M. Baggiolini, *J. Exp. Med.* **171,** 449 (1990).

[33] J.-M. Schroder and N.-L. M. Persoon, *J. Exp. Med.* **171,** 1091 (1990).

[34] A. Richmond, E. Balentien, H. G. Thomas, G. Flaggs, D. E. Barton, J. Spiess, R. Bordoni, U. Francke, and R. Derynck, *EMBO J.* **7,** 2025 (1988).

act as critical molecular regulators of the inflammatory response. They can be generated at virtually any site in the body by a wide variety of cell types and appear to resist clearance or inactivation. For example, unlike other agents such as C5a, *N*-formylmethionyl-leucyl-phenylalanine, and LTB_4, IL-8 (NAP-1) demonstrates a persistent chemoattractive effect following intradermal injection.[26] The apparent resistance of NAP-1 to oxidative or hydrolytic inactivation implies a possible fail-safe role for the recruitment of responder cells to areas of inflammation or infection.

Platelet-Activating Factor

PAF (1-*O*-alkyl-2-acetyl-*sn*-glyceryl-3-phosphorylcholine) demonstrates chemotactic activity for human PMNs and is produced by a variety of cells including neutrophils and endothelial cells.[35,36] Like other chemoattractants, PAF activates neutrophil respiratory burst and secretion functions, and similar to C5a, it also possesses anaphylatoxic properties.[21] PAF may play a role in stimulus–response coupling pathways as suggested by its ability to cross-desensitize PMNs for other secretagogues.[37] PAF receptors have been identified on neutrophil plasma membranes[38]; however, receptor structure remains unknown.

Exogenous Chemoattractants

Although it had been known for some time that most bacteria elaborate chemotactic factors in culture, Schiffman *et al.* first reported in 1975 that a series of *N*-formylmethionyl peptides, especially *N*-formylmethionyl-leucyl-phenylalanine (FMLP), were chemoattractive for PMNs.[39] Several years later, such peptides were identified and isolated from liquid cultures of various bacteria,[21] supporting their likely role as signals for recruitment of PMNs to sites of bacterial invasion. Following exposure to FMLP, PMNs undergo rapid transients in cytosolic calcium levels followed by granule secretion and increased oxidative metabolism.[40–42] The human

[35] E. J. Goetzl and W. C. Pickett, *J. Immunol.* **125,** 1789 (1980).
[36] R. N. Pinckard, R. S. Farr, and D. J. Hanahan, *J. Immunol.* **123,** 1847 (1979).
[37] J. T. O'Flaherty, C. J. Lees, C. H. Miller, C. E. McCall, J. C. Lewis, S. H. Love, and R. L. Wykle, *J. Immunol.* **127,** 731 (1981).
[38] A. G. Stewart and G. J. Dusting, *Br. J. Pharmacol.* **94,** 1225 (1988).
[39] E. Schiffmann, B. A. Corcoran, and S. M. Wahl, *Proc. Natl. Acad. Sci. USA* **72,** 1059 (1975).
[40] J. R. White, P. H. Naccache, T. F. P. Molski, P. Borgeat, and R. I. Sha'afi, *Biochem. Biophys. Res. Commun.* **113,** 44 (1983).
[41] E. L. Becker, H. J. Showell, P. M. Henson, and L. S. Hsu, *J. Immunol.* **112,** 2047 (1974).
[42] L. A. Boxer, M. Yoder, S. Bonsib, M. Schmidt, P. Ho, R. Jersild, and R. Baehner, *J. Lab. Clin. Med.* **93,** 506 (1979).

FMLP receptor is a membrane protein (55–75 kDa) containing 350 residues and is closely associated with a guanine nucleotide-binding protein (G protein).[43–45] Two isoforms, differing by a single amino acid, have been described and binding data have suggested the existence of high- and low-affinity states.[46,47] The relationship of affinity state to isoform is still unclear; however, work such as recent amplification of genomic DNA from HL-60 cells and human PMNs that produced a roughly 1-kb fragment identical in sequence to cDNA described for one of the isoforms may help resolve this question.[48]

Bacterial Effects on Neutrophil Chemotaxis

This section reviews the current understanding of the effect of bacterial toxins and other bacterial substances on neutrophil chemotaxis. Microbial products that reportedly affect chemotactic movement and/or related neutrophil functions can be organized into three major categories: (1) toxins, (2) metabolic products or structural components, and (3) complex mixtures such as culture filtrates and sonic extracts. These can be arranged as in Table I with respect to organismal derivation of the product(s).

Specific components such as the toxins can also be grouped according to their site or mode of action, as summarized in Table II. For convenience of discussion, this order is used within the subsequent text for both toxins and metabolic or structural components, followed by reports involving complex mixtures such as bacterial sonicates and extracts.

ADP-Ribosylating Toxins

Several bacterial toxins exert their effect by the ADP-ribosylation of proteins that are essential for normal cellular function. Three major toxins of this type have been shown to affect neutrophil chemotaxis. These are pertussis toxin, cholera toxin, and botulinum C2 toxin. As pseudomonas exotoxin A and staphylococcal leukocidin also possess ADP-ribosylation activity and affect PMN function, they are presented here as well. ADP-ribosylating toxins have been functionally divided into two domains, A

[43] J. Niedel, J. Davis, and P. Cuatrecassas, *J. Biol. Chem.* **255,** 7063 (1980).

[44] F. Boulay, M. Tradif, L. Brouchon, and P. Vignais, *Biochem. Biophys. Res. Commun.* **168,** 1103 (1990).

[45] P. G. Polakis, R. Uhing, and R. Synderman, *J. Biol. Chem.* **263,** 4969 (1988).

[46] C. Koo, R. Lefkowitz, and R. Snyderman, *Biochem. Biophys. Res. Commun.* **106,** 442 (1981).

[47] R. Snyderman, *Rev. Infect. Dis.* **7,** 390 (1985).

[48] E. De Nardin, S. J. Radel, N. Lewis, R. J. Genco, and M. Hammarskjold, *Biochem. Int.* **26,** 381 (1992).

TABLE I
BACTERIAL TOXINS AND OTHER COMPONENTS THAT INFLUENCE NEUTROPHIL CHEMOTAXIS

Toxins
Anthrax toxin
Cholera toxin
Clostridial toxins
Botulinum C2 toxin
Perfringolysin
Phospholipase C
Toxin A
Pertussis toxin
Pseudomonas toxins
Alkaline protease/elastase
Cytotoxin
Exotoxin
Staphylococcal toxins
α-Toxin
δ-Toxin
Staphylococcal leukocidin
Sphingomyelinase C
Metabolites/Structural Components
Ammonium
Short-chain fatty acids
Aminopeptidase inhibitors
Lipopolysaccharide
Complex Mixtures
Legionella
Oral gram-negative bacteria
Glycocalyces

and B. The A moiety contains the ADP-ribosylating activity and the B moiety mediates uptake through the cell membrane, via specific receptors that vary between toxins.

Cholera Toxin

Vibro cholerae produces a toxin which is an 84-kDa oligomeric protein comprising one A and 5 B subunits. The A subunit is 27 kDa and comprises two polypeptide chains, A1 (21 kDa) and A2 (6 kDa)[49–51] that are linked by a disulfide bond. On reduction, the enzymatically active A1 peptide

[49] D. M. Gill, *Biochemistry* **15,** 1242 (1976a).
[50] D. M. Gill, *Adv. Cyclic Nucleotide Res.* **8,** 85 (1977).
[51] D. M. Gill and R. S. Rapport, *J. Infect. Dis.* **139,** 674 (1979).

is released to induce toxic activity.[52-54] The B subunits constitute the membrane recognition unit necessary for cell intoxication.[55-60]

Although the main action of cholera toxin in the intestine is binding to the membrane of intestinal mucosal cells and activating adenylyl cyclase,[61] the mechanism of action in phagocytic cells is through a cAMP-independent mechanism. Cholera toxin, like pertussis toxin, inhibits chemotaxis by the ADP-ribosylations of guanine nucleotide-binding proteins, thereby uncoupling the receptor signal.[62] G proteins that can be ADP-ribosylated by both toxins, such as G_{i2} or G_{i3}, may be linked to chemotactic receptors[63,64]; however, at this time, it is not clear whether cholera toxin and pertussis toxin act through the same or independent G proteins.

Pertussis Toxin

Pertussis toxin is a product of the gram-negative bacterium *Bordetella pertussis,* the causative agent of whooping cough.[65] Its toxic activity is the result of ADP-ribosylation of signal-transducting guanine nucleotide-binding proteins. The functional consequence of G-protein ADP-ribosylation is the uncoupling of the receptor from the effector system of the cell. The transmembrane systems affected by pertussis toxins include adenylate cyclase, stimulation of phospholipases A_2, C, and D, stimulation of cyclic guanosine monophosphate (cGMP) phosphodiesterase in the retina, and the regulation of ion channels.[66,67]

[52] D. M. Gill and C. A. King, *J. Biol. Chem.* **250,** 6424 (1975).
[53] J. J. Makalanos, R. J. Collier, and W. R. Romig, *J. Biol. Chem.* **254,** 5855 (1979).
[54] J. Moss, S. J. Stanley, and M. C. Lin, *J. Biol. Chem.* **254,** 11993 (1979).
[55] P. Cuatrecasas, *Biochemistry* **12,** 3547 (1973).
[56] P. Cuatrecasas, *Biochemistry* **12,** 3558 (1973).
[57] P. Cuatrecasas, *Biochemistry* **12,** 3577 (1973).
[58] J. Holmgren, I. Lönnroth, and L. Svennerholm, *Infect. Immun.* **8,** 208 (1973).
[59] J. Holmgren, J-E. Mänsson, and L. Svennerholm, *Med. Biol.* **52,** 229 (1974).
[60] W. E. Van Heyningen, C. C. J. Carpenter, N. F. Pierce, and W. B. Greenough III, *J. Infect. Dis.* **124,** 415 (1971).
[61] D. V. Kimberg, M. Field, J. Johnson, A. Henderson, and E. Gershon, *J. Clin. Invest.* **50,** 1218 (1971).
[62] R. R. Aksamit, P. S. Backlund, Jr., and G. L. Cantoni, *Proc. Natl. Acad. Sci. USA* **82,** 7475 (1985).
[63] P. Gierschik, D. Sidiropoulos, and K. Jakobs, *J. Biol. Chem.* **264,** 21470 (1989).
[64] M. Verghese, R. J. Uhing, and R. Synderman, *Biochem. Biophys. Res. Commun.* **138,** 887 (1986).
[65] A. A. Weiss and E. L. Hewlett, *Annu. Rev. Microbiol.* **40,** 661 (1986).
[66] P. Gierschik, D. Sidiropoulos, K. Dieterich, and K. H. Jakobs, *in* "Transmembrane Signaling, Intracellular Messengers and Implications for Drug Development" (S. R. Nahorski, ed.), pp. 73–89, Wiley, Chichester, 1990.
[67] L. Birnbaumer, J. Abramowitz, and A. M. Brown, *Biochim. Biophys. Acta* **1031,** 163 (1990).

TABLE II
BACTERIAL PRODUCTS THAT AFFECT NEUTROPHIL CHEMOTAXIS

Bacterial product	Binding site or substrate	Mode of action	Effect on chemotaxis
ADP-ribosylating toxins			
Cholera toxin	B—Ganglioside G^{M1} A—G-protein(s)	ADP-ribosylation of G protein, receptor signal uncoupling	Inhibition
Pertussis toxin	B—Glycoproteins, glycolipids A—G-protein(s)	ADP-ribosylation of G protein, receptor signal uncoupling	Inhibition
Botulinum C2 toxin	C_{2II}—cell membrane (?) C_{2I}—G-actin	ADP-ribosylation of intracellular G-actin	Inhibition
Exotoxin A (*Pseudomonas aeruginosa*)	Cell membrane (?) Elongation factor 2	ADP-ribosylation of elongation factor 2, inhibition of protein synthesis	Unknown
Staphylococcal leukocidin	S component—ganglioside G_{M1} F component—lysophospholipids ADP-ribosylation substrate(s) (?)	Increased PLA_2 activity and phosphoinositide metabolism, increased membrane permeability, ADP-ribosylation	Inhibition (?) (S-component chemotactic)
Cytolytic toxins			
Staphylococcal α-toxin	Cell membrane (?)	Monomer binding → conformational rearrangement to hexameric toxin channels → increased permeability, increased phospholipid and arachidonate metabolism	Inhibition
Staphylococcal δ-toxin	Cell membrane (?)	Pore formation by toxin oligomers, increased phospholipid metabolism, increased permeability	Inhibition
Perfringolysin	Membrane cholesterol	Pore formation, increased permeability	Inhibition
Cytotoxin (*Pseudomonas aeruginosa*)	Cell membrane(s) (cytoplasmic, nuclear, granule)	Unknown	Inhibition

Other toxins			
Anthrax toxin	ATP	Protective antigen (PA) binds to cell surface, → promotes binding of edema factor (EF), EF exhibits adenylate cyclase activity → increased cAMP (in CHO cells)	Enhancement
Toxin A (*Clostridium difficile* enterotoxin)	Unknown	Unknown, increases intracellular [Ca^{2+}]	Unknown (chemotactic, chemokinetic)
Phospholipase C	Membrane phospholipids, especially phosphatidylcholine	Membrane destabilization, increased permeability	Inhibition
Sphingomyelinase C	Sphingomyelin	Release of phosphorylcholine membrane rearrangement, increased permeability	Inhibition
Alkaline protease/elastase (*Pseudomonas aeruginosa*)	Membrane proteins	Cleavage of receptors (?)	Inhibition
Bacterial Metabolites/ Components			
Aminopeptidase inhibitors Actinonin Amastatin	Aminopeptidases	Enzyme inhibition	Enhancement (actinonin)
Ammonium	Pancellular (?)	Variety of effects due to alteration of cellular pH, reduces actin polymerization/depolymerization	Inhibition
Short-chain fatty acids	Pancellular (?)	pH-dependent, proton shuttle activity capable of reducing cellular pH	Inhibition
Lipopolysaccharide	CD14	Unknown, induces release of neutrophil inhibition factor (?)	Inhibition

Chemotaxis to chemoattractants such as FMLP, LTB_4, and C5a is mediated by G protein-coupled receptors on the neutrophil surfaces.[68-70] Evidence that initially implicated G proteins in this process came from experiments in which cells exposed to pertussis toxin ceased to be chemotactically active. As pertussis toxin was found to ADP-ribosylate the α subunit of the G proteins, it was clear that the signal resulting from chemoattractant receptor occupancy must pass through or be transduced by these proteins. Since then, pertussis toxin has also been shown to inhibit neutrophil phagocytosis, degranulation, and oxidative burst.[68,71-73]

Botulinum C2 Toxin

Botulinum C2 toxin possesses ADP-ribosyltransferase activity toward actin. It is a binary toxin, consisting of components C2I and C2II. C2II binds to the eukaryotic cell membrane and facilitates intracellular access for C2I which selectively ADP-ribosylates nonmuscle G-actin at arginine-177.[74] This modification strongly inhibits the rate of G-actin polymerization. It also converts isolated G-actin into a capping protein that binds to the barbed end of actin filaments, blocking polymerization at this site. As depolymerization at the pointed end of the filaments is not affected, the net effect is progressive F-actin cleavage.

Exposure of isolated PMNs to botulinum C2 (0.4 μg/ml C2I, 1.6 μg/ml C2II) for 60 min completely blocked random migration and directed movement with several chemotactic stimulants.[75] Similarly to other agents that block actin polymerization (i.e., cytochalasins), C2 toxin treatment increased superoxide anion (O_2^-) production by PMNs stimulated with FMLP, PAF, and concanavalin A (Con A) but not phorbol myristate

[68] K. R. McLeish, P. Gierschik, T. Schepers, D. Sidiropoulos, and K. H. Jakobs, *Biochem. J.* **260,** 427 (1989).

[69] T. E. Rollins, S. Siciliano, S. Kobayashi, D. N. Cianciarulo, V. Bonilla-Argudo, K. Collier, and M. S. Springer, *Proc. Natl. Acad. Sci. USA* **88,** 971 (1991).

[70] M. W. Verghese, L. Charles, L. Jakoi, S. B. Dillon, and R. Snyderman, *J. Immunol.* **138,** 4374 (1987).

[71] H. D. Gresham, L. T. Clement, J. E. Volanakis, and E. J. Brown, *J. Immunol.* **139,** 4159 (1987).

[72] A. J. Feister, B. Browder, H. E. Willis, T. Mohanakumar, and S. Ruddy, *J. Immunol.* **141,** 228 (1988).

[73] G. J. Spangrude, F. Sacchi, H. R. Hill, D. E. Van Epps, and R. A. Daynes, *J. Immunol.* **135,** 4135 (1985).

[74] F. Grimminger, U. Sibelius, K. Aktories, N. Suttorp, and W. Seeger, *Mol. Pharmacol.* **40,** 563 (1991).

[75] J. Norgauer, E. Kownatzi, R. Seifert, and K. Atories, *J. Clin. Invest.* **82,** 1376 (1988).

acetate (PMA). Enhanced release of both primary and secondary granule markers was also noted. FMLP receptor binding and dissociation were not affected by the bacterial toxin, and ligand–receptor endocytosis was only mildly reduced, favoring postreceptor events as the primary site(s) of action.[76]

Recently, Grimminger *et al.* demonstrated amplification of ligand-induced lipid mediator generation by C2 toxin that exhibited differential, concentration, and stimulant-dependent effects.[74] In response to soluble stimulants, FMLP and PAF, low-dose C2 (50/100 to 100/200 ng/ml, C2I/II) favored 5-lipoxygenase activity with resultant increases in LTB_4, LTA_4 and 5-hydroxyeicosatetraenoic acid (5-HETE). With increasing toxin concentration, relative 5-HETE production increased as leukotriene generation declined. Further increases in toxin concentration resulted in elevated levels of inositol phosphates and PAF, a feature not seen in the absence of C2. The authors' interpretation of these data suggested that higher toxin doses resulted in a general amplification of PMN signal tranduction events and second messenger production, possibly as a result of suppressed cytoskeletal (actin-associated) rearrangement normally linked to receptor occupancy. Enhanced 5-lipoxygenase activity seen at lower C2 levels may result from initial toxin effects on an actin subset or pool near the cell membrane that serves to regulate enzyme activation. Consistent with the toxin's ability to prevent actin-dependent cell functions such as migration and phagocytosis,[75] C2 toxin also exerted concentration-dependent inhibition of leukotriene and PAF generation by PMNs in response to opsonized zymosan.

Exotoxin A (Pseudomonas aeruginosa)

Exotoxin A has been shown to inhibit eukaryotic protein synthesis by inactivation of elongation factor 2 through ADP-ribosylation, similar to the action of diphtheria toxin.[77,78] Inhibition of human myeloid stem cell proliferation by exotoxin A has been demonstrated, as have alterations in macrophage morphology in tissue culture and a priming effect on monocyte oxidative burst activity.[79–81] By contrast, PMN morphology, priming of oxidative burst response and phagocytosis were virtually unchanged by

[76] T. W. Doebber and M. S. Wu, *Proc. Natl. Acad. Sci. USA* **84,** 7557 (1987).
[77] B. H. Igleuski and D. Kabat, *Proc. Natl. Acad. Sci. USA* **72,** 2284 (1975).
[78] G. Döring and E. Müller, *Microbial Pathogenesis* **6,** 287 (1989).
[79] J. Patzer, H. Nielsen, and A. Kharazmi, *Microbial Pathogen.* **7,** 147 (1989).
[80] M. Pollack and S. E. Anderson, *Infect. Immun.* **19,** 1092 (1978).
[81] R. K. Stuart and M. Pollack, *Infect. Immun.* **38,** 206 (1982).

purified exotoxin A at similar concentrations.[79,82] Presently, information concerning toxin effect on chemotaxis is lacking. The finding that both PMN elastase and the MPO/H_2O_2/halide system are potent inactivators of exotoxin A ADP-ribosyltransferase activity could explain, in part, the relative resistance of PMNs to intoxication.[78]

Staphylococcal Leukocidin

Staphylococcal leukocidin consists of two protein components termed S (slow-eluted) and F (fast-eluted) on the basis of their separation by carboxymethyl-cellulose chromatography.[83] These components act synergistically to induce cytotoxic changes within a narrow spectrum of mammalian cells and especially human neutrophils. Although the exact mechanism of toxicity remains unclear, specific binding of S component to surface ganglioside G_{M1} has been shown to stimulate phospholipase A_2 activity in toxin-sensitive cells, lysophospholipid formation with subsequent increased binding of F component, and increased membrane permeability.[84–86] Both S and F components exhibit ADP-ribosyltransferase activity, catalyzing the transfer of ADP-ribose to rabbit PMN membrane proteins of 37 and 41 kDa, respectively.[87] In addition, stimulation of rabbit PMN cell membrane phosphoinositide metabolism by F component alone or both S and F components has been demonstrated.[88] The genes for both components have recently been cloned and sequenced.[89,90] Amino acid sequence homology between S and F was found to be 31%; however, no other significant homology was detected. Sublytic concentrations of leukocidin reportedly inhibit neutrophil chemotaxis, but definitive experiments using purified components have not been published.[91] Interestingly, S component demonstrated potent chemoattractant activity for rabbit PMNs at 10^{-10} to 10^{-8} *M*.[92]

[82] M. Bishop, A. Baltch, L. Hill, R. Smith, F. Lutz, and M. Pollack, *J. Med. Microbiol.* **24,** 315 (1987).
[83] M. Thelestam, *in* "Staphylococci and Staphylococcal Infections, 2" (C. S. F. Easomon and C. Adlam, eds.), p. 705, Academic Press, New York, 1983.
[84] M. Noda, I. Kato, T. Hirayama, and F. Matsuda, *Infect. Immun.* **29,** 678 (1980).
[85] M. Noda, I. Kato, F. Matsuda, and T. Hirayama, *Infect. Immun.* **34,** 363 (1981).
[86] M. Noda, T. Hirayama, F. Matsuda, and I. Kato, *Infect. Immun.* **50,** 142 (1985).
[87] I. Kato and M. Noda, *FEBS Lett.* **225,** 59 (1989).
[88] X. Wang, M. Noda, and I. Kato, *Infect. Immun.* **58,** 2745 (1990).
[89] A. Rahman, K. Izaki, I. Kato, and Y. Kamio, *Biochem. Biophys. Res. Commun.* **181,** 138 (1991).
[90] A. Rahman, H. Nariya, K. Izaki, I. Kato, and Y. Kamio, *Biochem. Biophys. Res. Commun.* **184,** 640 (1992).
[91] P. C. Wilkinson, *Rev. Infect. Dis.* **2,** 293 (1980).
[92] M. Noda, I. Kato, T. Hirayama, and F. Matsuda, *Infect. Immun.* **35,** 38 (1982).

Cytolytic Toxins

Staphylococcal α-Toxin

α-Toxin is secreted by most strains of *Staphylococus aureus* and is an important virulence factor with membrane-damaging properties. It is considered to be a membrane channel-forming protein with intriguing similarities to eukaryotic molecules such as complement C9 and the lymphocytic perforins.[83,93] Because of the exquisite sensitivity of rabbit red blood cells to α toxin, toxic activity is usually assayed in hemolytic units (HU). Although relatively resistant to the lytic effects of α-toxin,[94] neutrophil chemotaxis to FMLP was shown to be depressed at doses of 5 HU or greater.[95] Heat treatment of α-toxin (60°, 30 min) abolished this inhibitory effect as did preincubation of toxin with specific rabbit antiserum.

Pretreatment of neutrophils with low doses of α-toxin (<10 HU) was actually shown to increase phagocytosis and killing of opsonized staphylococci. Higher toxin doses, however, resulted in decreased ingestion and bactericidal activity, neutrophil aggregation, and some membrane damage as measured by trypan blue uptake.[96] More recent work has demonstrated the ability of α toxin to stimulate arachidonate metabolism and LTB_4 release from rabbit neutrophils.[97] This effect was associated with a rapid increase in membrane permeability to Ca^{2+}, but not inulin or dextran, consistent with the formation of discrete transmembrane channels. Increased cytosolic free Ca^{2+} was shown to result from the influx of extracellular Ca^{2+} and not mobilization of internal, membrane-associated Ca^{2+} stores.[98]

Staphylococcal δ-Toxin

Staphylococcal δ-toxin, in contrast to the α-toxin and leukocidin, is cytotoxic with a wide variety of mammalian cells.[83] It shares sequence homology with the bee venom toxin, melittin, and exhibits similar surface-active "detergent-like" properties and biological effects, such as activation of membrane phospholipase A_2 (PLA_2) and formation of various lipid intermediates. Although its precise mode of action has yet to be defined, transmembrane pore formation by toxin oligomers has been postu-

[93] M. Thelestam and L. Blomqvist, *Toxicon* **26,** 51 (1988).

[94] J. Jeljaszewicz, *in* "The Staphylococci" (J. O. Cohen, ed.), p. 249, Wiley, New York, 1972.

[95] D. Schmeling, C. Gemmell, P. Craddock, P. Quie, and P. Peterson, *Inflammation* **5,** 313 (1981).

[96] C. Gemmell, P. Peterson, D. Schmeling, and P. Quie, *Infect. Immun.* **38,** 975 (1982).

[97] N. Suttorp, W. Seeger, J. Zucker-Reimann, L. Roka, and S. Bhakdi, *Infect. Immun.* **55,** 105 (1987).

[98] N. Suttorp and E. Habben, *Infect. Immun.* **56,** 2228 (1988).

lated.[99] This event could explain the rapid influx of extracellular calcium observed with human PMNs following exposure to δ-toxin.[100] In the same study, δ-toxin caused enhanced production of PAF from exogenously added lyso-[^{3}H]PAF. Such data indicate that in addition to PLA_2 activation, δ-toxin may enhance acetyltransferase activity in the PMN membrane. Although a dose-dependent inhibition of human PMNs and monocyte chemotaxis toward casein (1 mg/ml) by purified δ-toxin has been reported,[91] detailed characterization of this effect is lacking.

Perfringolysin (Clostridium perfringens θ-Toxin)

Perfringolysin is an oxygen-labile cytolysin similar in structure and biological effect to streptolysin 0.[91] At subcytolytic concentrations it has been shown to inhibit both PMN random migration and chemotaxis to a variety of chemotactic factors.[101,102] Monocyte chemotaxis was also reportedly inhibited by perfringolysin, but the effect was relatively less than that seen with PMNs.[83] The inhibitory action of θ-toxin can be neutralized by the addition of cholesterol. Stevens *et al.* found that crude preparations of θ-toxin blocked both PMN chemotaxis and chemiluminescence.[103] Purified θ-toxin (12.75 HU/2.5 $\times$ 10^5 cells) was extremely toxic, with only 9% viability at 20 min. A biphasic chemiluminescent response was noted with enhancement at low toxin concentrations and dose-dependent inhibition at higher toxin concentrations. Similarly, very low doses of θ-toxin (0.02–0.06 HU/2.5 $\times$ 10^5 PMNs) increased random migration, whereas concentrations greater than 0.08 HU/2.5 $\times$ 10^5 PMNs significantly reduced random migration as well as chemotaxis toward FMLP.

Cytotoxin (Pseudomonas aeruginosa)

Cytotoxin exhibits lytic activity against most eukaryotic cells studied. Although its primary effect appears to be directed at the cell membrane, thymidine incorporation into DNA by murine splenocytes was decreased in the presence of cytotoxin.[104] Toxicity appears to be intimately related to Ca^{2+}-dependent membrane systems.[82,105] Baltch *et al.* reported that although exhibiting no chemoattractant qualities itself, 30 min of preexposure to cytotoxin at a subcidal concentration (4 μg/ml) caused significant

[99] J. H. Freer and T. H. Birbeck, *J. Theor. Biol.* **94,** 535 (1982).
[100] S. Kasimir, W. Schonfeld, J. Alour, and W. König, *Infect. Immun.* **58,** 1653 (1990).
[101] P. Wilkinson, *Nature* **255,** 486 (1975).
[102] P. C. Wilkinson and R. B. Allan, *Mol. Cell. Biochem.* **20,** 25 (1978).
[103] D. L. Stevens, J. Mitten, and C. Henry, *J. Infect. Dis.* **156,** 324 (1987).
[104] T. G. Obrig, A. L. Baltch, T. P. Moran, S. P. Mudzinski, R. P. Smith, and F. Lutz, *Infect. Immun.* **45,** 756 (1984).
[105] T. Hirayama and I. Kato, *Infect. Immun.* **43,** 21 (1984).

inhibition of PMN chemotaxis (to 10^{-7} *M* FMLP) and bactericidal activity.[106] Transmission electron microscopy revealed increased membrane permeability to ruthenium red at cytotoxin concentrations as low as 0.1 μg/ml and detectable ultrastructural changes at concentrations of 2 μg/ml and greater. Exposure of PMNs to 4 μg/ml cytotoxin for 2 hr caused marked clumping and cell destruction, nuclear blebbing, release of granules, and diminished phagocytic activity.[82]

Other Toxins

Anthrax Toxin

Anthrax toxin is produced by *Bacillus anthracis* and consists of three protein components: protective antigen (PA), edema factor (EF), and lethal factor (LF). PA bound to the surface of susceptible cells forms a receptor for EF and LF. EF possesses adenylate cyclase activity and acts in combination with PA to produce cAMP in susceptible cells. LF is identified by the finding that intravenous administration of LF and PA is lethal to many animal species; however, its specific mode of action is currently unknown. Although none of the components of anthrax toxin alone affected either random migration or chemotaxis, Wade *et al.* indicated that combinations of EF and PA, LF and PA, or EF and LF and PA significantly enhanced chemotaxis (directed migration) without affecting random migration.[107] The toxins themselves exhibited no chemoattractant activity. Moreover, although toxin combinations containing EF and PA were associated with increased levels of neutrophil cAMP, cAMP levels with LF and PA were equivalent to those of control cells. In a separate study, phagocytosis of killed, avirulent *B. anthracis* was shown to be reversibly inhibited by EF and PA, but not by LF and PA.[108] A similar pattern of inhibition was demonstrated with chemiluminescence induced by opsonized *B. anthracis* or PMA, which activates protein kinase C directly. Wright and Mandell examined the effect of anthrax toxin components on the release of superoxide anion (O_2^-) following FMLP stimulation.[109] Neutrophils isolated in the absence of LPS (<0.1 ng/ml) (See below) released minimal O_2^-. Pretreatment with anthrax toxin had little effect. Neutrophil priming by exposure to 3 ng/ml LPS (1 hr, 37°) or

[106] A. Baltch, M. Hammer, R. Smith, T. Obrig, J. Conroy, M. Bishop, M. Egy, and F. Lutz, *Infect. Immun.* **48,** 298 (1985).

[107] B. Wade, G. Wright, E. Hewlett, S. Leppla, and G. Mandell, *Proc. Soc. Exp. Biol. Med.* **179,** 159 (1985).

[108] J. O'Brien, A. Friedlander, T. Dreier, J. Ezzel, and S. Leppla, *Infect. Immun.* **47,** 306 (1985).

[109] G. Wright and G. Mandell, *J. Exp. Med.* **164,** 1700 (1986).

100 ng/ml muramyl dipeptide (MDP, a synthetic glycopeptide mitogen analogous to bacterial peptidoglycans) increased O_2^- release fivefold. Pretreatment of neutrophils with EF and PA or LF and PA inhibited LPS or MDP priming of O_2^- release up to 90%. If, however, LPS or MDP priming preceded toxin exposure, inhibition was markedly diminished. In contrast to the effect on phagocytosis,[108] anthrax toxin exhibited no inhibition of O_2^- release following PMA stimulation. As neither LPS nor MDP enhanced or primed neutrophil release of O_2^- in response to PMA, the authors suggested that anthrax toxin acts to prevent priming of membrane activation pathways at a point proximal to protein kinase C.

Toxin A (Clostridium difficile Enterotoxin)

Toxin A of *Clostridium difficile* was shown by Pothoulakis *et al.* to possess both chemokinetic and chemotactic activity for human PMNs.[110] Directed migration was observed at concentrations ranging from 10^{-7} to 10^{-6} *M* (25 to 200 μg/ml). Response at 100 μg/ml toxin A was greater than 90% of the response to 10^{-7} *M* FMLP. Although no superoxide anion production was detected following exposure to toxin A, intracellular Ca^{2+} was transiently increased, an effect blocked by PT. These data suggest that the toxin A effect is mediated, at least in part, by the guanine nucleotide-binding protein (G_i).

Phospholipase C (Clostridium perfringens α-Toxin)

Phospholipase C (PLC) is a Ca^{2+}-dependent enzyme with broad substrate specificity. It causes a reduction of both random migration and chemotactic movement of monocytes and PMNs, with relatively greater inhibition of monocyte motility.[101] More recent evidence, though, indicates that although crude preparations blocked both chemotaxis and chemiluminescence, purified α-toxin did not affect PMN viability or chemotactic responsiveness and moderately enhanced opsonized zymosan-induced chemiluminescence.[103] Furthermore, antitoxin activity has been demonstrated by either whole or sonicated PMNs, but not mononuclear cells or various cell lines.[111] This activity colocalized with the primary granule marker, myeloperoxidase, and enzyme inhibitor data were suggestive of a major role by neutral proteases, such as cathepsin G.

Sphingomyelinase C (β-Toxin)

Sphingomyelinase C (β-toxin), a staphylococcal phospholipase C that acts primarily on membrane sphingomyelin, was also shown to inhibit

[110] C. Pothoulakis, R. Sullivan, D. A. Melnick, G. Triadafilopoulos, A-S. Gadenne, T. Meshulam, and J. T. LaMont, *J. Clin. Invest.* **81,** 1741 (1988).

[111] H. Larson, G. Smith, and L. Shah, *Br. J. Exp. Pathol.* **66,** 243 (1985).

leukocyte migration. This inhibition was primarily directed against monocytes with little or no effect against PMNs, even at 10^2 *M* or greater concentrations.[91,101]

Alkaline Protease and Elastase (Pseudomonas aeruginosa)

Coincubation of cells with both alkaline protease (AP) and elastase from *Pseudomonas aeruginosa* was shown to inhibit PMN chemotaxis to *E. coli* culture filtrates in a dose-dependent manner.[112] Of the two, elastase exhibited more potent activity, with nearly 40% inhibition at 0.025 μg/ml compared with similar inhibition by AP at 0.5 μg/ml. If, however, the PMNs were preincubated with either AP (5 μg/ml) or elastase (2.5 μg/ml) for 1 hr and then washed prior to assay, AP inhibition was virtually unchanged whereas inhibition of chemotaxis by elastase was reduced by more than 50%. Oxygen consumption induced by PMA was unaffected by either AP or elastase; however, opsonized zymosan-stimulated O_2 consumption was inhibited 41% by 1 μg/ml AP but not by elastase (up to 25 mg/ml). Conversely, elastase was shown to inhibit the reduction of cytochrome C by superoxide anion in PMA-stimulated PMNs, whereas AP had little effect. In another report, AP and elastase (25 μg/ml) strongly inhibited myeloperoxidase-mediated PMN chemiluminescence.[113] Inhibition of monocyte chemotactic activity toward both FMLP and zymosan-activated serum and chemiluminescent response to opsonized zymosan was recently reported with elastase concentrations as low as 1 μg/ml.[114]

Bacterial Metabolites/Structural Components

Aminopeptidase Inhibitors

The action of bacterial protease inhibitors on neutrophil chemotaxis was reported by Matsuda *et al.*[115] Actinonin and amastatin, low-molecular-weight aminopeptidase inhibitors derived from *Actinomycetes,* had no effect on neutrophil random migration at 10^{-6} *M*. Responsiveness to FMLP (2×10^{-8} *M*) was significantly enhanced in the presence of both agents at submicromolar concentrations. Enhancement was detected with actinonin from 10^{-8} to 10^{-6} *M,* whereas amastatin was ineffective at these concentrations. Neutrophil responsiveness remained significantly elevated with exposure to either agent at 10^{-6} *M,* even if the cells were prewashed prior to assay. Regression line analysis demonstrated increased

[112] A. Kharazmi, G. Döring, N. Høiby, and N. Valerius, *Infect. Immun.* **43,** 161 (1984).
[113] A. Kharazmi, N. Høiby, G. Döring, and N. Valerius, *Infect. Immun.* **44,** 589 (1984).
[114] A. Kharazmi and H. Nielsen, *APMIS* **99,** 93 (1991).
[115] Matsuda, Y. Katsuragi, Y. Saiga, T. Tanaka, and M. Nakamura, *Biochem. Int.* **16,** 383 (1988).

total cell migration with both inhibitors, whereas actinonin alone was associated with an increased rate of migration. Bestatin [(25,3*R*)-3-amino-2-hydroxy-4-phenyl-butanoyl-L-leucine], a well-characterized immunomodulator produced by *Streptomyces olivoreticuli* that also inhibits aminopeptidase activity, has similarly been shown to augment both chemotactic and chemokinetic movement.[116]

Ammonium

Weak bases can alter the cytoplasmic pH of eukaryotic cells, an effect known to modify a variety of cellular functions. With leukocytes these include chemotaxis, phagocytosis, granule secretion, and bactericidal mechanisms.[117] Ammonium, a weak base produced by bacteria as part of urea metabolism, exhibited concentration-dependent inhibition of PMN migration in response to IL-8 (NAP-1) and FMLP.[118] Migration to IL-8 was reduced more than 80% by 15 m*M* ammonium chloride. More recently, evidence was presented by Brunkhorst and Niederman linking the inhibitory effects of ammonium to interference with cytoskeletal actin polymerization/depolymerization reactions.[117]

Short-Chain Fatty Acids

Short-chain fatty acids (SCFAs) are a major metabolic by-product of anaerobic bacteria. Succinic acid, a ubiquitous SCFA among *Bacteroides* species, has been found in particularly high concentrations within mixed infections containing *Bacteroides in vivo* (up to 31 m*M*) and *in vitro* (up to 36 m*M*) in *Bacteroides* culture filtrates. Rotstein *et al.* have reported that succinate or *Bacteroides fragilis* culture filtrate inhibits numerous neutrophil functions, including chemotaxis, bacterial killing, and respiratory burst activity.[119–122] Chemotaxis to both FMLP and zymosan-activated serum was significantly reduced following exposure to as little as 10 m*M* succinate for 20 min. Activity was strictly pH dependent, with marked inhibition noted at pH 5.5, but not at neutral pH, that paralleled a rapid, significant reduction of intracellular pH. These features were consistent with a model of inhibition mediated by undissociated succinate (pK_a = 5.57 and 4.19) that could enter the cell and serve as a proton shuttle to lower cytoplasmic pH levels. In support of this model, the

[116] C. Jarstrand and H. Blomgren, *J. Clin. Lab. Immunol.* **5,** 67 (1981).

[117] B. Brunkhorst and R. Niederman, *Infect. Immun.* **59,** 1378 (1991).

[118] A. Samanta, J. Oppenheim, and K. Matsushima, *J. Biol. Chem.* **265,** 183 (1990).

[119] O. Rotstein, T. Pruett, V. Fiefel, R. Nelson, and R. Simmon, *Infect. Immun.* **48,** 402 (1985).

[120] O. Rotstein, T. Pruett, J. Sorenson, V. Fiegel, R. Nelson,and R. Simmons, *Arch. Surg.* **121,** 82 (1986).

[121] Rotstein, P. Nasmith, and S. Grinstein, *Infect. Immun.* **55,** 864 (1987).

authors reported that ^{14}C-labeled succinate coincubated with neutrophils at pH 5.5 was found to accumulate rapidly intracellularly. Similar inhibitory activity was shown with several other SCFAs produced by *Bacteroides* species, including butyrate, isobutyrate, and propionate. Adipic acid, which is not produced by *Bacteroides,* was also shown to inhibit neutrophil function.

Lipopolysaccharide

Lipopolysaccharide, or endotoxin, is a major structural component of the outer membrane of gram-negative bacteria and has profound effects on PMNs both *in vivo* and *in vitro*. It commonly occurs as three covalently linked components: an O-polysaccharide (outer) core, an inner polysaccharide core, and lipid A. The minimal LPS structure consistent with bacterial growth contains only lipid A and two unique monosaccharide moieties of 3-keto-2-deoxyoctonate (KDO), yet this structure (termed Re LPS) retains full biological activity.[123,124] Besides its direct effects on PMNs, LPS is also a potent inducer of cytokine production and release by a variety of cells. Mononuclear phagocytes, in particular, respond to LPS stimulation by secretion of tumor necrosis factor-α (TNF-α), IL-1β, IL-6, and prostaglandin (PG)E_2.[125] These cytokines, in turn, serve as powerful amplification signals for the production of inflammatory mediators such as IL-8 (NAP-1). Recent work has shown that CD14, a 55-kDa glycoprotein, serves as an LPS receptor on PMNs and monocytes.[126] The receptor recognizes complexes of LPS with the serum protein, LPS-binding protein (LBP),[124,126,127] and can promote phagocytic uptake of bacteria or LPS-coated particles. LPS–LBP complexes also dramatically enhance TNF-α secretion by monocytes over equivalent LPS concentrations alone, responses that can be blocked by monoclonal antibodies to CD14.[126]

Because of the complex nature of not only the cellular and physiological interactions of LPS with host tissues and cells but also intrinsic variation among LPS chemotypes and even LPS preparations, the discussion is subdivided into minor headings that, it is hoped, will help guide the reader

[122] Rotstein, T. Vittorini, J. Kao, M. McBurney, P. Nasmith, and S. Grinstein, *Infect. Immun.* **57,** 745 (1989).

[123] C. R. H. Raetz, *Annu. Rev. Biochem.* **59,** 129 (1990).

[124] C. R. H. Raetz, R. J. Ulevitch, S. D. Wright, C. H. Sibley, A. Ding, and C. F. Nathan, *FASEB J.* **15,** 2652 (1991).

[125] D. C. Morrison and J. L. Ryan, *Annu. Rev. Med.* **38,** 417 (1987).

[126] S. D. Wright, R. A. Ramos, P. S. Tobias, R. J. Ulevitch, and J. C. Mathison, *Science* **249,** 1431 (1990).

[127] S. D. Wright, P. S. Tobias, R. J. Ulevitch, and R. A. Ramos, *J. Exp. Med.* **170,** 1231 (1989).

through an even more complex literature concerning endotoxin effects on neutrophil chemotaxis.

SYSTEMIC EFFECT. Experimental human endotoxemia, achieved by intravenous injection of *Pseudomonas* endotoxin (0.1 μg/kg) into healthy volunteers, resulted in a transient, circulating neutropenia at 1 hr, followed by neutropenia that peaked 2–4 hr postinjection.[128] Functional assays of PMN phagocytosis, killing, glucose metabolism, and random migration were normal. Chemotaxis toward zymosan-activated serum (ZAS), however, was significantly depressed 1 hr after injection with gradual recovery by 4 hr. In a rabbit model, Rosenbaum *et al.* reported that heterophils isolated from animals challenged 24 hr previously by intravenous injection of 100 μg of *E. coli* 055:B5 LPS exhibited reduced migration and granule release in response to activated plasma or partially purified C5a.[129] By contrast, chemotaxis and degranulation with FMLP was normal or elevated.

STEROID EFFECT. *Escherichia coli* endotoxin (1 ng/ml) inhibited PMN chemotaxis to C5a by 85%, with a small, but significant reduction in random migration at concentrations ranging from 0.05 to 50 ng/ml.[130] This inhibition could be reversed by both polymyxin B and glucocorticoids, but through different mechanisms. Polymyxin B, known to bind to the lipid A-KDO region of endotoxin,[131] inactivated endotoxin directly, a function resistant to dialysis. By contrast, the glucocorticoid effect apparently occurred through the PMNs themselves. Polymyxin B and methylprednisolone together acted synergistically to block endotoxin inhibition. In an effort to determine the basis for steroid inhibition of endotoxin-mediated chemotaxis inhibition, Issekutz *et al.* examined a number of compounds likewise known to increase intracellular cAMP levels.[132] Prostaglandin E_1, isoproterenol, dibutyryl-cAMP, and cholera toxin were shown to similarly antagonize endotoxin inhibition of chemotaxis toward C5a in dose-dependent fashion, at concentrations below those necessary for their own inhibitory effects on PMN chemotaxis.

An expanded modulatory role for prostaglandins in PMN response to endotoxin challenge was suggested by the work of Hayashi *et al.*[133] Using

[128] M. Territo and D. Golde, *Blood* **47,** 539 (1976).

[129] J. Rosenbaum, K. Hartiala, R. Webster, E. Howes, Jr., and I. Goldstein, *Am. J. Pathol.* **113,** 291 (1983).

[130] A. C. Issekutz and W. D. Biggar, *J. Lab. Clin. Med.* **92,** 873 (1978).

[131] D. C. Morrison and D. M. Jacobs, *Immunochemistry* **13,** 813 (1976).

[132] A. C. Issekutz, M. Ng, and W. D. Biggar, *Infect. Immun.* **24,** 434 (1979).

[133] T. Hayashi, H. Iwata, T. Hasegawa, M. Ozaki, H. Yamamoto, and T. Onodera, *Comp. Pathol.* **104,** 161 (1991).

a model system of mice chronically infected by lactic dehydrogenase virus (LDV), a pathogen that affects certain macrophage subpopulations, intraperitoneal injection of 30 μg of LPS (*E. coli* 026:B6) was shown to elicit significantly fewer peritoneal PMNs in infected animals compared with controls. Yet chemotactic responsiveness of these PMNs to FMLP *in vitro* was equivalent. As LPS is a known inducer of IL-1 production,[125,134] thioglycolate-elicited peritoneal macrophages were harvested and treated with 100 ng/ml LPS for 24 hr. LDV-infected mouse macrophages produced significantly less IL-1 than control cells and intraperitoneal injection of the cell supernatants into uninfected mice resulted in significantly greater PMN influx with control (IL-1-rich) supernatants. To determine if PGE generation might be responsible for the suppressed IL-1 response in infected animals, indomethacin, a potent inhibitor of cyclooxygenase activity, was tested. Although having no effect alone, addition of indomethacin (10^{-6} *M*) to LPS significantly increased IL-1 production by both control and infected macrophages. Indomethacin pretreatment of mice prior to LPS challenge also enhanced recruitment of peritoneal PMNs *in vivo* with no significant difference between control and infected animals. These results suggest that enhanced cyclooxygenase activity and prostaglandin production in LDV-infected mice may lead to decreased IL-1 production and reduced PMN recruitment following LPS challenge. The implications for other types of chronic infectious disease are intriguing.

INHIBITION FACTOR. Although most studies indicate that LPS acts as an inhibitor of PMN chemotaxis,[125,135,136–138] the precise mechanism for this effect is unknown. As initially reported by Goetzl and Austen,[139] incubation of human PMNs or monocytes with endotoxin (*E. coli* 026:B6, 0.2 μg/ml) induced release of a preformed inhibitor to PMN chemotaxis termed *neutrophil-immobilizing factor* (NIF), a proteolytically sensitive molecule of approximately 5 kDa that remains obscure.

Subsequently, culture filtrates (CFs) from *E. coli* or *Staphylococcus epidermidis* and immune-complex activated serum (AS) were both shown to induce chemotaxis of human PMNs. Preexposure to AS or CF resulted in deactivation to subsequent migration toward AS but not to CF.[140] Endotoxin (*E. coli* 0111-B4) did not exhibit chemoattractant qualities alone

[134] W. P. Arend and R. J. Massoni, *Clin. Exp. Immunol.* **64,** 656 (1986).
[135] D. C. Morrison and R. J. Ulevitch, *Am. J. Pathol.* **93,** 617 (1978).
[136] I. Ginsburg and P. Quie, *Inflammation* **4,** 301 (1980).
[137] I. Ginsburg, J. Goultchin, A. Stabholtz, N. Neeman, M. Lahav, L. Landstrom, and P. Quie, *Agents Actions, Suppl.* **26,** 240 (1980).
[138] M. E. Wilson, *Rev. Infect. Dis.* **7,** 404 (1985).
[139] E. J. Goetzl and K. F. Austen, *J. Exp. Med.* **136,** 1564 (1972).
[140] A. C. Issekutz and W. D. Biggar, *Infect. Immun.* **15,** 212 (1977).

(0.1–1000 μg/ml) but inhibited migration toward AS (presumably C5a) in a concentration-dependent fashion, with complete inhibition at 2 μg/ml. No inhibition by endotoxin was noted for chemotaxis toward CF (presumably FMLP). In addition, 0.2 μg/ml endotoxin did not detectably alter hexose monophosphate shunt activity in either resting or phagocytosing PMNs. Similar results were seen when PMNs from rabbits injected 18–24 hr previously with *E. coli* 055:B5 LPS demonstrated depressed chemotaxis to C5a, platelet-derived growth factor (PDGF), LTB_4, and PAF, but not FMLP.[141] Secretory response was diminished to the endogenous peptide factors, C5a and PDGF, but not to the lipid mediators, LTB_4 and PAF, or FMLP. A plasma-derived chemotaxis inhibitor(s) was suggested by inhibition of control rabbit PMN migration to LTB_4 and C5a following exposure to plasma from LPS-injected rabbits.

Haslett *et al.*, however, demonstrated that exposure of human PMNs to *E. coli* K235 LPS (10 mg/ml) for 1 hr at 37° caused a significant reduction in chemotaxis to 10^{-8} *M* FMLP.[142] Additionally, LPS induced significant morphological changes in resting PMNs and increased both superoxide anion production and granule secretion following FMLP stimulation. The authors cautioned that trace contamination by LPS of density gradient media used for PMN purification could act to "prime" cells and influence subsequent measures of cell function. The relationship of these findings to NIF or some other substance as well as the actual mechanism of inhibition remains to be determined.

Chemokinetic/Chemotactic Activity. Rather than inhibiting cell migration, both naturally produced lipid A and various synthetic analogs exhibited chemokinetic activity with human PMNs and monocytes at concentrations ranging from 0.1 to 100 ng/ml.[143,144] LPS prepared from several different bacterial strains was also chemokinetic for PMNs; however, monocytes were not affected. A direct chemoattractant role for LPS was suggested by Creamer *et al.*[145] When relatively high concentrations (450–1000 μg/ml) of LPS B from *Salmonella typhosa* were employed, significantly enhanced neutrophil migration was noted toward LPS compared with controls. In addition, lipid A (*Salmonella minnesota* Re 595) induced significant chemotaxis at 125 μg/ml. Chemotactic activity of the

[141] K. Hartiala, L. Langlois, I. Goldstein, and J. Rosenbaum, *Infect. Immun.* **50,** 527 (1985).

[142] C. Haslett, L. Guthrie, M. Kopaniak, R. Johnston, Jr., and P. Henson, *Am. J. Pathol.* **119,** 101 (1985).

[143] S. Kotani, H. Takada, M. Tsujimoto, T. Ogawa, Y. Mori, M. Sakuta, A. Kawasaki, M. Inage, S. Kusumota, T. Shiba, and N. Kasai, *Infect. Immun.* **41,** 758 (1983).

[144] S. Kotani, H. Takada, M. Tsujimoto, T. Ogawa, I. Takahasi, T. Ikeda, K. Otsuka, *et al., Infect. Immun.* **49,** 225 (1985).

[145] H. R. Creamer, N. Hunter, W. Bullock, and W. Gabler, *Inflammation* **15,** 201 (1991).

LPS preparation was reportedly heat stable and partially inhibited by polymyxin B. Neutrophil responsiveness to FMLP (2×10^{-8} M) was enhanced in the presence of LPS (125 μg/ml), suggesting a priming effect similar to that reported for other neutrophil functions.

ADHESION. Dahinden and colleagues examined the effect of endotoxin on PMN adhesiveness and function.[146] In parallel with increased adherence to plastic petri dishes, different *Salmonella* endotoxin preparations (5 μg/ml) induced release of the secondary granule components, vitamin B_{12}-binding protein and lysozyme. Release of primary granule marker, myeloperoxidase, was not measurable, yet glucose oxidation and superoxide (O_2^-) production were increased.[147] Extremely low levels of endotoxin inhibited PMN chemotaxis toward 10% inulin-activated plasma (IAP), with nearly 30% inhibition at 50 pg/ml of one rough-strain endotoxin preparation. Chemotaxis to FMLP could also be inhibited by and random migration reduced in the presence of endotoxin. The authors hypothesized that endotoxin exposure resulted in a generalized increase in PMN adhesiveness that reduced cellular motility. No chemoattractant activity was detectable with either endotoxin or lipid A over a broad concentration range (0.5 ng to 5 μg/ml).

In contrast to findings with adherent PMNs, exposure of PMNs in suspension to endotoxin had no detectable effect on glucose oxidation, granule release, or O_2^- production in the presence or absence of cytochalasin B.[147] Unlike FMLP, which stimulated both adherence and cell aggregation in suspension, no aggregatory response was detectable with endotoxin up to 5 μg/ml; however, Bryant *et al.* reported that 10 μg/ml endotoxin promoted PMN aggregation, an effect blocked by 10 mM sodium ethylenediaminetetraacetic acid.[148] In addition, if endotoxin exposure was performed at 4° and the cells were washed, random migration was not reduced as seen at 37°. This suggested to the authors that the inhibitory action of endotoxin required initial phago/endocytosis by the target cells.

LIPOPOLYSACCHARIDE STRUCTURE. In examining the effects of LPS structure on PMN function, LPS from *E. coli* J5, a UDP-galactose-4-epimerase-deficient mutant of *E. coli* 0111B4, was compared with the parent strain LPS.[149] As the mutation affects polysaccharide side-chain production, J5 LPS contains primarily lipid A. A 2-hr exposure to 0111B4 LPS (250 μg/ml) had no significant effect on PMN migration toward zymo-

[146] C. Dahinden, C. Galanos, and J. Fehr, *J. Immunol.* **130,** 857 (1983).

[147] C. Dahinden and J. Fehr, *J. Immunol.* **130,** 862 (1983).

[148] R. E. Bryant, R. M. Des Pres, and Y. De Rogers, *J. Biol. Med.* **40,** 192 (1967).

[149] P. Henricks, M. Van der Tol, R. Thyssen, B. van Asbeck, and J. Verhoef, *Infect. Immun.* **41,** 294 (1983).

san-activated serum, whereas profound inhibition was seen with J5 LPS. Similar findings were reported at concentrations up to 16 μg/ml. Other functions, such as phagocytosis and metabolic activation, were also diminished by J5 LPS, an effect that could be blocked by polymyxin B or an oxygen radical scavenger. By contrast, 0111B4 LPS had minimal effects. Separation of lipid A and the polysaccharide fractions from 0111B4 LPS revealed inhibitory activity with the lipid A fraction alone. As both J5 LPS and lipid A from 0111B4 were shown to stimulate PMN superoxide production directly, the authors hypothesized that their inhibitory actions could stem from radical-induced damage to the PMN membrane. Pugliese *et al.* demonstrated that in the absence of serum, both lipid A and LPS from smooth-strain (wild-type) *Salmonella minnesota,* as well as most rough-strain mutants, inhibited random migration in a dose-dependent manner (0.01–1.0 μg/ml).[150] At the same time, significant stimulation of chemiluminescence was uniformly noted with these substances at 1.0 μg/ml. With respect to migration, increasing the number of carbohydrate residues attached to lipid A resulted in decreased inhibition of random motility in a nearly linear fashion.

Further examination of the relationship between LPS composition and modulatory effects on human PMN function by Kharazmi *et al.* revealed that LPS from most but not all tested strains of *Pseudomonoas aeruginosa* inhibited chemotaxis at 10 μg/ml.[151] Similar results were demonstrated with LPS enhancement of PMN chemiluminescence following FMLP stimulation. LPS alone at concentrations up to 10 μg/ml had no chemotactic activity. Alanine content of the LPS core region was the major difference between the most potent LPS preparation (174-0:9, 391 nmol alanine/mg LPS) and the least active (1118-0:3, 110 nmol alanine/mg LPS), suggesting a possible key to chemotactic inhibitory activity. Bignold *et al.* tested the effect of endotoxin preparations (*E. coli, Klebsiella pneumoniae, Vibrio cholerae, Shigella flexneri, Salmonella typhosa,* and *P. aeruginosa*) on the chemotactic response of human PMNs.[152] At 100 μg/ml all endotoxins significantly inhibited chemotaxis to IL-8 (100 ng/ml) but not to FMLP (10^{-7} *M*). Only *S. flexneri* endotoxin was found to inhibit random migration.

Recent work in our laboratory has revealed that the neutrophil susceptibility to LPS may be dependent on the species of bacteria and/or prior infection of the host by the bacteria. *Porphyromonas* (*Bacteroides*) *gingi-*

[150] C. Pugliese, M. LaSalle, and V. DeBari, *Mol. Immunol.* **25,** 631 (1988).

[151] A. Kharazmi, A. Fomsgaard, R. Conrad, C. Galanos, and N. Høiby, *J. Leukocyte Biol.* **49,** 15 (1991).

[152] L. Bignold, S. Rogers, T. Slaw, and J. Bahnisch, *Infect. Immun.* **59,** 4255 (1991).

valis LPS is known to be structurally quite different from enteric LPS (little or no core polysaccharide, KDO, and heptose and markedly different lipid A structure) and to possess less toxin activity in limulus lysate or neutrophil priming assays.[153] Neutrophils from subjects infected with *P. gingivalis* are profoundly inhibited by *P. gingivalis* LPS compared with noninfected individuals, whereas the response to *E. coli* LPS is noninhibitory and equivalent in both cases (data not shown). Moreover, the inhibition is enhanced in the presence of serum of infected individuals, an effect also seen with normal cells. A role for these results support serum factors such as NIF, LPS antibodies or the presence of increased levels of LBP.

Complex Mixtures

Legionella

A sonic extract of *Legionella pneumophila* serogroup 1 (Lp1 Ag) demonstrated direct chemotactic activity for neutrophils as well as monocytes over a broad (15.6–1000 mg/ml) concentration range.[154] Neutrophil cytotoxicity was also detected, as viability decreased 10–40% at concentrations from 125 to 4000 mg/ml over a 2-hr exposure; however, no toxic effect was noted with monocytes. Similarly, pretreatment of monocytes with Lp1 Ag (31.25 μg/ml) did not influence their chemotactic response to a variety of agents, whereas neutrophil responsiveness was inhibited. Inhibition was partially reversed by heat treatment of Lp1 Ag, suggesting the presence of both heat-labile and heat-stable components. Interestingly, although heat treatment (100° or 121°, 30 or 60 min) had minimal effect on the direct chemoattractive activity of the sonicate for neutrophils, monocyte recruitment was markedly elevated with 121° treatment compared with 100° or untreated Lp1 Ag.

Oral Gram-Negative Bacteria

Inhibition of leukocyte migration by sonic extracts of dental plaque bacteria, including *Bacteroides melaninogenicus, Fusobacterium nucleatum, Veilonella parvula,* and *Actinomyces viscosus* was reported, although protein concentrations of the extracts varied considerably.[155] Among other oral bacterial species, culture filtrates and/or sonicates of *Actinobacillus actinomycetemcomitans, Bacteroides* species, *Capnocytophaga* species, and *Fusobacterium nucleatum* specifically inhibited PMN chemotaxis to

[153] S. C. Holt and T. E. Bramanti, *Crit. Rev. Oral Biol. Med.* **2,** 177 (1991).

[154] C. Rechnitzer, A. Kharamzmi, and H. Nielsen, *Eur. J. Clin. Invest.* **16,** 368 (1986).

[155] E. Budtz-Jörgensen, J. Kelstrup, T. Funder-Nielsen, and A. Knudsen, *J. Periodontal Res.* **12,** 21 (1977).

FMLP, a feature lost following dialysis.[156] Except for *Capnocytophaga* species, these soluble products inhibited binding of [^{3}H]FMLP to the PMN surface, suggesting direct competition with the FMLP receptor. Similar inhibition of chemotaxis to endotoxin-activated serum (EAS), however, implied a more general blocking effect. Sonic extract of *B. melaninogenicus* was neither chemotactic nor chemokinetic in studies performed by Gale *et al.*, although PMN migration with sonicate and 2% human serum albumin (HSA) was significantly greater than with HSA alone.[157] In a separate analysis, dialyzed sonic extracts from oral bacteria demonstrated significant chemoattractant activity and the capacity for serum activation.[158] The greatest response was seen with *Streptococcus sanguis* and *Bacteroides intermedius* extracts (1 mg protein/ml) and was largely attributed to formylmethionyl oligopeptide-like materials.

Culture supernatants from *P. gingivalis* reportedly diminished binding of [^{3}H]FMLP and suppressed O_2^- generation by FMLP-stimulated PMNs, although a protein concentration of 1 mg/ml was required for significant effect.[159] *P. gingivalis* ATCC 33277 was shown to produce chemotactically active C5a des-Arg from purified C5 through the action of bacterial protease(s).[160] In an earlier report, both culture filtrates (CFs) and outer membrane preparations (OMs) from several *Bacteroides* species demonstrated chemotactic, but not chemokinetic, activity with rabbit PMNs. Interestingly, coincubation of guinea pig complement with either CFs or OMs blocked chemotaxis but resulted in enhanced lysosomal granule release over the preparations alone.[161] It could be postulated that proteolytic activity within the *Bacteroides* CFs and OMs, similar to that described for *P. gingivalis* ATCC 33277, could produce sufficient C5a to inhibit chemotactic movement and elicit degranulation.

Glycocalyces

Bacterial glycocalyx, which includes both the tightly bound capsule and the more loosely associated slime layer, has been proposed as an important virulence factor among pathogenic strains. The glycocalyces of *Bacteroides fragilis, Bacteroides thetaiotaomicron,* and *Staphylococcus epidermidis* were found to inhibit opsonophagocytosis and the luminol-

[156] T. Van Dyke, E. Barthomomew, R. Genco, J. Slots, and M. Levine, *J. Periodontol.* **53,** 502 (1982).
[157] K. M. Gale, R. N. Powell, and G. J. Seymour, *J. Periodontal Res.* **18,** 119 (1983).
[158] D. Lareau, M. Herzberg, and R. Nelson, *J. Periodontol.* **55,** 540 (1984).
[159] K. Maeda, T. Hirofuji, N. Chinju, K. Tanigawa, Y. Iwamoto, T. Hatekeyama, and M. Aono, *Adv. Dent. Res.* **2,** 315 (1988).
[160] H. Schenkein and C. Berry, *J. Periodontal Res.* **23,** 1 (1988).
[161] S. Adamu and J. Sperry, *Infect. Immun.* **33,** 806 (1981).

dependent chemiluminescent response of PMNs to PMA (10 ng/ml).[162,163] Chemiluminescence following exposure to opsonized, heat-killed *Staphylococcus aureus* was not affected. Although the authors reported inhibition of PMN chemotaxis by glycocalyx preparations, use of nonactivated normal serum as a chemoattractant is unusual and suggests that random migration was actually measured. Inhibition of PMN motility by *B. thetaiotaomicron* and *S. epidermidis* glycocalyces probably represented local complement activation by these preparations. No inherent chemotactic activity was detected with any of the glycocalyx preparations.

Implications for Future Studies

The study of the effect of bacterial products on neutrophil function is important on two levels. First is the obvious importance of these products as virulence factors in the pathogenesis of specific infectious diseases. The second, however, relates to the use of these molecules as *in vitro* probes for the examination of specific neutrophil pathways involved in various cell functions. For example, both cholera toxin and pertussis toxin were critical tools in the study and elucidation of G-protein pathways. Our understanding of the complex biochemical pathways that lead to chemotaxis is incomplete. It is anticipated that as the mechanisms of action of specific toxins and bacterial products are learned, these properties can be exploited to dissect the intracellular pathways that underlie and control cell motility, chemotaxis, and other cellular functions.

Acknowledgments

We express our thanks to Ms. Kimberly Ladley and Miriam Hunt for excellent clerical assistance and our editors, Patrik Bavoil and Virginia Clark, for their encouragement and support.

[162] D. A. Ferguson, Jr., E. M. Veringa, W. R. Mayberry, B. P. Overbeek, D. W. Lambe, Jr., and J. Verhoef, *Microbios* **69,** 53 (1992).

[163] E. M. Veringa, D. A. Ferguson, D. W. Lambe, Jr., and J. Verhoef, *J. Antimicrob. Chemother.* **23,** 711 (1989).

Section II

Phagocytosis and Killing Assays

[8] Measurement of Nonopsonic Phagocytic Killing by Human and Mouse Phagocytes

By RICHARD F. REST and DAVID P. SPEERT

Introduction

The goal of this article is to present detailed methods for the quantitative measurement of phagocytosis and killing of bacteria by human and mouse neutrophils, monocytes, and macrophages. Methods that have been used and proven in the current literature, and generally accepted by those in the "bacterial killing" community (including reviewers and editors), are presented. Comments regarding the proper handling of reagents, phagocytes, and bacteria; expected results; and proper data interpretation are offered. The article does not describe the theoretical bases for the chemistry or cell biology on which the assays and their interpretations are based, nor does it present alternate assays for complex or unusually difficult systems or systems intended to screen large numbers of samples. See also Vol. 132 of this series (Immunochemical techniques, Part J, Phagocytosis and Cell-Mediated Cytotoxicity).

As students we were taught that antibody and each of the complement pathways work independently or in concert to opsonize, i.e., prepare pathogenic microbes for phagocytosis. Specific examples most notably included the group A streptococci with their antiopsonic M proteins[1] and the pneumococci with their antiphagocytic capsular polysaccharides.[2] When an immune response is mounted, specific antibody together with complement lead to opsonization of the invaders and to quick resolution of infection and disease; however, it has now become apparent that phagocytosis and killing of several microbes readily occur in the absence of antibody, complement, or other opsonins.[3,4] Such nonopsonic, i.e., opsonin-independent, phagocytosis has been observed for neutrophils, monocytes, and macrophages, interacting with essentially all types of microbes, including viruses,[5] bacteria,[3,4,6] *Actinomyces*,[7]

[1] V. A. Fischetti, *Sci. Am.* **264,** 58 (1991).
[2] G. A. Bruyn, B. J. Zegers, and R. van Furth, *Clin. Infect. Dis.* **14,** 251 (1992).
[3] I. Ofek, R. F. Rest, and N. Sharon, *ASM News* **58,** 429 (1992).
[4] I. Ofek and N. Sharon, *Infect. Immun.* **56,** 539 (1988).
[5] K. L. Hartshorn, A. B. Karnard, and A. I. Tauber, *J. Leukocyte Biol.* **47,** 176 (1990).
[6] A. Catanzaro and S. D. Wright, *Infect. Immun.* **58,** 2951 (1990).
[7] A. L. Sandberg, L. L. Mudrick, J. O. Cisar, J. A. Metcalf, and H. L. Malech, *Infect. Immun.* **56,** 267 (1988).

fungi,[8,9] and parasites.[10,11] Nonopsonic phagocytosis occurs *in vivo*[12] and is most probably important in protection against infection and disease progression. Nonopsonic phagocytic systems can be placed in two major functional categories, defined by the chemistry of the interaction between bacterial and host ligands: the first involves protein–protein interactions, and the second involves protein–carbohydrate interactions and has been called *lectinophagocytosis*.[4,13–18]

Isolation and Handling of Human Blood Neutrophils

Two common aqueous media are used to isolate neutrophils from blood: Percoll (a suspension of polyvinylpyrrolidone-coated colloidal silica developed by Pharmacia) and a mixture of Ficoll (a nonionic, synthetic sucrose polymer developed by Pharmacia that forms a dense solution) and Hypaque (a dense solution of a polysubstituted carboxylic acid developed by Winthrop Pharmaceuticals, division of Sterling Drug, Inc., N.Y., NY) (see also this series, Vol. 162 [44]). The method presented here is a modification of the Ficoll–Hypaque method, as described by Ferrante and Thong.[19,20] It is rapid, has a high recovery (≥80%), and yields very pure (≥93%) resting (minimally stimulated) neutrophils. Contaminating cells are mostly eosinophils (depending, of course, on the donors' eosinophil counts), with an occasional monocyte or lymphocyte. The method can be readily used for volumes of blood of 200 ml or less (yielding $\leq 5 \times 10^8$ neutrophils), which should be enough for phagocytic killing assays performed in one day. For isolation of substantially larger numbers of neutrophils refer to this series, Volume 162 [44] or 163 [28].

[8] L. Marodi, H. M. Korchak, and R. B. Johnston, Jr., *J. Immunol.* **146,** 1783 (1991).

[9] W. E. Bullock and S. D. Wright, *Adv. Exp. Med. Biol.* **239,** 45 (1988).

[10] C. B. Palatnik, J. O. Previato, L. Mendonça-Previato, and R. Borojević, *Parasitol. Res.* **76,** 289 (1990).

[11] M. Nakao and E. Konishi, *Parasitology* **103,** 23 (1991).

[12] W. Bernhard, A. Gbarah, and N. Sharon, *J. Leukocyte Biol.* **52,** 343 (1992).

[13] A. Gbarah, C. G. Gahmberg, I. Ofek, U. Jacobi, and N. Sharon, *Infect. Immun.* **59,** 4523 (1991).

[14] J. M. Antal, J. V. Cunningham, and K. J. Goodrum, *Infect. Immun.* **60,** 1114 (1992).

[15] A. Athamna, I. Ofek, Y. Keisari, S. Markowitz, G. G. Dutton, and N. Sharon, *Infect. Immun.* **59,** 1673 (1991).

[16] M. E. Conly and D. P. Speert, *Biol. Neonate* **60,** 361 (1991).

[17] W. M. Shafer and R. F. Rest, *Annu. Rev. Microbiol.* **43,** 121 (1989).

[18] D. P. Speert, F. Eftekhar, and M. L. Puterman, *Infect. Immun.* **43,** 1006 (1984).

[19] A. Ferrante and Y. H. Thong, *J. Immunol. Methods* **36,** 109 (1980).

[20] A. Ferrante and Y. H. Thong, *J. Immunol. Methods* **48,** 81 (1982).

Neutrophils have evolved as heavily armed, trigger-happy defenders. They are easily stimulated or aggravated. For reproducible measurements to be taken over the lifetime of a research project (generally several months), isolated neutrophils should be consistently in the same state of rest or minimal activation. Some basic rules help in achieving this goal. Use the highest purity water available, assiduously avoiding endotoxin contamination. Use only polypropylene plasticware, including pipettes, new and disposable, if possible. Do not use glassware. (If it is used, all glassware should be treated with a commercially available siliconizing solution, as directed by the manufacturer.) Neutrophils stick to glass, which leads to neutrophil activation and poor yields. Expose the neutrophils to as few temperature fluctuations as possible and do not temperature-"shock" the neutrophils; e.g., do not repeatedly move tubes of neutrophils from the ice bath to a bench-top rack and back again. Keep the neutrophils on ice, in calcium- and magnesium-free buffer, until ready to use. Do not vortex, pipette, or mix neutrophil suspensions vigorously. Do not aerate neutrophil suspensions; i.e., do not generate bubbles while pipetting or vortexing. The same considerations generally hold true for monocytes and macrophages, whose isolation and handling are described later. Neutrophils prepared by this method are very minimally stimulated. Indeed, they may act or respond more slowly, or to a lesser degree, than neutrophils obtained by other published methods.

Materials and Reagents

9% (w/v) Ficoll. Using a powder funnel, add 18 g of Ficoll (400,000 MW, Sigma, St. Louis, MO, No. F 4375) to 160 ml of very rapidly stirring water in a 500-ml Erlenmeyer flask. When the Ficoll is completely dissolved, fill to 200 ml. Nine percent Ficoll can be stored in the refrigerator for months, and can be autoclaved if desired. In work with neutrophils, the Ficoll–Hypaque solution (described below) need not be sterile, as the neutrophils will not be cultured for any length of time.

Hypaque-M, 75%, diatrizoate meglumine and diatrizoate sodium injection (Winthrop Pharmaceuticals, Division of Sterling Drug Inc.).

Ficoll–Hypaque solution. Mix 13 parts of 9% Ficoll with 3 parts of Hypaque-M, 75%. The two solutions are very dense so they must be thoroughly mixed. The Ficoll–Hypaque solution can be kept in the refrigerator for months. [Several companies sell sterile solutions of Ficoll–Hypaque for neutrophil, monocyte, and lymphocyte isolation. The density of these premade solutions (1.077, 1.083, etc.) is not the same as the density (1.114 g/ml) of the Ficoll–Hypaque solution described here, and so these solutions cannot be used for the separation method described below.]

Blood drawing supplies and reagents, including heparin. (For the methods described here, it is much easier to use 20- or 50-ml syringes with 18- to 20-gauge needles than it is to use several vacutainer tubes. For volumes of blood greater than 50 ml, use an 18- to 20-gauge butterfly setup and change syringes.)

15- and 50-ml polypropylene, sterile, disposable, conical, screwcap tubes (or equivalent).

A 20- to 50-ml syringe with an 18-gauge needle to which has been attached about 3 in. of thin-walled plastic tubing. Instead of the needle and tubing, an 18- to 20-gauge "butterfly" setup can be used, with the needle cut off, leaving about 3 in. of tubing remaining attached to the luer hub.

Dulbecco's phosphate-buffered saline (DPBS) (without calcium or magnesium) with 1 mg/ml gelatin, pH 7.3 (DPBSG). Heat but do not boil the buffer to dissolve the gelatin. The buffer contains, per liter, 8 g NaCl, 0.2 g KCl, 1.15 g Na_2HPO_4, and 0.2 g KH_2PO_4. The buffer can be made up (or purchased) at 10× concentration, without gelatin, filter-sterilized, and stored for months at room temperature (it will crystallize in the cold) in a tightly stoppered bottle. The 10× bottle can be used to quickly make up the 1× buffer with added gelatin.

Distilled and/or deionized water, stored in the refrigerator.

3.6% (w/v) NaCl, stored in the refrigerator.

Methods

Warning! Human blood should be handled with extreme caution because of the possible presence of human immunodeficiency virus type 1 (HIV-1) and other viral pathogens. Be acquainted with and use Universal Precautions. Wear gloves. Use needles and other "sharps" only when absolutely necessary, and discard needles in proper needleproof containers, not in the normal waste or trash. Do not recap needles prior to disposal, as this enhances risk of puncture injury. All materials (beakers, flasks, pipettes and pipette tips, tubes, syringes, gauze, "benchtop diapers," etc.) that have been in contact with blood or blood products (including plasma, serum, and blood cells) should be sterilized (preferably by autoclaving or by soaking or treating with 10% bottled bleach for several hours) before being discarded or reused. In addition, if human blood is handled on a regular basis (even if only infrequently) the researcher should be immunized against hepatitis B.

Slowly layer 7 to 8 ml of whole, freshly drawn anticoagulated (10 U/ml heparin, final concentration) blood over 5 ml of room-temperature (not cold) Ficoll–Hypaque in 15-ml conical screwcap tubes. For 50 ml of blood, seven tubes are needed. With a little practice, the blood can be layered over the Ficoll–Hypaque directly from the syringe used to draw the blood (with the needle removed). For larger volumes of blood, layer 20 to 25 ml of blood over 20 ml of Ficoll–Hypaque. (*Note:* It is the column height of Ficoll–Hypaque through which the blood cells travel that is important in their separation, not the absolute volume of Ficoll–Hypaque.[20]) Tightly cover the tubes (do not centrifuge open tubes of blood), and centrifuge for 30 min at 200 *g* at 20° (not 3°), with the break set off or low.

Place the centrifuge tubes on ice. The neutrophils will have migrated through the Ficoll–Hypaque as a diffuse (cloudy) band above the erythrocyte pellet and below the Ficoll–Hypaque/plasma interface (where the mononuclear cells are located, see below). Remove the neutrophils from each tube with a syringe and tubing. Place the tip of the tubing in the middle of the neutrophil "cloud" and, while slowly swirling the syringe, withdraw as much of the Ficoll–Hypaque layer as possible, without contaminating the neutrophils with erythrocytes from the pellet or monocytic cells from the Ficoll–Hypaque/plasma interface. At this point, there will be some erythrocyte and platelet contamination of the neutrophils. Before collecting the neutrophils from each subsequent tube, wipe the plastic tubing with a tissue, to minimize contamination of the neutrophils with mononuclear cells. Add about 15 ml of the neutrophil suspension to as many 50-ml tubes (on ice) as needed. (To avoid bubbles and to avoid activating or damaging the neutrophils, dispense the neutrophils from the syringe slowly, down the sides of the 50-ml tubes.) Add about 30 ml of cold DPBSG to each tube, mix well but gently, and centrifuge for 10 min at 200 *g* at 3°C. After centrifugation, the neutrophil pellet(s) should be red because of the contaminating erythrocytes. The supernatant(s) may be cloudy as a result of the presence of platelets. Any platelets remaining in the neutrophil pellet(s) will be eliminated during erythrocyte lysis (see below). [If you choose not to lyse the erythrocytes, repeat the washing procedure until the supernatant(s) is clear.] Carefully, in one rather slow motion, decant and discard the supernatant(s) from the neutrophil pellet(s). Be careful because the pellet(s) may be loose. Do not drain the tubes well; leave about 0.5 to 1 ml of liquid.

Contaminating erythrocytes are removed by hypotonic lysis. Perform the lysis on each 50-ml tube separately. With slow vortex mixing (by hand, not vortex mixer), gently and thoroughly resuspend the neutrophil pellet (and contaminating erythrocytes) in the small amount of liquid remaining in the tube. Rapidly add 12 ml of cold water, down the side of the tube, and vortex gently for 10 sec. This will lyse the erythrocytes, yet minimally harm the neutrophils. Immediately add 4 ml of cold 3.6% (w/v) NaCl, down the side of the tube, and vortex gently; this restores isotonicity. Centrifuge the neutrophils and lysed erythrocytes for 8 min at 125 *g* at 3°. The supernatant should be pale red and the pellet a light tan (with a greenish tinge, if no erythrocytes are present). Carefully decant and discard the supernatant. If there is still a red color to the neutrophil pellet, the hemolysis may be repeated once. [Although phagocytes do not lyse on short-term exposure to water, they are not left unscathed. Avoid exposing phagocytes to hypotonic conditions. Additionally, lysing the erythrocytes may not be absolutely necessary, depending on the system used. As an example, piliated gonococci (*Neisseria gonorrhoeae*) adhere avidly to human erythrocytes; thus, the erythrocytes must be lysed before the phagocytes are used in killing assays, as the erythrocytes compete with the phagocytes to bind gonococci. On the other hand, nonpiliated gonococci, or *Pseudomonas* species, generally do not adhere to human erythrocytes, so their presence in the phagocytosis assay is probably of little consequence.]

After the erythrocytes are lysed, pool the neutrophil pellets in a volume of cold DPBSG equal to one-tenth the volume of blood drawn, e.g., 5 ml of DPBSG if 50 ml of blood was drawn. The purified neutrophils can be resuspended either by gentle vortexing or by gentle mixing with a disposable plastic transfer pipette, preferably with a large bore to avoid damaging or activating the neutrophils. To determine the concentration of neutrophils dilute the suspension 1 : 10 (e.g., 10 μl in 90 μl) in 3% acetic acid (room temperature). This lyses erythrocytes (if not lysed already) and, with a little practice, allows clear distinction between neutrophils and contaminating monocytic cells on examination with light or phase-contrast microscopy. Immediately charge a hemocytometer according to the manufacturer's directions, wait about 2 or 3 min to allow the neutrophils to settle, and quantitate the number of neutrophils. Generally, resuspending the purified neutrophils in PBSG equal to one-tenth the volume of blood drawn yields about 1 to 2 $\times$ 10^7/ml, i.e., 50 ml of blood yields about 5 to 10 $\times$ 10^7 neutrophils. Add enough cold DPBSG (without calcium or magnesium) to bring the suspension to 1 to 2 $\times$ 10^7/ml.

Cell viability can be determined by trypan blue "exclusion" (also see this series, Vol. 132 [3]). On a microscope slide, mix 1 part (e.g., 10 μl) of neutrophil suspension with 1 part of a solution of 0.25% trypan blue in 0.85% NaCl for 5 min at room temperature, place a coverslip over the cells, and observe under 200× to 400× magnification with a light microscope. A hemocytometer can be used, if desired. Viable neutrophils are clear, whereas dead neutrophils are light blue. Do not let the neutrophils sit in the trypan blue for any length of time, as it is toxic and an unrepresentative number of neutrophils will turn blue.

Isolation and Handling of Human Blood Monocytes

Materials and Reagents

Isotonic Percoll. Mix 1 part sterile 10× Hanks' balanced salt solution (HBSS) with 9 parts of Percoll (Pharmacia, Piscataway, NJ).

HBSS (1×), sterile, containing, per liter, 8 g NaCl, 0.4 g KCl, 0.048 g Na_2HPO_4, 0.06 g KH_2PO_4, 0.06 g $MgSO_4 \cdot 7H_2O$, 0.14 g $CaCl_2$, 1 g glucose, 0.1 g $MgCl_2$, and 0.35 g $NaHCO_3$, pH 7.2.

Material and reagents to collect sterile, autologous human serum.

RPMI 1640 cell culture medium, complete, endotoxin "free," containing 100 units/ml penicillin and 100 μg/ml streptomycin.

Disposable, wide-mouth, plastic transfer pipettes (2–5 ml) or 20-ml syringes, sterile.

40-ml Oakridge (or similar) centrifuge tubes, sterile.

50-ml polypropylene, sterile, conical, disposable, screwcap centrifuge tubes, sterile (or equivalent).

Density marker beads (Pharmacia).

Methods

Also see this series, Vol. 132 [7]. To obtain autologous serum, draw 20 ml of blood, using no anticoagulant, and transfer to a sterile glass tube. (If the blood is drawn in vacutainer tubes, it need not be transferred to a new tube.) Let the blood clot at 37° for 1 hr, gently rimming the blood with a sterile wooden stick or transfer pipette at 30 and 60 min. Place the tube(s) of blood on ice for 2 hr, and centrifuge the clotted blood for 10 min, at 1000 *g*, at 3°C. Carefully remove the serum, leaving all erythrocytes and clot behind, and transfer to a fresh sterile tube. Keep the serum on ice. Properly sterilize and dispose of the clot. Serum can be frozen at −70° in aliquots.

One can isolate about 20% as many monocytes as neutrophils from human blood (about 4 to 8 × 10^7 monocytes from 100 ml of blood). The method described here yields mononuclear cells that are 90 to 95% monocytes. Prepare blood (≥100 ml) exactly as described above for neutrophil isolation, using sterile reagents and aseptic technique. With a pipette, carefully transfer the mononuclear cell layers from the Ficoll–Hypaque/plasma interfaces to 50-ml plastic centrifuge tubes, being careful to aspirate as little of the Ficoll–Hypaque solution (which contains neutrophils) as possible. Wash the mononuclear cells twice (10 min, 200 *g*, 3°), each time with about 40 ml of cold HBSS, and resuspend to 2 to 5 $\times 10^7$ mononuclear cells/ml in cold HBSS. (Ficoll–Hypaque is toxic to monocytes and should be removed as quickly as possible.)

To each of two 40-ml Oakridge-type centrifuge tubes, dispense 22 ml of isotonic Percoll, 1 ml of autologous serum, and 14.7 ml of HBSS, and mix well. To tube 1, add 10 μl each of the various beads from the density marker kit. Add an equal volume of HBSS to tube 2. Tube 1 is used as a reference to mark the position of the monocytes to be isolated in tube 2. To form a Percoll gradient, centrifuge both tubes, without cells, in a fixed-angle rotor, for 15 min, at 30,000 g, at 3°, with the brake off. When the gradients are ready to load, gently but thoroughly resuspend the washed mononuclear cells prepared above and layer them (in a volume of about 5 ml) onto the preformed gradient in tube 2. Add an equivalent volume of HBSS to tube 1, and centrifuge both tubes in a swinging bucket rotor for 20 min, at 1200 g, at 3°. After centrifugation, the monocytes band at a density of 1.076 (marker 5, blue beads), just above the lymphocytes. Using a pipette or 20-ml syringe with attached plastic tubing, carefully transfer the monocytes to a sterile, 15-ml polypropylene, conical, screwcap centrifuge tube and dilute with 5 or more vol of cold HBSS. Wash the monocytes twice for 10 min, at 200 *g*, at 3°. After the second centrifugation, resuspend the monocytes in 5 ml of RPMI 1640 plus penicillin and streptomycin, and determine their concentration, viability, and purity (by Wright stain). Purity can also be determined by cytochemical staining (see this series, Vol. 132 [8]). Adjust the suspension to 1 $\times 10^7$ monocytes/ml.

Preparation of Human Monocyte-Derived Macrophages

Materials and Reagents

Materials for isolating monocytes and autologous serum, as described above.

Sterile (autoclavable) Teflon beakers or screwcap Erlenmeyer flasks, 20–60 ml, (e.g., Teflon FEP flasks or Teflon PTFE beakers, Cole Parmer Scientific, Chicago, IL).

Methods

Isolate monocytes and autologous serum, as described above. Supplement the RPMI 1640 culture medium with 15% autologous serum and dispense the monocytes to the carefully washed sterile Teflon beakers or flasks. The number of monocytes per beaker or flask can be varied significantly. As a guide, dispense 1 to 2 × 10^6 monocytes in 2 to 4 ml of culture medium in 20-ml flasks, or 5 × 10^6 monocytes in 10 ml of culture medium in 50- to 60-ml flasks. Incubate for 4 to 7 days at 37° with 5% (v/v) CO_2, and "feed" with an additional 5% autologous serum every 3 days; the medium need not be changed. Monocytes can be removed after cooling the flasks or beakers on ice for 15 min and vigorously (with no air bubbles) pipetting the monocytes with a plastic pipette. Observe for morphologic changes by light microscopy and by cytochemical staining (see this series, Vol. 132 [8]) and for viability by trypan blue exclusion (see above).

Isolation and Handling of Mouse Peritoneal Macrophages

Materials and Reagents

- Thioglycolate broth, sterile, prepared according to the manufacturer's directions.
- 5- and 20-ml plastic, sterile, disposable syringes; 20-gauge needles.
- Metophane, ether, or other appropriate anesthetic.
- 70% (v/v) and 95% (v/v) ethanol.
- Dissecting scissors and forceps.
- Disposable plastic transfer pipettes, 2–5 ml, sterile.
- 15- or 50-ml sterile, polypropylene, conical, screwcap centrifuge tubes (or equivalent).
- HBSS, sterile (see above).
- Complete RPMI 1640 culture medium, endotoxin "free," with 10% fetal calf serum (FCS).

Methods

One mouse yields about 1 to 2 × 10^7 macrophages. Inject mice with 1 ml sterile thioglycolate broth each, intraperitoneally, and return to cage for 3 to 6 days. [Be careful to inject the mice intraperitoneally, not subcutaneously, intramuscularly (e.g., in the thigh), or into the lumen of the intestines. Consult the animal care committee for the proper procedure.] On the day of the experiment sacrifice each mouse, individually, by cervical dislocation (after anesthesia with metophane, ether, or other appropriate anesthetic, to prevent the mice from experiencing pain). Dip the mouse in 70% ethanol and lay it on its back on a paper towel. Being careful not to nick or cut the peritoneum, cut the skin over the peritoneum (the length of the mouse) with sterile scissors (dipped in 95% ethanol and flamed). Grasp the peritoneum with sterile forceps and inject 5 ml of HBSS into the peritoneal cavity, being careful to avoid internal organs. Massage the peritoneum rather vigorously, with thumb and forefinger, for about 60 sec to dislodge the infiltrated and adherent macrophages. With sterile scissors, make a small slit in the peritoneum and, with a 20-ml syringe (without needle) or pipette, withdraw as much liquid (containing the macrophages) as possible. Pool the macrophages in sterile, polypropylene tubes of the appropriate size, and immediately cool on ice. Wash the macrophages in cold RPMI 1640 medium by centrifuging for 10 min, at 200 *g*, at 3°. Decant and discard the supernatant fluid(s) and resuspend the pelleted macrophages to 1×10^7/ml in cold RPMI 1640 containing 10% (v/v) FCS. Keep on ice until ready to use. Keeping macrophages cold minimizes their adherence to plastic.

Isolation and Handling of Mouse Peritoneal Heterophiles (Neutrophils)

This method is identical to that for isolation of peritoneal macrophages, except that the cells are harvested 24 hr (instead of several days) after instillation of thioglycolate (see immediately above). Resuspend isolated neutrophils in HBSS to 1×10^7/ml. Because these are inflammatory neutrophils, they are activated; they have diapedesed through endothelial cells and basement substrates of the vasculature, have degranulated some of their specific granule contents, and have been exposed to cytokines in the tissue and peritoneum. This makes them very different from neutrophils isolated from blood, which are "resting" or minimally stimulated. As such, they may act quite differently (more aggressively) than neutrophils obtained from blood.

Determination of Opsonin-Independent Bacterial Killing by Neutrophils and Monocytes

In this section, we present methods used to investigate phagocytic killing of *Neisseria gonorrhoeae*[17,21–23] and *Pseudomonas aeruginosa*.[16,18] The methods should work equally well for almost any bacteria, with a little optimization of the assay. Phagocytic killing of bacteria is measured in a tumbling tube system, where the bacteria and phagocytes are tumbled

[21] S. H. Fischer and R. F. Rest, *Infect. Immun.* **56,** 1574 (1988).
[22] J. V. Frangipane and R. F. Rest, *Infect. Immun.* **60,** 989 (1992).
[23] F. L. Naids and R. F. Rest, *Infect. Immun.* **59,** 4383 (1991).

together, and bacterial viability is measured over time by removing, diluting, and plating aliquots.

For phagocytic killing assays, the ratio of bacteria to phagocytes is usually kept rather low, from 0.1 to 2 or 3. Because phagocytes have a limited capacity to internalize and kill their prey, they can become saturated (or satiated). For example, in an assay with 2×10^6 each of bacteria and neutrophils (a 1 : 1 ratio), the neutrophils may internalize and kill about 1.9×10^6 bacteria in 2 hr; i.e., the neutrophils may kill about 95% of the inoculum; however, at a ratio of 10 : 1 bacteria to neutrophils, i.e., with 20×10^6 bacteria and 2×10^6 neutrophils, the neutrophils may engulf and kill about 16×10^6 bacteria. In this case, even though the neutrophils phagocytically killed more than eight times the number of bacteria than in the first assay (where the ratio was 1 : 1), the percentage bacteria killed is only 80% of the inoculum. The percentage killed decreases even more as the ratio of bacteria to neutrophils increases. In addition, as the concentration of phagocytes increases, so does killing, within limits. Thus, for statistical analysis, it is best to optimize, not maximize, the number of bacteria killed. Another way to state this is that it is best to maximize the percentage bacteria killed, not the number of bacteria killed. In this manner, relatively small differences in phagocytic killing between strains, mutants, assay conditions, and so on, can be more easily detected.

Bacteria may interact with phagocytes quite differently depending on bacterial growth conditions and growth phase. Anaerobically and aerobically grown bacteria, log- and stationary-phase bacteria, agar- vs broth-grown bacteria, and bacteria grown in the presence or absence of specific nutrients might be killed to quite different degrees by the same phagocyte preparation, on the same day. In addition, monocytes, macrophages, and neutrophils can vary tremendously in their ability to kill a specific bacterial species or strain; generally, monocytes kill less efficiently than do neutrophils.

Materials and Reagents

Phagocytes at 1 to 2×10^7/ml in DPBSG or HBSS (prepared as above).
Bacteria at 1 to 2×10^8/ml in DPBSG or HBSS (prepared as above).
DPBSG (see above) with 1 m*M* calcium and 1 m*M* magnesium or HBSS, sterile (see above).
1.5-ml conical, microcentrifuge tubes or 3- to 4-ml snap-cap, round-bottom polypropylene tubes, sterile.
2.5 mg/ml cytochalasin B or D in dimethyl sulfoxide (DMSO). This

stock should be kept in small aliquots, frozen at $-20°$, and thawed and used only once.

2% (w/v) Saponin (Sigma) in water. This can be stored at room temperature for several months.

Sterile distilled and/or deionized water.

Appropriate sterile bacteriologic broth for dilutions.

Agar medium appropriate for growing bacterial species of interest.

Methods

Each phagocytic killing assay consists of an experimental tube containing 2.5×10^6 phagocytes, 2.5×10^6 bacteria, buffer to 500 μl, and one or two control tubes (see below). Assays can be made proportionally larger (and 3- to 4-ml snap-cap tubes used) for long assays from which several samples are to be drawn, or for assays in which multiple samples will be drawn, for instance, in preparing samples for light or electron microscopy. The shape of the tube makes no difference, as long as the phagocytes tumble freely. The size of the tube should be no larger than five times the volume of the assay mixture. Cap the tubes tightly and tumble them end-over-end, at about 15 rpm, in a 37° incubator. If no rotator is available, the tubes may be rocked end-to-end. (Be conscious of the pH of the medium during the phagocytosis assay. DPBSG contains no bicarbonate or glucose, whereas HBSS contains both. The two buffers have quite different buffering capacities in air and a CO_2 atmosphere. Measure the pH of a few tubes after the assay is completed; it should remain between pH 7.2 and pH 7.4.)

Immediately after starting the tubes tumbling, and before the first time point (e.g., at 30 or 45 min), quantitate the exact number of bacteria added to the phagocytic killing assay by appropriately diluting and plating the original bacterial suspension used for the killing assay (not the bacteria mixed with the phagocytes).

At various time intervals (e.g., 30–45 min) over 2 to 3 hr, remove duplicate 10-μl aliquots from the phagocytosis mix and the two control tubes (see below); dilute each 1 : 1000 (1 : 100 then 1 : 10) in the appropriate liquid bacteriologic medium (or saponin or water, see below); and plate 100 μl each (of the 1 : 1000 and the 1 : 100 dilution) on properly labeled, relatively dry agar plates prepared with the appropriate bacteriologic medium. This should yield 250 colonies on the 1 : 1000 dilution plate, if no killing occurs.

The question arises whether to lyse the phagocytes (to release internalized bacteria) before diluting and plating assay samples. We have not observed major differences between lysed and nonlysed cells with gonococci or *Pseudomonas* when normal human neutrophils are used; however, monocytes and macrophages generally need to be lysed, and your system may act differently. Three lysis methods are offered: First, phagocytes can be effectively lysed by adding an equal volume (500 μl) of 2% saponin for 15 min, and then diluting the assay mixture appropriately and plating. (Saponin generally does not kill bacteria at this concentration for this period of time; however, this should be verified for particular bacterial strains.) Second, phagocytes can be broken by sonication. An exact setting cannot be given in a general method like this. The setting on a particular sonicator, the particular cup or bath used, and the sensitivity of the bacterial strain and phagocytes to sonication must be determined empirically. A cup or bath sonicator, not a tip sonicator, should be used, to avoid aerosolization of pathogens. Third, dilutions of phagocytes and bacteria may be made in sterile water, to lyse the phagocytes. Which lysis method to choose (if indeed one is needed) depends on which phagocyte is used and how sensitive the bacterial strain is to water, saponin, or sonication.

The phagocyte bactericidal assay system should be tailored or optimized for each individual species of bacteria. This assay is simple enough that optimization can easily be done within a week or two. For instance, better killing may be achieved with a particular buffer (HBSS, DPBSG, or a more complete cell culture medium). Also, the concentration of phagocytes and the ratio of bacteria to phagocytes should be varied. The time of the assay can be extended to longer than the 3 hr suggested above, especially for monocytes and macrophages. Neutrophils, on the other hand, do not remain viable for long periods, and results from assays longer than 3 hr should be viewed with caution. For this reason, and because some bacteria are toxic to phagocytes, it is mandatory to determine phagocyte viability at the end of all phagocytosis assays, at least the first few times a new strain or variant is used.

Two control tubes are generally needed and sufficient. The first contains 5 μg/ml cytochalasin (i.e., 1 μl of a 2.5 mg/ml stock in the 500-μl assay), phagocytes, bacteria, and buffer. The second contains bacteria and buffer only. Cytochalasins inhibit microfilament polymerization and phagocytosis, and generally inhibit phagocytic killing by almost 100%. Cytochalasin activity is reversible; thus, it must be present in the assay buffer at all times, and should be added to the phagocytes about 10 min prior to the addition of bacteria. Extracellular killing of bacteria in the presence of cytochalasins, i.e., in the absence of phagocytosis, has been reported,[24] but may be a rare event. The cytochalasin control need be performed only the first few times a bactericidal assay is done with a

[24] J. Weiss, L. Kao, M. Victor, and P. Elsbach, *J. Clin. Invest.* **76,** 206 (1985).

particular bacterial species, to ensure that phagocytic killing is (or is not) occurring.

The "bacteria alone" control indicates whether the bacteria used in the assay are growing, remaining static, or dying during the assay, in the absence of phagocytes. [Clumping of bacteria in medium can occur, giving the impression that the bacteria are dying in the "bacteria only" controls when in fact they are not. If decreased bacterial numbers in the "bacteria alone" control are observed, put about 10 μl on a slide under a coverslip and observe under 400× to 600× phase-contrast microscopy to observe for clumping. Alternatively, gram stain about 10 μl of sample that has been gently distributed (to about the size of a dime) and heat-fixed on a slide.] The bacteria in the "bacteria only" control should remain viable or increase in number over time. If the viable count decreases, the assay buffer should be changed. The value or biological relevance of determining phagocytic killing of bacteria that are also losing viability in the absence of phagocytes is highly questionable.

The assay is quantitated by determining the percentage of viable bacteria remaining at different time points, by following this simple equation:

$$\%\ \text{viable} = \frac{\text{number of bacteria recovered from assay at } T_n \times 100}{\text{number of bacteria added to assay at } T_0}$$

T_0 is time zero, i.e., the time the bacteria and phagocytes are first mixed, and T_n is a time after T_0 when viability is determined. Note that the denominator is not the number of bacteria recovered at T_0 in the assay tube, is not the number of bacteria recovered in the "bacteria alone" control tube at T_n, and is not the number of bacteria in the cytochalasin control tube at T_0. None of these is acceptable, because they are all variables; i.e., they may differ for different bacteria, different buffers, different bacterial growth conditions, or different phagocytes. If anything but the actual T_0 input number of bacteria is used as the denominator, improper conclusions may be drawn.

Determination of Opsonin-Independent Bacterial Killing by Macrophages

Materials and Reagents

Please see above.

Methods

The methods used to assess the bactericidal capacity of mononuclear phagocytes are essentially the same as those for neutrophils; however, macrophages generally kill bacteria substantially less efficiently than do neutrophils. Furthermore, macrophages adhere avidly to plastic (unlike neutrophils). Therefore, hypotonic lysis (or sonication or saponin treatment) is essential prior to diluting the phagocytes and bacteria for determination of viable counts. Alternatively, phagocytic killing assays can be performed in Teflon flasks or beakers.

Differentiation of Adherent vs Internalized Bacteria

See also this series, Vol. 132 [6]. There are times when it is of interest (or necessary) to determine whether the bacteria in the phagocytic killing assay are indeed being phagocytized, for instance, in the case where a parental bacterial strain resists phagocytic killing whereas a mutant is readily killed. There are at least three explanations for such an observation: First, the mutant could adhere to the phagocytes substantially more than the parent and thus be phagocytized and killed to a greater extent. As a result, more mutant than parent bacteria would be associated with each phagocyte. Second, the mutant could associate with phagocytes to the same degree as the parent but be phagocytized and killed more efficiently, i.e., to a greater extent. As a result, a higher percentage of mutant bacteria would be intracellular (vs adherent) compared with the percentage of parent bacteria that are intracellular. Third, the mutant could associate with phagocytes to the same degree as the parent, but be more readily killed intracellularly. As a result, the mutant and parent would behave similarly regarding adherence and phagocytosis, but differ in the end result, i.e., killing. In many cases, these three situations can be objectively differentiated.

Quantitation of Phagocyte-Associated (Adherent and Ingested) Bacteria

Materials and Reagents

Same materials and reagents as described above for the phagocytic killing assay.

Cytocentrifuge (Cytospin, Shandon Southern Instruments, Sewickley, PA).

Wright stain, glass slides, light microscope, immersion oil.

Methods

Determine the number of phagocyte-associated bacteria by incubating bacteria with cytochalasin-treated phagocytes, similarly to the phagocytic killing assay described above; however, use 2×10^7 bacteria and 5×10^5 phagocytes (a ratio of 40 : 1 bacteria to phagocytes). At 30 to 60 min, deposit 150 μl of assay mixture on a glass slide using a cytocentrifuge. Stain the "dot" with Wright stain and quantitate the number of associated bacteria per 100 phagocytes by oil immersion (1000× magnification) light microscopy. Although the assay can (and should initially) be done without cytochalasin, we have found that it is easier to quantitate associated bacteria when cytochalasin is used. This may be due to the degradation of ingested bacteria when cytochalasin is not used. In our hands, results are the same with or without cytochalasin, at assay times less than 30 min. When viewing preparations from phagocytosis assays, you cannot determine whether bacteria are adherent or ingested, regardless of what you might think they look like under the microscope. A technique is described below to quantitate adherent vs ingested bacteria.

Although slightly more tedious to quantitate, for several reasons this assay is much preferred to radioactive determinations (where bacteria are radiolabeled, mixed with phagocytes, incubated, separated from phagocytes by centrifugation, and counted). First, depending on the bacteria and the phagocytes (especially neutrophils, which are denser than monocytes or macrophages), it may not be possible to separate bacteria from phagocytes quantitatively by centrifugation. This is certainly true for gonococci (personal experience). Second, important observations can be made by light microscopy that cannot be made by radioactive determinations concerning the association of bacteria with phagocytes. These include the state of the phagocytes (clumped, activated, or lysed), and the absolute numbers of bacteria per phagocyte (even or uneven distribution of bacteria per phagocyte).

In the scenario given above, if more mutants than parents are associated with phagocytes, then it is quite probable that the mutants have lost an "antiphagocytic" component or property (or conversely, but less likely, have gained a "prophagocytic" component). If, on the other hand, the mutants and parents associate equally well with phagocytes, further studies should be done, as outlined below. In this scenario, it is unlikely that the mutants will associate with phagocytes less than do the parents.

Quantitation of Adherent vs Ingested Bacteria

In this procedure, first described by Hed for yeast,[25] bacteria are prelabeled with fluorescein and incubated with phagocytes, aliquots of the phagocytosis mixture are mixed with trypan blue, and the mixture is

[25] J. Hed, *FEMS Lett.* **1,** 357 (1977).

viewed with fluorescence and phase-contrast microscopy. Intracellular bacteria fluoresce whereas the fluorescence of adherent (but not ingested) bacteria is quenched by the trypan blue. With the technique, adherent and intracellular bacteria can be quantitated, as can the percentage ingested bacteria. The assay originally described the use of crystal violet as a quenching agent, but Hed has now reported that crystal violet, being lysosomotropic, can enter phagolysosomes and quench the fluorescence of intracellular bacteria as well (see this series, Vol. 132 [6]).

There are drawbacks to this assay. Due to the physicochemical makeup of the cell surface, the bacterial strain(s) used might be resistant to significant fluorescein isothiocyanate (FITC) conjugation, or FITC conjugation may decrease or destroy the binding function of specific surface molecules. Some of these problems may be insurmountable, making this technique less useful.

Materials and Reagents

Same materials and reagents as described above for the phagocytic killing assay.
FITC, 0.1 mg/ml in 0.1 *M* carbonate buffer, pH 9.5, freshly prepared.
Trypan blue, 2 mg/ml in 0.15 *M* NaCl in 0.02 *M* citrate or acetate buffer, pH 4.4.
Fluorescence microscope, microscope slides.

Methods

To label bacteria with fluorescein, incubate 1×10^9 bacteria in 1 ml of 0.1 *M* carbonate buffer, pH 9.5, containing 0.1 mg/ml FITC for 30 min at 37°C. Wash the fluorescein-conjugated bacteria with DPBSG three times by centrifugation and resuspend to 5×10^8/ml in DPBSG.

To determine total adherent and intracellular bacteria, incubate bacteria with phagocytes exactly as indicated in the phagocytic killing assay described above, except use 5×10^7 fluorescein-labeled bacteria (to give a ratio of 20 : 1 bacteria to phagocyte). At time points determined empirically by initial phagocytic killing assays, deposit 5 μl of assay mixture on a glass slide, add 5 μl of trypan blue solution, mix gently, cover with a coverslip, and immediately observe with phase and fluorescence microscopy. It is important to observe intracellular and extracellular bacteria immediately after their removal from the phagocytosis assay, because phagocytosis may continue on the glass slide. Intracellular bacteria fluoresce, whereas adherent (extracellular) bacteria do not fluoresce. Under phase-contrast microscopy, intracellular bacteria are bright, whereas adherent bacteria are dark. By working with the numbers of blue (adherent) and fluorescent (intracellular) bacteria, "total associated" and "percentage ingested" bacteria can be easily determined. Care should be taken in interpreting data from time points taken more than 1 hr after phagocytosis, as the fluorescein might be degraded, released from the surface of the bacteria, or otherwise quenched.

In the scenario presented, where more mutants than parents are killed, it is likely that a higher percentage of mutant (compared with parent) bacteria are ingested (and subsequently killed).

[9] Measurement of Opsonic Phagocytosis by Human Polymorphonuclear Neutrophils

By JOHN R. KALMAR

Introduction

Since the original descriptions of the phagocytic process by Metchnikoff, phagocytosis has been recognized as a primary and critical cell function in host defense. In higher vertebrates, such as humans, this function is performed largely by "professional" phagocytes like the polymorphonuclear neutrophil (PMN). Although phagocytic ingestion of foreign particles and certain microorganisms is known to occur in the absence of exogenous factors or molecules (nonopsonic phagocytosis, see Chapter 8), it has long been recognized that several serum-derived substances could serve as powerful promoters of the ingestion process. Termed opsonins (from the Greek *opsonion,* "a relish, seasoning, or sauce"), these humoral factors bind to the surface of foreign material and, in turn, engage receptors on the phagocytic cell surface that initiate the endocytic (phagocytic) process.

Classically, much of the opsonic activity derived from serum of test animals or subjects has been ascribed to antibody (immunoglobulin) and complement molecules. Among the immunoglobulins, IgG1 and IgG3 are considered the predominant opsonic mediators. Recent evidence by Anderson *et al.* indicates that freshly isolated PMNs express only FcγRII (CD32) and FcγRIII (a nontransmembrane form of CD16 linked to glycosyl-phosphatidylinositol)[1]; however, exposure to interferon-γ results in

[1] C. L. Anderson, L. Shen, D. M. Eicher, M. D. Wewers, and J. K. Gill, *J. Exp. Med.* **171,** 1333 (1990).

the additional expression of FcγRI (CD 64). FcγRI and FcγRII were found to promote phagocytosis, whereas FcγRIII promotes binding, but not ingestion. Once considered possible dysopsonins,[2] IgA and secretory IgA have also demonstrated opsonic activity with PMNs through interaction with surface FcαR whose functional expression, like FcγRI, can be enhanced with either granulocyte–macrophage colony-stimulating factor (GM-CSF) or granulocyte colony-stimulating factor (G-CSF).[3,4]

The complement system is a series of glycoproteins that exist within the blood in a nonactive state. Once activated, these proteins undergo a rapid, sequential cascade of proteolytic and organizational events that serve to amplify the original signaling event and produce numerous biologically active by-products. Activation can occur following antibody–antigen interactions (classic pathway) or on the membranes and cell walls of microorganisms (alternative pathway). In terms of phagocytosis, complement component C3b apparently serves as the primary ligand for complement receptor 1 (CR1, CD35), found on several cell types including PMNs.[5,6] CR1 also shows affinity for C4b and weak affinity for iC3b, a split product of C3b. CR3 (CD11b/CD18), a member of the integrin family, has been shown to bind iC3b, but not C3b, and also exhibits an epitope with lectinlike activity that binds noncomplement ligands.[7] Although controversy has existed for years concerning the ability of complement (C3 fragments) alone to mediate phagocytosis, recent work has suggested such mediation is possible but target specific.[8] It is, however, clear that the combination of antibody and complement is strongly synergistic in promoting binding and ingestion of opsonized particles or microorganisms. In addition, a number of other serum factors,[2] including fibronectin, C-reactive protein, and more recently the lipopolysaccharide-binding proteins (LBPs)[9] and septin,[10] demonstrate opsonic activity and may also act synergistically with antibody, complement, or each other to promote phagocytosis.

[2] D. R. Absolom, this series, Vol. 132, p. 281.

[3] A. Gorter, P. S. Hiemstra, P. C. J. Leijh, M. E. van der Sluys, M. T. van den Barselaar, L. A. van Es, and M. R. Daha, *Immunology* **61,** 303 (1987).

[4] R. H. Weisbart, A. Kacena, A. Schuh, and D. W. Golde, *Nature* **332,** 647 (1988).

[5] R. Boackle, *in* "Introduction to Medical Immunology" (G. Virella, ed.), 3rd ed., p. 135, Marcel Dekker, 1993.

[6] I. Roitt, *in* "Immunology" (L. van den Berghe, ed.), 2nd ed., p. 232. Gower Medical, London, 1989.

[7] G. D. Ross, J. A. Cain, and P. J. Lachmann, *J. Immunol.* **134,** 3307 (1985).

[8] M. K. Hostetter, *in* "The Natural Immune System, Humoral Factors" (E. Sim, ed.), p. 177, IRL Press at Oxford University Press, New York, 1993.

[9] S. D. Wright, P. S. Tobias, R. J. Ulevitch, and R. A. Ramos, *J. Exp. Med.* **170,** 1231 (1989).

[10] S. D. Wright, R. A. Ramos, M. Patel, and D. S. Miller, *J. Exp. Med.* **176,** 719 (1992).

Measurements of phagocytosis of microbial parasites by phagocytic cells such as PMNs have employed a wide variety of techniques that can be broadly divided into two main categories: (1) direct measures of phagocyte–target interactions and (2) indirect measures that quantitate metabolic events or biochemical markers associated with ingestion and cellular activation, such as lysosomal enzyme release, superoxide anion (O_2-) production, and chemiluminescence.[11] Although possessing some advantages, the latter techniques are susceptible to various nonphagocytic (soluble) stimuli that can lead to cellular activation in the absence of phagocytosis. Furthermore, as indirect protocols measure postphagocytic events, abnormal values do not necessarily imply phagocytic dysfunction.[11] For these reasons, three direct techniques for the measurement of phagocytosis are presented within this article. The first is a novel flow cytometric assay that uses unfractionated whole blood. The second, a fluorescence microscopic method compatible with a wide variety of microorganisms, employs cytocentrifugation and cell fixation. The third is an adaptation of the oil red O method originally described by Stossel in 1973[12–14] to the user- (and technology-) friendly 96-well microtiter plate format. Additional protocols for the study of opsonic phagocytosis together with excellent discussion and comment may be found in Volume 132 of this series.

Isolation of Human Polymorphonuclear Neutrophils

As described, the one-step Ficoll–Hypaque technique provides rapid, simple fractionation of whole blood to isolate PMNs. Techniques employing other density gradient mediums such as Percoll (a colloidal suspension of polyvinylpyrrolidone-coated silica particles) are equally satisfactory.[15] Modification of the Ficoll–Hypaque method to create a discontinuous gradient is recommended.[16] Increased separation of leukocyte bands is achieved with resultant improvements in PMN yield and purity. This modification also permits the use of extremely small blood samples, such as with neonates or young children, and is compatible with a variety of anticoagulant formulations. (See Chapter 8.)

[11] R. Absolom, this series, Vol. 132, p. 95.
[12] T. P. Stossel, *Blood* **42,** 121 (1973a).
[13] T. P. Stossel, *J. Cell Biol.* **58,** 346 (1973b).
[14] T. P. Stossel, this series, Vol. 132, p. 192.
[15] D. Roos and M. De Boer, this series, Vol. 132, p. 225.
[16] J. R. Kalmar, R. R. Arnold, M. L. Warbington, and M. K. Gardner, *J. Immunol. Methods* **110,** 275 (1988).

Materials

Histopaque 1.077 brand of Ficoll–sodium diatrizoate, available from Sigma Chemical Company, St. Louis, Missouri, consists of 5.7% (w/v) Ficoll 400 and 9.0% (w/v) sodium diatrizoate with specific gravity 1.077 and osmolarity 290 mos*M*. Also, Hanks' balanced salt solution (HBSS) without calcium and magnesium (GIBCO BRL) can be substituted for Dulbecco's phosphate-buffered saline.

Methods

Before adding the fresh, heparinized blood to the 15-ml centrifuge tubes as described in Chapter 8, carefully layer 1 ml of Histopaque 1.077 (or equivalent) over the 5 ml of Ficoll–Hypaque (1.114 g/ml) mixture without mixing. Alternately, underlay the lighter solution with the heavier mixture. For larger blood volumes, scale up the amount of Histopaque 1.077 accordingly. Centrifugation at 500 *g*, rather than 200 *g*, for 30 min is recommended to ensure complete separation of the neutrophil band from the erythrocyte pellet, a feature that significantly reduces erythrocyte contamination.

As mentioned (in Chapter 8), minimal handling, stirring, and temperature changes of PMN suspensions is very important. Although there is evidence to suggest that functional decay is minimized at 4° compared with 37°,[17] maintenance of isolated cells at room temperature (20–24°) rather than 5° or 37° has been recommended to minimize alterations in cellular responsiveness.[18,19]

Experimental Protocols

Protocol 1: Flow Cytometry

The use of flow cytometry has facilitated the study of phagocytosis by permitting analysis of large cell populations, thus addressing potential populational heterogeneity of function. In addition, flow cytometric technology permits the kinetics of phagocyte–target interaction and ingestion to be examined at the individual cell level. Protocols that use purified populations of phagocytes have been presented in Volume 132[20,21] and

[17] A. Ferrante, L. J. Beard, and Y. H. Thong, *Clin. Exp. Immunol.* **39,** 532 (1980).
[18] T. A. Lane and B. Windle, *Blood* **54,** 216 (1979).
[19] J. McCullough, B. J. Weiblen, P. K. Peterson, and P. G. Quie, *Blood* **52,** 301 (1978).
[20] C. C. Stewart, B. E. Lehnert, and J. A. Steinkamp, this series, Vol. 132, p. 183.
[21] J. Hed, this series, Vol. 132, p. 198.

reviewed by others.[22,23] The method outlined here, by White-Owen *et al.*,[24] does not require a leukocytic purification step, uses whole blood samples as small as 100 μl, and permits delayed analysis through sample fixation.

Materials and Equipment

1. *Staphylococcus aureus* ATCC 25923 (or suitable replacement)
2. Pooled normal human serum, type AB (fresh or stored in aliquots, −70°)
3. Fluorescein isothiocyanate
4. Cytochalasin D
5. Immuno-Lyse (Coulter) with lysing agent
6. Microcentrifuge or medium-speed centrifuge with microtube adaptor(s)
7. Flow cytometer (filtered emission at 488 nm)

Bacteria

Culture *Staphylococcus aureus* ATCC 25923 for 16 hr at 37° in trypticase soy broth together with fluorescein isothiocyanate (FITC, Research Organics, Inc.) at 30 to 50 μg/ml in a light protected environment. Centrifuge (1500 *g*, 10 min), wash bacteria twice with Dulbecco's phosphate-buffered saline (DPBS, GIBCO Laboratories), and heat-inactivate at 60° for 30 min. Wash the bacteria again and resuspend in DPBS to a final concentration of 10^9 bacteria/ml. Examine and confirm uniformity of staining by fluorescence microscopy and flow cytometry. Aliquot and store bacterial suspension at −70°.

Blood Samples

Use freshly drawn, anticoagulated blood. Determine total white blood cell (WBC) counts with a hemacytometer following 3% acetic acid lysis of erythrocytes. If the WBC count exceeds 10^7/ml, dilute blood with DPBS to achieve a WBC count between 10^6 and 10^7 cells/ml.

Phagocytosis

Dispense 100 μl of whole or diluted blood into separate sterile 1.5-ml polypropylene microcentrifuge tubes for each time point and condition to

[22] R. Bjerknes, C-F Bassøe, H. Sjursen, O. D. Laerum, and C. O. Solberg, *Rev. Infect. Dis.* **11,** 16 (1989).

[23] C-F. Bassøe and R. Bjerknes, *Acta Pathol. Microbiol. Immunol. Scand. Sect. C* **92,** 51 (1984).

[24] C. White-Owen, W. Alexander, R. M. Sramkoski, and G. F. Babcock, *J. Clin. Microbiol.* **30,** 2071 (1992).

be studied. Centrifuge at lowest *g* and minimum time to permit blood cell pelleting (e.g., 1000 *g* for 15–30 sec or 200–250 *g* for 5 min); carefully remove plasma and wash twice with DPBS. Gently resuspend the cells in 100 μl of pooled normal human type AB serum. Briefly sonicate the stock FITC-labeled bacteria (3–5 sec, Branson Sonifier II with $\frac{1}{8}$-in. tapered microtip, setting 3) and add 10 μl (per 100 μl cell suspension) to each tube. For time zero control, immediately add cytochalasin D (Sigma Chemical Co., final concentration 5 μM). Incubate remaining tubes in a rotary incubator at 37° and stop phagocytosis at appropriate time points in each tube or 100-μl aliquot with cytochalasin D. Immediately following addition of cytochalasin D, place the tubes on ice and transfer the contents with ice-cold DPBS to 12 × 75-mm polypropylene to a total volume of 3 ml. A control tube without bacteria is recommended to evaluate background PMN fluorescence. [*Note:* Preopsonization may be useful for some microorganisms. This can be accomplished by incubating the FITC-labeled bacteria in an equal volume of pooled normal human type AB serum (50% serum v/v) for 15 min at 37°. Meanwhile, the washed PMN pellets are resuspended in 80 μl of fresh serum. Phagocytosis is initiated by adding 20 μl of the preopsonized, labeled bacteria to the PMNs.]

Erythocyte Lysis

Specimen erythrocytes can be removed with Immuno-Lyse (Coulter). Centrifuge and wash all experimental tubes twice with 3 ml cold DPBS. Dilute the Immuno-Lyse 1 : 25 with DPBS and add 1 ml to each tube. Vortex all tubes immediately and at 1 min after addition of lysing agent. Maintain tubes on ice and stop reaction with 250 μl of Coulter fixing reagent and vortexing. Centrifuge and wash tubes with cold DPBS to remove most free hemoglobin (usually two cycles). At this point, cells either can be fixed with 0.5 ml of cold 1% paraformaldehyde solution (containing 1% sodium cacodylate and 0.85% NaCl) or can be prepared for immediate analysis with addition of 0.5 ml cold DPBS.

Flow Cytometry

Perform the analysis using the blue-green 488-nm line of argon. Green fluorescence can be collected by using a 530 ± nm bandpass filter with linear amplification. Non-PMN debris may be excluded by using forward and 90°-angle light scattering. Details of data analysis and interpretation are presented in previously mentioned references[20,22] as well as others.[23,25,26]

[25] H. M. Shapiro, *in* "Practical Flow Cytometry" (H. M. Shapiro, ed.), 2nd ed., Alan R. Liss, New York, 1988.

[26] A. G. Fredrickson, C. Hatzis, and F. Sriene, *Cytometry* **13,** 423 (1992).

Comments

Some flow cytometric techniques are designed to distinguish between adherent particles or bacteria and those enclosed within completely internalized phagosomal structures. Quenching dyes such as crystal violet and trypan blue have been used to block the emission signal from the former, while the latter, inaccessible targets continue to fluoresce.[21,22] In development of the current assay, Immuno-Lyse was shown to reduce markedly the fluorescence of free FITC-labeled bacteria,[24] thereby obviating the need for a separate quenching agent. (See Comments under Protocol 2.)

As has been previously reported, acidification of developing phagolysosomes reduces the pH-sensitive emission from FITC.[22] This feature can be used to estimate intraphagolysosomal pH; however, as mentioned by Stewart *et al.*,[20] FITC-labeled bioparticles exhibit a high coefficient of variation in their fluorescence emission that can restrict their usefulness. Furthermore, not all microorganisms stain equally well, if at all, with FITC due to hydrophilic surface structures that block dye access to cell wall proteins.[21]

Protocol 2: Fluorescence Microscopy

Problems encountered with FITC labeling of certain microorganisms as mentioned by Hed[21] were among the factors that led to the use of 4′,6-diamidino-2-phenylindole (DAPI), a trypanocide derivative first synthesized by Dann *et al.*[27] for labeling bacterial targets. DAPI preferentially binds to A:T-rich regions of double-stranded DNA, forming a highly fluorescent complex not seen with either RNA or single-stranded DNA[28] and stable over a wide pH range (pH 4–11) and salt concentration (0–2.0 *M* NaCl).[29] Similar in many respects to the Hoechst dyes 33342 and 33258, DAPI exhibits less staining variability.[25] Previously, DAPI was used as a vital stain for free-living bacteria,[30,31] yeast,[32] and mycoplasma.[33]

In this assay, DAPI is used to vitally stain target microorganisms. Acridine orange (AO) is used as a counterstain to permit visualization of PMNs under fluorescent illumination. Cytocentrifugation concentrates the

[27] O. Dann, G. Bergen, E. Demant, and G. Volz, *Justus Liebig's Ann. Chem.* **749,** 68 (1971).

[28] M. L. Barcellona, R. Favilla, J. Von Berger, M. Avitabile, N. Ragusa, and L. Masotti, *Arch. Biochem. Biophys.* **250,** 48 (1986).

[29] J. Kapuscinski and B. Skoczylas, *Nucleic Acids Res.* **5,** 3775 (1978).

[30] A. W. Coleman, *Limnol. Oceanogr.* **25,** 948 (1980).

[31] K. G. Porter and Y. S. Feig, *Limnol. Oceanogr.* **25,** 943 (1980).

[32] D. H. Williamson and D. J. Fennell, *in* "Methods in Cell Biology" (D. M. Prescott, ed.), Vol. 12, p. 335, Academic Press, New York, 1975.

[33] W. C. Russel, C. Newman, and D. H. Williamson, *Nature* **253,** 461 (1975).

sample mixture onto a relatively small surface area and induces spreading of the PMNs that facilitates bacterial enumeration. As previously described,[34] cyanoacrylate fixation preserves cellular morphology and stain distribution and permits delayed sample analysis.

Materials and Equipment

1. DAPI stock solution, 10 mg/ml distilled sterile H_2O (warm slightly to dissolve, protect from light, stable at 4° for at least 6 months)
2. AO stock solution, 1.5–2.0 mg/ml HBSS (protect from light, stable at 4° for at least 6 months)
3. Pooled normal human serum (fresh or stored in aliquots, −70°)
4. Cytocentrifuge (Cytospin 2, Shandon)
5. Cyanoacrylate (Krazy Glue, Inc.)
6. Fluorescence microscope, mercury HBO 100 W with DAPI/Hoechst filter set
7. Sonicator (optional)

Bacteria

Prior to assay, subculture bacteria and grow in appropriate liquid culture media to a previously determined growth phase (typically mid- to late exponential). Harvest bacteria from broth by centrifugation (optimal *g* force and time will vary, however, 1500 *g* for 10 min works well with most) and resuspend in a liquid diluent such as distilled H_2O or 0.15 *M* NaCl. Alternately, bacteria can be harvested directly from solid media to produce a liquid suspension. Should clumping be observed in broth culture or liquid suspension, sonicate briefly at low-power setting to disperse. Add DAPI to the bacterial suspension (final concentration 1–3 μg/ml) and incubate in the dark at room temperature for 5–10 min. Centrifuge bacteria, wash twice, and resuspend to a final concentration of approximately 10^8 organisms/ml.

Opsonization and Phagocytosis

Add 150 μl bacterial suspension together with 50 μl autologous serum in a 1.5-ml polypropylene microcentrifuge tube and incubate at 37° for 15 min with constant, gentle agitation (rotary or rocking). Add 200 μl PMN suspension (10^7 cells/ml) to the bacteria and vortex mildly. For time 0, immediately remove 30 μl and add to 5 μl AO (1.5–2.0 μg/ml HBSS working solution) preloaded in a cytocentrifuge sample chamber (Cytospin 2). (*Note:* Adjustment of the AO concentration so that PMN granules appear very light yellow-green in fixed specimens under ultraviolet excita-

[34] W. Horn, C. Hansmann, and K. Federlin, *J. Immunol. Methods* **83,** 233 (1985).

tion is critical. This concentration serves the counterstain function yet is well below that required for metachromasia,[35] which can interfere with specimen evaluation.) Centrifuge on HI acceleration setting at 900 rpm (72.4 *g*) for 4 min. Remove all slide holders/chambers at once and place horizontally on countertop with specimen facing up. Remove slides from clip holders individually and let moisture evaporate (sheen will disappear in approximately 2–3 sec). Immediately place one drop of cyanoacrylate directly over the sample and affix glass coverslip without trapping air. (*Warning:* Cyanoacrylate can be hazardous to your fingers.) Repeat procedure for remaining conditions and time points. Slides can be examined immediately or stored in the dark at room temperature indefinitely.

Fluorescence Microscopy

Examine slides under ultraviolet excitation (preferably with a mercury arc lamp, HBO 100 W) and appropriate filter combinations for DAPI (excitation 365 nm, emission >420 nm). Bacteria will appear blue in contrast to the faint yellow-green PMN granules. Count the number of bacteria associated within the granule-delineated cytoplasmic outline of at least 100 random PMNs. Use of a 100× oil-immersion objective may be required; however, a 60–63× objective is optimal for most analyses. Prepared slides are extremely durable with preservation of cellular morphology, stain distribution, and intensity (specimens stored in the dark at room temperature for at least 4 years do not exhibit visually significant loss of signal) and resistance to photobleaching during analysis. Such features make these specimens well suited for automated video-based image analysis systems.

Comments

Although numerous investigators have developed strategies to differentiate adherent from ingested microorganisms, such discrimination has not been a problem at the suggested bacteria : PMN ratios (approximately 7.5–10 : 1). Furthermore, the concept of adherence versus ingestion is clouded by electron microscopic evidence that 20 to 40% of phagocytic vacuoles maintain some communication with the extracellular milieu (unsealed vacuoles).[36–38] Alternative controls for adherence such as exposure to cold (0–4°), 2-deoxyglucose, or cytochalasin D are recommended.[39] If necessary, quenching can be achieved in this assay with heat-inactivated

[35] J. Kapuscinski and Z. Darzynkiewicz, *Nucleic Acids Res.* **11,** 7555 (1983).
[36] Y. V. Jacques and D. F. Bainton, *Lab. Invest.* **39,** 179 (1978).
[37] P. Cech and R. I. Lehrer, *Blood* **63,** 88 (1984).
[38] P. Cech and R. I. Lehrer, *Blood* **64,** 147 (1984).
[39] S. D. Wright, this series, Vol. 132, p. 204.

target bacteria (60° for 30 min) using propidium iodide (PI), a cationic dye that binds both DNA and RNA and is excluded by intact cell membranes under isotonic conditions.[25] While positively stained with DAPI (blue), heat-killed bacteria are quenched in the presence of PI and appear red under ultraviolet fluorescence. As PI is excluded from sealed vacuoles, bacteria within such phagosomes retain their blue color. Preaddition of 2 μl PI stock solution (100 μg/ml HBSS) together with AO in the cytocentrifuge sample chamber provides adequate quenching and serves to monitor PMN viability. Addition of PI to bacterial suspensions immediately prior to phagocytosis can be used to assess bacterial killing as well as ingestion.[40–42]

Protocol 3: Oil Red O, Microplate Format

Stossel first described the use of oil droplets labeled with the lipophilic dye oil red O and emulsified with agents such as bacterial lipopolysaccharide (LPS) and denatured bovine serum albumin (BSA) that were capable of fixing complement-derived opsonins on the particle surface.[12–14] A modification reported by Rosen *et al.*[43] is presented here and features a 96-well microtiter plate format that facilitates simultaneous assessment of several experimental conditions and controls. Multiple centrifugation steps are eliminated, as well is the requirement for an oil extraction step to permit quantitation.

Materials and Equipment

1. Oil red O and glutaraldehyde (grade II, 25% solution, Sigma Chemical Co.)
2. Diisodecyl phthalate oil (ICN Pharmaceuticals)
3. Lipopolysaccharide (LPS, *Escherichia coli* serotype O26:B6, boivin preparation, Difco)
4. Pooled normal human serum (fresh or stored in aliquots, −70°)
5. Hanks' balanced salt solution (HBSS) without calcium and magnesium (GIBCO Laboratories)
6. 96-well polystyrene microtiter plates
7. Medium-speed centrifuge with microtiter plate carriers
8. ELISA reader with 540-nm bandpass filter capability
9. Multichannel pipettor (optional)
10. Vortexer/shaker with microtiter plate adaptors
11. Sonicator

[40] C. W. Cutler, J. R. Kalmar, and R. R. Arnold, *Infect. Immun.* **59,** 2097 (1991).
[41] C. W. Cutler, J. R. Kalmar, and R. R. Arnold, *Infect. Immun.* **59,** 2105 (1991).
[42] D. J. Uhlinger, D. N. Burnham, R. E. Mullins, J. R. Kalmar, C. W. Cutler, R. A. Arnold, J. D. Lambeth, and A. H. Merril, Jr., *FEBS Lett.* **286,** 28 (1991).
[43] H. Rosen, B. R. Michel, and A. Chait, *J. Immunol. Methods* **144,** 117 (1991).

Neutrophils

Isolate as described in Chapter 8, and resuspend at 2×10^7 PMNs/ml in HBSS.

Oil Red O-Labeled Diisodecyl Phthalate Oil

Dissolve oil red O in a minimal volume of $CHCl_3$ and combine with diisodecyl phthalate oil to a dye concentration of 40 mg/ml. Allow residual $CHCl_3$ to evaporate (room temperature or 56°) until odor is no longer detectable (up to several days). (*Note:* Residual $CHCl_3$ reportedly inhibits phagocytosis.) Labeled oil can be stored at room temperature, despite a tendency for precipitated dye to accumulate at the bottom of the storage container.

Lipopolysaccharide/Oil Drop Emulsions

Prepare a stock LPS solution (10 mg/ml) in HBSS with brief sonication to effectively disperse the LPS. Aliquots of the suspension can be frozen at −20° until use. Layer 1 ml of labeled oil over 3 ml of the LPS suspension and sonicate at 4° using 30-sec bursts (Branson Sonifier II, $\frac{1}{8}$-in. tapered microtip, setting 4) with the tip just below the LPS/oil interface until emulsified. Emulsions can be stored at −20° but must be resonicated (approximately 30 sec) after thawing.

Opsonization and Phagocytosis

Incubate emulsion with an equal volume of 50% normal human serum (or selected dilution made with HBSS) for 20 min at 37°. Use immediately or transfer onto ice until use. Prewarm microtiter plates to 37°. Likewise, preincubate PMN suspension at 37° for 5 min prior to assay. To each well add 5 μl of divalent cation solution (40 m*M* $CaCl_2$, 16 m*M* $MgCl_2$, in HBSS) and 50 μl of opsonized oil emulsion. Repeatedly mix emulsions on Vortex mixer during pipetting to ensure homogeneity. Phagocytosis can be initiated simultaneously in rowwise fashion by addition of 200 μl of PMN suspension per well using a multichannel pipettor. Mix contents with several pipettor strokes and incubate plate(s) at 37° with continuous mild vortexing using a microtiter plate adaptor. Phagocytosis is stopped by addition of 50 μl 3.7% formalin to each well, mixed with repeated, firm aspiration strokes. Time zero wells can be achieved by reversing the order of PMN–formalin addition.

Centrifuge plates at 1500 *g* for 20 min with no braking to separate cells from uningested oil and minimize disturbance of cells during deacceleration. Carefully aspirate the surface layer of uningested oil and discard. Next, add 50 μl of 25% glutaraldehyde for 20 min at room temperature to fix the cells to the plate. Wash plates by immersion in phosphate-buffered saline (PBS) twice, empty by inversion, and read absorbance at 540 nm.

Comments

To avoid confusion with standard absorbance units, multiply values by 1000 to generate "oil red O units." Assay limitations include high background absorption by the PMNs, although cell numbers can reportedly be reduced by 50% with acceptable results. In addition, disturbance of the PMN cell layer during fixation and washing may skew resultant absorbance values.

Although *E. coli* serotype O26:B6 LPS apparently enjoys a favored position among emulsifying agents for this technique, coemulsifications of LPS and bovine serum albumin or tetanus toxoid antigens produced by Rosen *et al.*[43] indicate a virtually limitless number of formulations and conditions could be examined. It would be of interest, for example, to determine the mechanism by which similarly emulsified *E. coli* serotype O11:B8 LPS can activate complement but not promote oil drop ingestion.[14] Careful selection and testing of such conditions could shed important light on the role played by various LPS structures or antigenic determinants in the phagocytic process.

[10] Measurement of Human Neutrophil Respiratory Burst Activity during Phagocytosis of Bacteria

By RICHARD F. REST

Introduction

In this article I describe detailed methods to measure the level of human neutrophil respiratory burst activity induced in response to bacterial challenge. Phagocyte activation can occur in the presence or absence of phagocytosis, as long as there is contact between bacteria and phagocytes. Although similar results may be obtained using granulocytes from animals, they can differ dramatically, both structurally and functionally, from human neutrophils. The methods are used and proven in current literature and accepted by those working in the phagocyte respiratory burst field. Specific comments are offered regarding proper handling of reagents, neutrophils, and bacteria; expected results; and proper data interpretation. The theoretical bases for the chemistry and cell biology on which the assays and their interpretation are based are presented only briefly. Alternate assays for complex systems or systems intended to screen large numbers of samples are not presented. See also Vol. 132 of this series (Immunochemical Techniques, Part J, Phagocytosis and Cell-Mediated Cytotoxicity).

Importantly, other good assays are available. This article presents guides to the best method(s) for a particular system, and for the particular biological, cellular, and molecular question(s) to be answered. The methods presented below may not necessarily be the best for your system, so experiment! The methods are presented as a set of assays to begin an organized approach to measure the oxidative burst induced by the bacterial strain(s) under investigation in the laboratory. Each bacterial strain yields slightly to dramatically different results, and major differences might be observed using unopsonized vs opsonized bacteria. (Methods for opsonizing bacteria are presented in [9]. The results of preliminary experiments should guide the direction of subsequent investigations. Because of inherent sensitivities, not all the assays described below use the same incubation conditions. Use the following methods as guides to proper approaches, not as procedures that should be followed unwaveringly. For example, if a bacterial strain induces the production of massive amounts of reduced oxygen products, decrease the number of neutrophils used in the assays or decrease the concentrations of bacteria used.

On contact with the neutrophil, most opsonized and some nonopsonized bacteria induce an oxidative burst, which comprises the assembly of a multicomponent system of enzymes and electron transport molecules that catalyzes a single-electron reduction of molecular oxygen to yield the superoxide anion. Superoxide, in turn, rapidly combines with itself and hydrogen ions to form hydrogen peroxide. In the presence of myeloperoxidase, lactoferrin, iron, and other cofactors found in the microenvironment of the neutrophil, superoxide anion and hydrogen peroxide combine to form several other reduced oxygen products, including hydroxyl radicals and singlet oxygen. The methods presented below include assays to measure two afferent and several efferent activities of the oxidative burst, including oxygen uptake, hexose monophosphate shunt activity, intracellular and extracellular production of superoxide, intracellular and extracellular production of hydrogen peroxide, luminol- and lucigenin-dependent chemiluminescence, and nitroblue tetrazolium (NBT) reduction.

To determine whether a bacterial strain(s) or mutant(s) induces an oxidative burst in human neutrophils, and if so, to what degree, some questions need to be answered. Is the oxidative burst stimulated? To what degree is the oxidative burst stimulated? Is the oxidative burst inhibited?

How does opsonization affect the ability of the bacteria to induce the oxidative burst? Is the burst limited to the phagolysosome? Is there an oxidative burst but no release of reduced oxygen products into the extracellular milieu? Is phagocytosis needed for induction of the burst, or is contact sufficient? To determine whether phagocytosis indeed occurs or whether the bacteria (regardless of their ability to induce an oxidative burst) are killed by neutrophils, see [8] and [9].

Isolation and Handling of Human Neutrophils

Also see this series, Vol. 108 [9], [28]–[32]; Vol. 132 [3], [8]; and Vol. 163 [28]. See [8] for methods discussing the isolation and handling of human neutrophils.

Measurement of Oxygen Uptake

Also see this series, Vol. 186 [443]–[448]. One of the initial events of the oxidative burst, along with increased hexose monophosphate shunt activity (see below for measurement of this metabolic pathway), is increased oxygen uptake. By definition, with no increased oxygen uptake, there can be no oxidative burst and no augmented secretion of reduced oxygen products. Reduction of 2 mol of oxygen by the oxidase yields 2 mol of superoxide. Depending on the system, however, this simple stoichiometry can rapidly decay as superoxide dismutes to hydrogen peroxide, hydroxyl radical, and other intermediates.

Materials, Reagents, and Equipment

Oxygen monitor, oxygen electrodes with microchamber adapters, and accompanying water-jacketed chamber assembly (e.g., Yellow Springs Instruments (Yellow Springs, OH), biological oxygen monitor, available through scientific distributors). Also, a thermostatted water bath/circulator and a chart recorder.

Microsyringes (e.g., Hamilton 700LT series) with 20-gauge, 2-in. needles (or equivalent).

Dulbecco's phosphate-buffered saline (DPBS) with 0.1% (w/v) gelatin (DPBSG), containing calcium and magnesium, at 37°. DPBS contains, per liter, 8 g NaCl, 0.2 g KCl, 1.15 g Na_2HPO_4, 0.2 g KH_2PO_4, 0.1 g $CaCl_2$, and 0.1 g $MgCl_2 \cdot 6H_2O$, pH 7.2–7.4. DPBSG contains 1 mg/ml gelatin which is added and solubilized with gentle heating; the buffer should not be boiled to dissolve the gelatin. DPBS can be made (or purchased) at 10× concentration, without gelatin, filter-sterilized, and stored several months at room temperature. The 10× buffer can be used to quickly make up the 1× buffer with added gelatin. (Alternatively, for any assay described below, Hanks' balanced salt solution (HBSS) containing 1 mg/ml gelatin, and buffered with 10 m*M* HEPES, pH 7.2–7.4, can be used. HBSS contains, per liter, 8 g NaCl, 0.4 g KCl, 0.048 g Na_2HPO_4, 0.06 g KH_2PO_4, 0.1 g $MgSO_4 \cdot 7H_2O$, 0.1 g $CaCl_2$, 1 g glucose, 0.1 g $MgCl_2$, and 0.35 g $NaHCO_3$.)

Neutrophils, 1 to 2 × 10^7/ml in DPBSG without calcium or magnesium, isolated from freshly drawn blood, stored on ice. Because this assay requires rather larger numbers of neutrophils, at least 100 ml of blood should be drawn.

Opsonized zymosan, used as a particulate positive control stimulus. Boil 10 ml of a 20 mg/ml suspension of zymosan for 10 min, wash twice in water (200 *g*, 10 min, room temperature), and resuspend in 10 ml DPBS. Mix the washed zymosan with 10 ml of fresh (or fresh-frozen) human serum (see [9] for methods), incubate at 37° for 20 min, wash once (200 *g*, 10 min, 3°), and resuspend in 10 ml cold DPBSG. [Do not opsonize for much longer than 20 min as the deposited, opsonic fragments of the third component of complement (C3b and iC3b) will gradually be degraded to less active or inactive opsonins.] Use the same day or aliquot and freeze immediately at −20° in a frost-free freezer. Frozen opsonized zymosan lasts several months.

Phorbol myristate acetate (PMA), used as a soluble, positive control stimulus. Dissolve 100 μg PMA per milliliter of dimethyl sulfoxide (DMSO). Store at −20° in a frost-free freezer, in 50-μl aliquots; thaw and use each aliquot only once. This frozen stock, which is 1000 to 10,000 times the concentration used to stimulate neutrophils, lasts several months in the freezer.

Bacteria, 2 × 10^8/ml in DPBSG. (Some bacteria retain higher viability with certain additions, such as 10 m*M* glucose.)

Methods

Each day equilibrate the oxygen monitor and confirm that the oxygen electrodes are working properly, closely following the manufacturer's directions. The semipermeable membranes used to cover the tips of the oxygen electrodes decay rapidly, clog easily, and are inexpensive; thus, they should be changed daily to avoid unnecessary delays or erroneous data. The Yellow Springs oxygen monitor accommodates two electrodes at once, whereas the chamber assembly holds four reaction vessels. In this manner, while two reactions are being monitored, two reaction vessels can be washed and readied for the next two samples.

To each of two reaction vessels, add 4×10^6 neutrophils (200 μl of 2×10^7/ml) and 1.3 ml of DPBSG, kept at 37°. Vigorously shake the tube or bottle of buffer for several seconds before each use, to ensure that it is saturated with air. With the stir bar not stirring, place the electrode in each chamber, being careful to avoid trapping air bubbles. Start the stir bars stirring and temperature-equilibrate the neutrophil suspension for about 5 min. Because it is important that each assay be run as similarly as possible, use the same equilibration time for all samples. Measure oxygen consumption using a chart recorder (or computer) attached to the oxygen monitor. There is generally a low level of oxygen uptake with no stimulus, 1 nmol/min/1×10^6 neutrophils or less. (If there is more than this minimal background, the neutrophils may have been primed or stimulated by improper handling or contamination with a stimulant, e.g., detergent. Data obtained with such neutrophils are open to question and interpretation.) With the microsyringe, add the bacteria or positive control stimuli [50 μl (1 mg) of opsonized zymosan or 1 μl (100 ng) of PMA, in a total volume of 50 μl DPBSG at 37°] through the vent hole in the probe chamber adaptor. Do not stop the stir bar or remove the electrode from the chamber during this step. Record the results for 5 to 20 min.

Measure the maximal linear portion of the oxygen consumption curve, where the maximal linear rate lasts for 1 min or longer. (Measuring the rate from a portion of the curve that is linear for less than 1 min may lead to inaccurate results.) Depending on the stimulus, there will be a variable lag period of a few minutes before the rate of oxygen consumption becomes linear, and the rate will remain linear for different periods of time. Generally, the higher the rate, the shorter the period of linearity, due to the limited oxygen availability in the chamber and the absolute ability of oxygen to diffuse across the membrane covering the oxygen electrode tip. Generally, the rate of oxygen uptake after 20 min is not reliable. Oxygen consumption is recorded by the oxygen monitor as percent O_2 consumed. At 100% saturation with air, at 37°, there is approximately 0.2 mM O_2, i.e., ~200 nmol oxygen/ml in DPBSG. Results are often reported as nmol O_2 consumed/min/1×10^6 neutrophils. Opsonized zymosan and PMA yield results of ~2 nmol O_2, consumed/min/1×10^6 neutrophils.

Samples should be run in duplicate, and the results should closely agree. Opsonized zymosan or PMA can be the positive control. Once a presumed linear rate of oxygen consumption is obtained, the reactions should be repeated, the same day, with half or twice as many bacteria; expected rates should be half or twice the previous rates. Common reasons for nonlinearity between data obtained with twofold dilutions of bacteria are that the oxygen electrode membranes are dirty; the neutrophils have been improperly handled or prepared; or the number of bacteria is too high and thus has exceeded the capacity of the system (neutrophils, oxygen electrodes, or both) to monitor oxygen uptake accurately.

As with all neutrophil assays, check the pH of the medium and the viability of the neutrophils (by trypan blue exclusion, see [8]) after the assays are completed. Viability should not decrease more than a few percent, and the pH should remain between pH 7.2 and pH 7.4. If viability decreases substantially, decrease the number of bacteria added and repeat the assay. It is possible that the bacteria are cytotoxic to the neutrophils.

Controls should be run with the same number of bacteria in the absence of neutrophils. Many bacteria use a substantial amount of oxygen; they may even undergo an oxidative burst of their own in the presence of neutrophils and compete with neutrophils for oxygen.[1,2] This problem may not be easily remedied. Several alternate approaches exist: (1) Use bacteria killed by different means such as ultraviolet light, antibiotics, or heat, to stimulate neutrophils; however, such treatment can alter the surface of the bacteria, thus altering their interaction with opsonins or neutrophils. (2) Use a buffer such as DPBSG, which does not contain glucose, instead of Krebs–Ringer phosphate, which contains glucose. As glucose may augment bacterial respiration, the lack of glucose may dampen it. Because intracellular stores of glycogen are present, the neutrophil oxidative burst is not immediately dependent on exogenously supplied glucose. Some bacteria, such as gonococci, use carbon sources in addition to glucose, including lactate which is produced and secreted by actively metabolizing neutrophils. (3) Add 1 m*M* KCN to the reaction mixture. This will not dramatically affect oxygen consumption by neutrophils as the oxidase is cyanide insensitive; however, this approach assumes that the bacterial oxygen consumption is progressing through respiration and that it is inhibited by KCN. This, of course, can be tested directly with the oxygen electrode. Another problem may be that if bacterial metabolism is in turn affecting the oxidative response of the neutrophil to the bacterium, adding KCN might cause effects that are uninterpretable.

Measurement of Hexose Monophosphate Shunt Activity

Also see this series, Volume 132 [17], [18]. The hexose monophosphate pathway or shunt (HMPS), also known as the pentose phosphate pathway,

[1] B. E. Britigan and M. S. Cohen, *Infect. Immun.* **52,** 657 (1986).

[2] B. E. Britigan, D. Klapper, T. Svendsen, and M. S. Cohen, *J. Clin. Invest.* **81,** 318 (1988).

metabolizes glucose to several smaller and larger products, and is the major source of NADPH (reduced nicotinamide–adenine dinucleotide phosphate) for the oxidase. Thus, it is an integral part, along with oxygen, in the grist for the oxidase mill. If there is no augmented HMPS activity, there is no oxidative burst. The second enzyme specific for the HMPS is 6-phosphogluconate dehydrogenase, which releases 1 mol of the C-1 carbon from 1 mol of 6-phosphogluconate by oxidative decarboxylation, and generates 1 mol of NADPH. (In total, 2 mol of NADPH are generated per 1 mol of glucose catabolized through the HMPS. One mole of NADPH reduces 2 mol of oxygen to superoxide.) Thus, during the respiratory burst, 1 mol of glucose metabolized through the HMPS optimally yields 4 mol of superoxide ion.

The HMPS is measured by incubating stimulated neutrophils with glucose radiolabeled in the C-1 carbon and following the release of radiolabeled CO_2. There is a low but detectable amount of CO_2 released from other glucose carbons, including C-6, because of continued (and sometimes slightly elevated) glycolytic activity. Glycolytic activity is determined by using parallel controls of glucose radiolabeled in the C-6 position. By knowing the specific activity of the added glucose, minimal, semiquantitative measurements can be made. The measurements are not quantitative for several reasons, the most important being that there are large glycogen stores (i.e., endogenous stores of glucose) in the neutrophil. These may "dilute" the exogenously added radiolabeled glucose; the extent of dilution is dependent on many variables, including the length of the assay, the state of neutrophil activation, and the stimulus. Also, different stimuli (in this case bacteria) induce different degrees and variations of the oxidative burst.

Some bacteria have an active HMPS. Thus, as with oxygen consumption, controls must be run with bacteria in the absence of neutrophils. If the bacteria do show an active HMPS, as indicated by substantial release of ^{14}C from the C-1 position, only stop-gap measures can be taken. Use bacteria killed by gentle methods, such as heat, antibiotics, or ultraviolet light, to avoid altering surface structure and function, or simply subtract the bacterial HMPS from the neutrophil-plus-bacteria HMPS. This is not always applicable, however, as the two cellular components of this system, bacteria and the neutrophils, might affect each other's metabolism. By affecting metabolism, artifacts may be introduced into the system.

Materials and Reagents

[1-^{14}C]Glucose, highest specific activity available.
[6-^{14}C]Glucose, highest specific activity available.

Glucose, 50 m*M* in water, filter-sterilized and stored (up to several months) in the refrigerator.

25-ml Erlenmeyer flasks with one-hole stoppers and one-armed plastic center wells (Kontes, Vineland, NJ).

No. 1 Whatman (Clifton, NJ) filter paper, 2 × 0.5 cm, folded into a fan to fit into the plastic center wells.

10% (w/v) KOH in water.

5 *N* H_2SO_4 in water.

10-ml syringe with 1½-in., 18- to 21-gauge needle.

Neutrophils, as described above.

DPBSG, as described above.

PMA or opsonized zymosan, as described above.

Bacteria, as described above.

Scintillation fluid and vials.

Methods

Add [1-^{14}C]- or [6-^{14}C]glucose to 50 m*M* glucose to a final radioactive concentration of 25 μCi/ml (yielding a specific activity of ~0.5 μCi/μmol) and store at −20° in a frost-free freezer until needed. Prior to use, the radiolabeled glucose stock is thawed and diluted fivefold in 50 m*M* glucose. Reactions are performed in 25-ml Erlenmeyer flasks and contain 20 μl of 50 m*M* [1-^{14}C]- or [6-^{14}C]glucose, stimulus and 2 × 10^6 neutrophils in a total of 1 ml DPBSG. Use 100 ng PMA or 1 mg opsonized zymosan as a positive control stimulus. Use bacteria at concentrations determined in preliminary experiments, usually in the range of 50 to 200 bacteria per neutrophil. Immediately after addition of neutrophils, which are added to the reaction mixture last, cap the Erlenmeyer flasks (with caps that are already "loaded" with center wells containing fluted filter paper and 200 μl of 10% KOH) and orbitally or reciprocally shake the flasks at ~100 rpm for 30 min at 37°.

To stop the reactions and to release $^{14}CO_2$ dissolved in the reaction buffer, inject 1 ml of 5 *N* H_2SO_4 through the rubber stopper into the reaction mixture, being careful not to get acid in the center well. Shake flasks for an additional 30 min. Being careful not to get any of the flask contents in the center wells, remove the rubber caps and center wells from the flasks. Place the center wells in scintillation fluid and count in a liquid scintillation counter. Results can be expressed as cpm of $^{14}CO_2$ released/2 × 10^6 neutrophils/30 min, for both C-1- and C-6-labeled CO_2, and can be presented in tabular form, or the C-6 results can be subtracted from the C-1 results and presented, as long as this is clearly indicated. Results may also be expressed as nanomoles of CO_2 released, by dividing the amount of radiolabeled CO_2 released (expressed in nCi) by the specific activity (nCi/nmol) of the original [1-^{14}C]- or [6-^{14}C]glucose, which should be about 0.1 nCi/mol. As indicated above, however, this may not reflect the true activity of the HMPS.

Measurement of Extracellular Superoxide Production

Also see this series, Vol. 105 [370]–[378], [393]–[398]; Vol. 132 [22]–[24]; Vol. 133 [449]–[493]; Vol. 186 [227]–[232], [567]–[575]. This assay measures the ability of superoxide anion generated by neutrophils to reduce extracellular, exogenously added ferricytochrome *c*, in the absence but not presence of superoxide dismutase. Reduction of cytochrome *c* is followed spectrophotometrically at an absorbance of 550 nm using rate assays, which are more revealing and less prone to misinterpretation, or end point assays, when many samples are to be measured and rate assays are impractical.

The assays described here efficiently measure superoxide released by neutrophils undergoing an oxidative burst, but not necessarily superoxide generated within phagocytic vacuoles or phagolysosomes. This is very important in data interpretation. For instance, even though *Neisseria gonorrhoeae* and *Bordetella pertussis* induce no measurable neutrophil superoxide release measurable by cytochrome *c* reduction, they do induce a substantial neutrophil oxidative response [measured as oxygen uptake, HMPS activity, intracellular oxidation of nitroblue tetrazolium, and intracellular hydrogen peroxide generation measured by the oxidation of 2′,7′-dichlorofluorescin (DCFH), see below].[3,4] Thus, if only cytochrome *c* (superoxide anion release) data were analyzed, one would erroneously conclude that gonococci and *B. pertussis* do not induce an oxidative burst in human neutrophils.

Rate Assay

Materials and Reagents

Cytochrome *c*, horse heart type VI (Sigma Chemical Co., St. Louis, MO), 10–12 mg/ml in buffer. This is stable for several weeks in the refrigerator.

Superoxide dismutase (SOD), 1 mg/ml in water. This is stable for several weeks in the refrigerator.

[3] F. L. Naids and R. F. Rest, *Infect. Immun.* **59,** 4383 (1991).

[4] L. L. Steed, E. T. Akporiaye, and R. L. Friedman, *Infect. Immun.* **60,** 2101 (1992).

Glass or disposable 1-ml spectrophotometer cuvettes.
Neutrophils, as described above.
DPBSG, as described above.
PMA or opsonized zymosan, as described above.
Bacteria, as described above.

Methods

In the rate assay, ferricytochrome *c* reduction (reflecting superoxide release) is measured continuously on a recording spectrophotometer. A multiple-sample, dual-beam thermostatted instrument is preferred, although any research spectrophotometer will suffice. Place 50 μl (5×10^5) neutrophils, 50 μl cytochrome *c*, and 850 μl of DPBSG (all at 37°) in each cuvette. Mix by gently inverting the cuvette(s) three times (using a piece of Parafilm and your thumb to cover the top). Prepare the reference cuvettes in the same manner, except add 10 μl of SOD. Place the cuvettes in their holders, equilibrate to temperature for 5 min, and add 50 μl of bacteria [or 1 μl (100 ng) of PMA or 50 μl (1 mg) of opsonized zymosan, for positive controls], with adequate mixing (with a small plastic paddle or fire-polished glass rod). The rate of the sample (−SOD) cuvette minus the rate of the reference (+SOD) cuvette yields the superoxide-dependent cytochrome *c* reduction. The rate of superoxide generation by unstimulated neutrophils should be minimal. The superoxide-independent rate of cytochrome *c* reduction is also generally very low. If it is high, check the system. With a double-beam instrument, superoxide-independent cytochrome *c* reduction and absorbance due to sample turbidity are automatically subtracted. A change in A_{550} of 1 equals 47.4 nmol of reduced cytochrome *c*; 1 nmol of superoxide reduces 1 nmol of cytochrome *c*.

End Point Assay

Materials and Reagents

Use the materials and reagents described above for the rate assay and, in addition, 1.5-ml conical, snap-cap microcentrifuge tubes.

Methods

Set up duplicate reactions exactly as described above, including the SOD controls, except use microcentrifuge tubes instead of cuvettes and incubate the tubes in a water bath instead of in a thermostatted spectrophotometer. At appropriate times (determined in preliminary experiments), centrifuge duplicate tubes for 30 sec in a microcentrifuge, immediately remove the supernatants, and read at 550 nm in a spectrophotometer. Subtract the readings of the supernatants containing SOD from the supernatants lacking SOD. This is done automatically using a double-beam instrument or manually with a single-beam instrument.

As with any end point assay, it is important to be cognizant of assay conditions. Superoxide release does not usually start immediately on addition of stimulant and varies for different stimuli. In addition, superoxide release may be linear for only a relatively short period. Thus, the rate of superoxide release cannot be determined by dividing the amount of superoxide released in a 20-min assay by 20. Such a number usually underestimates the maximal linear rate of superoxide release. This explanation has real consequences, which can lead to misinterpretation of results. For example, in a 20-min assay, assume a parent bacterial strain induces a respiratory burst with the following pattern: a lag of 2 min, a linear rate of 20 nmol/min/1 $\times$ 10^6 neutrophils for 10 min, and a rapidly decreasing rate for the next 8 min. Now assume that a particular mutant or variant of the parent induces the following pattern: a lag of 2 min and a linear rate of 10 nmol/min/1 $\times$ 10^6 neutrophils for 18 min. Although the two bacterial strains yield similar end point readings ($\sim$120 nmol and 90 nmol reduced cytochrome *c*, respectively), they have interacted with neutrophils quite differently. Therefore, when in doubt, do the rate assay.

Measurement of Intracellular and Extracellular Hydrogen Peroxide Production

Also see this series, Vol. 105 [393]–[398]; Vol. 132 [22]–[24].

Intracellular Assay

Bass and colleagues have introduced a flow cytometric assay to detect intracellular production of hydrogen peroxide in human neutrophils.[5,6] (Read the original articles to be better acquainted with some of the more involved interpretations of the method.) The assay involves loading neutrophils with 2′,7′-dichlorofluorescin diacetate (DCFH-DA), which can cross cell membranes. DCFH-DA is rapidly cleaved by cytosolic esterases to 2′,7′-dichlorofluorescin (DCFH), which cannot readily cross cell membranes. On neutrophil stimulation, hydrogen peroxide formed within phagocytic vacuoles or released into the neutrophil microenvironment

[5] D. A. Bass, J. W. Parce, L. R. DeChatelet, P. Szejda, M. C. Seeds, and M. Thomas, *J. Immunol.* **130,** 1910 (1983).

[6] P. Szejda, J. W. Parce, M. C. Seeds, and D. A. Bass, *J. Immunol.* **133,** 3303 (1984).

diffuses into the cytosol where it oxidizes the impermeant DCFH to the highly fluorescent 2′,7′-dichlorofluorescein (DCF), which also cannot readily cross cell membranes. Apparently this is a much faster reaction than that of peroxidases or catalase with hydrogen peroxide, as the assay is sensitive, quantitative, and stoichiometric with oxygen consumption and hexose monophosphate shunt activity. A possible drawback to the method is the need for a properly configured analytical flow cytometer.

Materials, Reagents, and Equipment

Analytical flow cytometer.
15-ml conical, polystyrene, screwcap centrifuge tubes.
Neutrophils, as described above.
DPBSG, as described above.
DPBSG, as described above, containing 2 m*M* ethylenediaminetetraacetic acid (EDTA).
PMA or opsonized zymosan, as described above.
Bacteria, as described above.
2′,7′-Dichlorofluorescin diacetate (DCFH-DA) (available from Eastman Kodak, Rochester, NY), 5 m*M* in ethanol, stored in the refrigerator.
Hydrogen peroxide, 30% (v/v), freshly diluted 100- to 1000-fold in distilled water. Concentration is determined by measuring absorbance at 230 nm, using an extinction coefficient of 81 cm^{-1} M^{-1}. Freshly prepare this solution daily.

Methods

Incubate 1×10^6 neutrophils/ml with 5 μM DCFH-DA in 5 to 10 ml DPBSG (without calcium or magnesium) for 15 min at 37° with gentle shaking, rocking, or tumbling. Wash the neutrophils once in warm DPBSG (150 *g*, 10 min, room temperature), and resuspend in 10 ml warm DPBSG containing 2 m*M* EDTA, which prevents neutrophil aggregation on stimulation. The washing step is optional. Determine the fluorescence profile for these unstimulated neutrophils. The green fluorescence emission is read at 510 to 550 nm, after excitation at 488 nm with an argon laser. Individual neutrophils are discerned by the combination of low-angle forward scattering and right-angle scattered laser light. Add the appropriate number of bacteria (varying from 1 : 1 to 200 : 1 bacteria to neutrophils) or, as controls, 100 ng/ml PMA or 1 mg/ml opsonized zymosan; shake for the appropriate length of time (up to 1 hr, but 15 or 20 min is usually sufficient); and record the fluorescence of 0.5- to 1-ml aliquots. Analyze 10,000 or more cells at each time point.

Extracellular Assays

The more commonly used assays for neutrophil hydrogen peroxide production involve detection of hydrogen peroxide produced on the outside surface of the cytoplasmic membrane and released directly into the neutrophil microenvironment; they probably do not detect very much hydrogen peroxide produced within enclosed phagolysosomes. (Neutrophils have active cytosolic peroxidases and catalase. It is thus unlikely that hydrogen peroxide produced within neutrophil phagolysosomes, or that has diffused into the cytosol, can survive long enough to exit the cell through the plasma membrane.) The method discussed here describes the use of horseradish peroxidase and scopoletin to measure released hydrogen peroxide. Hydrogen peroxide release is followed fluorometrically by measuring the decrease in fluorescence due to the oxidation of scopoletin.

Materials, Reagents, and Equipment

Spectrofluorometer.
Microcentrifuge and 1.5-ml snap-cap microcentrifuge tubes.
Neutrophils, as described above.
DPBSG, as described above.
PMA or opsonized zymosan, as described above.
Bacteria, as described above.
Scopoletin (7-hydroxy-6-methoxycoumarin), 400 μM in water. Scopoletin is barely soluble in water; the solution may need to be warmed for 1 or 2 hr to effect solution. Optionally, dissolve in 1/10 vol ethanol, and bring to a full volume with water. Stored in the refrigerator (in the dark); this stock will last several months.
Horseradish peroxidase (HPO), 1 mg/ml in DPBSG, stored in aliquots at -20°.

Methods

For each milliliter of reaction mixture, incubate 2×10^6 neutrophils, 8 nmol scopoletin, and 20 μg HPO in DPBSG for 5 min before adding stimulant, either 100 ng PMA, 1 mg opsonized zymosan, or bacteria, at ~10 : 1 to 200 : 1, bacteria to neutrophils. At various times thereafter (up to ~15 min), transfer 1-ml aliquots to 1.5-ml microcentrifuge tubes, and centrifuge for 30 sec, 6000 *g*, at room temperature, to stop the reaction. Read the fluorescence of the supernatants immediately (or put in a covered ice bath until ready to read several samples) at 460 nm, with an excitation wavelength of 350 nm. After an initial lag of a second to a few minutes depending on the stimulus, the rate of scopoletin oxidation (i.e., decreasing fluorescence) should be linear for at least several minutes. Record this linear rate. A slight background or resting rate should be subtracted from the stimulated rate.

To quantitate the amount of hydrogen peroxide released, measure the fluorescence of freshly prepared hydrogen peroxide-oxidized scopoletin standards, as follows. Dilute standard hydrogen peroxide (30%) 1 : 1000 in water, and measure its concentration spectrophotometrically at an absorbance of 235 nm, using an extinction coefficient of 81 cm^{-1} M^{-1} (the absorbance of a 10 m*M* solution will be 0.89). Dilute this solution another 50-fold to obtain a solution of about 0.25 m*M* hydrogen peroxide. Construct a standard curve by measuring the fluorescence of a set of tubes containing the assay reagents (8 μM scopoletin and 20 μg/ml HPO in DPBSG) and 0, 5, 10, and 20 μl of the ~0.25 m*M* hydrogen peroxide solution.

Balance the activity of the stimulus and the number of neutrophils used in the assay, both of which influence the amount of hydrogen peroxide secreted, with the amount of scopoletin used, reflecting the sensitivity of the assay. The more scopoletin used, the less sensitive the assay. Manipulating the concentrations of assay reagents (within a two- to fourfold range) is worth the extra effort it takes to optimize the system for the particular phagocytes, buffer, and bacterial strain(s).

Measurement of Luminol-Dependent Chemiluminescence as Semiquantitative Measure of Intracellular and/or Extracellular Release of Hydrogen Peroxide

Also see this series, Vol. 132 [33]. Luminol-dependent chemiluminescence (LDCL), perhaps one of the most widely used methods for measuring the oxidative burst, is also one of the least quantitative. The exact mechanisms of chemiluminescence are not fully understood, but appear to be due primarily to the oxidation of luminol by hydrogen peroxide and myeloperoxidase.[7,8] Results obtained with this method are only semiquantitative and generally faithfully reflect quantitative changes; thus, the results can be used to say that bacterial strain A induces a large oxidative burst, whereas strain B induces a much less vigorous one. Data cannot, however, be used to say that strain A induces 10 times (or whatever) more neutrophil superoxide or hydrogen peroxide production than does strain B.

[7] G. Briheim, O. Stendahl, and C. Dahlgren, *Infect. Immun.* **45,** 1 (1984).

[8] R. Lock and C. Dahlgren, *Acta Pathol. Microbiol. Immunol. Scand.* **96,** 299 (1988).

Measurement of Total (Intracellular and Extracellular) Chemiluminescence

Materials, Reagents, and Equipment

Scintillation counter or luminometer (e.g., LKB-Wallac, BioOrbit Model 1251 Luminometer).
Luminol, 1×10^{-3} *M* in dimethyl sulfoxide, kept at $-20°$, in a non-frost-free freezer. Protect from light.
Neutrophils, as described above.
DPBSG, as described above.
PMA or opsonized zymosan, as described above.
Bacteria, as described above.
Scintillation vials or luminometer cuvettes.

Methods

Thoroughly but gently swirl luminometer tubes or small scintillation vials containing 1×10^6 neutrophils, 1×10^{-5} *M* luminol, and the appropriate stimulus (100 ng/ml PMA, 1 mg/ml zymosan, or bacteria, at a ratio of 10 : 100 bacteria to per neutrophils) in a total of 1 ml of DPBSG. Generally, a ratio lower than 10 : 1 bacteria to neutrophils does not induce a sufficiently strong response. Measure chemiluminescence by placing the samples in the luminometer or scintillation counter (set to the tritium preset windows) for 1 to 2 hr, at least in initial studies. After experience has been gained with particular bacteria, other times may be more appropriate. Temperature and stirring affect the ability of stimuli to induce neutrophil LDCL. Some luminometers and scintillation counters stir samples and have a thermostatted chamber. If nonthermostatted equipment is used, repeat experiments under similar conditions on different days. The chambers of some counters warm up after prolonged use, which may affect the kinetics of the system.

Stimulation of LDCL generally follows a pattern: the higher the LDCL, i.e., the more vigorous the response, the more short-lived the response. On the other hand, the lower the response, the more protracted it is. For example, PMA-induced neutrophil LDCL usually peaks (at $\geq$50 mV on a luminometer or $\geq$100,000 cpm on a scintillation counter) at about 15 min, and decreases over the next 30 to 60 min to almost background. On the other hand, LDCL induced by some unopsonized gonococci[1] or low concentrations (10–100 μg/ml) of opsonized zymosan may still be rising at 120 min, and not reach more than 10 to 20 mV, or 10,000 to 20,000 cpm. Because of these variations in kinetics, data must be carefully and objectively recorded and reported. Recording just peak or just time-to-peak data is insufficient. Generally, recording the complete response curve or reporting the peak LDCL and the time-to-peak LDCL is appropriate. You can also report the initial rate of LDCL, which is automatically calculated on some luminometers and scintillation counters, or can be calculated manually from the response curve data. Be aware that some stimuli induce a bimodal response. This should be recorded and reported. For more information on the complexity of such kinetics, refer to articles by Briheim *et al.*[7] and Lock and Dahlgren.[8]

Measure neutrophil viability after assays are run. Some bacteria produce cytotoxins that are leukotoxic. Such bacteria can induce several different neutrophil oxidative responses, ranging from rapid, vigorous responses to no response. Thus, if no LDCL response is observed, it may be that the bacteria are killing the neutrophils, are inhibiting neutrophil oxidative responses, or are not inducing such responses. To begin to determine which of these is occurring (after determining that neutrophils are viable) costimulate (or poststimulate) neutrophils with bacteria and PMA or opsonized zymosan.

Measurement of Intracellular or Extracellular Chemiluminescence

Materials and Reagents

Reagents for measurement of chemiluminescence, described in the previous section.

Lucigenin, 5×10^{-3} *M* in water, or luminol, 1×10^{-3} *M* in DMSO, stored at $-20°$, in a non-frost-free freezer. Protect both stocks from light.

Catalase, 200,000 units/ml in water, stored at 3°.

Sodium azide, 100 m*M* in water, stored at 3°.

Horseradish peroxidase, freshly prepared or diluted to 4000 U/ml in PBSG, and kept on ice.

Methods

Run assays as indicated in the previous section, with a few additional reagents. Lucigenin-dependent luminescence reportedly measures mostly extracellular release of superoxide, in contrast to LDCL, which reportedly measures intracellular and extracellular production of hydrogen peroxide. Therefore, using lucigenin can give results similar to or very different from those obtained using luminol, depending on the stimulus. If a stimulus causes mostly intracellular (intraphagolysosomal) release of reduced oxygen products, like gonococci,[1] then luminol might give a larger signal than lucigenin. On the other hand, if a stimulus causes significant extracellular release of reduced oxygen products, like opsonized zymosan, then both luminol and lucigenin would be expected to give similar signals. In myeloperoxidase-deficient neutrophils, or in systems where azide is present (see below for an explanation of the role of azide in measuring LDCL), LDCL is inhibited whereas lucigenin-dependent luminescence is little affected.[9]

Assuming that the present knowledge is correct concerning the role of myeloperoxidase and hydrogen peroxide in LDCL, one can separately measure extracellular and intracellular (intraphagolysosomal) release of reduced oxygen products on stimulation of neutrophil oxidative metabolism. Addition of 2000 U/ml catalase (MW ~250,000) to the LDCL assay system allows measurement of intracellular hydrogen peroxide. Because of its size and activity, catalase scavenges (or degrades) extracellular, but not intraphagolysosomal, hydrogen peroxide before the hydrogen peroxide can be acted on by the degranulated myeloperoxidase to oxidize luminol. On the other hand, addition of a combination of 4 U/ml horseradish peroxidase and 1 m*M* sodium azide allows measurement of released, but not intracellular, reduced oxygen products. Because of its size and activity, azide inhibits both extracellular and intraphagolysosomal myeloperoxidase, but not added horseradish peroxidase, which is insensitive to azide inhibition.

Measurement of Nitroblue Tetrazolium Reduction to Formazan

Also see this series, Vol. 132 [3], [24]. Nitroblue tetrazolium (NBT) is colorless to yellow in solution. On reduction *in situ* (i.e., within the phagolysosome) by reduced oxygen intermediates, NBT forms formazan, which is insoluble in aqueous solutions and dark blue. There are many variations of the NBT assay. Two are described here. The first is semiquantitative and measures, by light microscopy, the percentage of neutrophils that have the capacity to phagocytose bacteria, undergo the oxidative burst, and consequently reduce NBT. The second is more quantitative and measures, by spectroscopy, organic extracts of neutrophils that have phagocytosed and reduced NBT.

Materials, Reagents, and Equipment

- NBT, 1 mg/ml in DPBSG, filter-sterilized, stored at 3°. Protected from light, this is stable for several months.
- 1.5-ml microcentrifuge tubes.

[9] H. Aniansson, O. Stendahl, and C. Dahlgren, *Acta Pathol. Microbiol. Immunol. Scand.* **92,** 357 (1984).

Neutrophils, as described above.

DPBSG, as described above.

Opsonized zymosan, as described above.

PMA/endotoxin. As opposed to the PMA solution mentioned several times above, this one contains 100 μg/ml PMA and an additional 100 μg/ml endotoxin, which reportedly complexes with NBT and acts as a stimulus for endocytosis. This solution is used as a positive control stimulus.

Bacteria, as described above.

Microscope slides, absolute methanol, and safranin (2.5% (w/v) in 95% ethanol).

Cytocentrifuge (e.g., Cytospin, Shandon Southern Instruments, Sewickley, PA).

Methods

Semiquantitative, Light Microscopy Nitroblue Tetrazolium Assay. Neutrophils (2.5×10^5/ml), 50 μg/ml NBT, and the appropriate stimulus (100 ng/ml PMA plus 100 ng endotoxin/ml, 1 mg/ml opsonized zymosan, or bacteria at a ratio of 10 : 1 to 50 : 1, bacteria to neutrophils) are mixed and incubated in a total of 500 μl DPBSG for 20 to 30 min at 37°. Longer incubation can lead to artifactual results. The neutrophils (150 μl) are then deposited on a glass slide by cytocentrifuge, air-dried, fixed with methanol for 1 min, stained with safranin for 1 min, air-dried, and observed with a light microscope (1000×, oil immersion). A positive result is observed as a dark blue/purple area around phagocytized bacteria or zymosan. When PMA is used as a positive control, small punctate areas of dark purple formazan can be seen within endocytic vacuoles within the cytoplasm. Surprisingly, little extracellular reduction of NBT is observed, regardless of the phagocytosis system or stimulus.

Quantitative Nitroblue Tetrazolium Extraction Assay. The same reagents and incubations as described immediately above are used for this assay; however, after incubation, the neutrophils are centrifuged in a microcentrifuge (30 sec) and extracted for 10 min, at room temperature, with 1 ml of dioxane. The extract is centrifuged (1000 *g*, 10 min), the extracted formazan (in dioxane) is removed to a cuvette, and its absorbance is measured at 560 nm. A relative (but not absolute) standard curve can be made by extracting neutrophils that have been stimulated by increasing concentrations of opsonized zymosan or PMA/endotoxin, as described in the preceding paragraph. Such a curve will indicate the lower and upper limits of the assay.

These methods can be modified for large numbers of samples, as discussed by Pick and co-workers (see this series, Vol. 123 [24]).

[11] Complement-Mediated Bacterial Killing Assays

By DANIEL P. MCQUILLEN, SUNITA GULATI, and PETER A. RICE

Introduction

Numerous studies have demonstrated that measurement of serum bactericidal activity provides an appraisal of the functional character of the humoral immune response to bacterial pathogens and correlates well with protection from certain infections. In the case of *Neisseria gonorrhoeae,* infection induces serum bactericidal activity.[1] Serum bactericidal activity against *N. meningitidis* is inversely related to the incidence of disease; in addition, individuals who developed meningitis with this organism had low baseline serum bactericidal titers.[2] The presence of serum bactericidal activity also correlates with protection from *Haemophilus influenzae* meningitis.[3] Bacteremia caused by *Escherichia coli* isolates that are sensitive to killing by normal human serum is less often associated with shock and death than bacteremia caused by serum-resistant *E. coli.*[4]

Although serum resistance has been correlated with the virulence of various organisms, the precise role of serum resistance at different stages of infection is incompletely defined. Membrane attack complex (MAC, C5b-9) insertion fully inhibits oxidative respiration by the inner membrane of serum-sensitive *E. coli,* but only transiently inhibits oxygen uptake of serum-resistant *E. coli,* implicating as yet undefined mechanisms that reverse the inhibitory effects of the MAC.[5] Recent observations indicate that the lipooligosaccharide of most *N. gonorrhoeae* is sialylated *in vivo,*[6] leading to phenotypic serum resistance in serum bactericidal assays. Therefore, although the presence or absence

[1] D. L. Kasper, P. A. Rice, and W. M. McCormack, *J. Infect. Dis.* **135,** 243 (1977).

[2] I. Goldschneider, E. C. Gotschlich, and M. S. Artenstein, *J. Exp. Med.* **129,** 1307 (1969).

[3] L. D. Fothergill and J. Wright, *J. Immunol.* **24,** 273 (1933).

[4] W. R. McCabe, B. Kaijser, S. Olling, M. Uwaydah, and L. A. Hanson, *J. Infect. Dis.* **138,** 33 (1978).

[5] J. R. Dankert, *J. Immunol.* **142,** 1591 (1989).

[6] M. A. Apicella, R. E. Mandrell, M. Shero, M. Wilson, J. M. Griffiss, G. F. Brooks, C. Fenner, C. F. Breen, and P. A. Rice, *J. Infect. Dis.* **162,** 506 (1990).

of *in vitro* serum resistance of *N. gonorrhoeae* correlates with defined clinical syndromes,[7] its exact relationship to mechanisms of pathogenesis remains uncertain.

We have selected three variations of the numerous assays of complement-mediated serum bactericidal activity. Each is easily adaptable for use in most clinical or experimental laboratory situations where serum killing of bacteria is of interest. The methods differ in the serum : organism ratios employed as well as their use of endogenous versus exogenous complement sources, yet all are reliable functional indicators of serum bactericidal activity. In addition, we describe two adaptations of the basic bactericidal assay that allow an assessment of the target specificity of bactericidal antibodies.[7] The specificity of bactericidal antibody can be ascertained by preincubation of organisms with antibody monospecific to bactericidal targets prior to addition of bactericidal antibody or preabsorption of bactericidal antibody with purified bacterial antigen(s) prior to use in the assay. An otherwise bactericidal antibody response can be subverted by blocking antibodies that act either by steric effects or by diversion of complement to nonbactericidal sites.[8,9] Blocking antibody specificity can be ascertained by preincubation of organisms with antibody monospecific to blocking targets prior to addition of bactericidal antibody or preabsorption of blocking antibody with purified bacterial antigen(s) prior to use in the assay. These bactericidal inhibition assays permit analysis of the antigenic target specificity and function of bactericidal and blocking antibodies.

Methods

Equipment

- 37° humidified 5% CO_2 incubator or a candle extinction jar in a 37° incubator
- Rotary shaker water bath
- Vortex mixer
- Plate rotator
- Glass rod spreaders

[7] P. A. Rice and D. L. Kasper, *J. Clin. Invest.* **70,** 157 (1982).
[8] P. A. Rice, H. Vayo, M. Tam, and M. S. Blake, *J. Exp. Med.* **164,** 1735 (1986).
[9] K. A. Joiner, R. Scales, K. A. Warren, M. M. Frank, and P. A. Rice, *J. Clin. Invest.* **76,** 1765 (1985).

Supplies and Reagents

Preparation of Complement

Phosphate-buffered saline (PBS)
Glutaraldehyde
Lysine

Growth Media and Nutrients

Chocolate agar plates are prepared by boiling 334 g GC II Agar Base (BBL/Becton Dickinson, Cockeysville, MD) until dissolved in 4500 ml deionized, distilled water. Next, 100 g dried bovine hemoglobin (BBL) is dissolved in 4500 ml deionized, distilled water. Each solution is divided into 500-ml aliquots, a magnetic stir bar is added to each hemoglobin flask, and the solutions are autoclaved for 30 min. After the solutions are cooled to 56° in a water bath, 500 ml of GC Agar solution is added to each 500 ml of hemoglobin solution. One tube (10 ml) of Isovitalex equivalent (see below) is added to each 1000-ml mixture, which is then stirred for 5 min. The mixture is poured into plastic petri dishes using sterile technique and allowed to solidify and cool prior to use. Plates are stored at 4° until use.

Gey's balanced salt solution (GIBCO, Grand Island, NY).

Solution A is prepared by dissolving 1.5 g proteose peptone No. 3 (GIBCO), 0.4 g potassium phosphate (dibasic), 0.1 g potassium phosphate (monobasic), 0.5 g sodium chloride, and 0.1 g soluble starch in 100 ml deionized, distilled water; autoclaved for 15 min; cooled to room temperature; and stored at 4° until use (shelf life 2 weeks).[10]

Solution B is prepared by dissolving 0.042 g sodium bicarbonate and 4.0 g glucose in 90 ml deionized, distilled water; filter-sterilized (0.22 μm Millipore filter); and stored at 4° until use.[10]

Isovitalex equivalent is prepared by first dissolving 1.1 g L-cystine, 25.9 g L-cysteine hydrochloride, 1.0 g adenine, and 10.0 g L-glutamine in 200 ml deionized, distilled water. A few drops of 6 *N* HCl are added if necessary for full dissolution. Separately, 0.1 g cocarboxylase, 0.25 g diphosphopyridine nucleotide (oxidized), 0.03 g guanine hydrochloride, 0.02 g $FeNO_3 \cdot 9H_2O$, 0.01 g vitamin B_{12}, 0.003 g thiamine hydrochloride, and 0.013 g *p*-aminobenzoic acid are dissolved in an additional 200 ml deionized, distilled water (using additional 6 *N* HCl if necessary). The

[10] S. A. Morse, S. Stein, and J. Hines, *J. Bacteriol.* **129,** 702 (1974).

two solutions are mixed, 100 g dextrose is added, and deionized, distilled water is added to a final volume of 1.0 liter.[10] The Isovitalex equivalent is filter sterilized (0.22-μm Millipore filter) and stored as 10-ml aliquots at −20° until use.

Preparation of Complement[11]

1. Clot freshly drawn human blood at room temperature for 15 min.
2. Centrifuge at 3000 *g* for 10 min, aspirate the serum, and immediately place it in an ice bath.
3. Transfer the test bacterial strain, after overnight growth on a heavily streaked culture plate, into a tube containing 1.0 ml of 0.1 *M* PBS. Wash the organisms three times in 1.0 ml of 0.1 *M* PBS.
4. Incubate the organisms in 1.0 ml of glutaraldehyde, 0.25% in PBS, for 30 min at 25°.
5. Wash the organisms three times with 1.0 ml of 0.1 *M* PBS.
6. Incubate the organisms in 1.0 ml of lysine (1 mg/ml in PBS) for 20 min at 25° to quench residual glutaraldehyde.
7. Wash the organisms three times with 1.0 ml of 0.1 *M* PBS.
8. Incubate the organisms in 5 ml of serum in an ice bath for 1 hr, with frequent stirring of the suspension.
9. Centrifuge the suspension to remove the organisms.
10. Sterilize the supernatant by filter sterilization through a 0.22-μm Millipore filter and store in aliquots at −70° until use as the complement source for a bactericidal assay involving the homologous bacterial strain against which it had been absorbed.
11. The level of complement activity after absorption should be determined by routine CH_{50}[12] and APH_{50}[13] assays and should be reduced by no more than 20%.

Preparation of Bacterial Suspension

1. Inoculate the test strain from frozen culture stock (−70°) onto a chocolate agar plate.
2. After overnight growth (or a few subpassages), inoculate a sterile swab full of organisms from the plate into 1 ml of growth medium (solution A) in a 12 × 75-mm sterile polystyrene tube. Vortex to evenly suspend the bacteria.

[11] K. A. Joiner, K. A. Warren, E. J. Brown, J. Swanson, and M. M. Frank, *J. Immunol.* **131,** 1443 (1983).
[12] W. Hook and L. Muschel, *Proc. Soc. Exp. Biol. Med.* **117,** 292 (1964).
[13] K. A. Joiner, A. Hawiger, and J. A. Gelfand, *Am. J. Clin. Pathol.* **79,** 65 (1983).

3. Inoculate the bacterial suspension, a few drops at a time, into a sterile sidearm flask containing 9 ml solution A, 1 ml solution B, and 0.1 ml Isovitalex equivalent until an OD_{650} of approximately 0.1 is attained.

4. Allow the culture to grow at 37°, stirring vigorously, to a mid-log-phase concentration of approximately 10^8 colony-forming units (cfu)/ml (OD_{650} = 0.2).

5. Perform four 10-fold dilutions of the liquid culture with solution A to obtain an inoculum with a concentration of approximately 10^5 cfu/ml. Incubate in a 37° rotary shaker water bath until use (maximum 10 min).

6. Add 0.025 ml of the final bacterial suspension into each reaction tube (0.15-ml total volume) in the bactericidal assay to give approximately 400 colonies per final culture plate.

Bactericidal Assay Using Serum Dilution (Method A)

Principle. The bactericidal assay method described below is our modification[1] of procedures described by Roberts[14] and Gold and Wyle[15] for *N. meningitidis*. In this assay, gonococci are incubated for 30 min with varying dilutions of human serum and a preabsorbed human complement source to quantitate the amount of bactericidal activity present in the test serum. Modifications in the type of growth medium employed may be made so that the assay can be easily adapted to use with multiple different bacterial species.

Procedures

1. Add 0.025 ml of diluent (Gey's balanced salt solution) to a sterile 12 × 75-mm sterile polystyrene capped tube.
2. Add 0.025 ml of bacterial suspension.
3. Add 0.05 ml of test serum.
4. Add 0.05 ml of the absorbed complement source serum.
5. Vortex to suspend the bacteria thoroughly.

Less than 0.05 ml of test serum may be used in the assay. Serial test serum dilutions may be used to delineate a dose–response effect of bactericidal activity. Less than 0.05 ml of complement source serum may be used in the assay; however, complement should be present at a minimum of 6–8% of the final reaction volume. Serial complement dilutions

[14] R. B. Roberts, *J. Exp. Med.* **126,** 795 (1967).

[15] R. Gold and F. A. Wyle, *Infect. Immun.* **1,** 479 (1970).

may be used to determine the appropriate amount of complement source serum (i.e., the dilution that gives the maximum complement effect in the absence of intrinsic bactericidal activity) for use in the assay. Diluent volume is precalculated to achieve a final reaction volume to 0.15 ml.

6. Inoculate duplicate chocolate agar plates with 0.025 ml of the reaction mixture immediately (within 15 sec) after complement addition to serve as baseline (zero minute incubation) growth controls.
7. Place the reaction tubes in a 37°C rotary shaker water bath.
8. Rotate the inoculated plates on a culture plate rotator and spread the reaction mixture inoculum thoroughly with a glass rod spreader.
9. Inoculate duplicate plates with 0.025 ml of the reaction mixture and repeat step 8 after a 30-min incubation at 37° (30-min incubation point).
10. Incubate the plates overnight at 37° in a 5% CO_2 atmosphere (candle extinction jar).
11. Count viable colonies on all plates after incubation.

Each experiment should include the following positive and negative controls. As a positive control for the presence of adequate complement activity, a serum with known bactericidal activity against the bacterial test strain should be used. Negative controls include an active complement control tube (containing active complement source without test serum) and a serum control tube [containing heat-inactivated (56° for 30 min) complement and test serum].

Analysis of Results. Bactericidal activity is expressed in terms of percentage of bacteria surviving in the 30-min reaction mixture divided by percentage surviving in the time zero control. The following criteria for killing should be met to validate results obtained in an individual experiment: (i) less than 50% survival in the 30-min sample compared with survival in the time zero sample (a killing end point between 50 and 90% may be selected); (ii) percentage survival in the test reaction mixture divided by percentage survival in the active complement control of less than 50%; and (iii) 90% or greater survival at 30 min in the active and heat-inactivated complement control mixtures. These criteria ensure that agglutination is not a significant factor in reduction of colony counts in the 30-min samples. Agglutination of organisms (e.g., if piliated or opacity-positive gonococci are used) may yield erratic assay results. Duplicate plating of 0.025-ml samples should not result in colony counts that vary more than 10 to 15%. Additionally, it should be noted that chamber counts yield numbers of organisms approximately tenfold higher than those obtained by colony counts.

Bactericidal Assay Using Undiluted Serum (Method B)

Principle. This method, described by McCutchan *et al.*,[16] is suited to demonstrating differences in susceptibility to undiluted serum. Serial dilutions of organisms are employed. Its advantages include the fact that it may more closely reproduce the interaction between intact human serum and bacteria during infection and avoids the possibility of introduction of endotoxin and other bacterial antigens into the complement source as a result of absorption. It is, however, essential to establish the presence of active complement in the serum used if an exogenous complement source is not added.

Equipment

37° humidified 5% CO_2 incubator or a candle extinction jar in a 37° incubator
Vortex mixer

Supplies and Reagents

Solid culture medium
Dulbecco's minimum essential medium (DMEM, GIBCO)
Normal and heat-inactivated serum

Procedures

1. Bacteria are grown overnight at 37° in 5% CO_2, suspended in DMEM, vortexed, and adjusted to 10^8 organisms/ml.
2. Serial 10-fold dilutions of the organisms are made.
3. Equal volumes (0.2 ml) of either normal or heat-inactivated serum and each 10-fold dilution of bacteria are mixed, then incubated for 1 hr at 37° in 5% CO_2.
4. Duplicate 0.1-ml aliquots of each mixture are spread on solid agar plates followed by overnight incubation at 37° in 5% CO_2.
5. Duplicate plates having between 10 and 100 colonies are counted and killing is expressed as the difference of the logarithm of survivors in heated and unheated serum.

Bactericidal Assay Using Pour-Plate Technique (Method C)

Principle. This method, described by Taylor,[17] uses a higher ratio of serum to organism suspension than does Method A and provides a measure

[16] J. A. McCutchan, S. Levine, and A. I. Braude, *J. Immunol.* **116**, 1652 (1976).
[17] P. W. Taylor, *Clin. Sci.* **43**, 23 (1972).

of bactericidal activity in a time-kill fashion. Like Method B, this method more closely approximates the concentrations of antibodies found *in vivo*.

Equipment

37° humidified 5% CO_2 incubator or a candle extinction jar in a 37° incubator
Vortex mixer

Supplies and Reagents

Liquid and solid culture media
0.06 *M* NaCl
0.05 *M* Tris(hydroxymethyl)aminomethane hydrochloride (Tris–HCl) buffer, pH 8.4

Procedures

1. Wash bacteria from a 90-min log-phase culture three times in 0.06 *M* NaCl.
2. Resuspend the organisms in 0.05 *M* Tris–HCl buffer, pH 8.4 and adjust to contain approximately 1×10^6 organisms/ml.
3. Add 1 ml of the bacterial suspension to 3 ml of test serum.
4. Incubate at 37°. Viable counts are obtained by the pour-plate technique at time zero and after 1, 2, and 3 hr of incubation.
5. Results are graded into six categories:

Grade 1: progressive decrease in the viable count at each hourly interval with a final count 10% or less of the initial inoculum.
Grade 2: progressive decrease in the viable count at each hourly interval with a final count 10% or greater of the initial inoculum.
Grade 3: viable counts at 1-, 2-, and 3-hourly intervals less than the initial inoculum but not showing an overall progressive decrease.
Grade 4: experiments in which viable counts greater or less than the initial inoculum are obtained.
Grade 5: viable counts at 1-, 2-, and 3-hourly intervals greater than the inoculum but showing an overall progressive increase or showing an overall progressive increase with a final count less than 200% of the initial inoculum.
Grade 6: progressive increase in the viable count at each hourly interval with at least a doubling of the inoculum after 3 hr.

Bactericidal Inhibition Assays[7]

Principle. The bactericidal inhibition assays described below permit more precise delineation of the target sites of bactericidal and blocking antibodies. Blocking antibodies can subvert an antibody response that might otherwise be bactericidal by steric and/or functional effects (i.e., diversion of complement to nonbactericidal sites on the organism). In these assays conditions and controls used are the same as those previously described in bactericidal assay Method A. Two different types of bactericidal inhibition assays can be performed. In the first (Method D) bactericidal antibody binding sites on the organisms are saturated by preincubation with a competitive antibody. In the second (Method E) antibody with specificity for the chosen antigen is absorbed out by preincubation of test bactericidal serum with purified bacterial membrane antigens (e.g., lipopolysaccharide or outer membrane proteins). Method D assesses the degree of competition of antibodies that share the same bactericidal target on the organism. It is most appropriate to use either a nonbactericidal antibody (usually a monoclonal antibody) or $F(ab')_2$ fragments of monospecific polyclonal antibody as the competing antibody. In Method E, absorption of the bactericidal antibody with increasing concentrations of the target antigen yields a decrease in bactericidal activity in the assay.

Alternatively, the binding target site(s) of the blocking antibody may differ from those sites to which bactericidal antibody is directed, as is the case for *N. gonorrhoeae*.[7–9] Complement-dependent killing of *N. gonorrhoeae* is blocked by IgG immunopurified from normal human serum.[7] This blocking IgG is specific for gonococcal outer membrane protein III (Rmp),[8] in contrast to bactericidal antibody, which is directed against gonococcal protein I (Por) and lipooligosaccharide. In addition, although complement is activated in the presence of blocking and bactericidal antibodies, deposition of a bactericidal MAC does not occur.[9] In Method D, blocking antibody sterically inhibits binding of bactericidal antibody leading to a decrease in bactericidal activity. In Method E, absorption of the putative blocking antibody with increasing concentrations of the blocking target antigen will yield an increase in bactericidal activity in the assay.

Procedure for Competitive Inhibition Assay (Method D)

1. Add 0.05 ml of competing antibody to 0.025 ml of bacterial suspension in sterile 12 × 75-mm sterile polystyrene capped tubes (use several dilutions of antibody in separate tubes). Incubate for 15 min in a 37° rotary shaker water bath.
2. Add 0.025 ml of bactericidal serum (dilution used is that which yields about 90% killing in a routine bactericidal assay).
3. Add 0.05 ml of prepared complement source serum.
4. Vortex thoroughly.
5. Immediately inoculate duplicate zero-minute incubation plates with 0.025 ml of the test reaction mixture.

6. Incubate the tubes for 30 min in a 37° rotary shaker water bath and follow steps 8 to 11 of Method A.

Procedure for Absorption Inhibition Assay (Method E)

1. Incubate different amounts of purified bacterial antigen with 0.05 ml of bactericidal (or blocking) serum in sterile 12 × 75-mm sterile polystyrene capped tubes for 15 min at 37° to saturate the binding capacity of the bactericidal (or blocking) antibody.
2. Add 0.025 ml of bacterial suspension.
3. Add 0.05 ml of prepared complement source serum.
4. Add diluent to a final volume of 0.15 ml, vortex thoroughly, and follow steps 6 to 11 of Method A.

Analysis of Results. For Methods D and E, killing in test mixtures is compared with killing resulting from controls that exclude either competing antibody (Method D) or purified antigens (Method E) from the reaction mixtures. In Method E, complement activity of serum mixed with bacterial antigen must be measured prior to use in the assay to ensure preservation of adequate complement activity.

Comments

The three bactericidal and two inhibition bactericidal assay methods described above should be useful and adaptable to most systems in which a functional correlate of a humoral immune response to a particular bacterial pathogen is desired. There are advantages and limitations to each of the bactericidal assay methods. Methods A and B offer the advantage of the ability to assess any changes in colonial morphology that may result from antibody action, whereas the use of the pour-plate technique in Method C precludes such an evaluation. Method A, by serum dilution, allows conservation of serum samples that are in limited supply. Although evaluation of killing at one time point (usually in the range of 1–2 hr) is most commonly employed, killing of some bacteria may not occur at a constant rate. In such cases it may be necessary to assess survival at several time points.

Inclusion of appropriate controls in bactericidal assay systems is of considerable importance. Such controls should always include (1) an organism growth control, (2) controls for both active and inactive complement and (3) a killing control (i.e., an antibody source with known bactericidal activity against the test strain). Variables that influence results obtained in bactericidal assays include the growth medium used, the number of times and conditions under which the organism has been subcultured, the type of diluent used, and the source of complement used. The addition of purified bacterial antigens (e.g., lipopolysaccharide) to serum (as in Method E) may lead to a decrement in complement activity. Therefore, the resultant complement activity must be measured prior to use of the serum in the assay. In addition, the use of broth-grown organisms versus plate-grown organisms has led to differences in results. These complex variables may make interpretation of results obtained in bactericidal assays done under different experimental conditions difficult. Finally, the relationship between resistance to complement-dependent serum bacterial killing measured *in vitro* and pathogenesis of various bacterial infections has been well established as discussed above. Therefore, *in vitro* complement-mediated bacterial killing assays represent a valid assessment of capability of the humoral immune response of the host to resist such infection.

Acknowledgment

This work was supported by Grants AI-32725 (P.A.R.) and Physician Scientist Award K11 AI-01061 (D.P.M.) from the National Institutes of Health.

[12] Measurement of Phagosome–Lysosome Fusion and Phagosomal pH

By THOMAS H. STEINBERG and JOEL A. SWANSON

Introduction

Most microorganisms that are ingested by eukaryotic cells are destroyed within the lysosomal compartment of the host cell. The low pH of this cellular compartment and the battery of acid hydrolases contained within it combine to make life difficult for ingested microbes. Nevertheless, some organisms survive and others thrive in the intracellular environment. They do so either by subverting the usual route that directs them to lysosomes, by blocking lysosomal acidification, or by adapting to the lysosomal environment.

Here we address two questions that are important to the student of intracellular parasitism. First, does the phagosome containing an organism fuse with lysosomes? Second, what is the pH of that phagosome?

Assessment of Phagosome–Lysosome Fusion

Some intracellular organisms survive by residing within membrane-delimited cytoplasmic organelles that do not fuse with lysosomes. How this occurs is not well understood: The organism may be ingested in a "normal" phagosome and may then either inhibit or fail to trigger phagosome–lysosome fusion. In some instances, the organism appears to be ingested within an unusual organelle that might not partake in the usual endocytic pathway. Whatever the mechanism, determining whether phagolysosome formation occurs is an important aspect of understanding intracellular survival tactics.

Phagosome–lysosome fusion has been assessed by several different methods. Most of these strategies involve either the identification of endogenous lysosomal markers in organelles containing ingested organisms or the colocalization of the phagocytosed organism with an exogenous marker endocytosed by the cell and sequestered within lysosomes. The former strategy makes use of the knowledge that a number of enzymes and several membrane components are found mostly in lysosomes after they are formed in the biosynthetic pathway. The latter strategy relies on the fact that many endocytosed markers can be chased into a slowly recycling compartment that comprises primarily lysosomes and a prelysosomal compartment. Either fluorescence or electron microscopy can be employed to assess phagosome–lysosome fusion in these experiments. We focus here on the use of fluorescence techniques, which use equipment that may be more readily accessible and is easier to operate.

Assay of Phagosome–Lysosome Fusion by Fluorescence Labeling of the Lysosome

Phagosome–lysosome fusion can be assessed by prelabeling the lysosomal compartment with an endocytosed marker that enables lysosomes to be identified by fluorescence microscopy. Transfer of the fluorescent marker from the lysosome to a phagosome is indicative of a phagosome–lysosome fusion event. A number of commercially available fluorescent molecules can be used for this purpose. We have used a Texas Red–ovalbumin conjugate to label the lysosomal compartment of J774 cells and mouse macrophages.[1] Labeling was performed as follows: Macrophages were plated on 13-mm No. 1 thickness glass coverslips in 24-well plates at 0.5 to 1×10^5 cells/coverslip and allowed to adhere for 6 hr or overnight. Texas Red–ovalbumin was dissolved in water as a stock solution at 10 mg/ml and added to the medium at a final concentration of 40 μg/ml (4

[1] T. H. Steinberg, J. A. Swanson, and S. C. Silverstein, *J. Cell Biol.* **107,** 887 (1988).

μl per 1 ml medium). The cells were incubated at 37° for 1 hr to allow endocytosis of the fluorescent probe. The coverslips were washed several times in buffered saline, fresh medium was added, and the cells were further incubated at 37° for 1 hr to chase the Texas Red–ovalbumin to the lysosomes. The cells were observed using epifluorescence microscopy at this time. The lysosomal network was brightly fluorescent and remained so for a considerable period thereafter (Fig. 1). It should be noted that because of their high constitutive rates of endocytosis, macrophages label far more efficiently with Texas Red–ovalbumin than do most other tissue culture cells. Higher concentrations or longer incubations with fluorophore are often needed to label lysosomes in other cell types.

The cells are then exposed to particles or microorganisms, and phagocytosis is allowed to occur. Several methods can be used to assess phagosome–lysosome fusion visually. In many cases, the organism can be viewed by phase microscopy. Simultaneous viewing of fluorescence and phase images often can be achieved by lowering the intensity of the transmitted light so that both images can be seen. Frequently, particles can be identified by fluorescence microscopy alone as nonfluorescent silhouettes within the fluorescent lysosomes. Using this approach, Eissenberg and

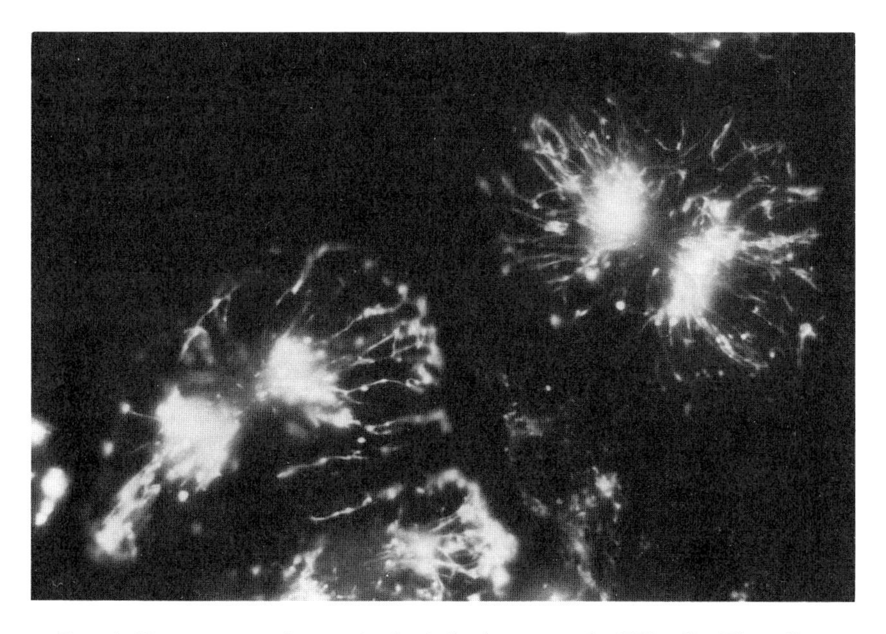

FIG. 1. Fluorescence micrograph of tubular lysosomes in J774 cells. The cells were incubated in Texas Red-conjugated ovalbumin, washed, and further incubated in medium without fluorophore as described in the text.

colleagues have employed fluorescein isothiocyanate–dextran conjugates to label lysosomes and demonstrate phagosome–lysosome fusion in macrophages infected with *Histoplasma capsulatum*.[2] Care must be taken in this sort of experiment to avoid subjecting the endocytic compartment to quantities of poorly degradable substances that may reduce the ability of lysosomes to fuse with other compartments.[3] High osmotic loads may also alter the morphology of the lysosomal compartment and convert tubular structures into more spherical ones.[4]

A related approach is to label fluorescently the organisms themselves with a fluorophore such as fluorescein isothiocyanate. Then, fluorescein-labeled bacteria and Texas Red-labeled phagosomes can be detected by using fluorescein and Texas Red (or rhodamine) filters set sequentially or by using specially designed dual-wavelength filter sets that can detect both dyes at the same time. Such filter sets are available from several vendors, including Omega Optical (Brattleboro, VT) and Chroma Optical (Brattleboro, VT), and can yield striking results. As derivatization of the microorganisms may affect their viability or intracellular trafficking, this method should not be used without independent confirmation.

The assay of phagosome–lysosome fusion, in addition to requiring little more than a fluorescence microscope, has the added benefit of allowing assessment of phagosome–lysosome fusion in either living or fixed cells. If the assay is performed on live cells, one can view a sample at different time points, especially if coverslips are mounted in a special chamber and viewed on an inverted microscope. For this purpose we use either a Sykes–Moore chamber (Bellco Glass Co.) or a heated Leiden Chamber (Medical Systems, Inc.). Fixable endocytic markers that possess amine groups that can be crosslinked to cellular proteins with paraformaldehyde can also be used.

Assay of Phagosome–Lysosome Fusion by Immunostaining of Lysosomal Markers

Antibodies that recognize lysosomal enzymes and lysosomal membrane proteins can be used in immunostaining assays to detect phagosome–lysosome fusion. Antigens that have been used in this fashion include the enzymes β-glucuronidase, cathepsin D, and cathepsin L[5];

[2] L. G. Eissenberg, P. H. Schlesinger, and W. E. Goldman, *J. Leukocyte Biol.* **43,** 483 (1988).
[3] R. R. Montgomery, P. Webster, and I. Mellman, *J. Immunol.* **147,** 3087 (1991).
[4] P. E. Knapp and J. A. Swanson, *J. Cell Sci.* **95,** 433 (1990).
[5] D. A. Portnoy, A. H. Erickson, J. Kochan, J. V. Ravetch, and J. C. Unkeless, *J. Biol. Chem.* **261,** 14697 (1986).

lactoferrin[6] and membrane markers such as Lamp-1 and Lamp-2. It should be noted that these markers do not distinguish late endosomes from lysosomes. An advantage of immunolocalization studies for phagosome–lysosome fusion over the methods discussed above is that ingestion and subsequent events are allowed to occur in cells that have had no prior manipulations that might affect the intracellular fate of the microorganism. For immunofluorescence staining in macrophages, we have used standard fixation techniques employing paraformaldehyde, methanol, or methanol/acetone. Autofluorescence can be a significant problem with these cells and may be less prominent with methanol or methanol/acetone fixation. We prefer a paraformaldehyde–lysine–periodate fixative adapted from McLean and Nakane[7] with modifications.[8] The cells are incubated in the fixative solution [70 m*M* NaCl, 5 m*M* KCl, 70 m*M* lysine hydrochloride, 5 m*M* $MgCl_2$, 2 m*M* ethylene glycol bis(β-aminoethyl ether) *N*,N′-tetraacetic acid (EGTA), 10 m*M* $NaIO_4$, 3.7% paraformaldehyde, 4.5% sucrose, 20 m*M* 2-morpholinoethanesulfonic acid (MES)] at room temperature for 90 min, and then washed three times in 150 m*M* NaCl, 4.5% sucrose, 20 m*M* Tris–HCl, pH 7.5.

For these experiments, ingestion of the microorganism is allowed to occur and the cells are fixed at appropriate times thereafter. Lysosomal components, and in some cases, organisms, can be identified by indirect immunofluorescence. Two-color secondary labeling to identify two different antigens can be performed if the primary antibodies are directly conjugated to fluorophores or if the primary antibodies have been obtained from different species. Confocal microscopy can aid in detecting and quantitating colocalization by minimizing the contribution of fluorescence from molecules out of the plane of focus.[9] Organisms can be stained with 4,6-diamidino-2-phenylindole (DAPI), which labels DNA and allows the bacteria to be visualized with an ultraviolet (UV) filter set. If a microscope is equipped with the three appropriate filter sets, the UV filter set (blue) can then be used to find the bacterium, and the fluorescein (green) and rhodamine (red) filter sets, to observe two additional markers.

It is worth pointing out that in some instances the point of contact between phagosome and the endocytic compartment may not be the lysosome, but may occur at an earlier stage. Thus Rabinowitz *et al.* interpret their recent studies[10] as suggesting that phagocytosed latex particles fuse

[6] M. E. Jaconi, D. P. Lew, J. L. Carpentier, K. E. Magnusson, M. Sjögren, and O. Stendahl, *J. Cell Biol.* **110,** 1555 (1990).

[7] I. W. McLean and P. K. Nakane, *J. Histochem. Cytochem.* **22,** 1077 (1974).

[8] E. L. Racoosin and J. A. Swanson, *J. Cell Biol.* **121,** 1011 (1993).

[9] R. R. Montgomery, M. H. Nathanson, and S. E. Malawista, *J. Immunol.* **150,** 909 (1993).

[10] S. Rabinowitz, H. Horstmann, S. Gordon, and G. Griffiths, *J. Cell Biol.* **116,** 95 (1992).

with a late endosomal "tubuloreticular compartment" rather than endosomes. This compartment expressed Lamp-1, Lamp-2, and Rab-7. Conversely, in some instances the initial fusion of phagosome and lysosome involves a lysosome that has not received endocytosed materials. Harding and Geuze,[11] in their studies of processing of *Listeria monocytogenes* by mouse macrophages, defined two classes of phagolysosomes, depending on whether or not the vacuole contained previously endocytosed bovine serum albumin (BSA)–gold in addition to cathepsin D and Lamp-1. Finally, recent studies have indicated that some markers that identify lysosomal membrane may report vesicle fusion earlier than markers that identify lysosomal contents. In studies of lysosome–macropinosome fusion, Racoosin and Swanson[8] detected lgp-A in macropinosomes before either Texas Red–Dextran, cathepsin L, or mannose 6-phosphate receptor.

Phagosome–Lysosome Fusion Assays Using Electron Microscopy

The approaches mentioned above can be used in conjunction with electron microscopic techniques. Electron-dense markers can be taken up by endocytosis and chased to the lysosomal compartment in experiments similar to those described for assays by fluorescence labeling. Reagents that have been used for this purpose include gold-conjugated BSA,[12] ferritin (either native ferritin as a fluid-phase marker or cationized ferritin as a marker that nonspecifically binds to membranes),[13] microperoxidase,[14] and thorium dioxide.[15,16] Alternatively, immunogold labeling can be used to identify phagosomal markers mentioned in the preceding section.[11,17] Lactoferrin has also been used in such studies.[15] Staining for acid phosphatase, one of the classic methods for identifying lysosomes, can also be done.[14,18,19]

Use of Acridine Orange in Studies of Phagosome–Lysosome Fusion

Acridine orange is a metachromatic dye that has been used extensively to assess phagosome–lysosome fusion. It has also been used to assess

[11] C. V. Harding and H. J. Geuze, *J. Cell Biol.* **119,** 531 (1992).
[12] J. W. Slot and H. J. Geuze, *Eur. J. Cell Biol.* **38,** 87 (1985).
[13] P. D. Hart and M. R. Young, *J. Exp. Med.* **174,** 881 (1991).
[14] J. Swanson, A. Bushnell, and S. C. Silverstein, *Proc. Natl. Acad. Sci. U.S.A.* **84,** 1921 (1987).
[15] N. A. Buchmeier and F. Heffron, *Infect. Immun.* **59,** 2232 (1991).
[16] M. A. Horwitz, *J. Exp. Med.* **158,** 2108 (1983).
[17] D. G. Russell, S. Xu, and P. Chakraborty, *J. Cell Sci.* **103,** 1193 (1992).
[18] U. Steinhoff, J. R. Golecki, J. Kazda, and S. H. Kaufmann, *Infect. Immun.* **57,** 1008 (1989).
[19] C. Frehel and N. Rastogi. *Infect. Immun.* **55,** 2916 (1987).

pH as discussed later in this article. At low concentrations acridine orange appears a dull green, but at high concentrations the dye exhibits a bright orange fluorescence. Furthermore, it is accumulated within acidic intracellular compartments and exhibits a bright orange fluorescence at the concentrations that it achieves in these compartments. Thus, in cells treated with acridine orange, the presence of bright orange staining of vacuoles containing phagocytosed organisms has been taken as evidence for phagosome–lysosome fusion.

Nevertheless, in several studies acridine orange has proven to be an unreliable reporter of phagosome–lysosome fusion. When acridine orange fluorescence and localization of Thorotrast (colloidal thorium dioxide) were compared as indicators of phagosome–lysosome fusion in the uptake of live *Candida albicans* by mouse macrophages,[20] Thorotrast was detected in almost all phagosomes, but most phagosomes did not appear to have fused with lysosomes as assessed by acridine orange. This discrepancy was not seen when dead yeast were used. The reason that acridine orange failed to detect phagosome–lysosome fusion was not clear in this study; however, the fact that its accumulation is pH dependent suggests one possibility: If phagolysosomal pH is increased by any means, acridine orange accumulation will be diminished and phagosome–lysosome fusion may be underestimated. Another possible explanation is that the osmotic load of the intravacuolar acridine orange might alter the ability of the lysosomes to fuse with other organelles, as mentioned above. These observations suggest that one must use caution in relying on acridine orange as a "lysosomal" marker in studies of phagosome–lysosome fusion.

Measurement of pH

The pH of phagosomes can be obtained simply or with great effort. Metchnikoff first described phagosome acidification by observing macrophages that had ingested litmus paper. The color change reported the pH change. Another colorimetric method of more practical value now is to add neutral red (10 μg/ml for 10 min) to preparations of living cells, either before or after ingestion of bacteria. Neutral red accumulates in acidic organelles, allowing them to be seen by bright-field microscopy. The principal difficulty encountered in this method will most likely be in identifying intracellular bacteria in these otherwise unstained preparations. The principal shortcoming is that the method cannot quantify pH with precision.

[20] N. Mor and M. B. Goren, *Infect. Immun.* **55,** 1663 (1987).

Fluorescent Probes of pH

Acridine Orange. Another simple microscopic method uses acridine orange, mentioned above. Acidic phagolysosomes can be identified using a rhodamine or Texas Red filter set (i.e., excitation 520–560 nm, emission 580–620 nm). Acridine orange also stains cells in various ways unrelated to pH. It binds readily to nucleic acids, for example. Therefore, caution is urged both in its use (it is a carcinogen) and in the interpretation of the fluorescent patterns it generates. One advisable control experiment to confirm that a red, fluorescent phagosome is indeed acidic is to raise the pH of all acidic organelles and see that the red fluorescence disappears. Exposure of acridine orange-labeled cells to 10 m*M* ammonium chloride in medium or saline for 10 min raises the pH of all acidic compartments. Neutralized organelles should no longer fluoresce red with acridine orange.

Fluorescein. Okhuma and Poole[21] first described the use of fluorescein to measure lysosomal pH. The principle of their method, and all the derivative methods, is simple: The fluorescence spectrum of fluorescein varies with the pH of its environment. Although the intensity of fluorescence emission is affected by dye concentration and by differences in the thickness of the cell from one location to another, under stable conditions the shape of the emission spectrum depends only on the pH of the local environment. This difference in the shape of the emission spectrum can be quantitated by measuring the ratio of the fluorescence emission intensities obtained when the fluorophore is excited at two wavelengths. Thus one can calculate $R_{450/490}$, the ratio of the intensity of the 520-nm fluorescein fluorescence obtained by exciting at 450 and 490 nm. This ratio varies according to pH, but is relatively independent of the quantity of fluorescein present. $R_{490/450}$ is lowest at pH 4.0 and increases with pH to a maximum at pH 8.0. Measurement of fluorescence ratios can therefore be used to obtain the pH of fluorescein environment.

Why not simply measure fluorescein fluorescence at one pH-sensitive wavelength, rather than obtain ratiometric data? For solutions or for populations of cells in a cuvette, such a method may be satisfactory. In the microscope, however, the fluorescence from any given cell is subject to other factors such as the absolute amount of fluorescein contained in the organelle, its concentration therein, and the path length, or thickness of the organelle itself. Obtaining the fluorescence at two wavelengths allows one to correct for these pH-independent factors.

Other pH-Sensitive Fluorophores. The fluorophore 8-hydroxypyrene 1,3,6-trisulfonic acid (HPTS or pyranine) also exhibits a pH-dependent

[21] S. Ohkuma and B. Poole, *Proc. Natl. Acad. Sci. U.S.A.* **75,** 3327 (1978).

excitation spectrum.[22] Fluorescence ratios obtained at 405 and 450 nm vary up to 20-fold between pH 4.0 and 8.0. This wider range of ratios makes HPTS a more sensitive indicator of pH than fluorescein. Nevertheless, there has been much less experience with HPTS than with fluorescein, and few useful conjugates of this dye are available.

Molecular Probes (Eugene, OR) has produced other pH-sensitive fluorophores, two of which, DM-NERF and Cl-Nerf, can be used like fluorescein or HPTS to measure pH ratiometrically. Because endocytic compartments are acidic, we have chosen here to describe dyes with maximal sensitivity (pK_a) in the pH range 3.0–7.0. Many other fluorophores are available for measurement of higher pH ranges.

Other pH-sensitive fluorophores have constant excitation spectra but pH-dependent emission spectra. Such probes have proven more useful with detection systems that must use a single excitation (ex) wavelength, such as laser-based flow cytometers and confocal microscopes. In this case, the ratiometric data are obtained using constant excitation light and collection at two emission (em) wavelengths.

Most of the dual-emission dyes currently in use report pH in the neutral or alkaline range. One that may prove useful for measuring pH of endocytic compartments is a dextran labeled with both fluorescein (pH-sensitive) and tetramethylrhodamine (pH-insensitive) (Molecular Probes). This combination of fluorophores on the same molecule permits ratiometric measurements based on either dual excitation, single emission (ex 495 and 520 nm, em 575 nm) or dual excitation, dual emission (ex 495/em 520 nm, ex 520/em 575 nm). An advantage of the latter system is that pH measurements may be obtained using a microscope equipped with a standard fluorescein and rhodamine filter set (see below).

Equipment for Fluorescence-Based Measurement of pH

Ratiometric measurement of pH requires a good fluorescence microscope and accessory equipment for collecting and measuring light (Fig. 2). Although many investigators now measure pH using digital image processing of video images from the microscope, simpler methods can suffice. For example, reliable measurements can be obtained with a fluorescence microscope equipped with standard fluorescein and rhodamine filter sets, a photomultiplier attached to one optical port, an aperture for restricting measurement to selected regions of the slide, an amplifier, and a chart recorder.

[22] K. A. Giuliano and R. J. Gillies, *Anal. Biochem.* **167,** 362 (1987).

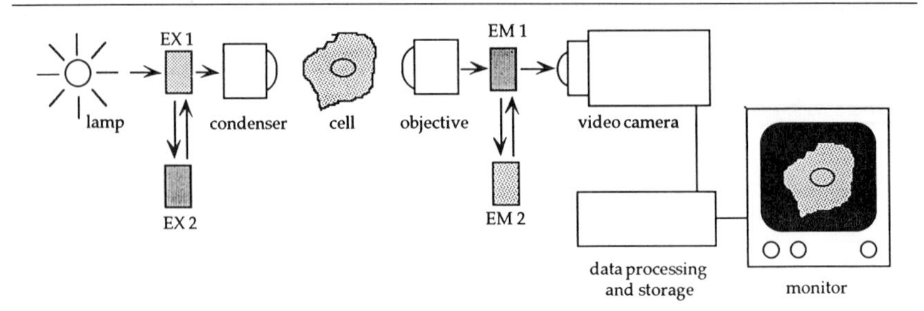

FIG. 2. Diagram of a fluorescence imaging system suitable for the measurement of vacuolar pH.

Microscopes. The optimal fluorescence microscope for ratiometric imaging allows control of the wavelengths of light reaching the specimen and similar control of the measured emitted light. Automated systems are available that change filter sets and that synchronize changing excitation light with image collection systems. Because pH data must be obtained from living cells, it is important to minimize light exposure and consequent photo damage. It is therefore preferable to have a shutter system that admits light to the specimen only when measurements are being taken.

In epifluorescence microscopy, the objective lens serves as both a light condenser and an imaging lens, and it is an important component of any pH measurement system. Optimal light gathering is obtained not so much from high magnification, as from a high-numerical aperture (N.A.). That is, a 63×, N.A. 1.40 lens is preferable to a 100×, N.A. 1.20 lens. We use a Zeiss, Thornwood, NY, IM-35 inverted microscope equipped with a 63× Neofluar lens, N.A. 1.4.

Observation Chambers. In epifluorescence microscopy, both excitation light and emission light cross the same coverslip. Because the high-numerical-aperture lenses used for measuring pH often have short focal lengths, the distance between the lenses and the cells must be kept short. This can be achieved by plating cells onto the upper coverslip, rather than on the lower surface of the chamber, and by using thin coverslips, No. 1 or 0 thickness.

Any of a variety of observation chambers may be used. Simple chambers can be constructed from 12-mm-diameter circular coverslips, with adherent cells, a glass slide, and Valap, a sealing wax.[23] More elaborate chambers are available commercially. We use Sykes–Moore chambers (Bellco Glass Co., Vineland, NJ), in which two large, circular coverslips are separated by a rubber gasket and are held together by a metal frame. These can permit perfusion of the sample and, with the appropriately

[23] J. A. Swanson, *Methods Cell Biol.* **29,** 137 (1989).

constructed stage, control of sample temperature. A more elaborate open-chamber system, PDMI-2 (Medical Systems Corp., Greenvale, NY), incorporates both temperature control and the ability to buffer culture medium in a controlled CO_2 environment.

Excitation and Emission Light. For ratiometric measurement of pH, one must obtain two fluorescent images of the cell, varying either the excitation or emission wavelength depending on the dye used. With fluorescein the emission filter is fixed (520 nm) but the excitation light must alternate between 450 and 490 nm. This can be accomplished in a number of ways: The two images may be acquired using two fluorescein filter sets in the microscope (450 ex/520 em and 490 ex/520 em), a filter wheel that switches between two different excitation barrier filters, or two different excitation light sources. If one chooses to alternate filter sets in the microscope, it is important to be sure that both resultant images are coincident; i.e., that the fluorescent cell occupies the same position in the microscope or video field. Otherwise, ratiometric data will be artifactually nonuniform.

Data Acquisition and Storage. Ratiometric measurement of pH by digital image processing requires a video camera capable of detecting dim signals and an image processing computer capable of digitizing and storing video images and of processing multiple images. A digitized video image is one in which the standard analog video signal, which is a linear sequence of voltage amplitudes that represents a linear scan of an image, is represented instead as a two-dimensional grid of small squares, typically 525 squares wide and 525 squares high. Each square, or pixel, contains a numerical value (0–255) corresponding to the light intensity of the corresponding position of the video image. Corresponding positions of two digitized images can be related mathematically, and consequently, one whole image can be subtracted from or divided by another image to yield a third image that reflects the quantitative relationship between those corresponding images. The spatial resolution of digital image processing allows measurement of fluorescence ratios and pH, from selected portions of the field, e.g., from individual bacterium-containing phagosomes.

Storage of a single digitized image requires considerable computer memory, typically greater than 200 kilobytes. Saving many images therefore requires large dedicated data storage devices.

Procedures for Fluorescence-Based Measurement of pH

The following is a procedure for measuring the pH of phagosomes using fluorescein–dextran as the pH probe, essentially as published in Aranda *et al.*[24] The procedure is applicable to other fluorophores as well,

[24] C. M. Aranda, J. A. Swanson, W. P. Loomis, and S. I. Miller, *Proc. Natl. Acad. Sci. U.S.A.* **89,** 10079 (1992).

with appropriate adjustments for spectral differences. A similar approach has been used by Eissenberg *et al.*[25]

Cells. Macrophages or other target cells are plated onto 25-mm-diameter, circular, No. 1 thickness coverslips and maintained in six-well plates containing appropriate culture medium. Infection with bacteria is done in these dishes. Bacteria may be labeled directly with fluorescein by covalent modification of the cell surface, or they may be introduced to host cells with fluorescein–dextran (5 mg/ml, Molecular Probes) added simultaneously to the medium. Use of the former method requires independent confirmation that fluorescein-labeled (or otherwise modified) bacteria remain viable and competent for infection. Because such resilience is often difficult to achieve, we recommend the latter method.

Trial and error determine the time and conditions for coincubation of cells, bacteria, and fluorescein–dextran necessary to obtain fluorescent phagosomes. For precise measurement of the time course of acidification, the period of infection must be shortened into a pulse. This can sometimes be done by binding bacteria to cells in the cold, then warming them in medium containing fluorescein dextran. After internalization, cultures are washed with saline, and medium is replaced with saline or medium lacking fluorescein–dextran. These infected cells may then be assembled into observation chambers or maintained in an incubator for a chase period before assembly and observation. The assembled chamber should contain a buffer that permits continued survival of both cell types. If possible, the chamber should be maintained at 37° with bicarbonate buffering of extracellular pH.

Image Collection and Analysis. Intense excitation light can damage cells, especially those containing fluorescein–dextran. Therefore, light exposure must be minimized by use of low-intensity excitation light and image intensification video, such as a silicon-intensified target (SIT) camera or a multichannel plate image intensifier coupled to a standard video camera or charge-coupled device (CCD) camera. The live video signal (analog) should be monitored through the image processing computer, so that a digitized image can be generated soon after an appropriately infected cell is identified. A bacterium in a fluorescein–dextran-containing phagosome should appear as a dark sphere or rod, 0.3 to 1.0 μm long, that displaces the fluorophore within the phagosome. The phagosome membrane may adhere closely to the bacterium or it may be more spacious (3–6 μm in diameter). The digitized video image is usually obtained as the average of a number of images; this reduces the effects of electronic noise in the intensified video image.

[25] L. G. Eissenberg, W. E. Goldman, and P. H. Schlesinger, *J. Exp. Med.* **177,** 1605 (1993)

One must obtain four images for each cell: the fluorescent image of the cell excited at 450 nm ($Cell_{450}$) and at 490 nm ($Cell_{490}$), and the background images, from nearby cell-free regions of the coverslip ($Background_{450}$ and $Background_{490}$). Background subtraction is performed digitally, $Cell_{450}$ minus $Background_{450}$ and $Cell_{490}$ minus $Background_{490}$, creating two background subtracted images, I_{450} and I_{490}, respectively. Then these two digital images are stored for later analysis. If filters are being changed manually, then care must be taken to ensure that the two images of the cell coincide. This is done by collecting $Background_{490}$ first, moving the cell into the field, collecting $Cell_{490}$, changing the filter set, collecting $Cell_{450}$, moving again to a cell-free region, and then collecting $Background_{450}$.

Image analysis can be performed after collecting and storing the experimental data. The two background-subtracted images of each cell are used to generate the ratio image, $R_{490/450}$, by dividing I_{490} by I_{450}. The fluorescence ratio for the fluorescent phagosome of interest, which is usually a small portion of the whole video field, can be obtained from the image $R_{490/450}$ by directly reading ratio values of the appropriate pixel positions or by creating binary masks that allow averaging of all pixels that correspond to the phagosome.

Calibration. Ratios obtained by digital image processing are converted to pH values by generating a standard curve *in situ*. This is done with microscope preparations of cells with fluorescein–dextran-containing phagosomes. The cells are incubated 10 to 15 min with the K^+/H^+ ionophore nigericin (10 μM) in a buffer containing 133 mM KCl, 1 mM $MgCl_2$, 15 mM HEPES, and 15 mM MES, pH 7.0.[26] The effect of the nigericin is to bring the pH of all intracellular compartments to that of the buffer (pH 7.0). Fluorescence images (I_{450} and I_{490} are obtained from several cells in this buffer; then the coverslip is perfused with a similar buffer, lacking nigericin, at a different pH. The *in situ* pH standard curve is generated by equilibrating cells for 10 min with buffers at pH 7.0, 6.5, 6.0, 5.5, 5.0, 4.5, and 4.0, and obtaining images of two or more cells at each pH. For pH 4.5 and 4.0, 30 mM citrate buffer is used instead of the HEPES/MES buffer combination. Average $R_{490/450}$ values are determined for the fluorescent organelles at each pH, generating the plot of $R_{490/450}$ vs pH, from which ratio values from experimental data can be translated into pH values.

Data Interpretation. Once it is certain that the pH of a bacterium-containing phagosome has been obtained, then other questions must be answered. The most pressing question is whether the phagosomes mea-

[26] G. R. Bright, G. W. Fisher, J. Rogowska, and D. L. Taylor, *J. Cell Biol.* **104,** 1019 (1987).

sured are those of viable or dead bacteria. If in general only a small percentage of bacteria internalized by cells successfully escape host defense mechanisms, then most of the fluorescent phagosomes will be those of the unsuccessful majority, usually the less interesting of the two groups. Independent measurements should be made of the efficiency of the infection process.

Phagosomal pH will most likely change with time, so some measure of the time course of acidification is appropriate. Macrophage phagosomes acidify shortly after formation. A change in the kinetics of acidification can be sufficient as a bacterial survival mechanism.

Finally, if it appears that bacterium-containing phagosomes alter the course of acidification, statistical tests of significance are required.

Acknowledgment

The authors thank Dr. Michael Koval for a critical reading of this manuscript and many helpful suggestions.

[13] Neutrophil Defensins: Purification, Characterization, and Antimicrobial Testing

By SYLVIA S. L. HARWIG, TOMAS GANZ, and ROBERT I. LEHRER

Introduction

Polymorphonucleated neutrophils (PMNs), the most abundant phagocytic cells of the blood, are classified as granulocytes because their cytoplasm contains thousands of small membrane-bound organelles (granules). One subset of these cytoplasmic structures, known as "primary" or "azurophil" granules, contains various hydrolases, myeloperoxidase, and a group of proteins and peptides with antimicrobial properties.[1] Among the antibiotic polypeptides of human PMNs are defensins, cathepsin G, azurocidin, bactericidal/permeability increasing (BPI) factor, and lysozyme (muramidase). This article describes the large-scale purification of the four defensins, HNP-1, -2, -3, and -4, found in human neutrophils[2,3]

[1] R. I. Lehrer and T. Ganz, *Blood* **76,** 2169 (1990).
[2] M. E. Selsted, S. L. Harwig, T. Ganz, J. W. Schilling, and R. I. Lehrer, *J. Clin. Invest.* **76,** 1436 (1985).
[3] R. I. Lehrer, T. Ganz, and M. E. Selsted, *Cell (Cambridge, Mass.)* **64,** 229 (1991).

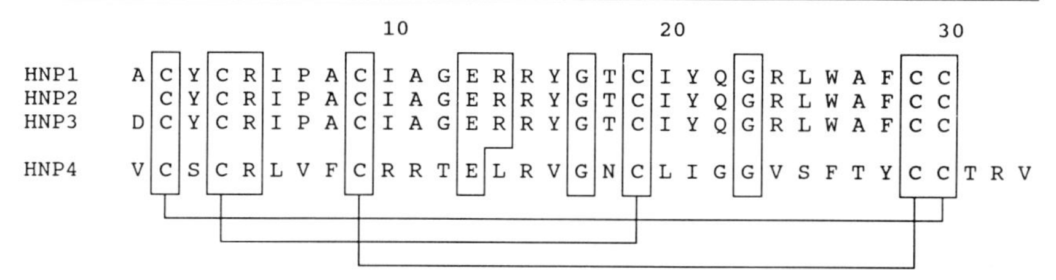

FIG. 1. Primary amino acid sequences of four human neutrophil defensins [M. E. Selsted, S. L. Harwig, T. Ganz, J. W. Schilling, and R. I. Lehrer, *J. Clin. Invest.* **76,** 1436 (1985); C. G. Wilde, J. E. Griffith, M. N. Marra, J. L. Snable, and R. W. Scott, *J. Biol. Chem.* **264,** 11200 (1989)]. The boxed residues include the six invariant cysteines that constitute the "defensin motif" and the other generally conserved or invariant residues [R. I. Lehrer, T. Ganz, and M. E. Selsted, *Cell (Cambridge, Mass.)* **64,** 229 (1991)]. The pattern of cystine–disulfide pairing is also shown [M. E. Selsted and S. S. L. Harwig, *J. Biol. Chem.* **264,** 4003 (1989)].

and describes procedures for testing their antimicrobial properties.[4,5] The primary amino acid sequences of these defensins are shown in Fig. 1. Smaller-scale procedures applicable to purifying antimicrobial defensins from the PMNs[6,7] or small intestinal tissues[8] of laboratory animals have been described elsewhere.

Isolation of Human Neutrophils

Materials

1. Phosphate-buffered saline (PBS) consisting of 0.135 *M* NaCl, 0.015 *M* sodium phosphate, 4 m*M* ethylenediaminetetraacetic acid (EDTA), pH 7.35.
2. Distilled H_2O with 4 m*M* EDTA.
3. NaCl 3.5% with 4 m*M* EDTA.
4. A single-donor leukaphoresis pack or pooled human neutrophils obtained from buffy coats or from fresh anticoagulated venous blood.

[4] T. Ganz, M. E. Selsted, D. Szklarek, S. L. Harwig, K. Daher, D. F. Bainton, and R. I. Lehrer, *J. Clin. Invest.* **76,** 1427 (1985).

[5] R. I. Lehrer, M. Rosenman, S. S. L. Harwig, R. Jackson, and P. B. Eisenhauer, *J. Immunol. Methods* **137,** 167 (1991).

[6] M. E. Selsted, D. Szklarek, and R. I. Lehrer, *Infect. Immun.* **45,** 150 (1984).

[7] P. B. Eisenhauer, S. S. L. Harwig, D. Szklarek, T. Ganz, M. E. Selsted, and R. I. Lehrer, *Infect. Immun.* **57,** 2021 (1989).

[8] P. B. Eisenhauer, S. S. L. Harwig, and R. I. Lehrer, *Infect. Immun.* **60,** 3556 (1992).

Equipment

1. Refrigerated centrifuge.
2. Aspirator.

Procedure

It is most convenient to obtain human neutrophils from normal donors by leukaphoresis. If leukaphoresis units are not available, neutrophils can be prepared from single-unit "buffy coats" or whole blood donations and then pooled. All donor blood should be tested for hepatitis virus and human immunodeficiency virus (HIV), and positive units should not be used for reasons of laboratory safety. A typical leukaphoresis procedure generates 1 to 2×10^{10} white cells, more than 90% neutrophils, in a final volume of 150 to 200 ml. In contrast, each "unit" (approximately 450 ml) of fresh whole blood typically provides approximately 1 to 1.5×10^9 neutrophils. Neutrophils should be processed within a few hours of harvest, with all processing steps performed at 0 to 4°.

To render the neutrophil preparation free of red cells and plasma, the leukaphoresis product is centrifuged at 1000 rpm (approximately 220 *g*) for 10 min and the supernatant plasma is carefully separated by aspiration and discarded. The residual plasma is removed by resuspending each pellet in 35 ml of PBS and repeating the centrifugation process. The pellets are again suspended in PBS and final volume is adjusted to 100 ml. Contaminating red cells are lysed by rapidly adding 300 ml of ice-cold H_2O with EDTA, mixing gently on ice for 45 to 50 sec, and then adding 100 ml of 3.5% NaCl with EDTA. The mixture is centrifuged at 1000 rpm for 10 min and the resulting white cell pellets (>90% viable by trypan blue) are ready for granule preparation. The red cell lysis step can be repeated one more time, if necessary.

Preparation and Extraction of Neutrophil Granules

Materials

1. Calcium-free Hanks' balanced salt solution (HBSS) with 2.5 m*M* $MgSO_4$, pH 7.4. This solution is prepared by dissolving 8 g NaCl, 0.4 g KCl, 0.06 g KH_2PO_4, 0.0477 g Na_2HPO_4, 0.35 g $NaHCO_3$, 0.098 g $MgSO_4$ and 1 g of glucose in 1 liter of H_2O.
2. Calcium-free HBSS with 5 m*M* ethylene glycol bis(β-aminoethyl ether) *N,N'*-tetraacetic acid (EGTA), pH 7.4.
3. Acetic acid 5%.

All reagents are kept at 4° and each step is performed in the cold.

Equipment

1. Bronwill Biosonik IV sonicator (VWR Scientific).
2. Refrigerated centrifuges, low speed and high speed.
3. Model 4635 Cell Disruption Bomb (Parr Instrument Company, Moline, IL).
4. Speed Vac Vacuum Centrifuge (Savant Instruments, Inc., Farmingdale, NY).

Procedure

The human leukocyte pellets are gently suspended in 100 ml of calcium and EGTA-free HBSS and transferred into a 400-ml beaker that is placed in the precooled (to 4°) Parr cell. After the "bomb" is pressurized with nitrogen to 750 psi and the suspension allowed to equilibrate for 20 min at 4°, with occasional gentle swirling, the pressure is slowly released and the now-disrupted cell suspension is collected dropwise, with gentle stirring, into a beaker that contains 100 ml of calcium-free HBSS with 5 m*M* EGTA. This mixture is centrifuged at 1200 rpm for 10 min and the postnuclear supernatant is further centrifuged at 16,000 rpm (27,000 *g*) for 20 min to sediment the granules, which are stored at −70° until their extraction.

Extraction is performed by suspending each granule pellet, typically derived from 2 to 3 × 10^9 neutrophils, in 5 ml of ice-cold 5% acetic acid and combining suspensions equivalent to approximately 4 × 10^{10} cells in a beaker. The total volume is adjusted to 100 ml with 5% (v/v) acetic acid and the solution is sonicated three times, 10 sec each, with 2 min of cooling on ice between cycles. The suspension is stirred overnight in the cold and the extract is cleared by centrifuging it at 27,000 *g*. This overnight extraction process is repeated twice more with 50 ml of ice-cold 5% acetic acid and the three supernatants are combined and concentrated by vacuum centrifugation to 8 to 10 × 10^8 cell equivalents ml^{-1}.

Purification of Defensins from Granule Extract

Sephacryl S-200 Size-Exclusion Chromatography

The concentrated granule extract of human PMNs is loaded on a 4.8 × 105-cm (V_t = 1900 ml) Sephacryl S-200 column (Pharmacia LKB Biotechnology, Piscataway, NJ) preequilibrated with 5% acetic acid and eluted with this solution at a flow rate of 108 ml hr^{-1} at 4°. The column effluent is monitored at 280 nm and 18-ml fractions are collected (Fig. 2). Fractions containing human defensin peptides (HNP-1 to -4) are identified by their electrophoretic mobility on 12.5% acid–urea polyacrylamide gel electrophoresis (AU-PAGE). HNP-1, -2, and -3 emerge with an elution volume considerably larger than the bed volume of the column, presumably reflecting nonspecific interactions between these peptides and the gel matrix. HNP-1 elutes first, followed by HNP-3 and HNP-2, which show considerable overlap. Although HNP-4 is only slightly larger (M_r 3709) than the other three defensins (HNP-1, M_r 3442; HNP-2, M_r 3371; HNP-3, M_r 3486), it elutes much earlier, within the calculated bed volume of the column. The column fractions are monitored by analytical AU-PAGE and early fractions containing only HNP-1 are pooled for subsequent purification by reversed-phase high-performance liquid chromatography (RP-HPLC). Fractions that contain mixtures of HNP-1 and -3 or HNP-2 and -3 are also pooled separately, concentrated, and desalted by RP-HPLC.

Reversed-Phase High-Performance Liquid Chromatography

Two different-size Vydac C_{18} columns (The Separations Group, Hesperia, CA) are used for RP-HPLC purification of defensin peptides. Concentrated Sephacryl S-200 fractions containing mainly HNP-4, the least abundant defensin peptide, are purified on a 4.6 × 250-mm C_{18} column, using a water–acetonitrile gradient that contains 0.1% trifluoro-

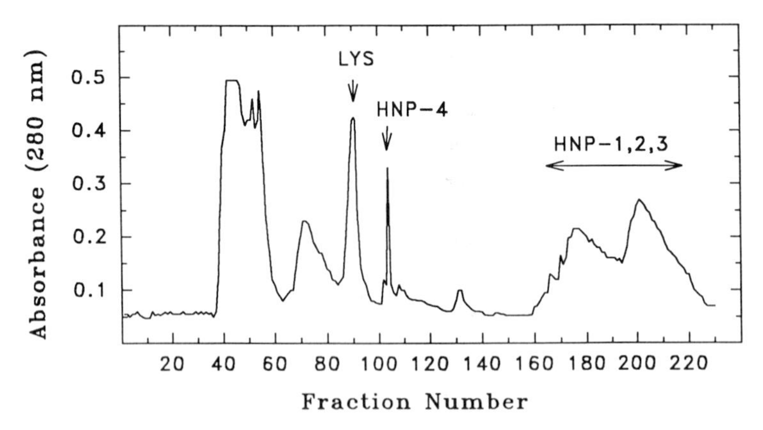

FIG. 2. Sephacryl S-200 chromatography of a human neutrophil granule extract. The fractions containing lysozyme (Lys) and the human defensin peptides HNP-1 through -4 are shown by arrows. Reprinted with permission from S. S. L. Harwig, A. S. K. Park, and R. I. Lehrer, *Blood* **79,** 1532 (1992).

acetic acid (TFA) in both solvents and that increases by 1% (v/v) per minute in acetonitrile. HNP-4 elutes consistently at 39% acetonitrile, and is well separated from human PMN lysozyme and the small amounts of the other defensin peptides or defensin precursors[9] that coelute with HNP-4 on a Sephacryl S-200 column.

Sephacryl S-200 fractions that contain HNP-1 and -3 or HNP-2 and -3 are initially purified on a 10 × 250-cm C_{18} column at a flow rate of 1.5 ml min^{-1}, with 0.1% TFA as an ion-pairing agent. Although HNP-1, -2 and -3 all elute at approximately 40% acetonitrile on this C_{18} column when a linear H_2O–acetonitrile gradient of 1% per minute is used, the step is effective in eliminating nondefensin contaminants.

Hydrophilic Interaction High-Performance Liquid Chromatography

Although defensins HNP-1, -2, and -3 differ only with respect to their N-terminal amino acid residue (see Fig. 1), the amino-terminal aspartic acid in HNP-3 renders it less cationic than HNP-1 or -2. We previously employed a Bio-Sil TSK-CM-3-SW cation-exchange HPLC column (Bio-Rad Laboratory, Richmond, CA) to separate HNP-3 from HNP-1 and -2, using 50 m*M* sodium phosphate, pH 6.5, with an increasing NaCl gradient. Although this was very effective, the column's performance deteriorated over time (months), presumably because of the inherent instability of the silica at near neutral pH. Consequently, we currently prefer using hydrophilic interaction (HI) HPLC with a 4.6 × 200-mm polyhydroxyethylaspartamide column (Poly LC, Columbia, MD). This technique uses a nonpolar mobile phase that retains peptides in proportion to their hydrophilicity, and subsequently elutes the peptides by increasing the polarity of the mobile phase. We have used this column to separate up to 3 mg of a mixture of HNP-2 and -3 or HNP-1 and -3. Peptides are suspended in 10 m*M* triethylamine phosphate (TEAP), pH 4.85 with 64% acetonitrile and loaded onto the column that has been well equilibrated with the same solvent at a flow rate of 1 ml min^{-1}. A decreasing acetonitrile gradient of 1% per minute is then used to elute the peptides. HNP-3 emerges first, followed by HNP-2 or -1. A representative HI-HPLC separation of HNP-2 and -3 is shown in Fig. 3. Fractions containing well-separated defensin peptides are concentrated by vacuum centrifugation to remove acetonitrile and subsequently desalted on a C_{18} column. Neither HI-HPLC nor cation-exchange HPLC allowed us to separate HNP-1 from its nearly identical congener HNP-2. The fractions containing HNP-1 and -2 can be used as such or recycled through the Sephacryl S-200 step to resolve them further. Figure 4 summarizes the overall purification process.

[9] S. S. L. Harwig, A. S. K. Park, and R. I. Lehrer, *Blood* **79,** 1 (1992).

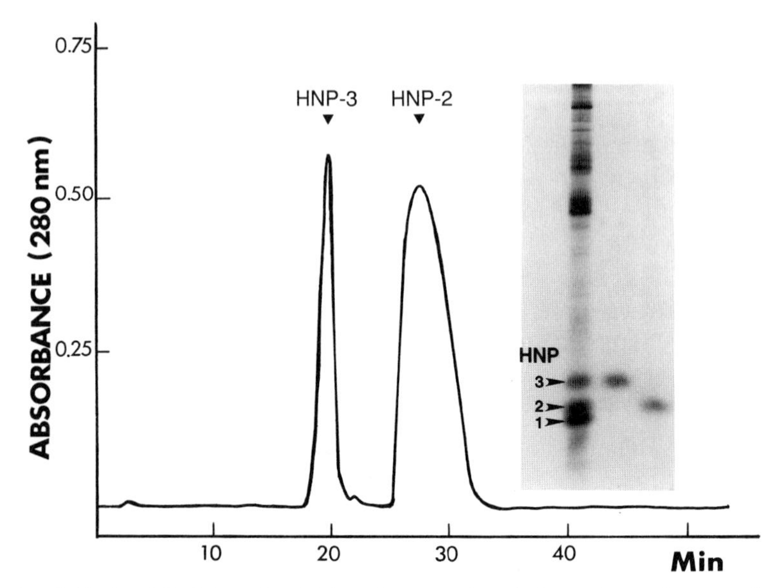

FIG. 3. Hydrophilic interaction high-performance liquid chromatographic separation of HNP-2 and HNP-3. Concentrated Sephacryl S-200 fractions containing only HNP-2 and HNP-3 were subjected to RP-HPLC purification, then lyophilized and dissolved in 10 m*M* TEAP, pH 4.85, with 64% acetonitrile and loaded onto a polyhydroxyethylaspartamide column equilibrated with the same solvent. The peptides were eluted with a 1% per minute gradient of acetonitrile (ACN), from 64% ACN to 0%, at a flow rate of 1 ml min^{-1}. The well-resolved peptides were desalted by RP-HPLC. Inset: a 12.5% AU-PAGE gel shows crude granule extract containing HNP-1, -2, and -3 (arrows) and HI-HPLC-purified HNP-2 and -3.

Identification and Quantitation of Purified Defensins

Peptide purity is assessed by 16.5% tricine sodium dodecyl sulfate–polyacrylamide gel electrophoresis (SDS–PAGE), using a minigel apparatus (Hoefer Scientific Instruments, San Francisco, CA), by 12.5% AU-PAGE, and by analytical RP-HPLC. The individual defensins can be identified by their characteristic migration on AU-PAGE (see Fig. 3) or by performing N-terminal amino acid sequencing. Quantitative amino acid analysis, performed by RP-HPLC on their phenylthiocarbamyl derivatives, also allows the precise identification of purified human defensins and provides a useful method for determining their concentrations by measuring the amounts of Ala, Glu, Phe, Gly, Arg, Thr, and Pro residues. We subject the defensins to a 40-hr vapor phase hydrolysis in 5.7 *N* HCl

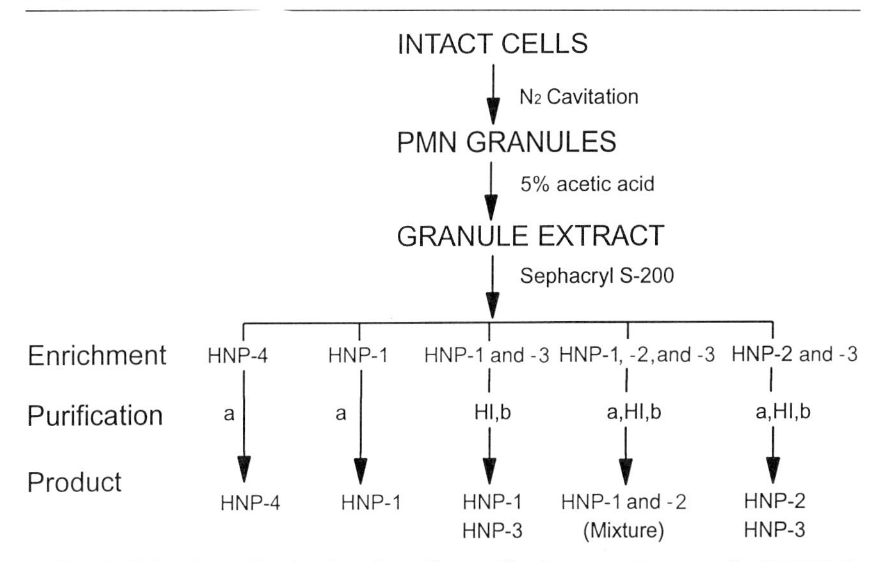

FIG. 4. Defensin purification flowsheet. For purification procedures: (a) C_{18} RP-HPLC; (HI) hydrophilic interaction HI-HPLC; (b) desalting by RP-HPLC. The relative amounts of defensins recovered in these fractions were approximately for HNP-4, 1%; for HNP-1, 20%; for HNP-1 and -3, 8%; for HNP-1, -2, and -3, 31%; and for HNP-2 and -3, 40%.

in vacuo, derivatize them with phenyl isothiocyanate,[10] and use the Pico tag column system (Millipore Waters, Milford, MA) for the amino acid analysis. Because we have observed that the slopes of serially diluted human defensin standards differ appreciably from those obtained with albumin or lysozyme standards, we cannot recommend either the BCA or Bradford method for quantitating defensins; however, the concentrations of purified defensins can be estimated by measuring their optical density at 280 nm[11] or by determining the difference between their optical densities at 215 and 225 nm.[12,13] Table I shows the theoretical and observed absorbances for solutions of purified human defensins prepared in our laboratory.

Antimicrobial Testing

As defensins have a broad spectrum of antimicrobial activity, various methods can be used to test them. We describe two procedures: conven-

[10] B. A. Bidlingmeyer, S. A. Cohen, and T. L. Tarvin, *J. Chromatogr.* **336,** 93 (1984).

[11] H. Edelhoch, *Biochemistry* **6,** 1948 (1967).

[12] W. J. Waddell, *J. Lab. Clin. Med.* **48,** 311 (1956).

[13] P. Wolf, *Anal. Biochem.* **129,** 145 (1983).

TABLE I
SPECTROPHOTOMETRIC MEASUREMENT OF DEFENSIN CONCENTRATIONS

Defensin	Concentration ($\mu g/ml^{-1}$)	A_{280}		Concentration ($\mu g/ml^{-1}$)	$A_{215} - A_{225}$	
		Calculated	Observed		Calculated	Observed
HNP-1	250[a]	0.794[b]	0.829	50	0.347[c]	0.296
HNP-2	250	0.811	0.909	50	0.347	0.316
HNP-3	250	0.784	0.816	50	0.347	0.302
HNP-4	250	0.128	0.143	50	0.347	0.446

[a] Determined by quantitative amino acid analysis.
[b] The theoretical absorbance at 280 nm (A_{280}) was calculated from the molar extinction coefficients of tryptophan, tyrosine, and cysteine, per the data.[11] Defensins were dissolved in 0.01% acetic acid for this measurement, but virtually identical results were obtained when the peptides were dissolved in 0.1 *M* ammonium acetate, pH 7.0, or 6 *M* guanidine hydrochloride, pH 4.7.
[c] Calculated by the method of Waddell.[12] Samples were dissolved in 0.01% acetic acid.

tional colony counting and a recently devised radial diffusion assay that is very sensitive and economical.[5] Although *Escherichia coli* ML-35 is used to illustrate the assays, both are more broadly applicable.

Defensins, whether purchased or prepared, are expensive reagents. Consequently, we have minimized their consumption by using small assay volumes and favorable conditions (near-neutral pH and low concentrations of divalent cations or salts). Human defensins show optimal activity against metabolically active organisms; thus, we also use mid-logarithmic-phase organisms to enhance sensitivity.

Colony Count Assay

Equipment

1. Inoculating turntable (Fisher 08-758-10).
2. Bacteriological incubator.
3. Autoclave.
4. Quebec Dark-Field Colony Counter (Scientific Products C8374-5).
5. Laboratory vortex mixer (Scientific Products S8223-1).
6. Bunsen burner.
7. Gilson Pipetman set (Models P-1000, P-200, and P-20, Rainin, Woburn, MA).
8. Shaking water bath (Model G76, New Brunswick Scientific, Edison, NJ).

Materials

1. Prepared trypticase soy agar (TSA) plates. These can be purchased from commercial suppliers or made from the powdered medium (BBL 11043, Beckton Dickinson, Cockeysville, MD).
2. Trypticase soy broth (TSB) BBL 11768, Beckton Dickinson, Cockeysville, MD). The full-strength broth contains 3.0 g of powdered medium per 100 ml of distilled water and is sterilized by autoclaving.
3. A bent glass rod.
4. Sodium phosphate buffer 10 m*M*, pH 7.4.
5. Sterile 12 × 75-mm capped plastic tubes.
6. Sterile 10 m*M* sodium phosphate buffer (NAPB), pH 7.4.
7. Pipette tips, autoclaved (Fisher Redi-tips, Catalog 21-197-8A and 21-197-8H).
8. Approximately 10 ml of acetone in a covered, glass petri dish.

Defensins

Stock solutions are prepared in sterile 0.01% acetic acid at 1 mg ml^{-1} in 12 × 75-mm sterile polystyrene tubes (Fisher Scientific 14-956-3D). The solutions retain activity for months at −20°. A solution of 0.01% acetic acid serves as a control.

Bacteria

Although we have most often used *E. coli* ML-35 in our studies,[14] *E. coli* K12 strains and *E. coli* ATCC 29648 (American Type Culture Collection, Rockville, MD) are suitable alternatives. The bacteria are cultured in full-strength TSB overnight (18–24 hr) at 37° in a shaking water bath, to obtain stationary-phase organisms. An aliquot, 50 μl, of the overnight culture is transferred to 50 ml of fresh full-strength TSB and incubated for 2.5 hr at 37° with shaking to obtain mid-logarithmic-phase cells. A portion, 25 ml, of the culture is centrifuged for 10 min at 2000 rpm (approximately 800 g) at 4°. After the supernatant is discarded, the bacterial pellet is resuspended in cold, sterile buffer and washed again at 800 g for 10 min at 4°. The pellet is resuspended in 5 ml of buffer, and 1 ml is transferred to a cuvette and its optical density is measured at 620 nm. If the OD_{620} is not between 0.3 and 1.0, the parent sample should be concentrated or another aliquot taken and diluted, so that the OD falls between these values. An OD_{620} of 0.20 corresponds to approximately 5×10^7 *E. coli* colony-forming units (cfu) ml^{-1}.

[14] R. I. Lehrer, A. Barton, and T. Ganz, *J. Immunol. Methods* **108,** 153 (1988).

Procedure

Dilute an aliquot of the washed, mid-logarithmic-phase bacteria to a concentration of 5×10^6 CFU/ml^{-1} in cold NAPB, keep on ice, and mix (vortex) before using. Prepare the assay medium by adding 100 μl of full-strength TSB to 6.9 ml of NAPB and warming this to 37° in a water bath. Prepare working defensin solutions in 0.01% acetic acid from the 1 mg ml^{-1} stock solution at 10 times the final desired concentrations; include a defensin-free control. Combine 35 μl of assay medium with 5 μl of defensin solution and 10 μl of bacterial suspension in a sterile 12×75-mm plastic tube, mix gently, and incubate for 2 hr in a 37° water bath. To stop the reaction, add 450 μl of ice-cold 0.15 *M* NaCl. This is called sample A. Prepare 10-fold serial dilutions by vortex mixing sample A and then transferring 50 μl to a tube containing 450 μl of cold buffer (sample B); repeat this process to prepare a further tenfold dilution (sample C). Place a TSA plate on the inoculating turntable, remove 25 μl from sample A, B, or C, and spread it evenly over the agar surface using the bent glass rod. Preparation of duplicate or triplicate plates from each dilution is recommended. Sterilize the rod between uses by dipping it into the acetone and passing it briefly through the Bunsen burner flame to ignite the acetone. The rod should *not* get hot. If the acetone in the petri dish accidently catches fire, merely cover it with its lid to extinguish the flames.

Incubate the plates (upside down) at 37° for 24 hr or until full colony development is obtained. Count only those plates containing up to 500 per plate. Calculate the mean number of colonies per plate and multiply this by the dilution to convert the data to colony-forming units per milliliter in the original 50-μl incubation mixture. With the suggested dilutions, the factors are for sample A, 400; for sample B, 4000; and for sample C, 40,000. Defensin-free controls (0.01% acetic acid) should be obtained at "zero time" and after 2 hr of incubation. The experimental results should be expressed as cfu ml^{-1} or as a percentage of *input* cfu ml^{-1} in the control. If the results are expressed as a percentage of the incubated control, information about input and control cfu ml^{-1} should always be provided.

Radial Diffusion Assay

Equipment

1. Laboratory hot plate/stirrer (Corning Model PC 351).
2. Autoclave.
3. Microwave oven.

4. A 37° bacterial incubator.
5. A 3-mm-internal-diameter agar punch or cork borer.

Materials

1. Agarose [Sigma, St. Louis, MO, Type I (low electroendosmosis)].
2. Na_2HPO_4 100 m*M*.
3. Citric acid 100 m*M*.
4. Trypticase soy broth powdered medium (BBL 11768, Beckton Dickinson, Cockeysville, MD).
5. Acetic acid 0.01% (sterile).
6. Tween 20 (Sigma, St. Louis, MO, Catalog P-1379).
7. Plastic petri dishes 100 × 100 × 15 mm square (Scientific Products, Catalog D-1936).
8. Glass media bottles with screw caps, Wheaton type, 125 and 250 ml (Scientific Products 87574-125 and 250).
9. Measuring magnifier, 7× (Bausch and Lomb, Catalog 81-34-35).
10. A small bubble level (obtainable from many local hardware stores).

Bacteria

Washed, mid-logarithmic-phase *E. coli* ML-35 (see above).

Overlay and Underlay Agars

Prepare these in advance. For each 100 ml of overlay agar, add 6 g trypticase soy broth powder and 1 g of agarose to 100 ml of distilled water in a foil-covered beaker, and stir on a heating plate until the contents dissolve. Place the agar into a Wheaton bottle and autoclave it. Store the sterile overlay agar at room temperature until needed. Melt it in a microwave oven prior to use and allow it to cool in a 43° waterbath so that it remains liquid.

Because it contains viable mid-logarithmic-phase bacteria, final assembly of the underlay agar is done immediately before use. The other components of the underlayer are assembled in advance, autoclaved, and dispensed into Wheaton bottles until needed. The ingredients of the underlay agar include dilute trypticase soy broth (30 mg of TSB powder/100 ml), 10 m*M* sodium phosphate or sodium citrate/phosphate buffer, and 1% (w/v) agarose. The pH of the agar should be adjusted prior to autoclaving. The following example describes the preparation of a pH 6.5 citrate/phosphate underlay agar.

Prepare stock Na_2HPO_4/citrate buffer at pH 6.0 by mixing 100 m*M* Na_2HPO_4 with 100 m*M* citric acid in a beaker while monitoring the pH.

For each 100 ml of agar, combine 10 ml of the stock buffer with 89 ml of distilled water in an uncovered beaker and add 1 ml of full-strength TSB solution and 1 g agarose. Adjust the final pH to 6.5 with 15% NaOH. Dissolve the agarose on a hot plate/stirrer or in a microwave oven, distribute it into 125-ml Wheaton-type bottles, and autoclave. Prior to the experiment, melt the underlay agar in a microwave oven and place it in a 43° water bath. If the addition of Tween 20 (polyoxyethylene sorbitan monolaurate) in the underlayer is desired, it should be added at this point by adding 200 μl of a filter-sterilized 10% Tween 20 solution per 100 ml of underlayer agar. Tween 20 (0.02% final concentration) is not antibacterial and it markedly enhances the bactericidal activity of low concentrations of HNP-4. Tween-containing agars cannot be remelted in the microwave without decomposing (color changes).

To prepare the assay plate, add 1×10^6 mid-logarithmic-phase bacteria (see above) to a sterile 15-ml culture tube. Then, add 10.0 ml of the bacterial underlay agar which has been maintained at 43°. Vortex the tube to mix the bacteria uniformly within the agar and pour the mixture into the square plastic petri dish, taking care that the dish remains level in both horizontal dimensions. Allow the agar to set (about 2 min). After tracing the outline of the petri dish on a piece of centimeter-ruled graph paper and placing 16 dots in an evenly spaced 4×4 array within its outline, place the marked graph paper template beneath the petri dish. Punch out the spots with a 3-mm gel punch and remove the cores by vacuum aspiration or with a bent needle.

Prepare fresh defensin dilutions (usually 50, 25, 10, and 5 μg ml^{-1}) in sterile 0.01% acetic acid from the 1 mg ml^{-1} stock solution. Add 5 μl of defensin samples or unknowns in 0.01% sterile acetic acid to the appropriate well and incubate the plate upside down for 3.0 hr at 37°. After 3 hr, pour 10.0 ml of the molten (43°) overlay agar on top of the bottom agar and allow it to solidify at room temperature, which occurs rapidly. Once firm, incubate the plate upside down overnight in an incubator at 37°.

The following day, after the surviving organisms have formed microcolonies, use a magnifying ruler to measure the diameter of the clear zone surrounding the well, in millimeters and subtract from this the diameter of the well (3.0 mm). The zone of clearing is expressed in units: 10 units = 1 mm. Units of clearing are related to defensin concentration in a log-linear manner. With *E. coli,* the assay easily detects the presence of less than 20 μg ml^{-1} defensins in a 5-μl sample (<100 ng of defensin). Other bacteria, such as *Mycobacterium fortuitum,* are substantially more sensitive indicators.

[14] Purification and Assay of Bactericidal/Permeability-Increasing Protein

By JERROLD WEISS

Introduction

The defense of the host against invading microorganisms, especially against bacteria capable of multiplying rapidly in extracellular fluids, requires the mobilization of polymorphonuclear leukocytes (neutrophils) (PMNs) to sites of infection. Among the essential properties of these highly specialized professional phagocytic cells is the elaboration of an array of cytotoxic substances that can arrest microbial multiplication and eliminate viable microorganisms. These substances include peptides and proteins stored in cytoplasmic granules after synthesis in the bone marrow during cell differentiation and oxidants formed *de novo* during the respiratory burst that accompanies PMN–microbe interactions.[1,2] On the one hand, it is likely that the optimal function of PMNs against a given microorganism depends on the integrated action of multiple antimicrobial systems. On the other hand, it is also apparent that the action of particular individual components of the antimicrobial arsenal of PMNs reflects their ability to target specific microorganisms. For example, phagocytosis of the gram-positive bacterium *Staphylococcus epidermidis* is accompanied by O_2 (oxidant)-dependent inhibition of bacterial viability (i.e., colony-forming ability in nutrient agar) and biosynthetic activity. In contrast, phagocytosis of many gram-negative bacteria (e.g., *Escherichia coli*) is accompanied by equally rapid arrest of bacterial multiplication without impairment of the bacterial metabolic machinery and without an apparent role of cytotoxic oxidants in this action.[1,3] Effects on viability of ingested bacteria are accompanied by discrete alterations of the gram-negative bacterial outer membrane, including an increase in permeability to small hydrophobic compounds that are normally excluded by the outer membrane.[4] PMN homogenates and extracts reproduce these effects. These findings

[1] P. Elsbach and J. Weiss, *in* "Inflammation: Basic Principles and Clinical Correlates" (J. I. Gallin, I. M. Goldstein, and R. Snyderman, eds.), p. 603. Raven Press, New York, 1992.

[2] S. J. Klebanoff, *in* "Inflammation: Basic Principles and Clinical Correlates" (J. I. Gallin, I. M. Goldstein, and R. Snyderman, eds.), p. 541. Raven Press, New York, 1992.

[3] B. A. Mannion, J. Weiss, and P. Elsbach, *J. Clin. Invest.* **86,** 631 (1990).

[4] S. Beckerdite, C. Mooney, J. Weiss, R. Franson, and P. Elsbach, *J. Exp. Med.* **140,** 396 (1974).

prompted the search for preformed constituent(s) of PMNs with bactericidal and/or permeability-increasing activity toward *E. coli* and other gram-negative bacteria and led to the isolation of the bactericidal/permeability-increasing protein (BPI) from both human and rabbit PMNs.

Isolation of Native Bactericidal/Permeability-Increasing Protein

Starting Material

To date, BPI has been detected only in PMNs and PMN precursors and has been isolated from human, rabbit, and cow PMNs. The BPI content of these cells is ca. 60 μg/10^8 PMNs.[5,6] Thus, greater than milligram quantities of BPI can be readily obtained from suitable starting material. Approximately 10^9 PMNs/animal can be collected in nearly homogeneous form from acute inflammatory peritoneal exudates of (New Zealand White) rabbits 6 to 15 hr after intraperitoneal injection of 250 to 300 ml of sterile physiological saline supplemented with glycogen (ca. 1–2 mg/ml).[7] PMNs can also be isolated from peripheral blood. Blood collected in the presence of anticoagulant (e.g., 14 units/ml of heparin) is mixed 1 : 1 (v/v) with 2% (w/v) dextran T-500 (Pharmacia, Piscataway, NJ). After sedimentation of red blood cells by gravity, the white blood cells (WBCs) are collected from the red cell-depleted upper phase and washed with physiological saline. Remaining red cells can be lysed by incubation of cells in 10 vol of 0.87% (w/v) ammonium chloride at 37° for 10 min; WBCs are recovered by sedimentation at ca. 100 *g* for 5 min and washed twice with saline. For purposes of purification of BPI, separation of PMNs from other WBCs is unnecesary. Recovery of PMNs is ca. 75% (approx 3 to 5 $\times$ 10^8 PMNs recovered/100 ml of blood). Leukaphoresis of patients with chronic myeloid leukemia can provide much larger numbers of BPI+ cells for purification (up to 10^{11}) provided the cells are at or beyond the promyelocytic stage of differentiation.[5]

Extraction of BPI and Purification by Gel Filtration, Ion-Exchange, and/or Reversed-Phase Chromatography

Like many other antimicrobial peptides and proteins of PMNs, BPI is highly cationic (p*I* > 9.5) and localized (mainly) in cytoplasmic granules.

[5] J. Weiss and I. Olsson, *Blood* **69,** 652 (1987).

[6] C. E. Ooi, J. Weiss, O. Levy, and P. Elsbach, *J. Biol. Chem.* **265,** 15956 (1990).

[7] P. Elsbach and I. L. Schwartz, *J. Gen. Physiol.* **42,** 883 (1959).

BPI can be extracted from isolated granules or from whole cell lysates.[8,9] Granule-rich fractions can be prepared by disruption of cells at 0° to 4° under osmotically protective conditions and differential centrifugation. Two different methods of cell lysis have been used: (1) Homogenization of PMNs in 0.34 *M* sucrose (up to 3×10^8 PMNs/ml) using a Potter–Elvehjem homogenizer with glass mortar and motor-driven Teflon-coated pestle[8]; (2) N_2 cavitation of PMNs (up to 2×10^8 PMNs/ml) equilibrated in a high-pressure bomb (Parr Inst. Co., Moline, IL) at 400 psi in relaxation buffer minus ethylene glycol bis(β-aminoethyl ether) *N,N'*-tetraacetic acid [EGTA: 100 m*M* KCl, 3 m*M* NaCl, 1 m*M* $ATP(Na)_2$, 3.5 m*M* $MgCl_2$, 10 m*M* PIPES, pH 7.3].[5] Disruption of cells is more uniform by the latter procedure and generally gives higher yields of intact granules. Low-speed pellets (400 g, 10 min) containing residual intact cells as well as heavy debris may be recycled to increase the yield of granules that are sedimented by centrifugation of the 400 *g* supernatant for 20 min at 27,000 *g*. BPI is more tightly membrane associated than other PMN granule proteins[5] and is not extracted efficiently by 0.05 *M* glycine hydrochloride buffer (pH 2.0), which has been used to extract other granule proteins.[10] Freeze–thawing (7×) of the granule-rich fraction followed by incubation with stirring for 3 hr at 0° to 4° with 0.2 *M* sodium acetate/acetic acid buffer, pH 4.0 (2×10^9 cell equivalents/ml), is most effective. Extracted proteins are recovered in the supernatant after centrifugation at 20,000 *g* for 20 min. At this stage, human BPI can be purified to apparent homogeneity in two steps: (1) gel filtration with Sephadex G-75 in the above buffer, and (2) ion-exchange chromatography of BPI-rich fractions on SP-Sephadex. BPI elutes from Sephadex G-75 just after the void volume corresponding to its apparent M_r of 50,000 to 60,000 and is eluted from SP-Sephadex with ca. 0.75 *M* NaCl buffered with 0.2 *M* acetate, pH 4.0. Eluted BPI can be readily detected by measurement of bacterial growth-inhibitory activity using a rough strain of *E. coli* such as *E. coli* J5 as test organism (see below). During both chromatographies, fractions containing BPI represent per unit protein by far (>10×) the most potent fractions. Unless noted otherwise, all purification steps are performed at 0° to 4°. Purified BPI is most stable by storage at 0° to 4° in weak acidic buffer (e.g., 20 m*M* acetate buffer, pH 4.0). Lyophilization and freeze–thawing of purified BPI should be avoided.

[8] J. Weiss, R. C. Franson, S. Beckerdite, K. Schmeidler, and P. Elsbach, *J. Clin. Invest.* **55,** 33 (1975).

[9] J. Weiss, P. Elsbach, I. Olsson, and H. Odeberg, *J. Biol. Chem.* **253,** 2664 (1978).

[10] J. E. Gabay, J. M. Heiple, Z. A. Cohn, and C. F. Nathan, *J. Exp. Med.* **164,** 1407 (1986).

Because recovery of granules in granule-rich fractions is incomplete, greater yields of BPI can be obtained by acid extraction of whole PMN homogenates. Whereas isolation of granule-rich fractions requires working with fresh cells, extraction of whole cell lysates can be done with stored (at $-20°$) cell pellets, permitting accumulation of large amounts of starting material before initiating purification. Most effective is homogenization of cells in distilled water (ca. 3×10^8 PMNs/ml) followed by addition of 2/3 vol of 0.4 N H_2SO_4 (0.16 N H_2SO_4, final concentration) and extraction with stirring for 30 min. Extracted proteins are recovered in the supernatant following centrifugation at 23,000 g for 20 min and equilibrated to higher pH by dialysis. Dialysis is accompanied by progressive accumulation of (protein) precipitates as the pH of the extract is raised. Rabbit PMN extracts can be neutralized to pH 7 by dialysis vs 1 mM Tris–HCl, pH 7.2, with little or no loss of BPI [or granule-associated "p15s" and phospholipase A_2 (PLA2); see below] but a fivefold reduction in total protein. The remaining soluble protein is separated from insoluble material by centrifugation at 23,000 g for 20 min (Sup II). The reduced complexity of this recovered soluble protein fraction and the unique properties of BPI permit isolation of rabbit BPI from Sup II in one step in greater than 70% yield by reversed-phase HPLC on a Vydac C_4 column (The Separations Group, Hesperia, CA) using a linear gradient of acetonitrile (0–95%, v/v) in 0.1% trifluoroacetic acid (TFA) developed over 30 to 60 min at a flow rate of 1 ml/min. Essentially all recovered *E. coli* growth-inhibitory activity is present in the last eluting protein peak (ca. 70% acetonitrile) which corresponds to purified BPI. Recovery of bioactive BPI requires prompt dialysis of eluted protein vs (dilute) acidic buffer.

In contrast to sulfuric acid extracts of rabbit PMNs, human PMN extracts can not be dialyzed to pH greater than 4.0 without losing BPI by coprecipitation with other precipitating substances during dialysis to higher pH. Purification of human BPI by conventional chromatographic procedures from this more complex protein mix is more difficult and less efficacious.

Purification of BPI by Using Escherichia coli as Affinity Adsorbent

These problems can be circumvented by taking advantage of the reversible, high-affinity binding ($K_d \sim 10$ nM) of BPI to rough strains of many gram-negative bacteria such as *E. coli*.[11] These binding properties reflect the strong attraction of BPI for gram-negative bacterial lipopolysaccharides (LPSs), i.e., both LPSs residing in the bacterial envelope and cell-

[11] B. A. Mannion, E. S. Kalatzis, J. Weiss, and P. Elsbach, *J. Immunol.* **142,** 2807 (1989).

free, isolated LPSs (see below). The restriction of LPSs to the unique outer membrane of gram-negative bacteria apparently accounts for the selectivity as well as the potency of BPI action toward gram-negative bacteria. The binding site for BPI likely includes anionic moieties in and adjacent to the lipid A region that is highly conserved in LPSs.[11-13] In the bacterial envelope, this site is at the membrane interface and less accessible to BPI when LPS molecules contain long polysaccharide chains (i.e., "smooth" strains)[12] (Fig. 1). Thus, BPI has lower affinity for these strains.[12,14] In contrast, the presence of an external but more porous capsular layer does not impede BPI interactions with the bacterial outer membrane.[15,16] The anionic groups in/near the lipid A region normally complex divalent cations (Mg^{2+} and Ca^{2+}) which are needed to reduce electrostatic repulsion between neighboring LPS molecules and thereby permit the high density of the LPS in the outer leaflet of the outer membrane of gram-negative bacteria. At saturation, up to 2×10^6 molecules of BPI are bound per bacterium within 15 min of incubation at 37°.[11] Optimal or near-optimal binding of BPI to *E. coli* occurs at 20° to 37° over a broad range of pH (4.0–7.5) and salt concentration (0 to ≤300 m*M* NaCl) and in the presence of physiological concentrations of Mg^{2+} and Ca^{2+} and high concentrations of other leukocyte proteins. Thus, BPI in dialyzed (pH 4.0) acid extracts of human PMNs or mixed leukocyte preparations (e.g., from a surgically removed spleen of a patient with chronic myeloid leukemia) binds readily to rough *E. coli* under these conditions and, in fact, at apparent saturation, is the only prominent leukocyte protein recovered in the bacterial pellet after washing to remove unbound or weakly bound proteins[11] (Fig. 2). Typically, we have used late-log-phase *E. coli* J5 (R_c chemotype of LPS), washed once with 20 m*M* acetate buffer, pH 4.0, and then resuspended with leukocyte extract diluted 1 : 1 with acetate buffer to a final bacterial concentration of 5×10^8 *E. coli*/ml. Incubations are carried out at 37° for 15 min with shaking. Bound and unbound proteins are separated by sedimentation of bacteria at 6000 g for 8 min and the bacteria are washed twice with 0.5 vol of acetate buffer. Treatment of bacteria (5×10^9 bacteria/ml) for 10 min at 37° with acetate buffer supplemented with 0.2 *M* $MgCl_2$ causes release of more than 80% of bound BPI with no detectable release of bacterial protein. To ensure purity, BPI in the

[12] J. Weiss, S. Beckerdite-Quagliata, and P. Elsbach, *J. Clin. Invest.* **65,** 619 (1980).

[13] H. Gazzano-Santoro, J. B. Parent, L. Grinna, A. Horwitz, T. Parsons, G. Theofan, P. Elsbach, J. Weiss, and P. J. Conlon, *Infect. Immun.* **60,** 4754 (1992).

[14] J. Weiss, M. Hutzler, and L. Kao, *Infect. Immun.* **51,** 594 (1986).

[15] J. Weiss, M. Victor, A. S. Cross, and P. Elsbach, *Infect. Immun.* **38,** 1149 (1982).

[16] J. Weiss, P. Elsbach, C. Shu, J. Castillo, L. Grinna, A. Horwitz, and G. Theofan, *J. Clin. Invest.* **90,** 1222 (1992).

a

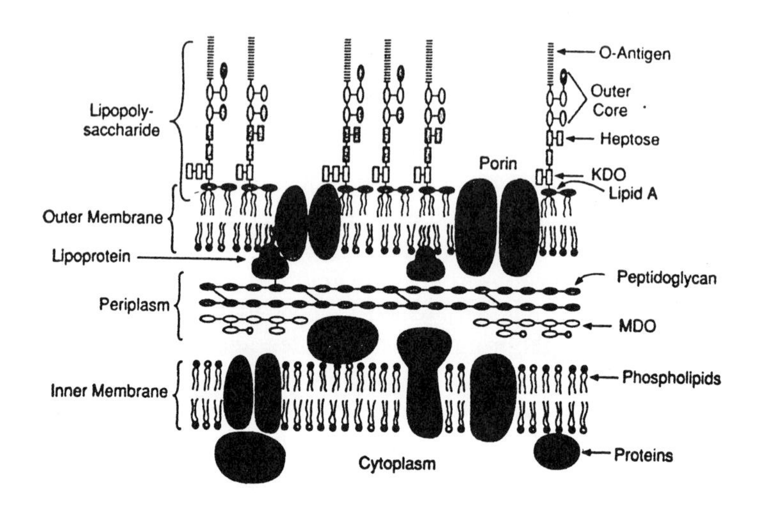

b

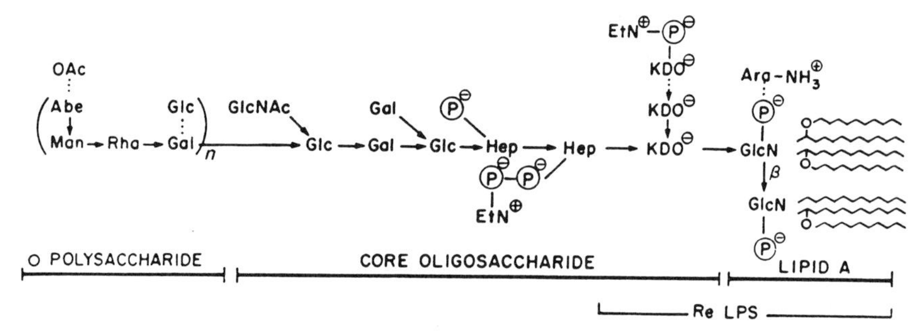

FIG. 1. Schematic representation of the envelope of gram-negative bacteria (a) and the chemical structure of gram-negative bacterial (*Salmonella*) lipopolysaccharide (b). KDO, Ketodeoxyoctanoate; MDO, membrane-derived oligosaccharides; GlcN, glucosamine; GlcNAc, *N*-acetylglucosamine; Ara-NH_3, aminoarabinose, EtN, ethanolamine; Hep, heptose; Glc, glucose; Gal, galactose; Rha, rhamnose; Man, mannose; Abe, abequose.

Mg^2 eluate is further purified by reversed-phase (C4) HPLC and promptly dialyzed as indicated above; human BPI elutes with ca. 65% acetonitrile. Affinity-purified BPI and BPI purified by gel filtration and ion-exchange chromatography are structurally identical and display equivalent bioactivi-

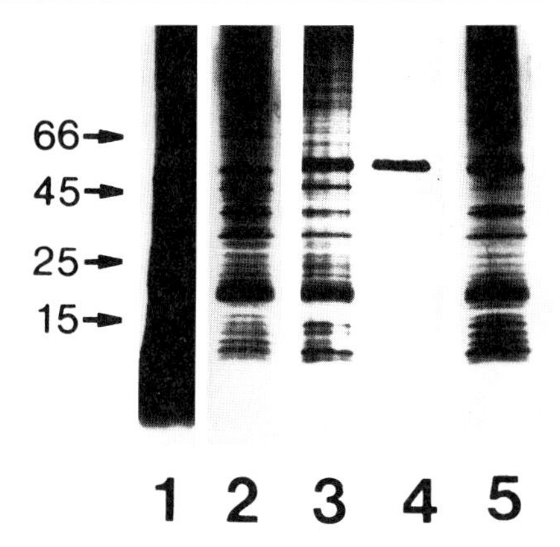

FIG. 2. One-step affinity purification of human BPI from crude leukocyte acid extracts by reversible binding to *E. coli* J5. Indicated samples were resolved by SDS–PAGE and stained with Coomassie blue. Lane 1 contains amount of leukocyte (WBC) extract added to the number of bacteria shown in lanes 2, 3, and 5. Lane 2, *E. coli* alone; lane 3, *E. coli* + extract, washed pellet; lane 4, 0.2 *M* $MgCl_2$ eluate of lane 3; 5, *E. coli* pellet after Mg^{2+}.

ties, including LPS-neutralizing activity (see below). No endotoxin is detectable (<10 pg/ml) in affinity-purified BPI.

This procedure has also been successfully employed to isolate rabbit[6] and cow BPI and should be useful for scanning other species/tissue sites in which BPI(-like proteins) have not yet been identified. Other proteins are present in leukocyte extracts that can bind *E. coli* when bacterial surface sites are not limiting. Thus, by manipulating the extract/bacteria ratio, other *E. coli*-binding proteins can be recovered. In this way, we have copurified novel p15 isoforms and the two most cationic (of six) defensin species from rabbit PMN extracts together with BPI. These proteins can then be separated by reversed-phase HPLC on a C4 column: elution of defensins NP-1 and -2, p15s A and B, and BPI occurs with 42, 45, 58, and 70% acetonitrile, respectively.

Isolation of (Bioactive) Fragments of Human BPI Generated by Limited Proteolysis

Some preparations of purified BPI show the gradual appearance of two fragments (ca. 25 and 30 kDa) during storage over months at 4° in very weak acidic buffer. This process is enhanced during concentration of the protein by ultrafiltration (Centricon 10, Amicon Corp., Lexington,

MA) and markedly accelerated by incubation at 37° in neutral pH buffer.[17] Treatment of purified BPI with the serine protease inhibitor phenylmethylsulfonyl fluoride (PMSF; 0.4 m*M* in 25 m*M* HEPES, pH 7) or the specific elastase inhibitor MeOSuc-Ala–Ala–Pro–Val-CH_2Cl (AAPVCK) prevents fragmentation. Incubation of PMSF-treated BPI with purified PMN elastase (as little as 1 part elastase/5000 parts BPI, mol/mol) at neutral pH and 37° reproduces this discrete fragmentation, suggesting that fragmentation of purified BPI during storage is due to trace amounts of contaminating PMN elastase. Fragmentation of BPI by small amounts of protease is seen only after reversed-phase chromatography, suggesting that this procedure causes subtle alteration of BPI conformation, greatly increasing the susceptibility of specific site(s) in BPI to proteolytic attack; no other alteration of BPI structure or function has been observed after reversed-phase chromatography.

The 25- and 30-kDa BPI fragments can be separated from residual holo-BPI and purified by reversed-phase HPLC (C4): The 25-kDa fragment elutes earlier (ca. 53% acetonitrile) and the 30-kDa fragment slightly after (ca. 70% acetonitrile) holo-BPI.[18] Complete separation of the 30-kDa fragment from holo-BPI requires repeated chromatography with a more shallow acetonitrile gradient (0–75% developed over 60 min at a flow rate of 1 ml/min). Edman degradation of the 25- and 30-kDa fragments shows that the NH_2 termini of these two fragments correspond to, respectively, the N terminus and residue 200 (or 203) of holo-BPI. Based on these findings and the apparent molecular weights of the fragments, we deduce that the 25- and 30-kDa fragments comprise residues 1–199 and 200(203)–456 (C terminus) of human BPI.[18]

By all functional criteria (see below), the N-terminal 25-kDa fragment is at least as active as holo-BPI on a molar basis and retains selectivity for gram-negative bacteria, indicating that the molecular determinants for LPS recognition, antibacterial cytotoxicity and endotoxin neutralization (see below), are encoded within this region of BPI.[16–18] In fact, toward smooth bacteria bearing LPSs with long polysaccharide chains, the fragment is up to 30 times more active than holo-BPI, reflecting the easier access of the fragment to binding sites at the membrane interface. The C-terminal fragment has no detectable cytotoxic activity toward bacteria but in certain assays (see below) shows some LPS-neutralizing activity.[18]

[17] C. E. Ooi, J. Weiss, P. Elsbach, B. Frangione, and B. A. Mannion, *J. Biol. Chem.* **262,** 14891 (1987).

[18] C. E. Ooi, J. Weiss, M. E. Doerfler, and P. Elsbach, *J. Exp. Med.* **174,** 649 (1991).

Expression and Purification of Recombinant BPI (Fragment)

To date, BPI mRNA has been detected only in precursors of PMNs. Full-length cDNA encoding human BPI was first cloned from a promyelocyte-like leukemic cell line (HL-60) pretreated for 4 days in culture with 1.3% (v/v) dimethyl sulfoxide to induce further neutrophilic differentiation.[19] Probes and primers derived from this cDNA have been used to isolate cDNA clones encoding rabbit and cow BPI.[20] Comparison of the primary (1°) structure of the three species of BPI deduced from the cDNA clones shows ca. 65% identity between each species and a closely similar overall structural organization. The N-terminal region of each protein corresponding to the bioactive 25-kDa fragment is highly cationic and globally amphipathic, the C-terminal region has little net charge and is very hydrophobic, and a central region (ca. residues 200–240) is rich in hydrophilic residues and prolines which may account for the selective fragmentation of BPI in this region.

Full-length human BPI cDNA (including signal sequence) encoding recombinant holo-BPI and a truncated cDNA with a stop codon inserted after codon 199 encoding recombinant BPI N-terminal fragment [created by mismatched primer polymerase chain reaction (PCR)] have been subcloned into a mammalian expression vector.[13] CHO-K1 cells (Celltech) have been transfected with these plasmids by the method of Wigler *et al.*[21] Secreted recombinant BPI (fragment) can be detected in the culture fluid by immunoblotting and purified in one step by cation-exchange chromatography. The functional properties of recombinant BPI (fragment) are indistinguishable from those of BPI (fragment) isolated from native sources, confirming that the molecular determinants of LPS recognition, cytotoxicity, and endotoxin neutralization are present within the region encompassing residues 1–199.

Assays of Antibacterial Cytotoxicity of BPI

The binding of BPI to susceptible gram-negative bacteria triggers almost immediate (≤1 min) inhibition of bacterial growth, discrete alterations of the bacterial outer envelope, but, initially, little or no disruption

[19] P. W. Gray, G. Flaggs, S. R. Leong, R. J. Gumina, J. Weiss, C. E. Ooi, and P. Elsbach, *J. Biol. Chem.* **264,** 9505 (1989).

[20] S. R. Leong and T. Camerata, *Nucleic Acids Res.* **18,** 3052 (1990).

[21] M. Wigler, S. Silverstein, L. S. Lee, A. Pellicer, Y. C. Cheng, and R. Axel, *Cell (Cambridge, Mass.)* **11,** 223 (1977).

of the intracellular energy-generating and energy-dependent machinery.[22] At this time, most (all?) of the bacterial damage is (potentially) reparable and bacterial growth inhibition is reversible. During longer incubations, bacterial injury extends to the cytoplasmic membrane and includes inhibition of metabolic activity, and bacterial growth inhibition becomes apparently irreversible. The potency of holo-BPI is up to 100-fold greater toward rough than toward smooth strains of susceptible gram-negative bacteria (LD_{90} values of, respectively, $\leq$10 and 1000 n*M* vs 10^6 bacteria/ml). These differences correlate closely with the greater affinity of (holo-)BPI for bacteria containing LPSs with short polysaccharide chains. Thus, the amount of bound BPI required to produce cytotoxic effects is apparently the same in more and less sensitive bacteria; however, the amount of bound BPI required to initiate cytotoxicity can vary appreciably (up to 20-fold) depending on pH[22] and the presence of other PMN granule proteins that while inactive alone can significantly reduce the BPI dose required to produce early sublethal damage.[6] The rate of progression from sublethal to lethal injury can also vary markedly, affected also by pH and the synergistic effects of other proteins (e.g., phospholipases A, late components of the complement system) with BPI that accelerate damage to the bacterial cytoplasmic membrane.[3]

Radiolabeling of BPI to Measure Binding

To facilitate measurement of binding of BPI to bacteria or to isolated LPSs (see below), BPI can be radioiodinated by incubation of 50 to 100 μg of protein for 10 min on ice with 0.2 to 4 mCi of carrier-free $Na^{125}I$ in 0.25 ml of phosphate-buffered saline (PBS, pH 6–7) in tubes coated with 75 μg iodogen (Pierce Chemicals Co., Rockford, IL).[11,13] The extent of labeling of BPI can be increased up to 10-fold (to 10 μCi/μg) by inclusion of 0.05% Tween 20 in the reaction mixture. This is recommended for preparing labeled BPI for use in assays that contain Tween 20 (e.g., LPS binding). Radioiodinated BPI can be separated from free $Na^{125}I$ by gel filtration on Sephadex G-25 equilibrated with 0.2 *M* acetate buffer, pH 4.0, or PBS + 0.05% Tween 20 and then stored at 4° with 1 mg/ml bovine serum albumin (BSA). Alternatively, BPI (fragments) can be radiolabeled with [^{35}S]methionine by *in vitro* translation of BPI mRNA obtained by *in vitro* transcription.[23] For this purpose, BPI cDNA is subcloned downstream of the T3 polymerase promoter site in pT3T7 18U (Pharmacia). The signal sequence encoded in human BPI cDNA can be deleted by digestion with *Dra*II followed by ligation, yielding a construct containing

[22] B. A. Mannion, J. Weiss, and P. Elsbach, *J. Clin. Invest.* **85,** 853 (1990).

[23] C. E. Ooi and J. Weiss, *Cell (Cambridge, Mass.)* **71,** 87 (1992).

codons (-31) to (-26) of the signal sequence linked in frame to codon 13 of the mature protein. *In vitro* transcription/translation reactions are carried out with commercially available kits (e.g., Stratagene; LaJolla, CA) following the instructions of the manufacturer. Typically, 1 μl of translation reaction mixture yields ca. 100,000 cpm precipitable with 10% trichloroacetic acid (TCA), the vast majority of which represents the desired (full-length) product.

Typical incubation mixtures to measure BPI binding contain 2×10^7 bacteria in 0.5 ml of buffered medium [e.g., nutrient broth supplemented with physiological saline and 20 mM phosphate buffer, pH 6–7.5 (buffered NB/NaCl)] in Eppendorf tubes.[11] Chloramphenicol (50 μg/ml) is added to samples containing added translation mixture to prevent bacterial incorporation of remaining free [^{35}S]methionine. Up to 0.2% (v/v) translation mixture/incubation can be added without affecting BPI binding or activity. Bound and unbound proteins are separated by sedimentation of bacteria followed by two washes with buffered medium. Binding may be measured simply by counting aliquots of the recovered supernatants and resuspended pellets or by analyzing these fractions by sodium dodecyl sulfate–polyacrylamide gel electrophoresis (SDS–PAGE)/autoradiography. Under these conditions, bound unlabeled BPI can be visualized by silver staining or immunoblotting. Thus the functional integrity of radiolabeled BPI preparations can be assessed by comparing the binding of labeled and unlabeled BPI in saturation curves, in the presence of increasing concentrations of $MgCl_2$ or $CaCl_2$ (5–100 mM) to inhibit binding, and to BPI-resistant bacteria (e.g., *Bacillus subtilis*).[24]

Effects of BPI (Intact PMNs) on Bacterial Growth (Viability)

Effects of antibacterial agents on bacterial viability (i.e., ability to multiply) can be assayed in liquid and in agar-based semisolid medium. Bacterial growth in liquid media can be measured as the time-dependent increase in turbidity (e.g., A_{550}) or by direct counting of bacteria in a bacterial (Petroff–Hauser) counting chamber. Bacterial growth in semisolid medium (e.g., nutrient agar) is measured as bacterial colony formation. Because colony formation requires longer than overnight incubation at the appropriate growth temperature, growth in liquid media provides a more direct assessment of the early effects on bacterial growth; however, because the antibacterial agent is present throughout the incubation, this method often cannot distinguish *reversible* inhibition of bacterial growth (bacteriostatic) from *irreversible* inhibition (bactericidal). Measurement

[24] G. In't Veld, B. Mannion, J. Weiss, and P. Elsbach, *Infect. Immun.* **56,** 1203 (1988).

of bacterial colony-forming ability typically involves dilution of bacterial samples from relatively high concentrations in the incubation mixtures (10^6–10^8/ml) to low concentrations in the petri dish ($<10^2$/ml) with a parallel dilution of the (soluble) antibacterial agent to concentrations far below its minimum effective dose. It is therefore generally presumed that inhibition of bacterial colony formation corresponds to an *irreversible* (bactericidal) effect on bacterial growth produced during the incubation period preceding sample dilution and plating. Incubation of susceptible bacteria with BPI leads to a rapid (within 1 min of incubation) loss of bacterial colony-forming ability on conventional bacteriological media (e.g., nutrient, trypticase soy, Mueller–Hinton agar). When, however, the media are supplemented with $\geq$0.5 mg/ml serum albumin most of the bacteria do form colonies, clearly demonstrating that these bacteria are *not* dead[22]! Increasing the time of BPI–bacteria incubation (in buffered NB/NaCl without albumin) before plating results in progressively fewer bacteria than can form colonies in albumin-supplemented media and a parallel reduction in bacterial uptake and incorporation of amino acids into bacterial protein (Fig. 3). Thus, the simple inclusion or exclusion of albumin in the growth medium (added to molten agar at 50°) permits a clear distinction between essentially undamaged bacteria (normal colony formation $\pm$ albumin), sublethally damaged bacteria (colony formation *only* in the presence of albumin), and dead bacteria (no colony formation $\pm$ albumin). In the same manner, it is possible to distinguish sublethally damaged from lethally injured *E. coli* that had been ingested by PMNs.[3] For both *E. coli* treated with purified BPI and *E. coli* ingested by PMNs, bacteria requiring albumin for growth manifest mainly outer envelope alterations, whereas bacteria unable to grow with or without albumin display structural and functional disruption of the cytoplasmic membrane as well[3,22]; (see Fig. 3). Albumin does not inhibit BPI binding or production of its early outer envelope alterations but somehow interrupts the progression from outer envelope (sublethal) to inner envelope (lethal) damage. Of the other proteins and polyanionic substances we have tested (e.g., immunoglobulins, lysozyme, glycosaminoglycans), only β-lactoglobulin is also able to retard the bactericidal action of BPI. Both albumin and β-lactoglobulin are fatty acid-binding proteins and their relative potency in rescue of BPI-treated bacteria parallels their relative free fatty acid binding affinities. Human and bovine serum albumin are equally effective as are fatty acid-poor, endotoxin-free, and HPLC-purified albumin preparations.

Despite the presence of high concentrations of albumin in blood (ca. 50 mg/ml), addition of nanomolar concentrations of BPI to whole blood *ex vivo* can cause the rapid *killing* of (encapsulated) gram-negative bacteria that otherwise survive the cellular and extracellular antibacterial systems

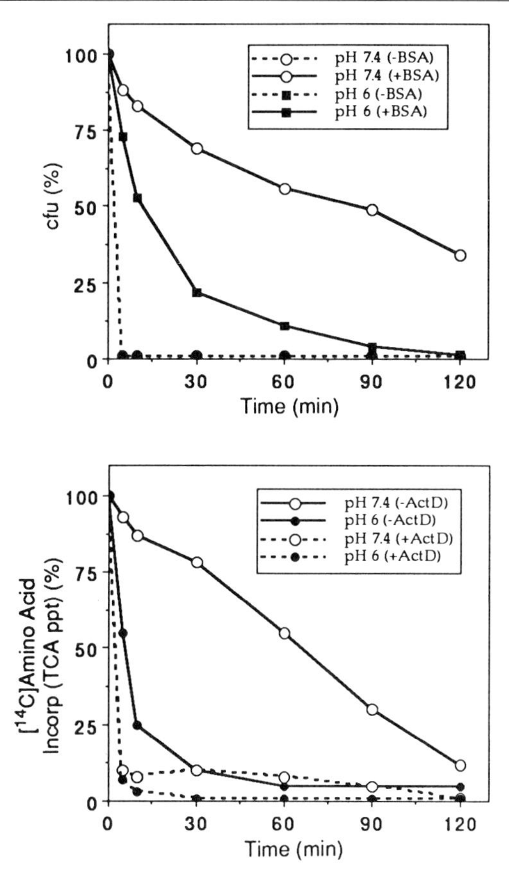

FIG. 3. Time- and pH-dependent action of BPI on *Escherichia coli* J5. *E. coli* were incubated at 37° in 20 m*M* phosphate-buffered NB/NaCl (pH 6.0 or 7.4) with BPI. At the indicated times, aliquots were taken for measurement of bacterial viability (cfu) in nutrient agar (± BSA) (upper) or pulsed for 15 min with ^{14}C-labeled amino acids ± Act D for assay of bacterial protein biosynthetic activity and outer membrane permeability (lower). All results are expressed as percentages of values obtained for aliquots of untreated bacteria taken at zero time. See the text and Mannian *et al.*[22] for further details.

in whole blood.[16] Bactericidal effects of added (extracellular) BPI are observed in blood collected with either buffered citrate or heparin as anticoagulant. Heparin may diminish the endogenous antibacterial activity of whole blood. Therefore, experiments are typically carried out in buffered citrate. BPI and bacteria (up to 10^6/ml, final concentration) are each added to blood in 2% (v/v) of the blood volume; BPI is added first. The ability of added BPI to cause bacterial killing in blood implies that other (extracellular) component(s) in blood are able to override the inhibitory effect of albumin on the bactericidal action of BPI. In fact, serum (complement)-resistant encapsulated *E. coli* treated with purified BPI in the presence of serum are killed more rapidly than when treated with BPI alone.[3,16] In contrast, C7-depleted serum (Quidel, San Diego, CA) acts like purified albumin and limits BPI action to the initial sublethal effects. Addition of purified C7 (Quidel) to the depleted sera reproduces the effect of normal serum, showing that the enhancement of bacterial killing by normal serum plus BPI is dependent on late components of the complement system acting synergistically with BPI. In a remarkably similar manner, use of normal serum to opsonize *E. coli* and facilitate phagocytosis by PMNs markedly accelerates killing of ingested bacteria, whereas opsonization with C7-depleted serum results in much more limited intracellular damage to ingested *E. coli*.[3] The striking similarities in the fate of *E. coli* exposed to purified BPI and to intact PMN, including the temporal separation of sublethal and lethal stages of action and the opposing effects of albumin and late components of the complement system, support the view that BPI plays an important role in the action of PMNs against ingested BPI-sensitive bacteria such as *E. coli*.

Effect of BPI (or Intact PMNs) on Bacterial Uptake of Radiolabeled Amino Acids and Incorporation into Bacterial Protein

As indicated, the loss of the ability of BPI-sensitive bacteria treated with either purified BPI or intact PMNs to grow and form colonies in albumin-supplemented growth media parallels inhibition of bacterial uptake and incorporation of added amino acids into protein. Thus, measurement of the latter provides a more rapid assay of progression of bacterial damage that appears linked to lethal injury. Incubation mixtures containing 1 to 5 $\times$ 10^7 bacteria/ml are simply supplemented with radiolabeled amino acids (e.g., ^{14}C-labeled L-amino acid mixture; 56 mCi/mmol; New England Nuclear, Boston, MA; 0.1 μCi/sample) and 0.1% (w/v) casamino acids (Difco Laboratories) and incorporation into protein is measured as radiolabeled material precipitable with ice-cold 10% TCA.[22] TCA precipitates are collected by vacuum filtration through a 0.45-μm Millipore (Millipore Corp., Bedford, MA) HAWP membrane, the filters are washed three times with 1 ml 10% ice-cold TCA, and radioactivity is measured by liquid scintillation counting. In experiments with intact PMNs, 0.5 m*M* cycloheximide is also added to inhibit protein synthesis by the PMNs.[3] To visualize the spectrum of bacterial proteins synthesized by (un)treated bacteria, [^{35}S]methionine is used for labeling: TCA precipitates are collected by centrifugation at 10,000 g for 10 min, washed twice with 5% TCA and then with ethanol/diethyl ether (1 : 1) to remove traces of TCA, resuspended in Laemmli sample buffer, and analyzed by SDS–PAGE/autoradiography. We have observed that an early effect of BPI (while overall bacterial protein synthesis is still intact) is the suppression of ompF synthesis and the reciprocal stimulation of ompC synthesis. These are two major outer membrane proteins that form aqueous channels (porins) permitting uptake of small hydrophilic molecules (e.g., nutrients) through the outer membrane. This effect of BPI on the synthesis of ompF/C mirrors the effect of hyperosmotic conditions and appears to be mediated by a signal transduction system encoded by *ompR/envZ*.[25]

To measure uptake of radiolabeled amino acids independent of protein synthesis, bacteria are preincubated with 100 μg/ml chloramphenicol (Parke–Davis Co., Morris Plains, NJ) for 10 min at 37° to block bacterial protein synthesis before adding BPI. Radiolabeled amino acids (4–20 μCi/ml) are added at various times after addition of BPI. After an additional 5-min incubation, the bacteria are collected by filtration onto cellulose nitrate membranes (0.45-μm pore size; Millipore) and washed once with 10 vol of 0.8% (w/v) buffered nutrient broth. The radioactivity entrapped in the filter is measured by liquid scintillation counting. To determine the amount of radioactivity binding nonspecifically to the bacteria and/or the filter, samples of *E. coli* are pretreated for 30 min with 10% (w/v) formaldehyde before adding radioactive amino acids. These values are subtracted from all samples to calculate the actual uptake of amino acids.

Effect of BPI (or Intact PMNs) on Bacterial Outer Membrane Permeability

Presumably due to the absence of typical phospholipid bilayer regions, the outer membrane of many gram-negative bacteria, unlike other biological membranes, is impermeable to small hydrophobic molecules such as actinomycin D (Act D; Merck, Sharp and Dohme, West Point, PA); thus, these bacteria are normally highly resistant to the toxic effects of this antibiotic.[4,26] An early effect of purified BPI (or intact PMNs) on bacterial outer membrane permeability can therefore be monitored by assaying bacterial sensitivity to Act D, measuring the effect of the added antibiotic (50 μg/ml) on either bacterial protein synthesis or colony formation in albumin-supplemented growth media, processes inhibited by intracellular Act D but not initially by BPI (or intact PMNs).[3,4,22] The I_{50} values of Act D for untreated and BPI-treated *E. coli* are 200 μg/ml and 100 ng/ml, respectively. Thus, under these experimental conditions, Act D has no

[25] S. Forst and M. Inouye, *Annu. Rev. Cell Biol.* **4,** 21 (1988).

[26] M. Vaara, *Microbiol. Rev.* **56,** 395 (1992).

effect on untreated bacteria and completely inhibits BPI-treated bacteria. The percentage of bacteria with a permeable outer membrane can therefore be calculated as

% bacteria

$$= 1 - \frac{\text{bacterial } ^{14}\text{C-labeled amino acid incorporation } (+ \text{ Act D})}{\text{bacterial } ^{14}\text{C-labeled amino acid incorporation } (- \text{ Act D})} \times 100$$

$$= 1 - \frac{\text{bacterial cfu in BSA } + \text{ Act D}}{\text{bacterial cfu in BSA } - \text{ Act D}} \times 100$$

Similar experiments can be done with rifampin but are not preferred because the intrinsic bacterial resistance to this antibiotic is significantly less ($I_{50} \leq 10\ \mu$g/ml).

Effect of BPI on Cytoplasmic Membrane Permeability

The passive permeability of the cytoplasmic membrane of *E. coli* can be monitored by measuring the hydrolysis of added *o*-nitrophenyl-β-D-galactopyranoside (ONPG; Sigma Chemical Co., St. Louis, MO; 0.5 mg/ml) at 37° by *E. coli* strains (e.g., ML-35) that contain cytoplasmic β-galactosidase but lack lactose permease.[22] Production of *o*-nitrophenyl (ONP) is measured by determination of A_{400} after addition at various times of 2 vol of 20 m*M* NaOH to the incubation mixture. Maximum rates of ONPG hydrolysis can be obtained by disruption of *E. coli* by sonication (15-sec pulses, four times on ice) using a microultrasonic cell disruptor (Kontes Co, Vineland, NJ) at 75% of full strength.

Activation by BPI of Bacterial Phospholipolysis by Endogenous (Bacterial) and Exogenous (Host) Phospholipases A

An effect of BPI on bacterial phospholipid (PL) turnover can be readily measured after prelabeling bacterial PL during growth by supplementing the growth medium with radiolabeled free fatty acids (FFAs). Addition of 1 μCi/ml [1-^{14}C]palmitic acid or [1-^{14}C]oleic acid (DuPont–New England Nuclear, Boston, MA, 60 Ci/mol) yields almost exclusive labeling of, respectively, the 1- or 2-acyl position of PL.[27–29] After growth to mid-logarithmic phase, a 30-min chase in fresh medium without labeled precursor and washing with medium supplemented with ≥0.5% BSA to remove

[27] P. Elsbach, J. Weiss, R. C. Franson, S. Beckerdite-Quagliata, A. Schneider, and L. Harris, *J. Biol. Chem.* **254,** 11000 (1979).

[28] J. Weiss, S. Beckerdite-Quagliata, and P. Elsbach, *J. Biol. Chem.* **254,** 11010 (1979).

[29] P. Elsbach and J. Weiss, this series, Vol. 197, p. 24.

cell-associated unesterified FFAs, 5×10^7 *E. coli* (ca. 1 nmol of PL) contains ca. 100,000 cpm, more than 97% of which is in PL. The distribution of radiolabeled PL corresponds to the chemical composition of envelope PL which in *E. coli* is typically ca. 75% phosphatidylethanolamine (PE), 20% phosphatidylglycerol (PG), and 5% cardiolipin (CL); approximately 40% of the total envelope PL is in the outer membrane, the rest in the inner membrane. To permit a more precise analysis of the nature of the phospholipolytic activities mediating bacterial PL degradation, samples are extracted with 6 vol of chloroform : methanol (1 : 2, v/v) 30 min or longer at 37° or overnight at room temperature. After addition of 2 vol of chloroform and 3 vol of 50 m*M* KCl, samples are mixed by vortexing and spun at 3000 g for 5 min to facilitate the separation of two phases. Undegraded PL and nearly all acyl-containing PL breakdown products are recovered in the lower chloroform phase. Recovery of monoacyl derivatives of PE (lyso-PE) is increased to more than 90% by washing the upper aqueous–methanol phase with chloroform and recovery of the chloroform phase; much less lyso-PG is recovered because of the more polar properties of the head group in PG. The lipids in the chloroform phase are concentrated by evaporation of chloroform under N_2 and separated by thin-layer chromatography on commercial silica gel F254 plates (Analtech, Inc., Newark, DE) using, for example, a solvent system consisting of chloroform/methanol/distilled water/glacial acetic acid, 65/25/4/1, v/v). Migration of lipid species is determined by cochromatography of purified lipid standards that are visualized after chromatography by exposure of the plates to iodine vapors. Thin-layer fractions of samples corresponding to migration of lipid standards are scraped off the plate and monitored by liquid scintillation counting to measure recovery of labeled lipid species. There is little or no accumulation of labeled lipid breakdown products (<2% of total labeled PL/hr) during incubation of [^{14}C]oleate (or [^{14}C]palmitate)-labeled *E. coli* alone, indicating little or no constitutive turnover of bacterial PL under these conditions. In contrast, incubation of labeled *E. coli* with purified BPI or intact PMNs promptly yields labeled FFAs and lyso compounds, indicating that activation of phospholipase(s) A (PLA) ± lyso-PLA is an early event in BPI (PMN) action.[27,28,30] Maximum recovery of labeled products requires the presence of albumin during the incubation (≥0.25% w/v with purified BPI, ≥1% with intact PMN) to complex released FFAs and lyso compounds and prevent recycling of these products back into intact PL. The ability of albumin to complex rapidly and quantitatively these breakdown products but not intact enve-

[30] G. W. Wright, J. Weiss, K. S. Kim, H. Verheij, and P. Elsbach, *J. Clin. Invest.* **85,** 1925 (1990).

lope PL can be exploited to simplify greatly the assay of bacterial PL degradation under these conditions. In brief, at the end of the incubation, samples are spun at sufficient *g* force to sediment the bacteria (e.g., highest speed in a microfuge for 3 min), and an aliquot of the recovered supernatant is measured by liquid scintillation counting. Labeled FFAs and lyso compounds are quantitatively recovered in the supernatant, whereas undegraded PLs are quantitatively recovered in the bacterial pellet.[29-31]

The role of specific PLA in the action of purified BPI and intact PMNs on *E. coli* can be determined by taking advantage of the availability of PLA-deficient mutant strains of *E. coli*. Essentially no PL degradation is triggered during BPI treatment of *pldA*$^-$ *E. coli,* indicating that the outer envelope PLA encoded by this gene is the principal bacterial PLA activated by BPI.[31] Bacterial PL degradation during phagocytosis of the *pldA*$^-$ strain is reduced but not absent, apparently reflecting the action of the granule-associated PMN PLA2.[30] Purified (granule-derived) 14-kDa PMN PLA2 (and a nearly identical enzyme from inflammatory fluid) are inactive toward untreated *E. coli* but highly active toward BPI-treated (*pldA*$^-$) *E. coli*.[27,28,32,33] These enzymes are members of a conserved family of 14-kDa PLA2s that share many common structural and functional properties yet can differ significantly in their action against specific biological targets.[29,31,34] The activity of various purified PLA2s toward BPI-treated *E. coli* can differ by up to 10,000-fold, whereas these enzymes show essentially identical activity toward purified dispersions of *E. coli* PLs. The testing of recombinant PLA2 variants expressed from mutagenized cDNAs toward BPI-treated *pldA*$^-$ *E. coli* has provided an extremely effective experimental setting in which to study the role of discrete "hypervariable" regions in the biological action of PLA2.[31,35]

Expression of the wild-type *pldA* gene in multicopy plasmids provides PLA-rich strains of *E. coli* that still show little or no constitutive turnover of PL but nearly two times greater PL degradation (up to 60% of total) during treatment with purified BPI or phagocytosis by PMNs. Under these conditions, the progression from sublethal to lethal injury is markedly accelerated (ca. fivefold).[30]

The bacterial and host PLAs that act on BPI-treated *E. coli* are Ca^{2+}-dependent enzymes that have millimolar Ca^{2+} requirements in assays with

[31] J. Weiss, G. W. Wright, A. C. A. P. A. Bekkers, C. J. van den Bergh, and H. M. Verheij, *J. Biol. Chem.* **266,** 4162 (1991).

[32] S. Forst, J. Weiss, P. Elsbach, J. Maraganore, I. Reardon, and R. L. Heinrikson, *Biochemistry* **25,** 8381 (1986).

[33] G. Wright, C. E. Ooi, J. Weiss, and P. Elsbach, *J. Biol. Chem.* **265,** 6675 (1990).

[34] F. F. Davidson and E. A. Dennis, *J. Mol. Evol.* **31,** 228 (1990).

[35] J. Weiss, M. Inada, P. Elsbach, and R. M. Crowl, *Clin. Res.* **41,** 328 (1993).

purified dispersions of PLs but are maximally active toward BPI-treated bacteria in media containing as little as 30 μM Ca^{2+}.[28,36] Addition of $\geq$10 mM Ca^{2+} actually inhibits PLA action by inhibiting BPI binding to bacteria. The Ca^{2+} requirements of these enzymes are apparently satisfied by LPS-bound Ca^{2+} that is locally released on binding of BPI.[36]

Assays of Lipopolysaccharide Binding and Endotoxin Neutralization by BPI

Lipopolysaccharides play an important role both in the structural properties of the bacterial outer envelope and in the interaction of gram-negative bacteria with the host. The lipid A region represents the endotoxic moiety of LPS capable of potently triggering a number of host responses.[37] Because this region of LPS is embedded in the outer membrane, it is generally assumed that release of LPS from an intact outer envelope is needed for LPS to serve this signaling function, but the precise physical form(s) of bioactive endotoxin (LPS) *in vivo* has not been well defined. BPI and the bioactive N-terminal fragment of BPI bind with high affinity to the lipid A region of LPSs presented in a variety of physical forms (e.g., purified LPSs dispersed in solution or immobilized on synthetic surfaces, shed outer membrane fragments, intact bacteria) and potently inhibit LPS (endotoxin) activity.[11,13,16,18,38,39] Host responses to LPSs are normally beneficial but, when excessive, can have life-threatening consequences as in gram-negative sepsis.[37] Thus, the combined antibacterial and antiendotoxic activities of BPI have spurred great interest in this protein (or derivatives) as a potential novel therapeutic agent. As a result, the number of bioassays in which effects of LPS or intact gram-negative bacteria are shown to be neutralized by BPI (fragment) are rapidly increasing and include several animal models in which lethal and sublethal host alterations induced by infusion of purified LPSs or live gram-negative bacteria can be blocked by injection of BPI (fragment). In the interest of brevity, we describe only a few examples, drawing mainly on *in vitro/ex vivo* assays done in our laboratories. Because of the great potency of LPSs (effects detectable at $\geq$10 pg/ml) and the prevalence of gram-negative bacteria, all recyclable vessels, etc. are baked at 180° for 4 hr and all consumable supplies are pretested in the chromogenic *Limulus* amebocyte

[36] P. Elsbach, J. Weiss, and L. Kao, *J. Biol. Chem.* **260,** 1618 (1985).

[37] E. T. Rietschel and H. Brade, *Sci. Am.* **267,** 54 (1992).

[38] M. N. Marra, C. G. Wilde, J. E. Griffith, J. L. Snable, and R. W. Scott, *J. Immunol.* **144,** 662 (1990).

[39] M. N. Marra, C. G. Wilde, M. S. Collins, J. L. Snable, M. B. Thornton, and R. W. Scott, *J. Immunol.* **148,** 532 (1992).

lysate assay before use in sensitive bioassays to ensure the absence of detectable contaminating LPSs.

Assay of BPI Binding to Isolated Lipopolysaccharide

Binding of BPI or bioactive fragments to the lipid A region of LPS has been demonstrated directly by making use of purified LPS or lipid A either available from commercial sources (e.g., List Biological Laboratories, Inc., Campbell, CA: Ribi ImmunoChem Research, Inc., Hamilton, Mont.; Calbiochem Corp., La Jolla, CA) or isolated from clinical isolates by the aqueous phenol procedure of Westphal and Jann.[40] Stock solutions of LPS (1–10 mg/ml) are prepared by dispersing powders of LPSs in sterile, pyrogen-free water either by sonication for 4 min at maximum output (rough forms and lipid A) or by vigorous vortexing for 2 min (smooth forms) and stored at 4°. Serial dilutions of LPSs for coating of wells (Immulon 2 Removawell strips; Dynatech, Inc., Chantilly, VA) are made in absolute methanol (R-LPS) or Dulbecco's PBS (D-PBS; S-LPS) with vortexing (15–30 sec). Wells are coated ± LPS by incubation with 50 μl of LPS (0.25–5 μg/ml) or buffer alone overnight. The wells are then blocked with 0.1% BSA in PBS for 3 hr at 37° and, after discarding the blocking solution, washed three or more times with PBS/0.2% Tween 20. Radiolabeled BPI (see above) is then added in a total volume of 50 μl containing 1 μl translation mix (where added) and $\geq$50 ng/ml unlabeled BPI (fragment) in PBS/0.2% Tween 20 and incubated in the well overnight at 4°. After the wells are washed (as above), the radioactivity remaining in the wells is counted in either a gamma counter ([^{125}I]BPI) or a liquid scintillation counter ([^{35}S]BPI). Binding of radiolabeled BPI to wells not coated with LPS is taken to represent nonspecific binding and in Tween (0.05–0.2%) is generally about 10 to 20% of optimal LPS-dependent binding. Nonspecific binding is up to two times higher in the absence of Tween. LPS-dependent binding of BPI is unaffected by Tween.

Interactions between BPI and LPS dispersions in solution can be assayed by measuring the ability of added LPS to inhibit binding of radiolabeled BPI to immobilized LPS. Dilutions of LPS are prepared in distilled water or buffered physiological salts solution and added to radiolabeled BPI before incubation in (LPS-coated) wells.

In contrast to BPI binding to intact bacteria, BPI binding to isolated LPS, either dispersed in solution or immobilized in wells, is not affected by the polysaccharide chain of LPS.[13] Binding is saturable, of high affinity ($K_d \leq 5$ nM) to natural and synthetic lipid A and to R- and S-forms of

[40] O. Westphal and K. Jann, *Methods Carbohydr. Chem.* **5,** 83 (1965).

LPS, demonstrable toward LPS from a wide variety of gram-negative species, and not inhibited by unrelated polyanionic substances (e.g., DNA). These findings indicate that lipid A provides a specific binding site for BPI in LPSs and suggest that the tight packing of LPS in the outer membrane accounts for the impaired access of (holo)BPI to the lipid A region of LPSs with long polysaccharide chains.

Assays of Lipopolysaccharide (Endotoxin)-Neutralizing Activity of BPI

Chromogenic Limulus Amebocyte Lysate Assay

A wide variety of LPS species can activate a procoagulant protease present in *Limulus* amebocyte lysates (LALs). Protease activity can be detected using a chromogenic substrate provided in a commercially available assay kit (Whittaker Bioproducts, Inc., Walkersville, MD), permitting sensitive and quantitative detection of LPSs (detection limit ca. 10 pg/ml) and, hence, measurement of LPS-neutralizing activity of BPI. Maximum neutralizing effects manifest in these assays after preincubation of LPS with 1 to 10 n*M* BPI 10 min or longer before addition of lysate.[18,38] Similar BPI dose requirements are demonstrated over a broad range of LPS concentrations (0.1–10 ng/ml).

Lipopolysaccharide Action on Isolated Polymorphonuclear Leukocytes

Among the important targets of LPS action in mammalian hosts are the PMNs. At nanogram-per-milliliter concentrations, LPS increases the adherence properties of PMNs in part by upregulating the surface expression of CR1 and CR3 receptors[38] and primes the cells for heightened metabolic (e.g., arachidonate, oxidative) responses triggered when the cells are exposed to a second (non-LPS) stimulus.[18,41,42] The sensitivity of PMNs to LPS can be increased about 100-fold by addition of a plasma LPS (lipid A)-binding protein, LBP, that is structurally related to BPI but that amplifies rather than inhibits host responses to LPS.[43,44] Thus, PMNs

[41] L. A. Guthrie, L. C. McPhail, P. M. Henson, and R. B. Johnston, Jr., *J. Exp. Med.* **160,** 1656 (1984).

[42] M. E. Doerfler, R. L. Danner, J. H. Shelhamer, and J. E. Parrillo, *J. Clin. Invest.* **83,** 970 (1989).

[43] R. R. Schumann, S. R. Leong, G. W. Flaggs, P. W. Gray, S. D. Wright, J. C. Mathison, P. S. Tobias, and R. J. Ulevitch, *Science* **249,** 1429 (1990).

[44] P. S. Tobias, J. Mathison, D. Mintz, J.-D. Lee, V. Kravchenko, K. Kato, J. Pugin, and R. J. Ulevitch, *Am. J. Respir. Cell Mol. Biol.* **7,** 239 (1992).

provide an experimental setting in which effects of BPI on LBP-dependent and LBP-independent host responses to LPSs can be measured (Fig. 4).

PMNs to be used in these experiments are isolated from venous blood of healthy human volunteers by conventional dextran and Ficoll–Hypaque separation procedures. Purified PMNs are resuspended to the desired concentration (10^7–10^8/ml) in Ca^{2+},Mg^{2+}-free Hanks' balanced salt solution [HBSS(−); GIBCO, Grand Island, NY] and used as is or after prelabeling for 30 min at 37° with [^{3}H]arachidonic acid ([^{3}H]AA, 0.4 μCi/ml). After labeling, the cells are washed three times with HBSS(−) containing 1.5% human serum albumin (HSA) and resuspended in HBSS(−). This labeling procedure does not alter PMN sensitivity to subsequently added stimuli. Purified LPS (usually from *Salmonella minnesota* Re595) is diluted in HBSS, preincubated where appropriate with BPI and/or LBP, and added to PMNs. Maximum upregulation of surface CR1 and CR3 requires incubation of PMNs for 30 min with 10 ng LPS/ml and is completely inhibited by preincubation of LPS with ≥4 n*M* BPI for 5 min or longer (inhibition is nearly complete without preincubation).[38] Surface CR1/CR3

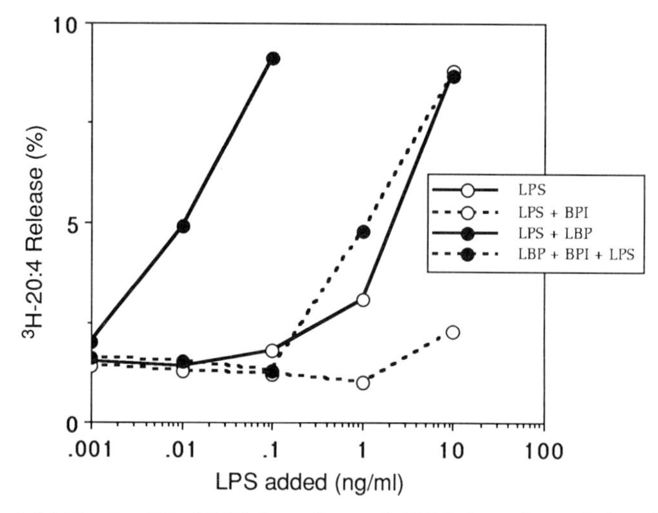

FIG. 4. Inhibition by BPI of LBP-dependent and LBP-independent priming of PMNs by (Re595) LPS. PMNs were prelabeled with [^{3}H]arachidonic acid (20 : 4), then preincubated with LPS as indicated for 45 min at 37°, followed by incubation with opsonized zymosan for 30 min. BPI (30 n*M*) and/or LBP (10–15 n*M*) were added just before addition of LPS to PMNs. Release of ^{3}H-20 : 4 during incubation with opsonized zymosan was measured by counting an aliquot of the extracellular supernatant collected at the end of the incubation and is expressed as a percentage of the total cellular radioactivity. No release of ^{3}H-20 : 4 was triggered during preincubation with LPS. Purified LBP was generously provided by Dr. Peter Tobias and Dr. Richard Ulevitch.

are monitored by fluorescence-activated cell sorting (FACS) after washing treated cells with PBS containing 0.05% sodium azide and 2% autologous plasma and then staining with commercially available fluorescent-labeled anti-CR monoclonal antibodies (e.g., Becton Dickinson, Mountain View, CA). Maximum priming of [^{3}H]AA release and leukotriene (B4) production stimulated by opsonized zymosan (2–40 particles/PMN), phorbol myristate acetate (0.03–1.0 μg/ml) or the Ca^{2+} ionophore A23187 (0.1 μg/ml) in the presence of 1 to 2 mM Ca^{2+} and Mg^{2+} requires incubation of PMNs for 45 min with 10 ng LPS/ml in the absence of LBP or with 0.1 ng LPS/ml in the presence of 10 to 15 nM LBP.[18,45] Maximum inhibition of LPS priming requires addition of $\geq$10 nM BPI to LPS at or before the time of addition of PMNs. At a given concentration of BPI, similar inhibition is observed over a broad range of LPS doses (0.1–100 ng/ml). Preincubation of LPS with LBP diminishes inhibition of LPS priming by BPI consistent with the view that part of BPI's effects involves competition with LBP for LPS. The specificity of the inhibitory effects of BPI can be tested by measuring the effect of BPI on other PMN priming and stimulatory agents [e.g., tumor necrosis factor α (TNF-α, 50 U/ml)].[18,38]

Cytokine-Inducing Activity of Isolated Lipopolysaccharide and (Live) Gram-Negative Bacteria in Whole Blood ex Vivo

The systemic toxicity of endotoxin is thought to be mediated at least in part by excessive production of cytokines such as TNF-α and various interleukins (ILs). Addition of purified LPS and of live gram-negative bacteria to anticoagulated whole blood *ex vivo* induces a dose- and time-dependent accumulation of TNF-α, IL-1, IL-6, and IL-8 in the extracellular medium.[16,18] Typically, incubations are carried out in 0.25 ml of whole blood for 4 to 6 hr at 37° with moderate shaking. At the end of the incubation, samples are diluted with 3 vol of tissue culture medium (RPMI) and extracellular cytokines are collected by centrifugation and measured by enzyme-linked immunosorbent assay (ELISA) using commercially available test kits (e.g., T Cell Sciences, Inc., Cambridge, MA). Cytokine accumulation is detected after incubation with as little as 10 pg/ml of purified LPS or 10^3 bacteria/ml (also containing ca. 10 pg of LPS/ml). Thus, per unit mass of LPS added, the cytokine-inducing activity of *E. coli* is at least as great as that of purified LPS. The activity of added gram-negative bacteria that are rapidly killed in blood and those that survive is not appreciably different and about 100 to 1000 times greater than that of the gram-positive bacterium *Staphylococcus aureus*.

[45] M. D. Doerfler, J. Weiss, J. Clark, and P. Elsbach *J. Clin. Invest.* (1994), in press.

The cytokine-inducing activity of both purified LPS and whole gram-negative bacteria can be potently inhibited by added nanomolar concentrations of BPI, demonstrating that the potent endotoxin-neutralizing effects of BPI can be exerted in the complex environment of whole blood containing other LPS-binding proteins (e.g., LBP, lipoproteins).[16,18] BPI is added to blood just prior to addition of LPS or bacteria; maximum inhibition does not require preincubation of BPI with LPS or bacteria. The inhibitory effect of the N-terminal fragment of BPI is at least as great as that of holo-BPI and, in fact, substantially greater than that of holo-BPI toward certain encapsulated strains containing long-chain LPSs.[18] This parallels differences in bactericidal potency and suggests that in these cases effects of BPI may depend on interaction with LPSs in the bacterial envelope. The isolated 30-kDa C-terminal fragment also exhibits inhibitory activity toward purified Re595 LPS in this assay.[16] There is no effect of BPI (fragments) on TNF-α production induced by heat-killed or live *S. aureus*.

Acknowledgments

This work was supported by USPHS Grants 5R37DK and AI-18571, and a Grant from XOMA Corp., Berkeley, CA.

[15] Regulation of Bacterial Gene Expression in Response to Oxidative Stress

By GISELA STORZ and MICHEL B. TOLEDANO

Introduction

Bacterial cells are able to sense and adjust to a great number of different environmental stresses including reactive oxygen species such as superoxide anion, hydrogen peroxide, and hydroxyl radicals. These oxidants are produced by the incomplete reduction of oxygen during respiration or by exposure to light, metals, radiation, and oxidation–reduction active drugs such as paraquat and menadione. As discussed in [10] in this volume, reactive oxygen species are also generated by phagocytic cells as an antibacterial defense. The reactive oxygen species constitute a stress because they lead to damage of almost all cellular components by causing mutagenesis, enzyme inactivation, and membrane damage.[1,2]

[1] B. Demple, *Annu. Rev. Genet.* **25,** 315 (1991).
[2] S. B. Farr and T. Kogoma, *Microbiol. Rev.* **55,** 561 (1991).

In recent years, considerable progress has been made in elucidating the mechanisms by which *Salmonella typhimurium* and *Escherichia coli* cells sense these reactive oxygen species and defend against their deleterious effects.[1,2] In this article, we present the approaches used to study the regulation of oxidative defense genes in *E. coli* and *S. typhimurium*. These general approaches should be useful in examining the response to oxidative stress in other organisms as well as in further elucidating the response in *E. coli* and *S. typhimurium*.

Oxidative Stress Conditions

Table I summarizes many of the conditions used to generate oxidative stress artifically. It should be emphasized that the majority of the studies on oxidative stress in bacteria have been conducted on cells grown in minimal glucose media during logarithmic growth; however, bacteria probably must respond to oxidants under a variety of growth conditions. The effects of oxidative stress on bacteria growing in rich media, under acidic or basic growth conditions, with varied carbon sources, during stationary or starved growth have only recently received more detailed attention.

Superoxide Anion

Increased levels of superoxide anion ($O_2^- \cdot$), the product of a one-electron reduction of oxygen, can be generated simply by treating cells with oxidation–reduction cycling drugs. Paraquat, menadione, and plumbagin are used most commonly.[3-6] The toxicity of these drugs is enhanced by the concomitant addition of iron and copper probably as a result of the increased oxidation–reduction cycling caused by the metals.[3,5] The effects of elevated superoxide anion concentrations have also been examined in *E. coli* strains lacking the two radical-detoxifying enzymes, manganese and iron superoxide dismutases. In logarithmically growing cultures, the concentration of superoxide anion for cells lacking the superoxide dismutases has been calculated to be 6.7 μM whereas the concentration for wild-type cells is 0.2 nM.[7]

Hydrogen Peroxide

Hydrogen peroxide (H_2O_2), the product of a two-electron reduction of oxygen, is fairly stable and readily diffuses into bacterial cells. The levels

[3] P. Korbashi, R. Kohen, J. Katzhendler, and M. Chevion, *J. Biol. Chem.* **261,** 12472 (1986).
[4] S. B. Farr, D. O. Natvig, and T. Kogoma, *J. Bacteriol.* **164,** 1309 (1985).
[5] J. T. Greenberg and B. Demple, *J. Bacteriol.* **171,** 3933 (1989).
[6] L. K. B. Walkup and T. Kogoma, *J. Bacteriol.* **171,** 1476 (1989).
[7] J. A. Imlay and I. Fridovich, *J. Biol. Chem.* **266,** 6957 (1991).

TABLE I
OXIDATIVE STRESS CONDITIONS

Oxidant[a]	Concentrations used for[b]					Ref.
	Adaptation	Killing	Zones of inhibition	2D gels	Mutant selections	
Paraquat (257.2) (methyl viologen)	1 mM[d]	0.25 mM + 50 μM Fe 0.25 mM + 1 μM Cu	0.074 M (10 μl) 0.1 M (20 μl)[d]	130–1300 μM 195 μM	5 μM[d]	3, 5, 6, 9, 14
Menadione[c] (172.2)	1.45 mM	100 mM + 50 μM Cu	0.58 M (10 μl)	145–1450 μM	8 μM	5, 9, 17
Plumbagin (188.2)	0.075 mM[d]	2 mM[d]		5–25 μM		4, 6
Hydrogen peroxide (34.01)	<0.06 mM	2.5–20 mM	0.88 M (10 μl) 0.88 M (20 μl)[d]	60–2000 μM	300 μM	5, 9, 10, 14
Cumene hydroperoxide[c] (152.2)			0.2 M (10 μl)	125 μM	200 μM	9, 11, 12a
tert-Butyl hydroperoxide (90.12)			0.078 M (10 μl)			9
N-Ethylmaleimide (125.1)			0.25 M (10 μl)	40 μM		5, 9
Cadmium chloride (183.3)			0.55 M (10 μl)	600 μM		9, 12

[a] Available from commercial sources. Formula weights given in parentheses.
[b] Grown in minimal glucose media except where indicated.
[c] More easily dissolved in dimethyl sulfoxide (DMSO).
[d] Grown in Luria broth.

of intracellular hydrogen peroxide are therefore easily manipulated by direct addition of hydrogen peroxide to the culture medium.[8-10] This oxidant most likely does not cause damage directly but is converted to the extremely reactive hydroxyl or ferryl radical in the presence of metals ($Me^{N+} + H_2O_2 \rightarrow Me^{(N+1)+} + HO\cdot + H_2O$).[10]

Other Oxidants

This article is focused mainly on studies of gene expression in response to superoxide anion- and hydrogen peroxide-generating conditions; however, other compounds and treatments may also be considered an oxidative stress and could be assayed for their effects on cells.[11,12] Butyl hydroperoxide and cumene hydroperoxide[9,12a] are two alkyl hydroperoxides that are available from commercial suppliers. *N*-Ethylmaleimide and iodoacetate affect the oxidation–reduction balance of cells by alkylating cysteine residues. Oxidants are also generated by treatment with cadmium chloride and by exposure to ultraviolet (UV) light. Finally, a shift between anaerobic and aerobic growth conditions or aerobic and anaerobic growth conditions or exposure to increased oxygen probably alters the cellular concentrations of reactive oxygen species. Each of these treatments, especially UV exposure and anaerobic–aerobic shifts, cause the induction of predominantly unique sets of proteins, suggesting that the underlying mechanisms for the responses to each of these types of stress are different.

Resistance Assays

The sensitivity of strains to oxidants has been assessed using several different approaches. The first evidence that *E. coli* cells possess a regulated defense response against oxidative stress came from growth curves of *E. coli*.[8] AB1157 cultures grown in K medium to a density of 5×10^7 cells/ml were treated with 0, 10, and 30 μM hydrogen peroxide for 30 min. The cells were then challenged with a lethal dose of 5 mM hydrogen peroxide, and aliquots of cells were removed after 20, 40 and 60 min, diluted, and spread on LB plates to determine the number of colony-

[8] B. Demple and J. Halbrook, *Nature (London)* **304,** 466 (1983).

[9] M. F. Christman, R. W. Morgan, F. S. Jacobson, and B. N. Ames, *Cell (Cambridge, Mass.)* **41,** 753 (1985).

[10] J. A. Imlay and S. Linn, *Science* **240,** 1302 (1988).

[11] R. W. Morgan, M. F. Christman, F. S. Jacobson, G. Storz, and B. N. Ames, *Proc. Natl. Acad. Sci. U.S.A.* **83,** 8059 (1986).

[12] R. A. VanBogelen, P. M. Kelley, and F. C. Neidhardt, *J. Bacteriol.* **169,** 26 (1987).

[12a] G. Storz, F. S. Jacobson, L. A. Tartaglia, R. W. Morgan, L. A. Silveira, and B. N. Ames, *J. Bacteriol.* **171,** 2049 (1989).

forming units. Cells that had not been pretreated with hydrogen peroxide were killed rapidly, whereas the cells pretreated with 10 or 30 μM hydrogen peroxide were resistant to the killing. The acquired resistance required protein synthesis, as it was blocked by prior treatment with chloramphenicol. Christman *et al.*[9] showed that *S. typhimurium* cells adapted to hydrogen peroxide by treating wild-type LT2 cells grown to an A_{650} of 0.2 in minimal glucose medium with 60 μM hydrogen peroxide for 60 min (in the presence or absence of 100 μg/ml chloramphenicol) and assaying the resistance of the cells to 10 mM hydrogen peroxide. In an experiment showing that *E. coli* cells could adapt to the superoxide-generating compound menadione, cells grown to 2×10^7 cell/ml were treated with 1.45 mM menadione for 1 hr and then pelleted, washed with M9 salts, and challenged with menadione in the presence of 50 μM $CuCl_2$.[5]

The sensitivity or resistance of wild-type or mutant strains has also been determined by zone of inhibition assays. For these simple assays, 0.1 ml of an overnight culture is plated on an agar plate in 2 to 3 ml of soft agar and a disk impregnated with an oxidant is placed in the center of the agar. The zone of killing or inhibition surrounding the disk is then measured after 12 to 24 hr of growth.[9] For this assay to be consistent the volume of the agar medium must be uniform. The growth of streaks of bacteria on plates containing a concentration gradient of a particular oxidant has also been used to determine the sensitivity of strains. Gradient plates can be made by pouring 25 ml of agar containing the oxidant into a 9-cm^2 petri dish set at an angle. Once the agar is hardened the dish is leveled and an additional 25 ml agar is added.[13]

Two-Dimensional Gels

Two-dimensional (2D) gel analysis of changes in total protein synthesis on treatment with hydrogen peroxide and superoxide-generating compounds has allowed for a further characterization of the adaptive responses. Two examples of the conditions used previously follow: Christman *et al.*[9] examined the proteins induced by hydrogen peroxide by labeling exponentially growing *S. typhimurium* cells. The cells (0.4-ml aliquots grown in minimal glucose medium) were labeled with 200 μCi/ml [^{35}S]methionine for 10- to 60-min intervals after the addition of 60 μM hydrogen peroxide. The labeling reaction was then quenched by the addition of 10 μl of 0.1 M methionine, and the proteins were separated on an isoelectric focusing gel (0.9% pH 3.5–10, 1.2% pH 4–6, 1.2% pH

[13] R. P. Cunningham, S. M. Saporito, S. G. Spitzer, and B. Weiss, *J. Bacteriol.* **168,** 1120 (1986).

6–8, and 1.2% pH 7–9 ampholytes) in the first dimension and a 10% sodium dodecyl sulfate–polyacrylamide gel electrophoresis (SDS–PAGE) gel in the second dimension. The proteins induced by the superoxide-generating agents were examined by Walkup and Kogoma.[6] Aliquots (0.5–1 ml) of exponentially growing cells were treated with 195 μM paraquat or 5–25 μM plumbagin and then labeled for 10–15 min with 100 μCi/ml $Na_2{}^{35}SO_4$ followed by a 2-min chase with 5 mM $MgSO_4$. The labeled proteins were then separated in the first dimension on an isoelectric focusing gel (1.6% pH 5–7 and 0.4% pH 3–10 ampholytes) and a 11.5% SDS–PAGE gel in the second dimension. These studies have shown that superoxide-generating agents and hydrogen peroxide induce distinct sets of proteins in both *E. coli* and *S. typhimurium,* indicating the existence of separate regulatory mechanisms. The analyses also demonstrated that some of the proteins are induced by multiple oxidative stress conditions as well as by nonoxidative types of stress such as heat shock and starvation.

Gene Fusions

Studies on the regulation of the genes encoding the oxidative stress-inducible proteins have been facilitated by promoter or in-frame gene fusions to the readily assayable *lacZ* reporter gene. The random insertion of *lacZ* genes within the bacterial genome has provided a powerful assay for the identification of stress-inducible genes. Kogoma *et al.*[14] isolated *lacZ* fusions to three uncharacterized genes that were induced by treatment with paraquat and plumbagin. Exponentially growing cultures were infected by Mu phage (MudX carrying a promoterless *lacZ* gene) and ampicillin transductants selected. The transductants were then replica-plated on indicator 5-bromo-4-chloro-3-indolyl-β-D-galactoside (X-Gal) plates in the presence and absence of 5 μM paraquat. Colonies that were more strongly blue in the presence than in the absence of paraquat were tested further for the inducibility of β-galactosidase. It is interesting to note that hydrogen peroxide-inducible genes have not been isolated using this approach of random fusions possibly because the induction is so transient or because hydrogen peroxide is detoxified so quickly.

lacZ fusions constructed directly with known genes have also allowed delineation of the sequences important for regulation. Manganese superoxide dismutase and catalase (encoded by *sodA* and *katG*) are both induced by oxidative stress. The genes corresponding to these activities were cloned by screens for plasmids giving rise to elevated superoxide dismu-

[14] T. Kogoma, S. B. Farr, K. M. Joyce, and D. O. Natvig, *Proc. Natl. Acad. Sci. U.S.A.* **85,** 4799 (1988).

tase or catalase activity. The cloned genes were then used to make promoter (transcriptional) or protein (translational) fusions between specific regions of the promoters and the *lacZ* gene. *sodA* fusions generated by Mu (MudIIPR13 carrying a promoterless *lacZ* gene) transposition showed that *sodA* expression is controlled by multiple regulators at both a transcriptional and a posttranscriptional level.[15] A 340-bp *katG* promoter fragment was subcloned directly into the promoter expression vector pRS415 to generate a *katG–lacZ* fusion.[16] The expression of the fusion construct was inducible by hydrogen peroxide, demonstrating that the hydrogen peroxide response element was contained within the 340-bp *katG* fragment.

Screens for Regulatory Mutants

The isolation of mutants that are altered in their responses to oxidative stress has been very instrumental in elucidating the mechanisms by which bacterial cells regulate gene expression in response to oxidative stress. In the following section we describe the approaches used to identify the predominant oxidative stress regulators described in *E. coli* thus far. Two approaches have been especially effective: (1) selections for mutants that have increased resistance to oxidants has yielded mutations that cause positive regulators to be constitutively active. Although they have not yet been isolated, null mutations that abolish the action of negative regulators should also be detected by this approach. (2) Screens for mutants that have altered expression of *lacZ* fusion constructs have been used to isolate constitutive and null mutations in both positive and negative regulators, as mutations giving rise to both increased and decreased β-galactosidase expression can be detected.

soxR and soxS Genes

The *soxRS* regulators of superoxide-inducible genes were identified by two different approaches. Greenberg *et al.*[17] isolated *soxR*^{c} (constitutive) mutants resistant to 1.4 mg/ml menadione on minimal A plates that arose spontaneously or on mutagenesis with nitrosoguanidine. Two-dimensional gels and enzyme assays then showed that these mutants had elevated levels of nine superoxide-inducible proteins. Tsaneva and Weiss[18] isolated

[15] A. Carlioz and D. Touati, *EMBO J.* **5,** 623 (1986).

[16] L. A. Tartaglia, G. Storz, and B. N. Ames, *J. Mol. Biol.* **210,** 709 (1989).

[17] J. T. Greenberg, P. Monach, J. H. Chou, P. D. Josephy, and B. Demple, *Proc. Natl. Acad. Sci. U.S.A.* **87,** 6181 (1990).

[18] I. R. Tsaneva and B. Weiss, *J. Bacteriol.* **172,** 4197 (1990).

mutations in the same *soxR* locus in a screen for mutations that affected the expression of a *nfo'* (encoding superoxide-inducible endonuclease IV)–*lacZ* operon fusion. Using this approach, one noninducible mutant (out of 4×10^4 cells) with reduced levels of β-galactosidase activity was isolated from transductants mutagenized with a mini-Tn*10* element.[18] β-Galactosidase-overproducing *soxR*^{c} strains were isolated in a hypermutagenic *mutL* strain background lysogenized with λ carrying the nfo'–lac fusion.

oxyR Gene

Salmonella typhimurium strains carrying mutations in the *oxyR* regulatory locus were isolated in a selection for diethyl sulfate mutants resistant to 300 μM hydrogen peroxide on minimal glucose plates.[9] Resistant *oxyR1* mutants constitutively overexpressed 9 of the 30 proteins induced by hydrogen peroxide. An *E. coli* mutant, *oxyR2*, with the same phenotype was isolated similarly. Although the *soxR* and *oxyR* mutants were isolated on plates, it should also be possible to isolate resistant mutants by selecting for growth in liquid cultures treated with oxidants. *oxyR* deletion mutations were obtained by selecting for imprecise excision of a Tn*10* linked to *oxyR* in both *S. typhimurium* and *E. coli*. These deletion mutants were recessive, hypersensitive to oxidants, and noninducible for the nine hydrogen peroxide-inducible proteins, demonstrating that *oxyR* is a positive regulator. *oxyR* was independently found to be a negative regulator of the Mu phage *mom* gene in a mutant screen for increased expression of a *mom* promoter fusion.[19] Deletion mutations of the *oxyR/momR* locus were also isolated as *mor* mutations which affected colony morphology and autoaggregation of the *E. coli* K12 strain in yet another independent screen.[20] Although the *oxyR* gene has been identified as a regulator in these different responses, the relationship between oxidative stress, Mu phage growth, and *E. coli* colony morphology is not yet understood.

katF Gene

katF, a third regulator of the oxidative defense activities was identified by several independent genetic screens. *katF* mutants were first identified as catalase-deficient mutants in a screen of colonies carrying random Tn*10* insertions.[21] *katF* was identified as *appR* in a second screen for the locus

[19] M. Bölker and R. Kahmann, *EMBO J.* **8,** 2403 (1989).

[20] S. R. Warne, J. M. Varley, G. J. Boulnois, and M. G. Norton, *J. Gen. Microbiol.* **136,** 455 (1990).

[21] P. C. Loewen and B. L. Triggs, *J. Bacteriol.* **160,** 668 (1984).

giving rise to different levels of periplasmic "pH 2.5 acid phosphatase activity" in different strains.[22] Finally, additional *katF/appR* mutants were isolated as *rpoS* mutants in a third screen for C-starvation-induced fusions.[23] It is noteworthy that although the *oxyR* and *katF* regulators were both identified as oxidative stress regulators, both genes were also identified by methods unrelated to oxidative stress.

The experiments described above illustrate how the isolation of mutants with increased resistance and mutants with altered *lacZ* fusion expression allowed characterization of the *soxRS, oxyR,* and *katF* regulators. The approach of isolating suppressors of existing mutations may allow the identification of additional regulators, although this strategy has only been partially exploited. In one study, all hydrogen peroxide-resistant mutants isolated in a hydrogen peroxide-sensitive, *oxyR* deletion background mapped to the *oxyR*-regulated *katG* and *ahp* genes, suggesting that no additional regulators are required for the *oxyR* response.[24]

Cloning of Genes Encoding Regulators

The *soxR, oxyR,* and *katF* genes were all cloned by complementation of the respective null mutations. Clones that complemented *soxR* deletion mutations were found to encode two proteins denoted SoxR and SoxS, and both genes were found to be required for appropriate regulation in response to superoxide.[25,26] Interestingly, the deduced protein sequences suggested specific functions for all of the regulators. The sequence comparisons showed that the SoxR protein was very similar to the MerR family of regulator proteins, particularly in regions encompassing helix–turn–helix binding domains and a cysteine-rich metal-binding domain.[26] SoxS also appears to have a helix–turn–helix DNA binding domain and is related to the AraC family of bacterial regulatory proteins.[25] The deduced protein sequence of the *oxyR* gene showed extensive homology to the LysR family of bacterial transcriptional regulators.[19,20,27,28] The *katF* gene product shares similarity with the RNA polymerase subunit sigma 70 suggesting a role for KatF as an alternate sigma factor.[29]

[22] E. Touati, E. Dassa, and P. L. Boquet, *Mol. Gen. Genet.* **202,** 257 (1986).
[23] R. Lange and R. Hengge-Aronis, *Mol. Microbiol.* **5,** 49 (1991).
[24] J. T. Greenberg and B. Demple, *EMBO J.* **7,** 2611 (1988).
[25] J. Wu and B. Weiss, *J. Bacteriol.* **173,** 2864 (1991).
[26] C. F. Amábile-Cuevas and B. Demple, *Nucleic Acids Res.* **19,** 4479 (1991).
[27] M. F. Christman, G. Storz, and B. N. Ames, *Proc. Natl. Acad. Sci. U.S.A.* **86,** 3484 (1989).
[28] K. Tao, K. Makino, S. Yonei, A. Nakata, and H. Shinagawa, *Mol. Gen. Genet.* **218,** 371 (1989).
[29] M. R. Mulvey and P. C. Loewen, *Nucleic Acids Res.* **17,** 9979 (1989).

For bacteria that are less well characterized than *E. coli* or *S. typhimurium*, it should generally be possible to clone the genes corresponding to recessive mutants by complementation with a wild-type library. Genes corresponding to dominant mutants may be cloned by moving libraries constructed with the DNA of the dominant mutant into a wild-type strain. Regulators of oxidative stress responses in other organisms may also be isolated on the basis of sequence similarity or immunological cross-reactivity with SoxRS, OxyR or KatF. To our knowledge, however, no regulators of oxidative stress responses have yet been identified using this approach.

Purification and *in Vitro* Assays

Genetic studies and sequence analyses have suggested functions for the oxidative stress regulators, but a complete understanding of their actions requires biochemical studies. OxyR was implicated as a positive regulator of oxidative defense genes and a negative regulator of itself and the *mom* gene. The following biochemical studies of the OxyR protein helped to elucidate the manner of OxyR action and the mechanism of OxyR activation. Overproduction of the OxyR protein was extremely useful for all of the biochemical studies and was achieved by subcloning *oxyR* behind the highly expressed *tac* or λP_L–P_R promoters together with deletions of the *oxyR*-negative regulatory element in the *oxyR* promoter.[19,30,31] A large percentage of the OxyR protein was in the cell pellets on lysis for all cases of overproduction. OxyR was purified from both the supernate and the pellet by three different approaches. Storz *et al.*[30] purified OxyR to near homogeneity by fractionation of the soluble fraction on heparin Sepharose, Mono S (Pharmacia, Piscataway, NJ), and an affinity column constructed with *katG* promoter sequences. Bölker and Kahmann[19] achieved 90% purification on extracting the cell pellet with 1 *M* NaCl. Tao *et al.*[31] resuspended the cell pellet in 500 m*M* NaCl and then obtained additional purification by a 60% ammonium sulfate precipitation followed by fractionation on a phosphocellulose column.

The similarity between OxyR and the LysR family of regulatory proteins suggested that OxyR could bind specifically to DNA. Indeed, extracts derived from OxyR-overproducing strains effectively protected *oxyR*, *ahpC*, *katF*, and *mom* promoter fragments from DNase I digestion in footprinting assays.[16,31] The protected sequences correlated with the pro-

[30] G. Storz, L. A. Tartaglia, and B. N. Ames, *Science* **248,** 189 (1990).

[31] K. Tao, K. Makino, S. Yonei, A. Nakata, and H. Shinagawa, *J. Biochem.* (*Tokyo*) **109,** 262 (1991).

moter regions found to be important for *oxyR* regulation as shown by the gene fusion studies described above. Mobility shift assays also allowed the OxyR–DNA binding constants to be determined.[31,32] Searches for "OxyR boxes" in the promoter regions of the *oxyR*-regulated genes proved to be misleading as the footprinting analysis showed that true OxyR binding sites share very little primary sequence similarity.

Polyclonal antiserum raised against an OxyR–β-galactosidase fusion protein was another important tool in the biochemical analysis of OxyR action.[30] The antibodies were raised on generation of the 150-kD fusion protein which could be separated from other *E. coli* proteins on a polyacrylamide gel. The gel containing the fusion protein was excised and the fusion protein was eluted and injected into rabbits. The polyclonal serum obtained was then used to examine the levels of the OxyR protein on Western blots of untreated and hydrogen peroxide-treated cells. The rate of OxyR synthesis could also be determined by pulse-labeling cells for 2-min intervals after treatment with hydrogen peroxide and then immunoprecipitating the labeled OxyR protein. The finding that neither the levels nor the synthesis of OxyR increased suggested that OxyR had to be activated by a modification of the OxyR protein.

The ability of OxyR to respond to oxidative stress and to activate transcription was examined in *in vitro* transcription–translation and *in vitro* transcription assays.[30] The coupled *in vitro* transcription–translation experiments were carried out with extracts of OxyR-overproducing strains. These extracts were incubated with NTP, amino acids, [^{35}S]methionine, and a plasmid template encoding the *oxyR*-regulated *ahpCF* genes and the control *bla* gene. The labeled protein products were then examined on polyacrylamide gels. For the *in vitro* transcription assays, purified OxyR and purified RNA polymerase holoenzyme were incubated with plasmid DNA. For these assays, the transcription products were examined by reverse transcriptase extension of a labeled primer complementary to the RNA. The results with both types of assays were qualitatively similar. OxyR activated expression of the *ahpC* gene but not of the control *bla* gene in the absence of hydrogen peroxide, suggesting that the OxyR protein was activated on release from the reducing environment of the bacterial cell. This conclusion was substantiated by *in vitro* experiments showing that OxyR reduced by treatment with 100 m*M* dithiothreitol (DTT) under aerobic conditions and 10 m*M* DTT under anaerobic conditions no longer activated expression. On removal of the 100 m*M* DTT by passage over a small Sephadex column, OxyR could again activate expression. In addition, critical footprinting experiments showed that

[32] L. A. Tartaglia, C. J. Gimeno, G. Storz, and B. N. Ames, *J. Biol. Chem.* **267,** 2038 (1992).

OxyR binds to DNA under both oxidizing and reducing conditions but that the footprints are different.

The biochemical techniques described above, overproduction and purification and *in vitro* assays to determine DNA binding or transcriptional activation, should be useful in evaluating the action of other regulators of oxidative stress genes. These assays should, however, take into consideration that the functions of other regulators could be affected by exposure to oxygen on lysis of the bacterial cells. The biochemical approaches may also be used to isolate regulators if the corresponding genes have not been identified and cloned using genetic strategies. First, a target gene that is regulated by oxidative stress must be identified and the *cis*-acting elements critical for regulation need to be delineated using gene fusion constructs described above. These *cis*-acting elements can then be used in mobility shift assays to identify proteins in crude cell extracts that bind to the regulatory regions. Once a specific mobility shift is detected, the mobility shift assay can be used to monitor purification of the factor. Finally, additional biochemical studies (chemical analysis of the modifications, structural studies) are also necessary to address the persisting questions (nature of the oxidative modification) about the known oxidative stress regulators.

Acknowledgments

We thank Dr. Richard Klausner and Dr. Paul Ross for helpful comments.

Section III

Adherence

[16] Bacterial Lectinlike Adhesins: Determination and Specificity

By JANINA GOLDHAR

Introduction

Bacterial lectinlike adhesins are proteinaceous structures located on the surface of bacterial cells that mediate the specific adhesion of the bacteria to the host cells. These recognizing proteins bind to the sugar components of glycoproteins or glycolipids on the surface of target cells. The capacity of bacteria to bind specifically to host cells is considered an important virulence factor involved in the initial step of infection.

The adhesins of various bacterial pathogens, mainly *Escherichia coli,* have been studied extensively during the last two decades. This article presents the experimental procedures employed in studies on *E. coli* lectinlike adhesins, focusing on models for testing adhesins and for determining glycoprotein receptors of the adhesins.

Most of the procedures have been used, or can be adapted, for studies on other bacterial adhesins. More extensive information about structure, biogenesis, biological function, and interaction with glycolipid receptors is discussed elsewhere in this volume and can be found in the scientific literature.[1-9]

Experimental Models for Detection and Characterization of Adhesins

Erythrocytes as Target Cells

Hemagglutination. The hemagglutination (HA) test has been widely used for the detection and characterization of lectins. The large natural

[1] J. P. Duguid and D. C. Old, *in* "Bacterial Adherence" (E. H. Beachey, ed.), Recept. Recognition, Ser. B, Vol. 6, p. 185. Chapman & Hall, London, 1980.

[2] I. Ofek and E. H. Beachey, *in* "Bacterial Adhesins" (E. H. Beachey, ed.), Recept. Recognition, Ser. B, Vol. 6, p. 2. Chapman & Hall, London, 1980.

[3] N. Sharon and I. Ofek, *in* "Microbial Lectins and Agglutinins" (D. Mirelman, ed.), p. 55. Wiley, New York, 1986.

[4] D. C. Old, *in* "The Virulence of *Escherichia coli*" (M. Sussman, ed.), p. 287. Academic Press, London, 1985.

[5] K. Jann and H. Hoschutzky, *Curr. Top. Microbiol. Immunol.* **151,** 55 (1991).

[6] F. K. de Graaf, *Curr. Top. Microbiol. Immunol.* **151,** 29 (1991).

[7] J. Hacker, *Curr. Top. Microbiol. Immunol.* **151,** 1 (1991).

variability of glycoproteins and glycolipids on the surface of erythrocytes (RBCs) of various animal species provides a tool for evaluating the specificity of bacterial adherence.

Duguid and Old[1,4] were the first to use the agglutionation of RBCs of different animal species as a basis for classification of *E. coli*. They found that the effect of experimental conditions on the results of HA depends on the type of hemagglutinin. Thus, the HA activity of group I strains of *E. coli* (later defined as carrying type I fimbriae) is best developed in a stationary phase of bacterial culture grown in liquid medium at 37°. The HA activity may be expressed, albeit less strongly, in a culture grown at 20° or on solid medium. These authors also observed that D-mannose and its derivatives inhibited HA, and these were designated mannose sensitive (MS). HA that was not inhibited by D-mannose was designated mannose resistant (MR). In contrast to MS adhesins, MR adhesins are best expressed on solid medium at temperatures above 18–20°. The MR hemagglutinins possess liability for elution at temperatures above 45°, and were designated mannose resistant eluting (MRE).[10] Thus, the choice of growth conditions of the bacteria for evaluating HA capacity is crucial. It is recommended that an *E. coli* tested for expression of type 1 fimbriae or of MR adhesins be subcultured a number of times under appropriate conditions (liquid or solid medium).[11]

Many *E. coli* clinical isolates can express two or more types of adhesins according to growth conditions, a possibility that must be considered when interpreting results.[1,11]

Evans *et al.*[12] used the MS HA and MR HA of RBCs of various animal species to detect and type clinical isolates of enterotoxigenic *E. coli* (ETEC). They proposed a typing system involving agglutination of human group A, bovine, chicken, and guinea pig RBCs in the presence or absence of D-mannose.

Procedure for Testing MR HA.[13] Bacteria are grown on a solid medium (CFA) containing 1% casamino acids, 0.15% yeast extract, 0.005% $MgSO_4$, 0.0005% $MnCl_2$, and 2% agar, pH 7.4, at 37° for 18–24 hr (this is one of the most suitable media for promoting expression of various MR

[8] N. Gilboa-Garber and N. Garber, *in* "Glycoconjugates: Composition, Structure and Function" (H. J. Allen and E. C. Kisailus, eds.), p. 541. Dekker, New York, 1992.

[9] H. Leffler and C. Svanborg-Edén, *in* "Microbial Lectins and Agglutinins" (D. Mirelman, ed.), p. 83. Wiley, New York, 1986.

[10] S. M. Ip, P. B. Crichton, D. C. Old, and J. P. Duguid, *J. Med. Microbiol.* **14,** 223 (1981).

[11] I. Ofek, J. Goldhar, Y. Eshdat, and N. Sharon, *Scand. J. Infect. Dis., Suppl.* **33,** 61 (1982).

[12] D. J. Evans, Jr., D. G. Evans, and H. L. DuPont, *Infect. Immun.* **23,** 336 (1979).

[13] D. G. Evans, D. J. Evans, Jr., S. Clegg, and J. A. Panley, *Infect. Immun.* **25,** 738 (1979).

adhesins of human isolates of *E. coli*), harvested into phosphate-buffered saline, pH 7.2 (PBS), washed once, and resuspended in PBS to required density (generally 1×10^{10} bacteria/ml). The RBC suspensions are prepared by drawing human group A, bovine, guinea pig, or chicken blood into an anticoagulant and diluted 1 : 4 with PBS or 1% D-mannose in PBS.

Bacterial suspension is mixed with a drop (20–50 μl) of an appropriate RBC suspension on an objective slide, and the resulting HA is recorded after 1–2 min of gentle rotation and another 2 min on the surface of ice. The HA was designated MR if the same degree of HA occurred in the presence and absence of D-mannose, and MS if it was reduced in the presence of D-mannose. This procedure used by various laboratories is recommended for screening individual colonies as well as a confluent growth of bacteria.

Modifications include the type of anticoagulant and concentration of RBCs (3–5%).[14,15]

Titration of HA capacity with good test standardization is performed with freshly prepared bacterial suspension (see above), twofold serially diluted and mixed (25–50 μl v/v) with RBC suspension. The HA titer will be the minimal concentration of bacteria causing HA (minimal HA unit, MHU).

The titration can be performed by slide agglutination or tube agglutination or in microtiter plates (U). The last method is the most convenient for titration and HA inhibition assays, although some types of bacterial adhesins (such as colonization factor antigens I and II carrying *E. coli* with bovine RBCs) and some types of RBCs (like chicken) cannot be tested in microtiter plate wells because the arrangement of the sediment on the bottom of the well complicates interpretation of the results. The optimal conditions for HA, including pH, temperature, type of microtiter plate, and agitation must be established for each type of adhesin. The interindividual differences in RBCs in the same species should also be taken into account.

HA is a secondary effect of adhesion and requires a critical density of adhesin molecules on the bacterial surface for the binding of adhesin-carrying bacteria or cell-free adhesin molecules to many erythrocytes (to be polyvalent). Indeed, *E. coli* carrying colonization factor antigen (CFA) fimbriae caused MR HA, whereas the isolated adhesins did not.[13] Disappearance of HA caused by bacteria at high temperatures (eluting hemagglutinins, as described by Duguid *et al.*[1,10]) may be due to release of the

[14] G. Kallenius, R. Mollby, S. B. Svenson, J. Winberg, A. Lundblad, S. Svensson, and B. Cedergren, *FEMS Microbiol. Lett.* **7,** 297 (1980).

[15] J. Goldhar, R. Perry, and I. Ofek, *Curr. Microbiol.* **11,** 49 (1984).

hemagglutinins from the bacterial surface, which causes them to become monovalent.

The hemagglutinins (adhesins) of *E. coli* are usually associated with fimbrial structures protruding from bacterial surfaces. Duguid and coworkers[16] were the first to observe on electron microscopy (EM) that not all *E. coli* strains causing HA are fimbriated. Subsequently, *E. coli* strains carrying nonfimbrial adhesins originating from both enteric and extraintestinal infections were isolated and characterized. In contrast to fimbrial hemagglutinins, the isolated nonfimbrial adhesins retained polyvalency and expressed HA activity.[5,15,17,18] Jann and Hoschutzky[5] reported that isolated and purified fimbria types 1, S, and P retained the adhesive capacity (optimal at pH 7.2), but HA was observed only at lower than neutral pH and in the presence of divalent ions. Salit and Gotschlich[19] reported that the isolated type 1 fimbriae caused HA only when they were crosslinked by specific antisera, causing the fimbriae to aggregate and become polyvalent. Jann and Hoschutzky[5] claimed that the isolated fimbriae were monovalent when the adhesin (the receptor-recognizing protein) and major fimbrial proteins were distinct subunits (P, S, type 1). They showed that when the major subunit of the adhesive polymer served as a receptor-recognizing protein, the isolated adhesin was polyvalent and caused HA, like nonfimbrial adhesins. The assembly of the subunits into rodlike structures may mask the receptor binding sites along the fimbriae, leaving only the terminal subunit free to bind to receptor, as described for CFA-I.[20]

Hemadhesion. The capacity to bind to RBCs of monovalent fimbriae isolated from bacteria causing HA can be tested using the hemadhesion test described by Hoschutzky *et al.*[21] Briefly, RBC suspension is distributed into the wells of a microtiter plate in which the test adhesin has been immobilized. After incubation, the nonadhering RBCs are gently washed out and the remaining (attached) erythrocytes are lysed with water. The amount of released hemoglobin is measured with an enzyme-linked immunosorbent assay (ELISA) reader, providing quantitation of the binding reaction.

Hemabsorption. In a bacterial population carrying gene complex encoding a particular lectin, not all bacterial cells express surface adhesin.

[16] J. P. Duguid, S. Clegg, and M. I. Wilson, *J. Med. Microbiol.* **12,** 213 (1979).

[17] I. Ørskov, A. Birch-Andersen, J. P. Duguid, J. Stenderup, and F. Ørskov, *Infect. Immun.* **47,** 191 (1985).

[18] J. Goldhar, R. Perry, J. R. Golecky, H. Hoschutzky, B. Jann, and K. Jann, *Infect. Immun.* **55,** 1837 (1987).

[19] I. E. Salit and E. C. Gotschlich, *J. Exp. Med.* **146,** 1169 (1977).

[20] T. Buhler, H. Hoschutzky, and K. Jann, *Infect. Immun.* **59,** 3876 (1991).

[21] H. Hoschutzky, F. Lottspeich, and K. Jann, *Infect. Immun.* **57,** 76 (1989).

Many lectin-producing bacteria undergo phase variation.[3,22] The percentage of bacterial cells expressing the lectin depends on the strain and growth conditions.[11] In a bacterial population containing a mixture of cells, a subpopulation rich in lectin-expressing bacteria may be separated from the non-lectin-expressing bacteria, using the ability of bacteria to elute HA at elevated temperatures.[17] This procedure was also employed to enrich an *E. coli* population carrying nonfimbrial adhesin type 1 (NFA-1).[18] Briefly, a bacterial suspension in PBS containing 2×10^9 cells/ml was mixed with 1 ml of human packed RBCs, and the mixture was incubated overnight at 4°. The bacterial suspension above the settled RBCs contained mostly cells lacking hemagglutinin, whereas the sediment contained RBCs with attached bacteria. After centrifugation at 4° (1000 *g*, 5 min), the sediment was switched to 40° for 30 min and the RBCs were separated from eluted bacteria by 10 min of centrifugation at 1000 *g*. The eluted bacterial suspension possessed higher HA activity than the original one. A correlation was found between the HA titer and the degree of adherence to tissue culture cells.[18] Such a correlation between HA and adherence was also found in other systems.[23,24]

Agglutination of Other Types of Target Cells or Receptor-Carrying Particles

Because erythrocytes express multiple receptors on their surfaces it is sometimes more convenient to use target cells with fewer receptors for the agglutination test. This is the reason why some laboratories replaced guinea pig erythrocytes with mannan-carrying yeasts in their studies on type 1 fimbriae[25–27]; they used the standardized yeast suspension and monitored and quantitated the reaction with a Payton aggregometer. The method is simple, reproducible, and specific in all cases requiring monitoring of MS.

Agglutination of latex beads coated with Gal(α1–4)Gal (sugar component of P-fimbria receptor) has been used to detect P fimbriae in clinical isolates of *E. coli*, providing a convenient experimental model for studying P fimbriae.[28]

[22] B. Nowicki, M. Rhen, V. Väisänen-Rhen, A. Pere, and T. Korhonen, *J. Bacteriol.* **160,** 691 (1984).

[23] A. Tavendale and D. C. Old, *J. Med. Microbiol.* **20,** 345 (1985).

[24] R. Freter and G. W. Jones, *Rev. Infect. Dis.* **5,** Suppl., S647 (1983).

[25] I. Ofek and E. H. Beachey, *Infect. Immun.* **22,** 247 (1978).

[26] T. K. Korhonen, *FEMS Microbiol. Lett.* **6,** 421 (1979).

[27] K. Jann, G. Schmidt, E. Blumenstock, and K. Vosbeck, *Infect. Immun.* **32,** 484 (1981).

[28] S. B. Svenson, G. Kallenius, R. Mollby, H. Hultberg, and J. Winberg, *Infection* **10,** 209 (1982).

Epithelial Cells

Since Ofek and Mirelman[29] first suggested that the MS *E. coli* adhesin (lectin) mediates the binding of bacteria to animal cells, it has become generally accepted that the capacity of bacteria to express various lectins is a determinant of their capacity to colonize the host tissues and cause infection. This, of course, makes the choice of target cells for adherence studies critical. Interaction of bacteria with epithelial cells may be examined by several methods.

Adherence of Bacteria to Epithelial Cells in Suspension. Ofek and Mirelman[29] used buccal epithelial cells to study type 1 fimbriae *in vitro*. Briefly, buccal cells were scraped from the oral cavity, washed and suspended in PBS, and incubated with the bacterial suspension. The nonadherent bacteria were separated from the epithelial cells with attached bacteria by differential centrifugation and washing with PBS. The epithelial cells were placed on a microscopic slide and stained with gentian violet, and the adherence was recorded by microscopic counting.

Svanborg-Eden *et al.*[30] described a similar procedure for testing adherence of *E. coli* responsible for urinary tract infections. Uroepithelial cells from the urine of healthy women were used, and the attachment was quantitated by phase-contrast microscopy. Other quantitation techniques for studying attachment to uroepithelial, and ileal epithelial cells include radiolabeling and EM.[31-33] The advantage of epithelial cells is their relevance to the natural site of infection. Most studies quantitate the adherence to epithelial cells compared with HA, a common reference for lectin activity. The disadvantage of epithelial cells in adherence studies is a variability in receptor expression. Differences have been observed in composition, availability, and density of particular receptors (1) among epithelial cells of various mucosal surfaces such as buccal vs urinary,[2,30] (2) among various types of cells (urinary squamous vs transitional), (3) between individuals (depending on blood group, secretors and nonsecretors), and (4) in the same individual at different times.[34] When bacterial adherence to epithelial cells is used to test binding capacity of the host cells, the bacteria must be standardized. When adhesins are characterized, the maximal stability of the target cells should be ensured.[2]

[29] I. Ofek and D. Mirelman, *Nature (London)* **265,** 623 (1977).

[30] C. Svanborg-Edén, B. Eriksson, and L.-A. Hanson, *Infect. Immun.* **18,** 767 (1977).

[31] C. F. Deneke, K. McGowan, G. M. Thorne, and S. I. Gorbach, *Infect. Immun.* **39,** 1102 (1983).

[32] C. P. Cheney and E. C. Boedeker, *Infect. Immun.* **39,** 1280 (1983).

[33] I. Ørskov, F. Ørskov, and A. Birch-Andersen, *Infect. Immun.* **27,** 657 (1980).

[34] H. Lomberg, B. Cedergren, H. Leffler, B. Nilsen, A. S. Carlström, and C. Svanborg-Edén, *Infect. Immun.* **51,** 919 (1986).

A technical problem of the adherence assay based on incubation of bacterial suspension with eukaryotic cell suspension (epithelial or other) is the separation of nonattached bacteria from animal cells with attached bacteria. Differential centrifugation[29] and filtration[31] have been used. The separation is more effective when the target cells are immobilized on slides or in microtiter plate wells prior to the addition of bacterial suspension.[35] Monolayers of tissue culture cells offer an advantage here, and various types of cell lines have been used for studying adherence of *E. coli* and other bacteria. These are discussed below.

Adherence to Tissue Culture Cell Lines and Quantitation of the Binding. Vosbeck and collaborators[36,37] used epithelioid cell line, intestine 407 (ATCC-CCL6) and a number of other cell lines to test adherence of fecal and urinary isolates of *E. coli*. Measuring the binding by viable counting or radiolabeling bacteria (with ^{14}C) enabled the authors to analyze the kinetics of adhesion and determine the linear relation between the numbers of added and bound bacteria. Analysis of the data by Scatchard plot revealed the specificity of binding to particular cell lines.[37] HEp-2 and HeLa epithelial cells have been widely used for testing adherence of *E. coli* representing various HA patterns originating from various clinical sources.[38,39] In these studies, bacterial binding was evaluated by microscopic examination of Giemsa-stained preparations. We examined the adherence of various *E. coli* isolates to the human kidney cell line (HK), which is genetically closely related to the HeLa cell line, and found it suitable for testing adherence of *E. coli* strains expressing both MS and certain MR adhesins.[15,18,40]

Testing Adherence of Bacteria to Cell Monolayers by ELISA. An ELISA-type assay was used to analyze adherence to the monolayers of tissue culture cell lines and to various types of cells (epithelial and polymorphonuclear leukocytes) immobilized in microtiter plate wells.[35,41] The procedure, which was used to analyze the adherence of *E. coli* expressing NFA-1 and NFA-2,[18] is described here: The suspension (100 μl of

[35] I. Ofek, H. S. Courtney, D. Schifferli, and E. H. Beachey, *Clin. Microbiol.* **24,** 512 (1986).

[36] P. S. Cohen, A. D. Elbein, R. Solf, H. Mett, J. Regos, E. B. Menge, and K. Vosbeck, *FEMS Microbiol. Lett.* **12,** 99 (1981).

[37] K. Vosbeck and U. Huber, *Eur. J. Clin. Microbiol.* **1,** 22 (1982).

[38] M. J. Bergman, W. S. Updike, S. J. Wood, S. E. Brown, III, and R. L. Guerrant, *Infect. Immun.* **32,** 881 (1981).

[39] A. Craviato, R. J. Gross, S. M. Scotland, and B. Rowe, *Current Microbiol* **3,** 95 (1979).

[40] A. Zilberberg, M. Lahav, R. Peri, and J. Goldhar, *Zentralbl. Bakteriol., Mikrobiol. Hyg., Abt. I, Orig. A* **254,** 234 (1983).

[41] L. Stanislawsky, W. A. Simpson, D. Hasty, N. Sharon, E. H. Beachey, and I. Ofek, *Infect. Immun.* **48,** 257 (1985).

HK cells, 3×10^5 cells/ml) was seeded to each of 96 wells of a polystyrene microtiter plate. After 48 hr at 37° in a 5% CO_2 atmosphere, the monolayers were fixed with glutaraldehyde (0.25% w/v in PBS) for 10 min and treated with 0.2% w/v glycine and 0.1% v/v bovine serum albumin at room temperature. After two washings with PBS, the bacterial suspension or purified adhesin was added (100 μl) and incubated for 30 min at 37°. Following three washings with PBS and fixation with methanol (10 min), the number of bacteria or the amount of adhesin bound to the HK cells was measured with antibacterial or antiadhesin rabbit serum in the ELISA setup, using horseradish peroxidase-conjugated anti-rabbit immunoglobulin and appropriate substrate as the detection system. Adhesion was measured as ELISA binding unit [optical density (OD) at 405 nm] net values after subtraction of OD values of controls. For testing inhibition of bacterial adhesion, the inhibitor to be tested was added to the wells after glycine treatment, incubated for 30 min at 37°, and washed; bacterial suspension at a concentration of about 10^8 cells/ml (previously determined to reach binding saturation) was added and the procedure was continued as described above. The percentage inhibition was calculated according to OD values in the presence of inhibitor compared with those without prior treatment. The assay proved to be reproducible and easy to perform and allowed the handling of a large number of test samples on the same day. It was successfully used in other laboratories for testing adherence of various bacteria.[42–44]

Using the ELISA for adherence studies necessitates certain technical preparations: (1) The antiserum used as detector must be strong enough to detect the bacteria bound. In our hands the minimal number of bacteria per well that could be detected was generally about 1×10^4, whereas the number of bacteria bound per well at the saturation point was at least 1×10^5. (2) A control should be performed to eliminate cross-reaction between components of the detection system and the eukaryotic cells. (3) Nonspecific sites of the plate should be blocked and the OD of the control wells containing the cells only should be determined. Some nonspecific binding might occur if the bottom of the well is not covered by cells. Too high a concentration results in detachment of the cells. (4) The fixation process must be elaborated and optimized for each system. (5) Each plate must contain all necessary control wells. (6) In evaluation of the results, if the study includes one antigen–antibody system (as in the inhibition experiment), the OD net values (ELISA binding units) can be used to

[42] R. Marre, B. Kreft, and J. Hacker, *Infect. Immun.* **58,** 3434 (1990).

[43] S. B. Olmsted and N. L. Norcross, *Infect. Immun.* **60,** 249 (1992).

[44] L. J. Forney, J. R. Gilsdorf, and D. C. L. Wong, *J. Infect. Dis.* **165,** 464 (1992).

quantitate the results and calculate the percentage inhibition, as described above. If different antigen–antibody systems are studied, the ELISA units should be "translated" into the number of bacteria (or amount of adhesin) using the reference curve,[45] as follows: Bacterial suspension containing 10^9 bacteria/ml is serially twofold diluted in sterile distilled water, and the duplicates of each concentration (range 10^9–10^4) are distributed (100 μl) into the wells. The bacteria are dried overnight at 37°, and ELISA is performed as described above. The OD values show a linear relation to the number of bacteria per well in the range 10^4 to 5×10^7 (the data vary according to the detection system). In all binding experiments measured by ELISA, the derived curve serves for calculation of the number of bacteria or amount of adhesin bound per well.

This procedure is reproducible and has also been used to determine the expression of adhesin on the surface of the bacteria.

Relevance of Cell Lines as Models for Studying Adherence

As mentioned above, the main purpose for studying adherence of clinical isolates is to determine the role of the adhesins (lectins) in the development of natural infection. Numerous authors have found a correlation between HA, adherence *in vitro* to particular cell lines, and bacterial pathogenicity. Epithelial cell lines are not differentiated, however, and the availability and density of receptors specific to particular pathogens (and their adhesins) do not necessarily correspond to natural conditions.

In an effort to improve the experimental model for studying specificity of attachment of diarrheogenic *E. coli,* Nesser *et al.*[46] used a cell line originating from human colon cancer (HT-29). Grown under suitable conditions (in the presence of galactose), the cells exhibited enterocytic differentiation and formed polarized epithelium with an apical brush border. The adherence of *E. coli* expressing CFA was measured by Giemsa staining, immunofluorescent antibodies (IFA), and radiolabeling of bacteria. This last technique provides reproducible and objective quantitation, but, in contrast to microscopic examination, does not pinpoint the site of binding. The HT-29 line was used in another study to test attachment of enteric pathogens and the specificity of binding of uropathogenic *E. coli* carrying MS and MR adhesins.[47] The methodology revealed that both types of adhesins mediated the binding, but with different binding patterns. Other

[45] I. Ofek and A. Athamna, *J. Clin. Microbiol.* **26,** 62 (1988).

[46] J. R. Neeser, A. Chambaz, M. Golliard, H. Link-Amster, V. Fryder, and E. Kolodziejczyk, *Infect. Immun.* **57,** 3727 (1989).

[47] A. E. Wold, M. Thorssen, S. Hull, and C. Svanborg-Edén, *Infect. Immun.* **56,** 2531 (1988).

differentiated cell lines have been used to test binding (and invasion) of diarrheogenic *E. coli,* including Caco-2, which expresses some markers of normal intestinal villus cells.[48,49]

In the search for an experimental model close to natural conditions, some authors used biostomy or biopsy tissues from relevant organs of animals and humans.[31,32] Virkola *et al.*[50] studied the binding of uropathogenic *E. coli* carrying various adhesins (P, S, type 1, type 1C, and 075X) to frozen tissue sections obtained from different sites in the urinary tract. Immunofluorescent staining enabled the authors to determine the distribution of specific receptors in various compartments of the urinary tract and the differences in the binding specificity of various adhesins. The disadvantages of this kind of specimen are difficulty in obtaining material and fluctuations in results because of interindividual differences.

Table I lists a number of experimental models for detection and quantitation of *E. coli* attachment.[50a,50b] In general, procedures should be chosen according to the relevance of the target cells to natural infection and the specificity of the tested adhesin, and the reliability and convenience of performance and quantitation of multiple samples.

Determination of Specificity of Adhesin

The above-described experimental models can be used for the following purposes[24]:

1. To compare attachment of a wild strain (clinical isolate) with its variants, mutants, or transformants that differ in expression of the adhesin
2. To provide evidence that the adhesin is expressed on the surface of the bacteria
3. To show that an isolated adhesin and the whole bacteria bind in a similar manner
4. To determine the inhibition of binding of the parent bacteria by the isolated adhesin
5. To determine the inhibition of binding of bacteria by adhesin-specific antibodies (polyclonal or monoclonal)

[48] B. B. Finlay and S. Falkow, *J. Infect. Dis.* **162,** 1096 (1990).

[49] S. Kerneis, G. Chauviere, A. Darfeuille-Michaud, D. Aubel, M.-H. Coconnier, B. Joly, and A. L. Servin, *Infect. Immun.* **60,** 2572 (1992).

[50] R. Virkola, B. Westerlund, H. Holthöfer, J. Parkkinen, M. Kekomaki, and T. Korhonen, *Infect. Immun.* **56,** 2615 (1988).

TABLE I
PROCEDURES FOR ESTIMATING ADHESION OF *Escherichia coli* TO EUKARYOTIC CELLS

Target cells	Procedure	Adhesin involved	References[a]
RBCs (guinea pig, human, bovine etc.)	Hemagglutination	Type 1	1, 11, 12
	Hemagglutination	Various MR (e.g., CFA, P)	12–15
	Hemadhesion	P	21
Yeasts	Agglutination	Type 1	25
Coated latex particles	Agglutination	P	28
Epithelial cells (buccal, urinary, intestinal)	Microscopic, stained	Type 1	29
	Microscopic, phase-contrast	Type 1, P	30, 32
	EM	Type 1, MR	30
	Radiolabeling	Type 1, CFA, MR	31, 47
	ELISA	Type 1, MR	35
Cell lines (HEp-2, HeLa, Int 407, HK)	Microscopic, stained	Type 1, various MR	23, 38
	Viable count	Type 1, various MR	37
	EM	Type 1, various MR	37
	Radiolabeling	Type 1, various MR	37
	ELISA	Type 1, various MR	18, 35
Differentiated cell lines (HT-29, CaCo-2)	EM	CFA	46
	Radiolabeling	CFA	46
	Immunofluorescence assay	CFA	46
	Microscopic, phase contrast	Type 1, P, other MR	47
Human and animal tissues (including experimental infection *in vivo*)	Radiolabeling	Type 1 and MR	50a
	Immunofluorescence assay	Type 1, type lC, P, S	13, 50
	Viable count	CFA	50b

[a] References to the literature numbered as per text footnotes.

Next are some examples of experimental procedures employed for determination of adhesin (lectin) specificity. Evans *et al.*[13] showed that following preincubation of *E. coli* carrying CFA-I with antibodies prepared against the purified adhesin, the HA caused by the bacteria was inhibited. Preexposure of intestinal mucosa of infant rabbits to purified CFA-I caused inhibition of adhesion of *E. coli* carrying CFA-I.

In another study conducted to determine P-fimbria receptor-specific interaction, Fab fragments of the antifimbria antibodies were used for inhibition of HA and bacterial adhesion to human uroepithelial cells.[51,52] In several structural and functional studies on adhesins, monoclonal anti-

[50a] A. Zilberberg, I. Ofek, and J. Goldhar, *FEMS Microbiol. Lett.* **23,** 103 (1984).
[50b] J. Goldhar, A. Zilberberg, and I. Ofek, *Infect. Immun.* **52,** 205 (1986).
[51] T. K. Korhonen, S. Eden, and C. Svanborg-Edén, *FEMS Microbiol. Lett.* **7,** 237 (1980).
[52] M. Rhen, P. Klemm, E. Wahlström, S. B. Svenson, G. Kallenius, and T. K. Korhonen, *FEMS Microbiol. Lett.* **18,** 233 (1983).

bodies were used instead of polyclonal antibodies specific for the adhesin.[20,53,54]

When the ELISA-type test described above was used to measure the inhibition of binding of bacteria by purified adhesins, the two nonfimbrial adhesins, NFA-1 and NFA-2, expressed by two *E. coli* urinary isolates were found to possess partial binding cross-reactivity.[18]

Determination of Receptor Specificity of Adhesins

As already established in plant lectins, HA is the most convenient test for studies on the receptor specificity of bacterial adhesins. The classic example is the type 1 fimbria, an MS lectin, first characterized by the inhibition of HA by D-mannose and related structures. D-Mannose is a component of the glycoprotein receptors found on various eukaryotic cell surfaces. In addition to MS lectin, the sugar binding specificity of a number of MR lectins of *E. coli* has been determined, including Gal(1–4)Gal, which is a component of the glycolipid receptor of P fimbriae (Pap adhesin[9,51]), and *N*-acetyl-D-glucosamine and NeuAc(α2–3)Gal, which are components of glycoprotein receptors of fimbriae G[55] and S,[56] respectively. Experimental approaches recommended for determination of receptors for bacterial lectinlike adhesins present on the surface of various eukaryotic cells include[57] (1) inhibition of binding by putative receptor components such as mono- and oligosaccharides; (2) inhibition of binding by competing lectins possessing known binding specificity; (3) reduction of binding after enzymatic or chemical modification of the putative receptor-containing complex; and (4) binding to target cells expressing the known receptor or to isolated immobilized receptor component.

Some of the experimental procedures for determination of receptors expressed on the erythrocytes are described below. Similar procedures have been used for determination of receptors on other types of target cells.

Hemagglutination Inhibition by Putative Receptor or Receptor Analog

Erythrocytes (RBCs) and hemagglutinin are prepared as described for the HA test, preferably using the suspension in PBS containing

[53] S. N. Abraham, J. P. Babu, C. S. Giampapa, D. L. Hasty, W. A. Simpson, and E. H. Beachey, *Infect. Immun.* **48,** 625 (1985).

[54] T. Moch, H. Hoschutzky, J. Hacker, K.-D. Kronke, and K. Jann, *Proc. Natl. Acad. Sci. U.S.A.* **84,** 3462 (1987).

[55] V. Väisänen-Rhen, T. K. Korhonen, and J. Finne, *FEBS Lett.* **159,** 233 (1983).

[56] T. K. Korhonen, V. Väisänen-Rhen, M. Rhen, A. Pere, J. Parkkinen, and J. Finne, *J. Bacteriol.* **159,** 762 (1984).

[57] G. W. Jones and R. E. Isaacson, *CRC Crit. Rev. Microbiol.* **10,** 229 (1983).

D-mannose (2%) for testing MR hemagglutinins. Bacterial suspension or isolated adhesin should be adjusted to the required concentration according to titration (see Hemagglutination Test), performed with the reagents volumes and experimental conditions used in the inhibition assay. A panel of monosaccharides, oligosaccharides, polysaccharides, or glycoconjugates is prepared in solution in PBS (pH 7.1–7.4), at a concentration of 100 m*M*.

HA Inhibition Test. The putative inhibitors are serially diluted and mixed with the bacterial suspension or the isolated adhesin (4–6 MHU, v/v). After 30 min of incubation at ambient temperature the RBC suspension is distributed into microtiter plate wells. Following incubation at the optimal conditions for each HA system, the inhibition titer is recorded as a percentage of the values of the control without inhibitor. Controls containing only the highest concentration of inhibitor and RBCs should be included.[15,58] An appropriate panel of oligosaccharides possessing various linkages or differing in a secondary or tertiary structure (linear, branched or aromatic) allows deduction of the binding site of the lectin that corresponds to the structure causing the greatest inhibition. In general, di- and trisaccharides are more active than monosaccharides; glycolipids and glycoproteins are most active but the results are difficult to interpret. Several studies have been successfully conducted by HA inhibition: (1) determination of the configuration of the binding site of MS lectin of *E. coli,* compared with MS lectin of *Salmonella* and concanavalin A (all recognize D-mannose[58,59]; (2) a set of neuraminic acid derivative oligosaccharides and polysaccharides used to determine the receptor of S fimbriae, recognizing (α2–3)-linked sialyl galactosides[56]; (3) experiments showing that sialic acid is the most potent inhibitor of binding of CS2 fimbriae[60]; and (4) the testing of various structural families of naturally occurring sialopeptides as potential inhibitors of HA mediated by CFA-1 and CFA-II, to estimate the structures of the receptors of these sialic acid-recognizing adhesins.[61]

HA inhibition experiments carried out for determination of carbohydrate receptor components of four different *E. coli* MR lectins are presented in Table II.

Despite the relative simplicity of the HA inhibition test, it can be problematic. Difficulties include the influence of some complex carbohydrates on pH and on the agglutinability of RBCs (then the high limit

[58] N. Firon, I. Ofek, and N. Sharon, *Infect. Immun.* **43,** 1088 (1984).

[59] I. Ofek and N. Sharon, *Curr. Top. Microbiol. Immunol.* **151,** 91 (1991).

[60] P. O. Sjöberg, M. Lindhal, J. Porathand, and T. Wadström, *Biochem. J.* **255,** 105 (1988).

[61] J.-R. Neeser, A. Chambaz, K. Y. Hoang, and H. Link-Amster, *FEMS Microbiol. Lett.* **49,** 301 (1988).

TABLE II
CARBOHYDRATE BINDING SPECIFICITY OF *Escherichia coli* MR ADHESINS BY INHIBITION OF HEMAGGLUTINATION

Inhibitor	Sugars and glycoconjugates used for inhibition of HA caused by[a]			
	G-fimbriae[b]	S-fimbriae[c]	CS2[d]	NFA-4[e]
N-Acetyl-D-glucosamine	2.1		>100	
D-Glucose	32		>100	
D-Mannose	130		n.d.	
L-Fucose	>130		>100	
D-Galactose	>130		>100	
N-Acetyl-D-galactosamine	>130		>100	
Orosomucoid	>1.9 mg/ml	0.001		>1 mg/ml
Asialoorosomucoid	>1.9 mg/ml	None		
Agalactosylorosomucoid	0.48 mg/ml			
NeuAc(α2–3)Gal(β1–4)Glc		0.3	6.2	>10.0
NeuAc(α2–3)Gal(β1–4)Glc NAc		0.3		
NeuAc(α2–6)Gal(β1–4)Glc		1.3		
NeuAc(α2–6)Gal(β1–4)Glc NAc		1.3		
NeuAc(α2–8)NeuAc(α2–3)Gal(β1–4)Glc		5.0		
NeuAc		>10.0	32.9	
Gal(β1–4)Glc		>10.0		
Colomic Ac		>10.0		
NeuGc			17.3	
Fetuin				>10 mg/ml
Glycophorin A^{MM}				<1 mg/ml
Asialoglycophorin A^{MM}				<1 mg/ml

[a] Minimal Inhibitory concentration (MIC, m*M*) required to prevent hemagglutination.
[b] Inhibition of hemagglutination of endo-β-galactosidase-treated erythrocytes.[55]
[c] Inhibition of hemagglutination by purified fimbriae.[56]
[d] Concentration of monosaccharides giving 50% hemagglutination of human RBCs by *E. coli* carrying CS2.[60]
[e] Inhibition of hemagglutination of erythrocytes MM by purified adhesin. Indicated are maximal concentrations of sugars tested or minimal concentrations that caused reduction of HA titer from 1 : 2000 to 1 : 10.[66]

concentration of the putative inhibitor must be lowered), the quantitation of results within the range of twofold dilution, and, most critical, the choice of an appropriate set of potential inhibitors.

Firon and co-workers[58,59] confirmed the specificity of the results by demonstrating binding inhibition by the same sugar to other target cells. The authors showed that yeast agglutination caused by fimbria type 1 was inhibited by a range of D-mannose derivatives, including di-, tri-, and aromatic saccharides. *p*-Nitro-*o*-chlorophenylmannose was the best inhibitor of yeast agglutination, as well as of the binding to guinea pig ileal epithelium mediated by the lectin.

One must be cautious in such experiments, however, as the sugar (carbohydrate) component specifically recognized by lectin may be part of various glycoproteins in different types of cells. Moreover, the accessibility of such a receptor varies with the type of tissue. Certain soluble glycoconjugates naturally occurring in body fluids, such as Tamm–Horsfall protein in urine[62] and glycoproteins containing *O*-linked carbohydrates in milk,[61] play an important role in inhibiting colonization by pathogenic bacteria *in vivo*. These can be used as additional tools for elucidation of receptor–lectin interaction.

Enzyme or Chemical Modification of Erythrocytes

Sodium Periodate Treatment. It is generally recommended that the procedure of receptor determination begin with an evaluation of the possible carbohydrate nature of the receptor.[15,29,63] For this purpose, RBCs are treated with sodium periodate: RBC suspension in PBS (5%) is mixed (v/v) with 10 m*M* $NaIO_4$ + 1 *M* glycine; the mixture is incubated in the dark at room temperature for 10 min (the control RBC suspension is incubated in PBS), washed three times with PBS, and adjusted to the required concentration (2.5–3%). Titration of HA is carried out in parallel with treated and untreated RBC. Disappearance of RBC agglutinability following periodate treatment suggests the carbohydrate nature of the receptor.

Neuraminidase Treatment. In the case of bacterial adhesins recognizing sialic acid residues (S fimbriae, CFA-I, and CFA-II), the first indication of specificity is the loss of agglutinability of RBCs after neuramindase treatment.[60,64,65] Modifications of the procedure have been introduced by various authors. According to Hoschutzky *et al.*[66] RBC suspension (10%) in PBS, pH 5, is incubated (v/v) with NANAse (*Clostridium perfringens*) 4 U/ml, at 37° for 1 hr. The subsequent steps—washings, control, titration—are as described above.

[62] J. L. Dunkan, *J. Infect. Dis.* **158,** 1379 (1988).

[63] I. Ofek, H. Lis, and N. Sharon, *in* "Bacterial Adhesion" (D. C. Savage and M. Fletcher, eds.), p. 71. Plenum, New York, 1985.

[64] J. Parkkinen, J. Finne, M. Achtman, V. Väisänen, and T. K. Korhonen, *Biochem. Biophys. Res. Commun.* **111,** 456 (1983).

[65] A. Faris, M. Lindahl, and T. Wadström, *FEMS Microbiol. Lett.* **7,** 265 (1980).

[66] H. Hoschutzky, W. Nimmich, F. Lottspeich, and K. Jann, *Microb. Pathog.* **6,** 351 (1989).

The interpretation of results depends on the specific procedure used. NANase in low concentration enhanced HA caused by P[67] and type 1 fimbriae.[19] HA mediated by S fimbriae was affected by 1 U/ml NANAse treatment, whereas HA mediated by NFA-3 (N blood group specific) was affected above 5 U/ml of NANAse.[68]

Endo-β-galactosidase Treatment.[55] Washed human RBCs (250 μl) were suspended in 650 μl 10 m*M* sodium phosphate (pH 7.4) containing 0.15 *M* NaCl. Ten milliunits of endo-β-galactosidase (from *Escherichia freundi,* Japan) in 100 μl of 50 m*M* sodium acetate buffer (pH 5.8) is added, and the mixture is incubated at 37° for 24 hr and washed. This procedure led to the appearance of terminal *N*-acetyl residues on RBCs.[69] The effect represents enzyme modification, which converts the nonagglutinable RBCs into agglutinable RBCs by a specific lectin. The specificity of the *N*-acetyl-D-glucosamine lectin (G fimbriae) was demonstrated by inhibition of HA of endo-β-galactosidase-treated RBCs by the sugar and by glycoproteins treated by the same enzyme.

Treatment by Proteolytic Enzymes. Treatment of RBCs with trypsin (0.2–1 mg/ml) and pronase (1 mg/ml), at 37° for 30 min provides an indication of the involvement of a protein component in the receptor.[15,66,68,70,71] Interpretation of the results requires treatment by a number of proteolytic enzymes at different concentrations.

In addition to target RBCs, similar chemical and enzymatic modifications can be made in isolated putative receptors, receptor components, glycoproteins, polysaccharides, and oligosaccharides used for inhibition of HA. Korhonen and co-workers[56,64] showed that sialoproteins (orosomucoid, fetuin, and glycophorin A) inhibited HA mediated by S fimbriae while the desialized glycoproteins did not, a finding that provided additional evidence of sialyl function as a receptor. HA mediated by NFA-4, nonfimbrial *E. coli* lectin recognizing blood group antigen MM, was equally well inhibited by glycophorin A^{MM} and its asialoderivative[66] (Table II).

Binding to Target Cells Possessing Known Antigenic Structure: Human Blood Group Antigens as Receptors for Bacterial Lectins. One of the analogies between a number of plant and bacterial lectins is their

[67] T. K. Korhonen, V. Vaisanen, H. Saxen, H. Hultberg, and S. B. Svenson, *Infect. Immun.* **37,** 286 (1982).

[68] J. Grünberg, R. Perry, H. Hoschutzky, B. Jann, K. Jann, and J. Goldhar, *FEMS Microbiol. Lett.* **56,** 241 (1988).

[69] J. Vitala and J. Finne, *Eur. J. Biochem.* **148,** 393 (1984).

[70] P. D. Issit and C. H. Issit, "Applied Blood Group Serology," 2nd ed., p. 183. Cooper Biomed. Inc., Malvern, PA, 1983.

[71] K. Wasniowska, C. Reichert, M. H. McGinniss, K. R. Schroer, D. Zopf, E. Lisowska, L. Messeter, and A. Lundblad, *Glycoconjugates* **2,** 163 (1985).

capacity to recognize blood group antigens. The first described was Pap (P-fimbria) adhesin recognizing P blood group antigen of the glycolipid nature.[9,67] Later, N, M, and Dr antigen-specific adhesins were described.[66,68,72,73]

One of the procedures employed to determine the blood group antigen specificity of bacterial adhesins is a panel of several types of erythrocytes expressing or not expressing particular blood group antigens. The panels are used by blood bank laboratories for detection of unusual antibodies.

The following example (Table III) shows the presumptive determination of NFA-3 binding specificity (N blood group-recognizing adhesin)[68] using the RBC panel (Data-Cyte Plus, Antigenic Construction Matrix, Dade), conducted according to the manufacturer's instructions. NFA-3 agglutinated only these RBCs that contained N antigen, similar to a plant lectin, *Vicia graminea* (Vg) known to recognize the N blood group antigen). Similarly, neither NFA-4 nor anti-M monoclonal antibody agglutinated RBCs lacking M antigen, compared with NFA-1,[18] which did not recognize any of the blood group antigens included in the panel (Table III).

Panels containing RBCs treated by enzymes that selectively destroy some antigens are also available for confirmation of results. For example, ficin-treated RBCs were used to identify N/M blood group specificity.[70] Nowicki *et al.*[73] employed a set of erythrocytes of known phenotypes expressing or lacking antigens of the IFC complex to demonstrate the receptor similarity of Dr, AFA-I, and AFA-III *E. coli* hemagglutinins.

Binding to Isolated Glycoprotein Receptors

To close the circle of receptor identification, it is advisable to demonstrate the specific binding of bacteria or isolated adhesin to the isolated, immobilized receptor glycoprotein compound. Two experimental procedures can be used: (a) immobilization of the substance in the microtiter plate wells and quantitation of the binding by any of the techniques described for estimating binding to target cells; (b) SDS–PAGE of a complex containing a putative receptor, electrophoretic transfer of the separated components onto nitrocellulose (blotting), and estimation of binding of bacteria or of isolated adhesin to the blot.

[72] M. Jokinen, C. Ehnholm, V. Väisänen-Rhen, T. K. Korhonen, R. Pipkorn, N. Kalkkinen, and G. Gahmberg, *Eur. J. Biochem.* **147,** 47 (1985).

[73] B. Nowicki, A. Labigne, S. Moseley, R. Hull, S. Hull, and J. Moulds, *Infect. Immun.* **58,** 279 (1990).

TABLE III

DETERMINATION OF BLOOD ANTIGEN AS RECEPTOR FOR *Escherichia coli* NONFIMBRIAL ADHESIN (NFA-3)[a]

L10078-E [Rev. 3/87]

Dade

Data-Cyte Plus
Antigenic Constitution Matrix

Lot No.: DC-519
Mfg. Date: 5 OCT 87
Exp. Date: 28 NOV 87

Vial no.	Rh	Donor code	Special typings	Rh								*MNS				P
				D	C	E	c	e	f	C^w	V	M	N	S	s	P_1
SCI*																
SCII*																
SCIII*																
1	rr	E366BA		0	0	0	+	+	+	0	0	+	+	0	+	+
2	rr	D938	KK	0	0	0	+	+	+	0	0	+	0	+	+	+
3	r′r	K512SE		0	+	0	+	+	+	0	0	+	+	+	+	0
4	r″r	4193		0	0	+	+	+	+	0	0	+	+	+	0	+
5	rr	K134BA		0	0	0	+	+	+	0	0	+	+	+	0	+
6	Ror	G814BM		+	0	0	+	+	+	0	+	0	+	0	+	+
7	R_1R_1	X556BA		+	+	0	0	+	0	0	0	+	0	+	+	+
8	R_1R_1	9026		+	+	0	0	+	0	0	0	+	+	+	0	+
9	$R_1R_1^w$	G228BA		+	+	0	0	+	0	+	0	0	+	+	+	+
10	R_2R_2	X557BA		+	$+^w$	+	+	0	0	0	0	+	+	+	+	0
11	Ror	D123	‡$Js(a^+b^-)$	+	0	0	+	+	+	0	0	+	+	+	0	+

[a] J. Grünberg and J. Goldhar, unpublished table. Grünberg *et al.*[68]

[b] HA of erythrocytes expressing designated antigens (*) by NFA-3. *Vicia graminea* lectin (Vg), NFA-4[66]; monoclonal antibodies anti-M, and NFA-1.[18]

TABLE III (*continued*)

Name ____________________

Pt. I.D. No. ____________________ Drawn ____________________

Blood Group __________ Rh __________ DAT __________ IoG __________ C3 __________

Interpretation ____________________ Results[b]

____

Technologist ____________________ Date Tested __________

Lewis		Lutheran		Kell				Duffy		Kidd			Vial no.	NFA-3	Anti-N(Vg)	NFA-4	Anti-MAb	NFA-1	Sex linked
Le^a	Le^b	Lu^a	Lu^b	K	k	Kp^a	Js^a	Fy^a	Fy^b	Jk^a	Jk^b	Xg^a							
													SCI*						
													SCII*						
													SCIII*						
+	0	+	+	0	+	0	0	+	0	0	+	0	1	+	+	+	+	+	
0	+	0	+	+	0	0	0	0	+	+	0	+	2	0	0	+	+	+	
0	+	0	+	0	+	0	0	+	+	0	+	0	3	+	+	+	+	+	
$+^w$	0	0	+	0	+	0	0	+	+	0	+	+	4	+	+	+	+	+	
+	0	0	+	0	+	0	+	+	0	+	+	+	5	+	+	+	+	+	
0	0	0	+	0	+	0	0	0	0	+	0	+	6	+	+	0	0	+	
+	0	0	+	0	+	0	0	0	+	+	0	0	7	0	0	+	+	+	
0	+	0	+	+	+	0	0	0	+	0	+	0	8	+	+	+	+	+	
+	0	0	+	0	+	0	0	+	0	+	0	+	9	+	+	0	0	+	
0	0	0	+	0	+	0	0	+	0	+	0	+	10	+	+	+	+	+	
0	+	0	+	0	+	0	+	0	0	+	+	+	11	+	+	+	+	+	
													Cord						

Detection of Binding by Radiolabeling. Parkkinen *et al.*[74] employed ^{125}I-labeled bacteria to identify the receptor of S fimbriae. They studied the binding to glycophorin A adsorbed onto microtiter wells. The bacterial binding was a function of glycophorin A concentration and was inhibited by the same sialyloligosaccharides and sialoglycoproteins that inhibited HA mediated by the bacteria. In another experiment, erythrocyte membrane proteins and purified glycophorin A were separated by SDS–PAGE, transferred onto nitrocellulose, and incubated with the radiolabeled bacteria. Bacterial binding was demonstrated by the radiolabeled band corresponding to a subunit of glycophorin A; pretreatment of the blot with NANAse inhibited the binding. This procedure is elegant and the results are clear-cut, but the technique requires skill.

Detection of Binding by ELISA. In our laboratory, binding of bacteria and isolated adhesins (NFA-1 and NFA-3) to immobilized glycophorin A was determined with an ELISA setup,[68] as described above, with some modifications: Glycophorin A and fetuin (used as control) are dried in microtiter plate wells (0.05–2.0 μg/well) overnight at 37°. (We prefer to dry the substance rather than coat it because drying is more economical.) The wells are filled with 2% bovine serum albumin (BSA) in PBS, incubated for 1 hr and washed once with PBS, and the bacterial suspension or the adhesin is then distributed into microtiter plate wells. Binding of various concentrations of bacteria to glycophorin A in various concentrations was estimated using the ELISA procedure described for binding to tissue culture cells.

Once the binding system was established and characterized, the complementary tests for determination of specificity could be performed. This included comparing binding of bacteria expressing and not expressing the adhesin (variants and mutants); inhibition of binding by antiadhesin antibodies, competing lectin, or receptor components; and testing of binding to glycophorin A pretreated with enzymes before addition of the bacteria.

Identification of Receptor Component by Immunoblotting. The procedure was employed to determine receptors for *E. coli* MS adhesin on human granulocytes.[75] The purified protein antigens or the granulocyte lysates were separated by SDS–PAGE and blotted onto nitrocellulose sheets. Binding of bacteria was tested by cutting the blots into strips (5 × 1.8 cm), incubating them in PBS–5% BSA at 4° for 12 hr, washing them

[74] J. Parkkinen, G. N. Rogers, T. K. Korhonen, W. Dahr, and J. Finne, *Infect. Immun.* **54,** 37 (1986).

[75] A. Gbarah, C. G. Gahmberg, I. Ofek, U. Jacobi, and N. Sharon, *Infect. Immun.* **59,** 4524 (1991).

with PBS–0.05% Tween 20, and incubating them for 12 hr at 4° in 15 ml bacterial suspension (10^8/ml of 2.5% BSA in PBS). The strips were washed four more times with PBS–Tween and incubated with rabbit anti-*E. coli* antibodies (1 : 210, in PBS–Tween–BSA) for 1 hr at 37°. The binding was visualized in a system containing the anti-rabbit antibody conjugated with alkaline phosphatase or horseradish peroxidase and appropriate substrates. The specificity of binding was confirmed by incubating the same strips with bacteria suspended in buffer containing D-mannose.

Each tested system has optimal experimental conditions, including concentrations of separated materials, bacterial suspension, and all reagents, as well as separation and transfer conditions.

Acknowledgments

I thank Izhak Ofek, Department of Human Microbiology, Tel-Aviv University, for constructive suggestions and comments during the preparation and preliminary review of the manuscript. Sincere appreciation is extended to Judith Rapoport for typing the manuscript.

[17] Possible Interaction between Animal Lectins and Bacterial Carbohydrates

By ROBERT E. MANDRELL, MICHAEL A. APICELLA, RAGHAR LINDSTEDT, and HAKON LEFFLER

Introduction

All living cells, whether in unicellular or multicellular organisms, surround themselves with a shell of complex carbohydrates. In mammals, mucosal epithelial surfaces, which are exposed to other organisms, are particularly rich in carbohydrates. Conversely, the bacteria that colonize the mucosal surfaces are also rich in carbohydrates. The carbohydrates are characterized by great structural diversity and variability related to species, individual, and cell type. It is not surprising then that complex carbohydrates appear to be frequently involved in the interaction between microbes and their hosts.

An important step in microbial pathogenesis is the interaction of the microbe with host cells and substances. Bacteria–host cell interactions involve multiple types of interactions that can contribute to the specificity of a pathogen for a particular host, tissue, and/or cell and for specific mechanisms of virulence. Many biologically important host–pathogen interactions appear to involve the specific recognition and binding of host carbohydrates by bacterial lectinlike proteins[1,2] (also reviewed elsewhere in this volume, see [16]). Recent evidence suggests that the specific recognition of bacterial carbohydrates by host lectins may also be important. This latter mode of interaction is the topic of this article.

It is not possible currently to present a general methodology for the study of the interaction between animal lectins and bacteria. This is because the animal lectins are a diverse group of proteins probably involved in many different steps in the host–pathogen interaction, and, so far, only a few systems have been studied at the molecular level. Therefore, in this article we present an overview of the types of animal lectins that have been described, the types of bacterial carbohydrates that might be likely to interact with these lectins, and examples of bacteria that bind to human lectins or lectinlike proteins. Finally, we present methodology for purification and characterization of one class of lectins, the S-Lac lectins, that is potentially applicable to future studies of lectins and bacterial carbohydrates.

Background

What Are Lectins?

Lectins are nonenzyme, nonimmunoglobulin proteins that have at least one carbohydrate binding domain[3]; they are not necessarily multivalent for carbohydrates nor must they act as agglutinins as required in an earlier definition.[4] The discoveries of carbohydrate binding domains associated with different types of other domains implies that binding of a protein to complex carbohydrates might be involved in a wider range of functions than thought previously.[2,5]

C-Type, S-Lac (S-Type, galectins), and Other Types of Animal Lectins

Lectins have been classified based on the properties and sequence of their carbohydrate binding domains.[5] The family of animal lectins studied most extensively are the C-type lectins. They can be soluble secretory proteins or integral membrane proteins and they include receptors involved in the uptake of plasma glycoproteins in the liver,[6] the selectins[7]

[1] K. Karlsson, *Annu. Rev. Biochem.* **58,** 309 (1989).

[2] N. Sharon and H. Lis, *Science* **246,** 227 (1989).

[3] S. H. Barondes, *Trends Biochem. Sci.* **13,** 480 (1988).

[4] I. J. Goldstein, R. C. Hughes, M. Monsigny, T. Osawa, and N. Sharon, *Nature* (*London*) **285,** 66 (1980).

[5] K. Drickamer, *J. Biol. Chem.* **263,** 9557 (1988).

[6] G. Ashwell and J. Harford, *Annu. Rev. Biochem.* **51,** 531 (1982).

[7] M. Bevilacqua, E. Butcher, B. Furie, M. Gallatin, M. Gimbrone, J. Harlan, K. Kishimoto, L. Lasky, R. McEver, J. Paulson, *et al.*, *Cell* (*Cambridge, Mass.*) **67,** 233 (1991).

involved in leukocyte homing,[8] and the serum mannan-binding protein involved in complement activation.[9] The C-type lectins require calcium for carbohydrate binding activity. Each carbohydrate binding domain has one or two shallow binding sites[10] with specificities that include a wide range of terminal monosaccharide residues (e.g., Man, Glc, Fuc, Gal, GalNAc, or GlcNAc). Higher-affinity binding appears to be conferred by multivalent interaction between a multimeric lectin and the precisely spaced antennae of complex saccharides.[11] The selectins, for example, recognize at varying affinities the common structure sialyl Lewis^x [NeuNAcα2–3Galβ1–4GlcNAc(α1–3Fuc)β1–3Gal],[12] a monosialylated, monofucosylated type 2 series carbohydrate.[13,14] The availability of the binding domain of selectins for carbohydrates on circulating cells is perhaps best exemplified by the role of the E-selectin in the endothelium in slowing the "rolling" of circulating leukocytes, presumably by binding to leukocyte sialyl Lewis^x glycoconjugates.[15]

Another group of animal lectins, the soluble lactose-binding (S-Lac) lectins,[16] also named S-type lectins,[5] were identified initially by their affinity for lactose and other β-galactosides[17] (Table I). The available sequences demonstrate conserved elements and allow the definition of a putative carbohydrate binding domain of about 130 amino acids (Oda *et al.*[18] list most of the known sequences).[18a] The most completely characterized mammalian S-Lac lectins are L-14-I and L-29, but their function still remains unknown (see Table I). All of the S-Lac lectins identified so far

[8] L. A. Lasky, *J. Cell. Biochem.* **45,** 139 (1991).

[9] K. Sastry and R. A. Ezekowitz, *Curr. Opin. Immunol.* **5,** 59 (1993).

[10] W. I. Weis, K. Drickamer, and W. A. Hendrickson, *Nature (London)* **360,** 127 (1992).

[11] K. G. Rice, O. A. Weisz, T. Barthel, R. T. Lee, and Y. C. Lee, *J. Biol. Chem.* **265,** 18429 (1990).

[12] C. Foxall, S. R. Watson, D. Dowbenko, C. Fennie, L. A. Lasky, M. Kiso, A. Hasegawa, D. Asa, and B. K. Brandley, *J. Cell Biol.* **117,** 895 (1992).

[13] M. L. Phillips, E. Nudelman, F. C. A. Gaeta, M. Perez, A. K. Singhal, S.-I. Hakomori, and J. C. Paulson, *Science* **250,** 1130 (1990).

[14] G. Walz, A. Aruffo, W. Kolanus, M. Bevilacqua, and B. Seed, *Science* **250,** 1132 (1990).

[15] T. A. Springer, *Nature (London)* **346,** 425 (1990).

[16] H. Leffler, F. R. Masiarz, and S. H. Barondes, *Biochemistry* **28,** 9222 (1989).

[17] S. H. Barondes, *Science* **223,** 1259 (1984).

[18] Y. Oda, J. Herrman, M. J. Gitt, C. Turck, A. L. Burlingame, S. H. Barondes, and H. Leffler, *J. Biol. Chem.* **268,** 5929 (1993).

[18a] The new name galectins has recently been adopted to designate the S-Lac family of animal lectins defined by their calcium independent affinity for lactose and other β-galactosides and by certain conserved sequence elements within their carbohydrate binding domains. Four mammalian galectins have been fully sequenced and there is evidence for multiple additional ones. Individual mammalian galectins are named by consecutive numbering. The new names of the mammalian S-Lac lectins mentioned in this paper are as follows: L-14I is galectin-1, L-14II is galectin-2, L-29 is galectin-3, and L-36 is galectin-4. [S. H. Barondes, V. Castronovo, D. N. W. Cooper, *et al.*, *Cell* **76,** 597 (1994)].

TABLE I
ANIMAL S-LAC LECTINS EXPRESSED IN MUCOSAL TISSUES

Name	Other names	Mucosal location		Other locations	Secretion	Ref.[a]
		Organ	Cell type			
L-14-I	Galaptin, L-14 bovine heart lectin, galectin-1	Intestine, lung	Submucosal muscle cells	Muscle, heart	From muscle cells	1, 2
L-14-II	Galectin-2	Lower small intestine	Unknown			3, 3a
L-29	Mac-2, CBP-35, L-34, εBP, galectin-3	Intestine, lung, kidney, other	Columnar and cuboidal epithelial cells	Macrophages	Cultured epithelial cells (apical) and macrophages	2, 4–6, 6a,b,c
L-36	Galectin-4	Small and large intestine	Unknown			5
CLL-II	Possibly C-14	Intestine (chicken)	Goblet cells	Embryonal tissues	From goblet cells	7
XL-16		Skin (frog)	Granule gland cells		Holocrin mechanism	8

[a] Key to references: (1) D. N. W. Cooper and S. H. Barondes, *J. Cell Biol.* **110,** 1681 (1990); (2) F. L. Harrison, *J. Cell Sci.* **100,** 9 (1991); (3) M. A. Gitt, S. Masa, H. Leffler, and S. H. Barondes, *J. Biol. Chem.* **267,** 10601 (1992); (3a) H. Leffler, R. Atchison, M. A. Gitt, M. Huflejt, R. Lindstedt, S. M. Massa, and S. H. Barondes, *J. Cell. Biochem.* **17A,** 366, Abst. CZ 019 (1993); (4) B. J. Cherayil, S. J. Weiner, and S. Pillai, *J. Exp. Med.* **170,** 1959 (1989); (5) Y. Oda, J. Herrman, M. J. Gitt, C. Turck, A. L. Burlingame, S. H. Barondes, and H. Leffler, *J. Biol. Chem.* **268,** 5929 (1993); (6) D. Brassart, E. Kolodziejczyk, D. Granato, A. Woltz, M. Pavillard, F. Perotti, L. G. Frieri, F. T. Liu, Y. Borel, and J. R. Neeser, *Eur. J. Biochem.* **203,** 393 (1992); (6a) D. K. Hsu, R. I. Zuberi, and F. T. Liu, *J. Biol. Chem.* **267,** 14167 (1992); (6b) R. Lindstedt, G. Apodaca, S. Barondes, K. Mostov, and H. Leffler, *J. Biol. Chem.* **268,** 11750 (1993); (6c) S. Sato and R. C. Hughes, *J. Biol. Chem.* **269,** 4424 (1994); (7) E. C. Beyer and S. H. Barondes, *J. Cell Biol.* **92,** 28 (1982); (8) P. Marschal, J. Herrmann, H. Leffler, S. H. Barondes, and D. N. W. Cooper, *J. Biol. Chem.* **267,** 12942 (1992).

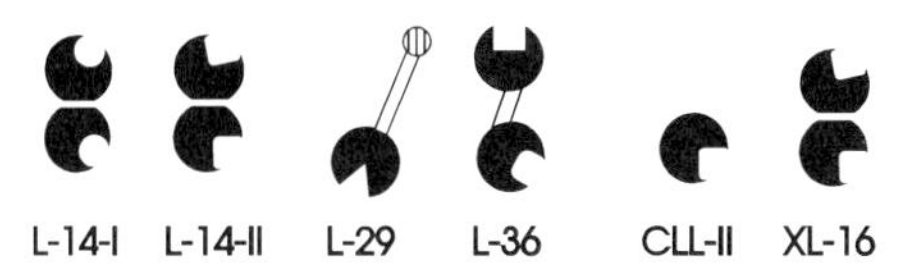

FIG. 1. Schematic structure of some S-Lac lectins. The different carbohydrate binding domains in various lectins are shown filled; other domains are depicted as open or hatched areas. L-14-I, L-14-II, and XL-16 are composed of noncovalent dimers and the carbohydrate binding domains within each lectin are identical, L-29 is a monomer with additional domains, L-36 is a covalently bound dimer with the two binding domains of a different specificity, and CLL-II is a monomer with one binding domain. Other information about the lectins and references is provided in Table I.

have been soluble proteins. Also, there are a variety of organizations for the carbohydrate and other binding domains of these lectins. Schematics of these depicted in Fig. 1 suggest the types of interactions that might occur with bacteria. Soluble lectins with two identical (e.g., L-14-I, L-14-II, XL-16) or two different (e.g., L-36) carbohydrate binding domains might function as bridges between bacterial and host cell carbohydrates. A lectin with a carbohydrate binding domain and another non-carbohydrate type of binding domain (e.g., L-29) could act similarly. Conversely, soluble lectins with a single carbohydrate binding domain (e.g., CLL-II) might block the binding of bacterial carbohydrates with other molecules. The mode of carbohydrate binding of S-Lac lectins differs from that of the C-type lectins in that calcium is not required for activity of the S-Lac lectins. They appear to interact with an epitope that spans two or more saccharides in a chain, and this epitope can be on terminal or internal residues.[19] All S-Lac lectins recognize lactose, but they have higher affinities for selected subsets of more complex β-galactoside-containing saccharides, especially saccharides containing *N*-acetyllactosamine[20]; accordingly, various polylactosaminoglycans are good ligands for the S-Lac lectins.[21]

There are also many other animal lectins that belong to neither the C-type nor S-type lectin family. These include a Kupffer cell glycoprotein hormone receptor,[22] a sialoadhesin (a sialic acid-specific lectin),[23] and various lymphokines. Theoretically, any of these lectins could interact with bacteria expressing the appropriate carbohydrates and present in tissues where the lectins are expressed. In recent years, a number of mucosal pathogens have been identified that express polysaccharides or glycolipids that are potential ligands for certain types of lectins.

[19] H. Leffler and S. H. Barondes, *J. Biol. Chem.* **261,** 10119 (1986).
[20] F. L. Harrison, *J. Cell Sci.* **100,** 9 (1991).
[21] C. P. Sparrow, H. Leffler, and S. H. Barondes, *J. Biol. Chem.* **262,** 7383 (1987).
[22] K. Drickamer, *Cell (Cambridge, Mass.)* **67,** 1029 (1991).
[23] A. S. McWilliam, P. Tree, and S. Gordon, *Proc. Natl. Acad. Sci. USA* **89,** 10522 (1992).

Structure of Bacterial Glycolipids and Polysaccharides

Lipooligosaccharides of Mucosal Pathogens (Potential Ligands for Human Lectins)

Studies in the last few years have shown that a number of gram-negative mucosal pathogens can synthesize glycolipids that are identical chemically or antigenically to carbohydrates of glycosphingolipids present in host cells or tissues.[24–26] Strains of *Neisseria gonorrhoeae, Neisseria meningitidis,* and some *Haemophilus* species synthesize heterogeneous types of lipooligosaccharides (LOSs) that mimic lactoneo series[24,27,28] or globo series[26,28,29] glycosphingolipids present in human cells (Table II).[30] The heptose-linked disaccharide that is a precursor for most of the major LOSs is lactose (Galβ1–4Glc); for a few gonococcal and meningococcal strains, the major branch terminates in only lactose.[31,32] It seems relevant that the lactoneo series oligosaccharide lacto-*N*-neotetraose (Galβ1-4GlcNAcβ1-3Galβ1-4Glc) is present in these LOSs (Table II), because this tetrasaccharide is not only the major precursor (paragloboside) of human erythrocyte blood group antigen glycolipids (ABHS),[33] but is also present in high concentrations in human neutrophils,[34,35] semen,[36] cervical mucin,[37] and myo- and endometrium.[38] Another interesting feature of this lactoneo series LOS is that it is the

[24] R. E. Mandrell, J. M. Griffiss, and B. A. Macher, *J. Exp. Med.* **168,** 107 (1988).

[25] R. E. Mandrell and M. A. Apicella, *Immunobiology* **187,** 382 (1993).

[26] R. E. Mandrell, *Infect. Immun.* **60,** 3017 (1992).

[27] H. J. Jennings, C. Lugowski, and F. E. Ashton, *Carbohydr. Res.* **121,** 233 (1983).

[28] R. E. Mandrell, R. McLaughlin, Y. A. Kwaik, A. Lesse, R. Yamasaki, B. Gibson, S. M. Spinola, and M. A. Apicella, *Infect. Immun.* **60,** 1322 (1992).

[29] M. Virji, J. N. Weiner, A. A. Lindberg, and E. R. Moxon, *Microb. Pathog.* **9,** 441 (1990).

[30] C. L. M. Stults, C. C. Sweeley, and B. A. Macher, this series, Vol. 179, p. 167.

[31] J. L. Di Fabio, F. Michon, J.-R. Brisson, and H. J. Jennings, *Can. J. Chem.* **68,** 1029 (1990).

[32] C. M. John, J. M. Griffiss, M. A. Apicella, R. E. Mandrell, and B. W. Gibson, *J. Biol. Chem.* **266,** 19303 (1991).

[33] S. Hakomori, *Semin. Hematol.* **18,** 39 (1981).

[34] M. N. Fukuda, A. Dell, J. E. Oates, P. Wu, J. C. Klock, and M. Fukuda, *J. Biol. Chem.* **260,** 1067 (1985).

[35] B. A. Macher, W. M. F. Lee, and M. A. Westrick, *Mol. Cell. Biochem.* **47,** 81 (1982).

[36] G. Ritter, W. Krause, R. Geyer, S. Stirm, and H. Wiegandt, *Arch. Biochem. Biophys.* **257,** 370 (1987).

[37] E. C. Yurewicz, F. Matsuura, and K. S. Moghissi, *J. Biol. Chem.* **257,** 2314 (1982).

[38] Z. Zhu, H. Deng, B. A. Fenderson, E. Nudelman, and Z. Tsui, *J. Reprod. Fertil.* **88,** 71 (1990).

major site for sialylation of LOSs of strains of meningococci,[39] gonococci,[40] *H. influenzae*,[28] and *Haemophilus ducreyi*.[41,42] The presence of the lactoneo series structure also may be related to the mechanism whereby sialylation of the LOSs prevents or decreases adherence[43] and decreases both opsonic and nonopsonic phagocytosis of gonococci and meningococci by human neutrophils.[44–46] Neutrophils contain similar linear lactoneo series glycolipids (both sialylated and nonsialylated).[34,35] The *N*-acetyllactosamine/polylactosamine specificity reported for many S-Lac lectins, including some present in human lung[21] and human leukocytes,[47–49] indicates that lectins may be available of an appropriate specificity and in the appropriate tissue or cells for binding to nonsialylated lactoneo series LOSs of bacteria.

Lactoneo Series Streptococcal Polysaccharides

The capsular polysaccharides of some Gram-positive bacteria also are composed of sialylated lactoneo series-like structures. The capsular polysaccharide of type XIV *Streptococcus pneumoniae* is similar antigenically to lacto-*N*-neotetraose,[50] and similarly, group B streptococcal polysaccharide types Ia, II, III, IV, and V contain the sialylated lactosamine NeuNAcα2–3Galβ1–4GlcNAc, either as a branching chain (Ia, IV, V) or as an internal unit (II, III).[51] If some of these streptococcal "lactosaminoglycan" units could become desialylated *in vivo* (acidic conditions or host neurmaninidase), then they might be potential receptors for epithelial cell lectins that bind terminal Gal or Gal–GlcNAc residues (Table I). The prevalence of this type of carbohydrate among pathogenic strains of gram-

[39] R. E. Mandrell, J. J. Kim, C. M. John, B. W. Gibson, J. V. Sugai, M. A. Apicella, J. M. Griffiss, and R. Yamasaki, *J. Bacteriol.* **173,** 2823 (1991).

[40] R. E. Mandrell, A. J. Lesse, J. V. Sugai, M. Shero, J. M. Griffiss, J. A. Cole, N. J. Parsons, H. Smith, S. A. Morse, and M. A. Apicella, *J. Exp. Med.* **171,** 1649 (1990).

[41] J. A. Odumeru, G. M. Wiseman, and A. R. Ronald, *J. Med. Microbiol.* **23,** 155 (1987).

[42] A. A. Campagnari, S. M. Spinola, A. J. Lesse, Y. A. Kwaik, R. E. Mandrell, and M. A. Apicella, *Microb. Pathog.* **8,** 353 (1990).

[43] R. F. Rest and J. V. Frangipane, *Infect. Immun.* **60,** 989 (1992).

[44] L. M. Wetzler, K. Barry, M. S. Blake, and E. C. Gotschlich, *Infect. Immun.* **60,** 39 (1992).

[45] J. J. Kim, D. Zhou, R. E. Mandrell, and J. M. Griffiss, *Infect. Immun.* **60,** 4439 (1992).

[46] M. M. Estabrook, N. C. Christopher, J. M. Griffiss, C. J. Baker, and R. E. Mandrell, *J. Infect. Dis.* **166,** 1079 (1992).

[47] A. Sharma, R. Chemelli, and H. J. Allen, *Biochemistry* **29,** 5309 (1990).

[48] A. Sharma, R. A. DiCioccio, and H. J. Allen, *Glycobiology* **2,** 285 (1992).

[49] B. J. Cherayil, S. Chaitovitz, C. Wong, and S. Pillai, *Proc. Natl. Acad. Sci. USA* **87,** 7324 (1990).

[50] B. Siddiqui and S.-I. Hakomori, *Biochim Biophys. Acta* **330,** 147 (1973).

[51] M. Wessels, J. L. DiFabio, V.-J. Benedi, D. L. Kasper, F. Michon, J.-R. Brisson, J. Jelinkova, and H. J. Jennings, *J. Biol. Chem.* **266,** 6714 (1991).

TABLE II

PROPOSED CARBOHYDRATE STRUCTURES OF LIPOOLIGOSACCHARIDES OF *Neisseria* AND *Haemophilus* SPECIES AND STRUCTURES OF SIMILAR HUMAN GLYCOSPHINGOLIPIDS[a]

Carbohydrate structure of LOS	Bacteria	Refs.[d]
Galβ1→4GlcNAcβ1→3Galβ1→4Glc-R	*N. meningitidis*	1–4
	N. gonorrhoeae	4, 5
	H. influenzae type b[b]	6, 7
	H. influenzae NT[c]	6
	N. lactamica	8
	N. cinereae	8
GlcNAcβ1→3Galβ1→4GlcHep-R	*N. gonorrhoeae*	7, 9–13
Galβ1→4GlcHep-R	(mutants), *N. meningitidis*,	
GlcHep-R	and *H. influenzae*[b]	
Hep-R		
Galβ1→4GlcNAcβ1→3Galβ1→4Hep-R	*H. ducreyi*	14, 15
NeuNAcα2→3Galβ1→4GlcNAcβ1→3Galβ1→4Glc-R	*N. meningitidis*	16, 17
NeuNAc-Galβ1→4GlcNAcβ1→3Galβ1→4Glc-R	*N. gonorrhoeae*	18
	H. influenzae type b[b]	6, 7
	H. influenzae NT[c]	6
	N. lactamica	19
Galα1→4Galβ1→4Glc-R	*N. meningitidis*	9, 20, 21
	N. gonorrhoeae	10, 20, 21
	H. influenzae type b[b]	6, 20
	H. influenzae NT[c]	6, 20
	N. lactamica	21
	B. catarrhalis	21
GalNAcβ1→3Galβ1→4GlcNAcβ1→3Galβ1→4Glc-R	*N. gonorrhoeae*	5
Carbohydrate structure of GSL similar to LOS listed above	**Common Name of GSL**	
NeuNAcα2→3(6) Galβ1→4GlcNAcβ1→3Galβ1→4Glcβ1→1Cer	Sialylparagloboside	22
Galβ1→4GlcNAcβ1→3Galβ1→4Glcβ1→1Cer	Paragloboside	22
GalNAcβ1→3Galβ1→4GlcNAcβ1→3Galβ1→4Glcβ1→1Cer	χ_2 and asialo-G_3	22
GalNAcβ1→4Galβ1→4Glcβ1→1Cer	Asialo G_{M2}	22
Galα1→4Galβ1→4Glcβ1→1Cer	P*K*	22
Galβ1→4Glcβ1→1Cer	LacCer	22

[a] Gal, galactose; GalNAc, *N*-acetylgalactosamine; Glc, glucose; GlcNAc, *N*-acetylglucosamine; NeuNAc, *N*-acetylneuraminic acid; Cer, ceramide; KDO, 3-keto-2-deoxyoctanoic acid. The structural analyses of some of the LOSs have been described previously (see references cited). Only the structure of the longer oligosaccharide bound to the KDO-linked heptose is shown. For the gonococcal and meningococcal LOSs, R is Galβ1→4Hep(GlcNAcα1→2Hep-α1→3)α1→5KDO. Substitutions of Glc or phosphatidylethanolamine occurring on the α1→3-linked heptose are not shown. The proposed structures for other LOS are based on the anticarbohydrate specificity of MAbs that bind the LOS.

[b] Structural analysis of a *H. influenzae* type b strain detected a Gal1→4GlcNAc1→3Hex1→4Glc1→Hep branch.[7]

[c] Structural analysis of one strain of *H. influenzae* NT did not confirm the presence of a terminal Galβ1→4GlcNAc moiety nor Gal–Gal–Glc.[13]

[d] Key to references: (1) F. Michon, M. Beurret, A. Gamian, J.-R. Brisson, and H. J. Jennings, *J. Biol. Chem.* **265,** 7243 (1990); (2) A. Gamian, M. Beurret, F. Michon, J.-R. Brisson, and H. J. Jennings, *J. Biol. Chem.* **267,** 922 (1992); (3) H. J. Jennings, C. Lugowski, and F. E. Ashton, *Carbohydr. Res.* **121,** 233 (1983); (4) R. E. Mandrell, J. M. Griffiss, and B. A. Macher, *J. Exp. Med.* **168,** 107 (1988); (5) R. Yamasaki, B. E. Bacon, W. Nasholds, H. Schneider, and J. M. Griffiss, *Biochemistry* **30,** 10566 (1991); (6) R. E. Mandrell, R. McLaughlin, Y. A. Kwaik, A. Lesse, R. Yamasaki, B. Gibson, S. M. Spinola, and M. A. Apicella, *Infect. Immun.* **60,** 1322 (1992); (7) N. J. Phillips, M. A. Apicella, J. M. Griffiss, and B. W. Gibson, *Biochemistry* **32,** 2003 (1993); (8) J. J. Kim, R. E. Mandrell, and J. M. Griffiss, *Infect. Immun.* **57,** 602 (1989); (9) J. L. DiFabio, F. Michon, J.-R. Brisson, and H. J. Jennings, *Can. J. Chem.* **68,** 1029 (1990); (10) C. M. John, J. M. Griffiss, M. A. Apicella, R. E. Mandrell, and B. W. Gibson, *J. Biol. Chem.* **266,** 19303 (1991); (11) D. E. Kerwood, H. Schneider, and R. Yamasaki, *Biochemistry* **31,** 12760 (1992); (12) N. J. Phillips, C. M. John, L. G. Reinders, J. M. Griffiss, M. A. Apicella, and B. W. Gibson, *Biomed. Mass Spectrosc.* **19,** 731 (1990); (13) N. J. Phillips, M. A. Apicella, J. M. Griffiss, and B. W. Gibson, *Biochemistry* **31,** 4515 (1993); (14) A. A. Campagnari, S. M. Spinola, A. J. Leese, Y. A. Kwaik, R. E. Mandrell, and M. A. Apicella, *Microb. Pathog.* **8,** 353 (1990); (15) W. Melaugh, N. J. Phillips, A. A. Campagnari, R. Karalus, and B. W. Gibson, *J. Biol. Chem.* **267,** 13434 (1992); (16) R. E. Mandrell, J. J. Kim, C. M. John, B. W. Gibson, J. V. Sugai, M. A. Apicella, J. M. Griffiss, and R. Yamasaki, *J. Bacteriol.* **173,** 2823 (1991); (17) R. Yamasaki, J. M. Griffiss, K. Quinn, and R. E. Mandrell, *J. Bacteriol.* **175,** 4565 (1993); (18) R. E. Mandrell, A. J. Lesse, J. V. Sugai, M. Shero, J. M. Griffiss, J. A. Cole, N. J. Parsons, H. Smith, S. A. Morse, and M. A. Apicella, *J. Exp. Med.* **171,** 1649 (1990); (19) R. E. Mandrell, J. M. Griffiss, H. Smith, and J. A. Cole, *Microb. Pathog.* **14,** 315 (1993); (20) M. Virji, J. N. Weiner, A. A. Lindberg, and E. R. Moxon, *Microb. Pathog.* **9,** 441 (1990); (21) R. E. Mandrell, *Infect. Immun.* **60,** 3017 (1992); (22) C. L. M. Stults, C. C. Sweeley, and B. A. Macher, this series, Vol. 179, p. 167.

positive and gram-negative bacteria suggests that this structure is part of a common mechanism of survival or invasion of these bacteria in humans.

Examples of Interaction between Animal Lectins and Microorganisms

Macrophage Lectins and Serum Mannan-Binding Protein

The mannan-binding animal lectins and their interactions with microbes have been the best characterized and the properties of the numerous known lectins of this type have been reviewed recently.[52] Mannan-binding proteins are present in granulocytes, macrophages, and serum (Table III) and all of them have been implicated as receptors for bacterial ligands. The serum mannan-binding protein activates complement in an antibody-independent manner when it encounters cognate carbohydrates on a microbial surface.[9] Evidence for the biological importance of this interaction is the fact that individuals with a genetic defect in the mannan-binding protein have increased susceptibility to certain infections.[53] Mycobacteria,[54,55] yeast,[56] and the host-derived oligosaccharides of viruses, including human immunodeficiency virus (HIV),[57] contain simple and complex mannose structures that also have been implicated in pathogenesis. It is likely that the interaction of these mannan-binding proteins with microbes is a late event in pathogenesis.

Animal Lectins at Mucosal Surfaces

Much less is known about the potential role of lectins present at the surface of mucosal tissue. The C-type lectins identified in mucosal epithelia include the lung surfactant proteins A and D in alveloar type II cells,[58] the hepatic asialoglycoprotein receptor minor subunits at the apical surface of intestinal epithelia,[59] and a putative lectin (identified based on sequence homology) induced in pancreatic cells during inflammation.[60] Of these,

[52] P. D. Stahl, *Curr. Opin. Immunol.* **4,** 49 (1992).

[53] R. J. Lipscombe, Y. L. Lau, R. J. Levinsky, M. Sumiya, J. A. Summerfield, and M. W. Turner, *Immunol. Lett.* **32,** 253 (1992).

[54] D. Chatterjee, S. W. Hunter, M. McNeil, and P. J. Brennan, *J. Biol. Chem.* **267,** 6234 (1992).

[55] D. Chatterjee, S. W. Hunter, M. McNeil, and P. J. Brennan, *J. Biol. Chem.* **267,** 6228 (1992).

[56] T. G. Pistole, *Annu. Rev. Microbiol.* **35,** 85 (1981).

[57] J. Mizuochi, M. W. Spellman, M. Larkin, J. Soloman, L. J. Basu, and T. Feizi, *Biochem. J.* **254,** 599 (1988).

[58] S.-F. Kuan, K. Rust, and E. Crouch, *J. Clin. Invest.* **90,** 97 (1992).

[59] C. B. Hu, E. Y. Lee, J. E. Hewitt, J. U. Baenziger, J. Z. Mu, K. DeSchryver-Kecskemeti, and D. H. Alpers, *Gastroenterol.* **101,** 1477 (1991).

[60] J. Iovanna, B. Orelle, V. Keim, and J. C. Dagorn, *J. Biol. Chem.* **266,** 24664 (1991).

TABLE III
BACTERIA RECOGNIZED BY HUMAN LECTINS OR LECTINLIKE MOLECULES

Source of lectin	Sugar specificity[a]	Bacteria	Ref.[b]
Macrophages	Man	*Pseudomonas aeruginosa*	1
Monocytes	Glc, Gal	*Staphylococcus albus*	2
Granulocytes	Gal	*S. albus*	2
Polymorphonuclear leukocytes	GalNAc, D-GalNH$_2$, D-GlcNH$_2$, Gal	*Neisseria gonorrhoeae*	3
Monocytes	GalNAc, D-GalNH$_2$, D-GlcNH$_2$, Gal, Fuc	*N. gonorrhoeae*	3
Macrophages (CR3, LFA-1, p150,95)	GlcNAc-GlcNAc	*Escherichia coli*	4
Serum	Man	*Salmonella montevideo*	5
Serum	L-Glycero-D-manno-Hep; GlcNAc	*E. coli*	6
Lung (16K)	GalNAc	*P. aeruginosa*	7
Airway secretions (65K)	?	*P. aeruginosa*	8
Macrophages	Man	*Mycobacterium*	9
Macrophages	Manα2–3Man-	*Klebsiella pneumoniae*	10
Granulocytes (CD11/CD18)	Oligomannose	*E. coli*	11
Alveolar type II and bronchiolar cells (surfactant-associated protein D)	Mal>Glc>Man>Fuc	*E. coli;* other Gram-negative bacteria	12

[a] Fuc, fucose; Gal, galactose; GalNAc, *N*-acetylgalactosamine; D-GalNH$_2$, D-galactosamine, Glc, glucose; GlcNAc, *N*-acetylglucosamine; D-GlcNH$_2$, D-glucosamine; Hep, heptose; Mal, maltose; Man, mannose. The designation, molecular weight, and other information about the lectins are shown in parentheses.

[b] Key to references: (1) D. P. Speert *et al., J. Clin. Invest.* **82,** 872 (1988); (2) E. Glass, J. Stewart, and D. M. Weir, *Immunology* **44,** 529 (1981); (3) D. F. Kinane, D. M. Weir, C. C. Blackwell, and F. P. Winstanley, *J. Clin. Lab. Immunol.* **13,** 107 (1984); (4) S. D. Wright and M. T. C. Jong, *J. Exp. Med.* **164,** 1876 (1986); (5) M. Kuhlman, K. Joiner, and R. A. B. Ezekowitz, *J. Exp. Med.* **169,** 1733 (1989); (6) N. Kawasaki, T. Kawasaki, and I. Yamashina, *J. Biochem. (Tokyo)* **106,** 483 (1989); (7) H. Ceri, W. S. Hwang, and H. Rabin, *Am. J. Respir. Cell Mol. Biol.* **5,** 51 (1991); (8) J. S. Hata and R. B. Fick, Jr., *J. Lab. Clin. Med.* **117,** 410 (1991); (9) D. Chatterjee, S. W. Hunter, M. McNeil, and P. J. Brennan, *J. Biol. Chem.* **267,** 6234 (1992); (10) A. Athamna, I. Ofek, Y. Keisari, S. Markowitz, G. G. Dutton, and N. Sharon, *Infect. Immun.* **59,** 1673 (1991); (11) A. Gbarah, C. G. Gahmberg, I. Ofek, U. Jacobi, and N. Sharon, *Infect. Immun.* **59,** 4524 (1991); (12) S.-F. Kuan, K. Rust, and E. Crouch, *J. Clin. Invest.* **90,** 97 (1992).

only surfactant protein D has been shown to interact with bacteria.[58] Other examples of lectins that bind to bacterial carbohydrates are presented in Table III. In this table we have only included examples of human cell lectins that bind bacteria or bacterial carbohydrate; additional examples of other animal lectins with similar properties are discussed in other reviews.[2,61–63] The carbohydrate specificity of the lectins, if known, is also presented, but it should be noted that it is based usually on monosaccharide or disaccharide inhibitions of the lectin–bacteria interaction and may not be representative of the specificity of the interaction *in vivo*.

Specificity of S-Lac Lectins versus Structure of Bacterial Carbohydrates

S-Lac lectins are good candidates for interacting with bacteria because their specificity for β-galactosides suggests that they might be able to interact with some of the bacterial carbohydrates mentioned above. Many of them are abundant in mucosal epithelial cells as outlined in Table I. Leukocyte lectins,[47–49] which might be involved in later stages of infection by mucosal bacteria, have not been included except for one present in epithelial cells.[49] A notable feature of S-Lac lectins is that they have structural properties of cytosolic proteins. One of these properties is the absence of a classic secretion signal; they appear to be secreted by a nonclassic pathway (see Table I).

Potential Interaction of Gonococci with Human Cells

Gonococci are a good example of a mucosal pathogen that has both lectins and carbohydrates that interact with human cells and contribute to pathogenesis L-29 is a potent agglutinin for certain bacterial strains (see below under binding assays). Numerous studies of gonococci have identified lectinlike proteins that bind to ligands on human cells.[64–67] For example, a recent study reported that a previously noninvasive *E. coli* strain will invade human endocervical epithelial cells when the *E. coli* express Opa, a gonococcal protein implicated in adherence of gonococci to leukocytes and responsible for colony opacity.[68] Different Opa proteins appear to be separately and directly responsible for the specificity of gonococci to adhere to and invade leukocytes or epithelial cells, although the inability of certain *E. coli* Opa recombinants to duplicate this adherence

[61] D. M. Weir, *Immunol. Today* **1,** 45 (1980).
[62] I. Ofek and N. Sharon, *Infect. Immun.* **56,** 539 (1988).
[63] D. M. Weir, *FEMS Microbiol. Immunol.* **47,** 331 (1989).
[64] J. Swanson, E. Sparks, B. Zeligs, M. A. Siam, and C. Parrot, *Infect. Immun.* **10,** 633 (1974).
[65] D. K. Paruchuri, H. S. Seifert, R. S. Ajioka, K. A. Karlsson, and M. So, *Proc. Natl. Acad. Sci. U.S.A.* **87,** 333 (1990).
[66] S.-I. Makino, J. P. M. van Putten, and T. F. Meyer, *EMBO J.* **10,** 1307 (1991).
[67] H. Perrollet and R. M. F. Guinet, *Lancet* **1,** 1269 (1986).
[68] D. Simon and R. F. Rest, *Proc. Natl. Acad. Sci. U.S.A.* **89,** 5512 (1992).

and invasion indicated that additional factors could be involved.[69] Although these studies suggest that certain Opa proteins play a major role in adherence events during pathogenesis, there is also evidence that LOSs may bind to human lectins; for example: (1) Purified gonococcal LOS damages cultured human fallopian tube epithelial cells, but not similar types of cells from other species[70]; (2) it binds to human ciliated epithelial cells[71] and, in the presence of serum, binds to human monocytes[72,73]; and (3) sugars present in gonococcal LOS, such as galactose, *N*-acetylgalactosamine, galactosamine, and glucosamine (see Table II) can inhibit the binding of gonococci to human monocytes or neutrophils (see Table III).[74] These observations have led to a hypothesis that genital epithelial cell lectins specific for LOSs contribute to the host and tissue tropism of gonococci.[75]

Evidence of Role for Human Lectins in Invasion of Hep-G2 Cells by Gonococci

The studies described above suggest that gonococci have multiple mechanisms for binding and invading human cells. In a preliminary study, the human hepatoma cell line Hep-G2 was used to evaluate whether putative gonococcal (the Opa proteins) and human cell lectins (the hepatic asialoglycoprotein receptor) are important in the invasion by gonococci. The presence of a family of different asialoglycoprotein receptors in the Hep-G2 cells has been reported[76] and several of the genes for these receptors have been cloned with this cell line by cDNA methods.[77] Two gonococcal strains were tested for intracellular uptake by this receptor in Hep-G2 cells: *N. gonorrhoeae* strain 1291, which contains a terminal

[69] E.-M. Kupsch, B. Knepper, T. Kuroki, I. Heuer, and T. F. Meyer, *EMBO J.* **12,** 641 (1993).

[70] C. R. Gregg, M. A. Melly, C. G. Hellerqvist, J. G. Coniglio, and Z. A. McGee, *J. Infect. Dis.* **143,** 432 (1981).

[71] M. D. Cooper, P. A. McGraw, and M. A. Melly, *Infect. Immun.* **51,** 425 (1986).

[72] N. Haeffner-Cavaillon, J.-M. Cavaillon, M. Etievant, S. Lebbar, and L. Szabo, *Cell. Immunol.* **91,** 119 (1985).

[73] P. J. Watt, M. E. Ward, J. E. Heckels, and T. J. Trust, *in* "Immunobiology of *Neisseria gonorrhoeae*" (G. F. Brooks, E. C. Gotschlich, K. K. Holmes, W. D. Sawyer, and F. E. Young, eds.), p. 253. Am. Soc. Microbiol., Washington, DC, 1978.

[74] D. F. Kinane, D. M. Weir, C. C. Blackwell, and F. P. Winstanley, *J. Clin. Lab. Immunol.* **13,** 107 (1984).

[75] Z. A. McGee, C. R. Gregg, A. P. Johnson, S. S. Kalter, and D. Taylor-Robinson, *Microb. Pathog.* **9,** 131 (1990).

[76] P. Weiss and G. Ashwell, *in* "Alpha 1-Acid Glycoprotein: Genetics, Biochemistry, Physiological Functions, and Pharmacology" (P. Braumann, ed.), p. 169. Alan R. Liss, New York, 1989.

[77] A. L. Schwartz, S. E. Fridovich, B. B. Knowles, and H. F. Lodish, *Cell (Cambridge, Mass.)* **61,** 1365 (1990).

Galβ1–4GlcNAc LOS residue (lactoneo series LOS), and an isogenic mutant 1291e, which has a truncated LOS that lacks this residue (see Table II). The phenotype of each of the strains was characterized (piliation, Opa proteins, and LOS) by colony morphology and by sodium dodecyl sulfate–polyacrylamide gel electrophoresis (SDS–PAGE)/immunoblot analysis with monoclonal antibodies (MAbs) MC02 (pili), 4B12 (Opa), and 3F11 (binds lactoneo LOS). This was necessary because variants of the mutant 1291e expressing Opa do not exhibit any visible signs of opaque colony morphology even though expression of an Opa can be demonstrated by MAb reactivity. The gonococci used in initial experiments were piliated and divided into four groups according to Opa expression and LOS type: *opa*$^+$/*3F11*$^+$, *opa*$^-$/*3F11*$^+$, *opa*$^+$/*3F11*$^-$, and *opa*$^-$/*3F11*$^-$. The assays were performed in 24-well microtiter plates coated with type I collagen. Hep-G2 cells (2×10^5) were seeded into each well and the wells were incubated 24 hr. After 24 hr, the monolayer of cells was washed with antibiotic-free tissue culture medium and gonococci (2×10^5 cfu) of each of the four phenotypes were added to the wells. After incubation for 6 hr (37° in a 5% CO_2), the wells were washed (5×) with fresh tissue culture medium; then tissue culture medium containing gentamicin (20 μg/ml) was added to the monolayers. The plate was incubated an additional 90 min and then the monolayer was washed twice with phosphate-buffered saline (PBS). The cells were removed from the surface of the plastic by agitation with 1 ml of PBS containing 5 m*M* ethylenediaminetetraacetic acid (EDTA). The cells were divided into two samples; 10-fold dilutions of one sample were placed in duplicate on supplemented GC agar and the other sample was processed for electron microscopy. Colony counts were made 24 and 48 hr after incubation (37° in a 5% CO_2). Bacterial adherence was assayed similarly except that the cell monolayers were removed prior to the addition of the gentamicin-containing medium and the 90-min incubation step.

The results of these experiments are shown in Fig. 2. Neither the Opa protein nor the lactoneo$^+$ LOSs are absolutely required for binding to the Hep-G2 cells (see *opa*$^-$/*3F11*$^-$), a result indicating that interactions among other host and bacterial components can also mediate binding. Gonococci with either the lactoneo$^+$ LOS or the Opa protein alone (*Opa*$^-$/*3F11*$^+$ and *Opa*$^+$/*3F11*$^-$) can invade Hep-G2 cells, but bacteria containing neither Opa nor LOS do not invade (*opa*$^-$/*3F11*$^-$). These results are similar to the results of Robertson *et al.* who reported that gonococcal mutants expressing an invasion-associated Opa and a truncated, lactoneo$^-$ LOS (due to lack of a functional UDPgalactose-4-epimerase) still could adhere well to two different types of human cells (Chang cells and an endome-

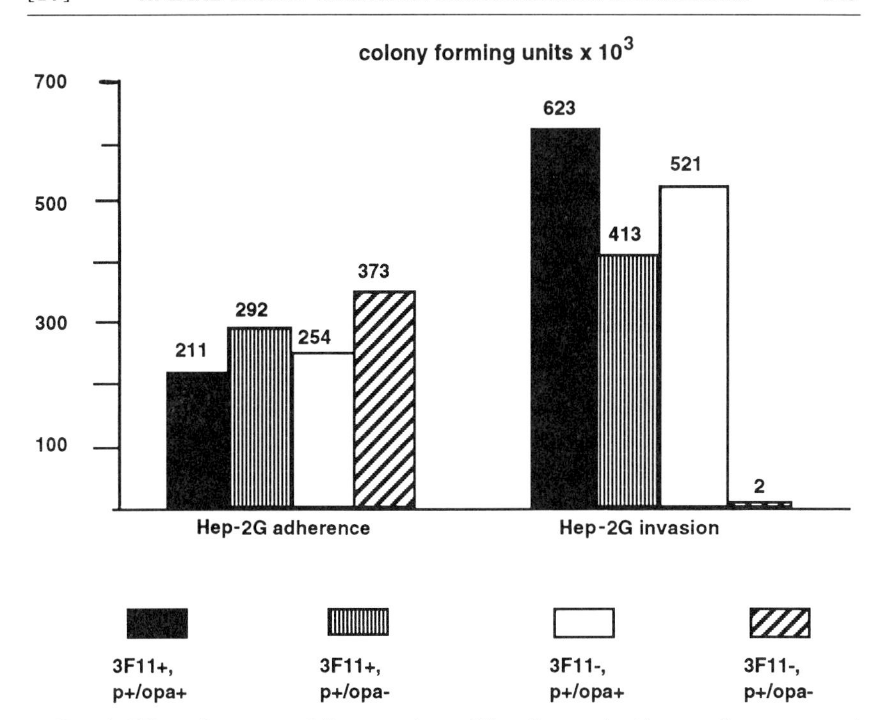

FIG. 2. Effect of gonococcal Opa proteins and lipooligosaccharides on adherence to and invasion of Hep-G2 cells by strain 1291e. The phenotypes of the bacteria were determined by SDS–PAGE and immunoblots with specific MAbs. MAb 3F11 binds to LOSs containing terminal Galβ1–4GlcNAc; the *p/opa* designation indicates whether pili or opacity protein is expressed.

trium-derived cell line), but that a lactoneo$^-$ LOS mutant could not invade the endometrial cell line.[78] These results suggested that adherence to and invasion of cells by Opa$^-$ gonococci can be mediated by human cell lectin–LOS interactions or by another type of bacterial lectin, but that specific LOS structures may be required for invasion of different cells, tissues, or hosts. These results are similar to those noted above indicating that *opa*$^-$*/3F11*$^-$ gonococci do not invade Hep-G2 cells, and suggesting that other gonococcal lectin–cell carbohydrate or cell lectin–LOS interactions were not sufficient for invasion of this cell type. One hypothesis is that invasion of Hep-G2 cells by *opa*$^-$*/3F11*$^+$ gonococci may be due partly to interaction with β-galactoside-binding host lectins such as the asialoglycoprotein receptor or to S-Lac lectins.

[78] B. D. Robertson, M. Frosch, and J. P. M. van Putten, *Mol. Microbiol.* **8,** 891 (1993).

Gonococci also express outer membrane lectins that bind to lactoneo and ganglio series glycosphingolipids,[79,80] glycolipids that contain carbohydrates similar or identical to those also present in LOSs (see Table II). These results seemed consistent with earlier observations by Blake who reported that purified gonococcal Opa could bind homologous LOS, an interaction that was thought to be responsible for the aggregation and clumping of some gonococcal strains.[79] However, the gonococcal lectin mediating the binding to glycosphingolipids *in vitro* apparently is not related to any identified Opa protein nor to pili.[65] Similarly, non-Opa lectin may have been responsible for the binding of *opa*$^-$/*3F11*$^-$ gonococci to Hep-G2 cells as described above (see Fig. 2). Although all the evidence points to Opa being of major importance in invasion by gonococci, LOS appears to have a subtle role depending on the human cell type and the gonococcal strain. These separate observations together reflect the potential complexity of the network of interactions that can occur among gonococci and host cells *in vivo* and the need for additional studies of Opa and LOS under *in vivo* conditions. Some of the methods provided in this article should be helpful for studying this part of the biology of the gonococcus.

Identification of Lectins in Epithelial Cells

Theoretical Considerations

An important criterion for identifying candidate lectins is the ability to bind the bacteria, but other criteria depend on the biological context in which the lectin is thought to be involved. Previously, agglutination of a suitable cell (e.g., erythrocytes or bacteria) was the most common way to screen for the presence of lectins in extracts. As mentioned above, however, this criterion may cause one to miss certain lectins.

Many lectins or lectinlike proteins are present in epithelial cells, but there is limited information about how these mammalian lectins are involved in pathogenesis at the molecular level. Lectins that bind bacteria might have varied properties as adherence can occur in so many different ways. It is therefore important for investigators to appreciate that lectins or lectinlike molecules might also be present in the cells or tissues relevant to the initial stages of adherence or invasion by their pathogen. Some of the methods for identifying these lectins are presented below.

[79] N. Strömberg, C. Deal, G. Nyberg, S. Normark, M. So, and K.-A. Karlsson, *Proc. Natl. Acad. Sci. U.S.A.* **85,** 4902 (1988).

[80] C. D. Deal and H. K. Krivan, *J. Biol. Chem.* **265,** 12774 (1990).

Affinity Chromatography

Affinity chromatography is an efficient procedure for identifying lectins. A tissue extract is passed over affinity columns containing immobilized saccharides and then bound lectins can be eluted with the same type of saccharides in soluble form. This approach is limited by the availability of suitable saccharides. The only animal lectins that have been identified in this way are those that have the ability to bind inexpensive saccharides such as mannose, galactose, fucose, and lactose. Almost all of these lectins have much higher affinities for more complex saccharides.[11,21] This suggests that in tissues or secretions, other lectins might be present that are not isolated by affinity chromatography on simple saccharides because they can bind only to more complex saccharides. Therefore, to screen adequately a mucosal tissue for lectins would require that multiple extraction protocols be used with and without various detergents, with and without Ca^{2+}, and by affinity chromatography with different immobilized saccharides.

Amino Acid Sequence Homologies

Similar amino acid sequences and other structural similarities among lectins have led to the identification of additional putative animal lectins. This was possible because consensus sequence elements have been identified that are typical of families of lectins with similar carbohydrate binding properties.[5] The lung surfactant proteins A and D discussed above are examples of this. In addition, monoclonal and polyclonal antibodies that bind to conserved epitopes and genetic probes that bind to conserved sequences will be very useful in identifying similar types of lectins in other tissues.[16,81]

Isolation of S-Lac Lectins from Rat Intestine and Cultured Epithelial Cells

Isolation of Lectins

The general procedures for isolation of S-Lac lectins are essentially as described for lectins from rat intestine.[16] To isolate lectins from a tissue sample, approximately 150 g of tissue is extracted with 300 ml of 75 m*M* KH_2PO_4/Na_2HPO_4, 75 m*M* NaCl, pH 7.4 (PBS), containing 4 m*M* 2-mercaptoethanol, 2 m*M* EDTA (MEPBS), and 150 m*M* lactose in an Omnimixer (Omni International, Gainesville, VA). The extract is centri-

[81] M. A. Gitt, S. Masa, H. Leffler, and S. H. Barondes, *J. Biol. Chem.* **267,** 10601 (1992).

fuged first at 5000 *g* and then at 100,000 *g* to remove debris and particulate material. The clear supernatant is dialyzed against MEPBS to remove lactose. The dialyzate is centrifuged again after dialysis to remove precipitated material and then loaded on a 100-ml column with immobilized lactose (lactosyl-Sepharose). After the column is washed with MEPBS until the protein content of the effluent is less than 2 mg/ml, lectin [as measured by the Bio-Rad protein assay (Bio-Rad, Richmond, CA)] is eluted with 150 m*M* lactose in MEPBS. This typically yields a fraction containing about 10 mg of a mixture of lactose-binding lectins. The lectins are purified further by ion-exchange chromatography. The lectin fraction is dialyzed against 5 m*M* Na^+/K^+ phosphate buffer, pH 7.4, and loaded on a DEAE-Sepharose column. The flowthrough and fractions eluted with increasing concentrations of NaCl are collected. The flowthrough is adjusted to pH 7.0 by adding KH_2PO_4, loaded on a column of CM-Sepharose, and then eluted with increasing concentrations of NaCl.

A similar protocol can be used to isolate lectins from cultured epithelial cells [e.g., Madin–Darby canine kidney line, (MDCK)], but without including lactose in the extraction buffer. Confluent cells on 1–10 tissue culture plates are lysed in MEPBS containing 2.5% Triton X-100 and centrifuged at 500 *g*. The supernatant is passed over 5–10 ml of lactosyl-Sepharose equilibrated in MEPBS and the lectin is eluted with MEPBS containing 150 m*M* lactose. This procedure yields about 50 μg of lectin per 10-cm tissue culture plate.[82]

Lactosyl-Sepharose

Lactosyl-Sepharose can be made exactly as described by Levi and Teichberg,[83] except that lactose obtained commercially (Sigma) was used without further purification. This procedure should be done in a fume hood because of the toxicity of divinyl sulfone. Lactosyl-Sepharose is stable for years if stored under conditions preventing microbial growth and the same materials can be reused for lectin purification many times. It is best to use a separate lactosyl-Sepharose preparation for each source of lectin to avoid cross-contamination by proteins remaining bound to the column.

Lactosyl-Sepharose has weak anionic properties. Therefore lectins that are negatively charged at neutral pH, like L-14-I, may be eluted from the column with salt-free buffers or water, as suggested previously.[83a] Conversely, lectins with a positive charge at neutral pH (e.g., RI-F and

[82] R. Lindstedt, G. Apodaca, S. H. Barondes, K. Mostov, and H. Leffler, *J. Biol. Chem.* **268,** 11750 (1993).

[83] G. Levi and V. I. Teichberg, *J. Biol. Chem.* **256,** 5735 (1981).

[83a] V. I. Teichberg, D. A. U. Erez, and E. Pinelli, *J. Biol. Chem.* **263,** 14086 (1988).

RI-H) will be slightly retarded on the column at very low salt concentrations, even in the presence of lactose. Nevertheless, lactose elutes all of these lectins efficiently if more than 10 m*M* NaCl is present.

Previously, asialofetuin (ASF) linked to Sepharose via CNBr activation was the affinity matrix used most commonly for purification of S-Lac lectins,[84] however, this matrix is much more cumbersome and expensive to make and is less stable. Also, the "stickiness" of ASF causes tissue proteins to accumulate and this limits the number of times each preparation can be used. Theoretically, ASF should bind some lectins much better than does lactose, but the high concentration of lactose on lactosyl-Sepharose apparently compensates for the presumed lesser affinity of the interaction with a simple disaccharide. This appears to be the case for all of the lectins we have studied so far.

Reducing Agents

Reducing agents were required for the isolation and preservation of the first discovered S-Lac lectin, L-14-I. Therefore, a requirement of a reducing agent for lectin activity has come to be regarded as a characteristic feature of S-Lac lectins and underlies the alternative name S-type lectins.[5] L-29 and L-36 do not, however, require reducing agents and one of the carbohydrate binding domains of L-36 does not contain any cysteine.[18]

Degradation of Lectins by Proteolysis during Purification

Structural analysis of some of the S-Lac lectins from rat intestine has revealed that they represent degradation products of larger proteins. For example, RI-F is the carbohydrate binding domain (CBD) of L-29 and RI-H is domain II of L-36.[18] This degradation occurs despite the presence of mercaptoethanol, EDTA, and phenylmethylsulfonyl fluoride (PMSF) in the extraction buffer. Whereas the CBDs appear to be stable to proteolysis, other domains of L-29 and L-36 are very sensitive to proteolysis. Purification of L-29 from lung or from cultured cells has resulted in much less degradation. Therefore, stronger protease inhibitors or detergents might be required for isolation of lectins from some tissues.

Assays for Interaction between S-Lac Lectins and Glycoconjugate Ligands or Carbohydrate-Containing Particles

Assays for the interaction of S-Lac lectins with bacteria have not been described in detail. In this section, therefore, we review binding assays

[84] S. H. Barondes and H. Leffler, this series, Vol. 138, p. 510.

developed for other ligands that could be adapted for bacteria or components isolated from bacteria. Lectins can be used labeled with ^{125}I or fluorescein or they can be used unlabeled either immobilized on a matrix, or free in solution.

Many S-Lac lectins have been labeled successfully with ^{125}I with the Bolton–Hunter reagent (New England Nuclear/Dupont, Boston, MA).[16,19,21] This has produced active lectin of specific activities of 1–10 μCi/μg. In cases where this reagent destroys the activity of a lectin, the chloramine-T method is an alternative procedure.[85] L-14-I is an example of a lectin where the protection of sulfhydryl groups by reaction with iodoacetamide before labeling is necessary,[19] but for other lectins (e.g., L-29),[86] this procedure is not necessary. The labeled active lectin is repurified by loading the whole labeling reaction mixture on a lactosyl-Sepharose column, washing the column, and then eluting the active bound lectin with lactose. The remaining lactose is removed by dialysis. Alternatively, the lactosyl-Sepharose column is eluted with 10 m*M* lactose and the eluted lectin is diluted at least 100-fold before assays to decrease the lactose to a noninhibitory concentration.[19]

S-Lac lectins can be fluoresceinated with iodoacetamido-fluorescein (Molecular Probes, Eugene, OR) as described for L-14-I[87] or with fluorescein isothiocyanate (Molecular Probes) as for L-29.[88] The labeled lectin can be repurified on lactosyl-Sepharose to remove reagents and inactivated protein and then dialyzed to remove lactose.

Binding Assays

Microscale Assay for Comparing Lectin Affinities for Oligosaccharides in Solution

The relative affinity of a lectin for soluble oligosaccharides is measured by a microscale inhibition assay as described previously.[19] Briefly, 5 μl of the labeled lectin (in PBS containing 2% albumin) is added to a 1.5-ml microfuge tube with 5 μl of lactosyl-Sepharose (diluted 1 : 30 in PBS) and 5 μl of PBS or PBS containing the saccharide to be tested. After 1 hr at room temperature, the sample is overlayed with 100 μl of oil (adjusted to a density higher than that of the soluble sample and lower than that of

[85] P. Marschal, J. Herrmann, H. Leffler, S. H. Barondes, and D. N. W. Cooper, *J. Biol. Chem.* **267,** 12942 (1992).

[86] S. M. Massa, D. N. W. Cooper, H. Leffler, and S. H. Barondes, *Biochemistry* **32,** 260 (1993).

[87] D. N. W. Cooper, S. M. Massa, and S. H. Barondes, *J. Cell Biol.* **115,** 1437 (1991).

[88] R. Lindstedt, G. Apodaca, K. Mostov, S. H. Barondes, and H. Leffler, *J. Biol. Chem.* **268,** 11750 (1993).

lactosyl-Sepharose) and centrifuged in a microfuge (approximately 13,000 *g*) for 2 min. The droplet of solution containing unbound material is removed, the radioactivity in the remaining pellet is counted, and this value is used to calculate the percentage bound of the total amount of added lectin. This type of assay could be used to test the interaction of lectin with any solubilized component of bacteria.

Binding of Labeled Lectin to Immobilized Ligands

A sample containing the target (bacteria or components of the bacteria) is immobilized in wells (e.g., Immulon II, Removawells, Dynatech Laboratories, Inc., Chantilly, VA) as described for soluble glycoproteins[86] and for membrane glycoproteins dissolved in Triton X-100 or particulate material dispersed by sonication.[89] The unoccupied sites are blocked by incubation with 2% albumin (radioimmunoassay grade should be used because it does not contain significant amounts of glycoprotein contaminants that can bind the lectins). Labeled lectin is added, the wells are incubated and washed, and the amount of ^{125}I-labeled bound lectin is measured in a gamma counter. Different times and temperatures of incubation and washing should be tested to optimize the conditions of the assay.[86] This assay can be adapted for any bacterial product that can be immobilized in a well.

Ligands Immobilized on Nitrocellulose or Silica

Bacterial products bound to nitrocellulose can be probed with lectin as described for laminin.[86] Similarly, glycolipids separated by thin-layer chromatography can also be probed with lectin.[90] Bacterial glycoconjugates such as lipooligosaccharides[91,92] or polysaccharides[93] can be separated in a gel, transferred to nitrocellulose,[94] and probed with a tagged lectin.

Binding of Labeled Lectin to Particles

Simple sugars such as lactose can be bound to Sepharose beads (lactosyl-Sepharose) and then used to assay the binding of a lectin. The beads

[89] P. Rafiee, H. Leffler, J. C. Byrd, F. J. Cassels, E. C. Boedeker, and Y. S. Kim, *J. Cell Biol.* **115,** 1021 (1991).

[90] J. C. Solomon, M. S. Stoll, P. Penfold, W. M. Abbott, R. A. Childs, P. Hanfland, and T. Feizi, *Carbohydr. Res.* **213,** 293 (1991).

[91] H. Schneider, T. L. Hale, W. Zollinger, R. C. Seid, Jr., C. A. Hammack, and J. M. Griffiss, *Infect. Immun.* **45,** 544 (1984).

[92] R. E. Mandrell, H. Schneider, M. Apicella, W. Zollinger, P. A. Rice, and J. M. Griffiss, *Infect. Immun.* **54,** 63 (1986).

[93] S. Pelkonen and J. Finne, this series, Vol. 179, p. 104.

[94] W. N. Burnette, *Anal. Biochem.* **112,** 195 (1981).

containing bound lectin can be separated from unbound lectin by centrifugation (see above). Oil can be added to separate the phases if it is necessary to use small volumes. Lectin binding to whole gonococci has been assayed in this type of assay (R. E. Mandrell and H. Leffler, unpublished, 1987).

Fluorescein-Conjugated Lectin

Fluoresceinated lectin has been used to reveal ligands in tissue sections or cultured cells by fluorescence microscopy.[87] Analogous procedures could be developed to stain ligands on bacteria.

Purification of Ligands with Lectin Immobilized on Sepharose

Solubilized ligands have been isolated by affinity purification on lectin–Sepharose columns.[95] This type of matrix could be used for purification of solubilized ligands and for determining whether whole bacteria can interact with a lectin. Soluble ligands could be purified by this approach, but integral membrane molecules would first have to be solubilized in a detergent and the detergent-solubilized ligands would have to bind the lectin in the presence of detergent for this approach to be successful.

Soluble S-Lac Lectins

S-Lac lectins, in many cases, can be used as agglutinins directly.[17] This includes the monovalent lectin L-29, which at high concentration appears to aggregate in the presence of certain ligands.[86,88] Agglutination provides a simple method for identifying a potential bacterial ligand that binds a mammalian lectin. L-29 was recently found to act as a potent agglutinin for certain strains of *Neisseria* as well as *Haemophilus influenzae* (Lindstedt, R., Kim, J. J., Barondes, S. H., and Leffler, H., unpublished, 1993); 200 ng/ml of lectin induced microscopic agglutination of the bacteria. The lectin also co-agglutinated bacteria with neutrophils and leukocytes.

Specific Probes for Analysis of Bacterial Carbohydrates that Bind Animal Lectins

The precise *in vivo* role of many mammalian lectins is not known; however, *in vitro* studies with mammalian lectins have been helpful in determining the types of carbohydrates they can recognize.[8,16] Complete structural analysis of a putative carbohydrate receptor is beyond the capability of most laboratories, but investigators interested in defining mamma-

[95] R. K. Merkle and R. D. Cummings, *J. Biol. Chem.* **263,** 16143 (1988).

TABLE IV
CARBOHYDRATE-SPECIFIC REAGENTS

Type or source of reagent[a]	Refs.[b]
Monoclonal antibodies	1–3
Lectins	
Bacterial	4–6
Mammalian	See Table I
Plant, other	7
Viruses	8
Bacterial toxins	9
Endo-/exoglycosidases	10
Glycosyltransferases	11

[a] Some of these reagents are available from commercial sources (see references).

[b] *Key to references*: (1) J. L. Magnani, *Chem. Phys. Lipids* **42,** 65 (1986); (2) R. Kannagi and S. Hakomori, *in* "Handbook of Experimental Immunology" (D. M. Weir, ed.), p. 117. Blackwell, Oxford, 1986; (3) R. E. Mandrell, *Infect. Immun.* **60,** 3017 (1992); (4) D. M. Weir, *FEMS Microbiol. Immunol.* **47,** 331 (1989); (5) N. Sharon, *FEBS Lett.* **217,** 145 (1987); (6) K.-A. Karlsson, this series, Vol. 138, p. 212; (7) A. M. Wu, S. Sugii, and A. Herp, *in* "The Molecular Immunology of Complex Carbohydrates" (A. M. Wu and L. G. Adams, eds.), p. 819. Plenum, New York, 1988; (8) J. C. Paulson, J. E. Sadler, and R. L. Hill, *J. Biol. Chem.* **254,** 2120 (1979); (9) L. D. Heerze and G. D. Armstrong, *Biochem. Biophys. Res. Commun.* **172,** 1224 (1990); (10) F. Maley, R. B. Trimble, A. L. Tarentino, and T. H. Plummer, Jr., *Anal Biochem.* **180,** 195 (1989); (11) M. Basu, T. De, K. K. Das, J. W. Kyle, H.-C. Chon, R. J. Schaeper, and S. Basu, this series, Vol. 138, p. 575.

lian lectin–bacterial carbohydrate interactions can take advantage of the many carbohydrate-specific reagents that have been developed for characterizing mammalian tissues, cells, and molecules (Table IV). Some preliminary information about the carbohydrate receptor would be an advantage in selecting the appropriate reagents for analyzing a putative carbohydrate receptor or the interaction between bacteria and lectins. Some of these reagents (e.g., MAbs, lectins, glycosidases, glycosyltransferases, viruses] can be obtained from commercial sources or culture collections (see references), and as the interest in glycobiology increases, new reagents will become available.

Conclusions

Although many examples of bacterial lectins binding host carbohydrates have been identified, there is a growing appreciation of the existence of host lectins that can bind to microbial carbohydrates. There is a diverse array of animal lectins and other types of carbohydrate-binding proteins that have multiple specificities for carbohydrates,[96] and it has been proposed that these are part of a network of intracellular and intercellular interactions among bacteria and mammalian cells.[62] Complementary interactions between bacterial lectins and carbohydrates with host cells could contribute to host specificity; it can be speculated that LOS plays a role in the specificity of gram-negative mucosal pathogens for certain hosts, cells, and/or tissues. In many cases, a host lectin–bacterial carbohydrate interaction does not appear to be of primary importance in attachment and invasion under the conditions of the *in vitro* assays, but it should be kept in mind that the conditions encountered by a pathogen *in vivo* might be more relevant to these interactions. For example, the concentration of soluble S-Lac lectins *in vivo* might be much different than the amount secreted by cell lines grown *in vitro*, a factor that could be important in the interactions of the bacteria with particular cells. Some of the recent information related to animal lectins and bacterial carbohydrates has provided the basis for some interesting concepts; it is hoped this will stimulate future studies of this type of interaction.

Acknowledgments

Some of the studies that led to this report were supported by Public Health Service Grants AI-18384, AI-24616, and NHRSA T31AG-00212 and by funds provided by the Cigarette and Tobacco Surtax Fund of the State of California through the Tobacco-Related Disease Research Program of the University of California (Grant 279).

[96] T. Feizi and R. A. Childs, *Biochem. J.* **245,** 1 (1987).

[18] Coaggregation between Bacterial Species

By SUSAN A. KINDER and STANLEY C. HOLT

Introduction

Bacterial species in nature have a strong tendency to colonize surfaces. The subsequent bacterial multiplication, adherence of additional bacterial species, and production of extracellular polymers results in the formation of a complex microbial community on the colonized surface, referred to as a biofilm.[1] Examples of such microbial ecosystems include the biofilms found in natural aquatic environments such as on the rocks or other surfaces of the ocean or a stream; in industrial aquatic environments such as the metal surfaces in heat exchangers or bioreactors; and in animal host environments such as in the oropharyngeal, gastrointestinal (GI), and vaginal tracts or on medical prostheses.[1] Biofilms constitute a specialized, sequestered microenvironment offering selective advantages over growth as individual (planktonic) cells in solution. The basis for, and examples of, selective advantages associated with physiological and structural properties conferred by biofilm growth are briefly reviewed below.

Enhanced metabolic activity of the bacteria in biofilms versus planktonic bacteria have been documented, for example, an increase in glutamate metabolism[1] in biofilm cells. In addition, the localization of biofilm bacteria in close proximity to specific substrates and other bacterial species undoubtedly facilitates metabolic interactions. This is reflected by the preferential colonization of specific surfaces, such as the adherence of cellulolytic bacteria to cellulose surfaces.[1] Other examples include the nutritional interrelationships between bacteria found in biofilms on human teeth. For instance, a protoheme produced by *Campylobacter*(*Wolinella*) *rectus* functions as a growth factor for *Porphyromonas*(*Bacteroides*) *gingivalis,* and formate produced by *Prevotella* (*Bacteroides*) *melaninogenica* stimulates the growth of *C. rectus*.[2] The localization of bacterial cells within a biofilm in an animal host may additionally constitute a strategy for gaining access to host tissue cells for the delivery of tissue toxic substances.

From a structural standpoint, biofilm bacteria are embedded in a hydrated extracellular matrix, consisting of bacterial exopolymers and trapped environmental macromolecules. This matrix functions as a permeability barrier, enriching the local nutrient supply by limiting the diffusion of nutrients derived from biological substrates and bacterial metabolism. The biofilm matrix structure also accounts for protective advantages, concentrating defensive substances as well as limiting access of harmful substances and predatory cells to bacteria within the biofilm matrix. For example, antibiotic-degrading enzymes (e.g., β-lactamases) produced by the bacteria are thought to be concentrated in the biofilm matrix.[3] The penetration of specific antibiotics (e.g., tobramycin) into biofilms is thought to be limited, as bacterial cells within biofilms are more resistant

[1] J. W. Costerton, K.-J. Cheng, G. G. Geesey, T. I. Ladd, J. C. Nickel, M. Dasgupta, and T. J. Marrie, *Annu. Rev. Microbiol.* **41,** 435 (1987).

[2] D. Grenier and D. Mayrand, *Infect. Immun.* **53,** 616 (1986).

[3] B. Giwercman, E. T. Jensen, N. Hoiby, A. Kharazmi, and J. W. Costerton, *Antimicrob. Agents Chemother.* **35,** 1008 (1991).

to the antibiotic than planktonic cells of the same strains.[1,4,5] Biofilm bacteria may be similarly protected from fluctuating environmental conditions as well as potentially harmful substances in a host environment such as complement, antibody, and toxin. Finally, biofilm bacteria in an animal host are thought to be protected from the host cellular defenses. It has been shown, for example, that *Pseudomonas aeruginosa* biofilms are less efficient than planktonic cells in stimulating an oxidative burst response in human polymorphonuclear leukocytes (PMNs).[6] In an animal host, it has further been speculated that biofilm bacteria are less susceptible to cellular defenses, such as phagocytosis by PMNs, because of limited access of the PMN or other host cell to bacteria within the biofilm matrix.[1] Thus, the sheltered life within a biofilm appears to offer many advantages to a bacterium. An understanding of the processes by which bacterial species colonize and emerge in biofilms is critical to our understanding of normal ecology, as well as the ability of bacteria to function as a pathogen in these environments.

The initial colonization of a surface involves bacteria–substrate interactions, whereas subsequent colonization leading to the formation of biofilms also involves bacterial adherence to extracellular polymers or other bacterial cells present on the substrate.[7–9] The adherence of cells of different bacterial species or strains is referred to as *coaggregation,*[9] in contrast to *autoaggregation,* which refers to the adherence of cells of the same bacterial strain. Extensive studies on the coaggregation of oral bacteria have been reported.[9,10] Interestingly, with the exception of coaggregation of urogenital bacteria,[11] there is essentially no information on bacteria–bacteria adherence involved in biofilm development in natural and industrial aquatic environments, on the plastic, metal, and rubber surfaces of medical devices, or in the GI and vaginal tracts of animal hosts. Physiological interactions between bacteria in these environments have been described,[1,8] but the focus of adherence studies has been on the interaction of bacteria with extracellular polysaccharides of the biofilms.[1,8] It is not

[4] H. Anwar, M. Dasgupta, K. Lam, and J. W. Costerton, *J. Antimicrob. Chemother.* **24,** 647 (1989).

[5] H. Anwar, T. van Biesen, M. Dasgupta, K. Lam, and J. W. Costerton, *Antimicrob. Agents Chemother.* **33,** 1824 (1989).

[6] E. T. Jensen, A. Kharazmi, K. Lam, J. W. Costerton, and N. Hoiby, *Infect. Immun.* **58,** 2383 (1990).

[7] R. J. Gibbons and J. van Houte, *J. Periodontol.* **44,** 347 (1973).

[8] M. Fletcher, *Adv. Microb. Physiol.* **32,** 53 (1991).

[9] P. E. Kolenbrander, *CRC Crit. Rev. Microbiol.* **17,** 137 (1989).

[10] P. E. Kolenbrander, *Annu. Rev. Microbiol.* **42,** 627 (1988).

[11] G. Reid, J. A. McGroarty, P. A. G. Domingue, A. W. Chow, A. W. Bruce, A. Eisen, and J. W. Costerton, *Curr. Microbiol.* **20,** 47 (1990).

clear from the literature that coaggregation is in fact lacking in other environments or if it has not been investigated. If coaggregation of bacteria does contribute significantly to the development of biofilms in nonoral environments, it will be interesting to see if the principles that have emerged from the examination of oral species apply.

Studies on the coaggregation of oral bacteria have involved more than 700 bacterial strains representing over 15 genera.[9] These studies have revealed that coaggregation is a highly specific process which often involves the interaction of complementary bacterial surface molecules functioning as adhesins and receptors.[9,10] Examples of bacterial structures or molecules involved in coaggregation include fimbriae from numerous oral species, such as *Streptococcus sanguis, Actinomyces viscosus,* and *P. gingivalis*[9,10]; protein adhesins on the surface of *Capnocytophaga gingivalis* and *Fusobacterium nucleatum*[9,10,12–15] and a carbohydrate receptor on *S. sanguis*.[9,10] Efforts have focused on the isolation of coaggregation adhesins at a molecular level.[9,10]

In this review, we discuss assays used to measure bacterial coaggregation, focusing on the advantages of the different assay systems and variables of importance in conducting this type of experimentation.

Types and Uses of Coaggregation Assays

Assays used to examine bacterial coaggregation measure either the adherence of bacterial cells of different species to one another in suspension (referred to as coaggregation of bacterial cells in suspension) or the adherence of one bacterial species to cells of a second species that have been immobilized on a solid substrate (referred to as solid-state coaggregation assays). The first type of assay relies on the formation and precipitation of bacterial aggregates in suspension, which occur secondary to the adherence of the individual bacterial cells. This type of assay system may be used to generate semiquantitative measurements yielding ordinal data as well as quantitative measurements yielding interval data. The solid-state assay systems, in contrast, directly measure the adherence of cells of one bacterial species to another. Coaggregation assays have been used in the characterization of adherence properties of individual bacterial species, delineation of the mechanisms of adherence, identification and isolation of the bacterial surface molecules responsible for mediating ad-

[12] S. A. Kinder and S. C. Holt, unpublished data.

[13] S. A. Kinder and S. C. Holt, *Infect. Immun.* **57,** 3425 (1989).

[14] P. Lancy, Jr., J. M. DiRienzo, B. Appelbaum, B. Rosan, and S. C. Holt, *Infect. Immun.* **40,** 303 (1983).

[15] P. Lancy, Jr., B. Appelbaum, S. C. Holt, and B. Rosan, *Infect. Immun.* **29,** 663 (1980).

herence, and isolation of adherence-deficient mutant strains. Each assay system offers specific advantages depending on the purpose for which the assay is being used, as reviewed below.

Coaggregation of Bacterial Cells in Suspension

Principle

As stated above, coaggregation of bacterial cells in suspension is assayed based on the formation and precipitation of large bacterial aggregates. The phenomena of aggregate formation and precipitation occur secondary to the adherence of individual bacterial cells. The semiquantitative macrocoaggregation assays use a visual assessment of the extent of coaggregation. The quantitative assays include the microcoaggregation, filter-retention, and spectrophotometric assays. These systems measure the extent of coaggregation by determining the incorporation of cells into coaggregates using radiolabeling of cells or turbidity measurements.

Basic Parameters Which May Influence Coaggregation of Bacterial Cells in Suspension

Bacterial Cell Age and Viability. Bacterial surface molecules are known to vary during growth phases,[16] and in specific instances the growth of bacterial cells has been shown to have a profound effect on bacterial adherence to eukaryotic tissue cells.[17] Significant alterations in prokaryotic coaggregation based on growth phase have, however, not been observed.[10,13,18,19] Similarly, the chronological age of bacterial cultures (number of *in vitro* passages) is known to influence the expression of surface molecules. For example, the loss of surface structures with *in vitro* passage has been reported for the capsules of pneumonocci[20] and *Bacteroides fragilis,*[21] the fimbriae of *Neisseria gonorhoeae,*[22] and the S layer of

[16] L. V. Stamm, R. L. Hodinka, P. B. Wyrick, and P. J. Bassford, Jr., *Infect. Immun.* **55,** 2255 (1987).

[17] C. A. Lee and S. Falkow, *Proc. Natl. Acad. Sci. U.S.A.* **87,** 4304 (1990).

[18] G. Bourgeau and B. C. McBride, *Infect. Immun.* **13,** 1228 (1976).

[19] R. J. Gibbons and M. Nygaard, *Arch. Oral Biol.* **15,** 1397 (1970).

[20] C. M. MacLeod and M. R. Krauss, *J. Exp. Med.* **92,** 1 (1950).

[21] D. L. Kasper, A. B. Onderdonk, B. G. Reinap, and A. A. Lindberg, *J. Infect. Dis.* **142,** 750 (1980).

[22] J. Swanson, S. J. Kraus, and E. C. Gotschlich, *J. Exp. Med.* **134,** 886 (1971).

C. rectus.[23] Chronological age has not, however, been found to be a significant variable in the interbacterial adherence observed in oral strains.[10] Further, coaggregation is thought to occur regardless of cell viability.[9,10,12] The storage of bacterial cells in buffer with 0.02% (w/v) sodium azide at 4° for up to 10 years with no loss in coaggregation properties has been reported.[9,10] Thus studies on the significance of cell age and viability indicate that they do not alter bacterial coaggregation, but their dramatic influence on the adherence of bacteria to eukaryotic cells suggests that these variables should continue to be examined in future studies of interbacterial adherence.

Culture Medium Used in Bacterial Growth. In general, the medium used to culture bacterial strains has not been found to influence coaggregation properties,[10] but isolated reports of an effect of culture media exist. For example, *S. sanguis* and *Streptococcus mutans* grown in the presence of sucrose coaggregate with *A. viscosus*, whereas the same species grown in the presence of glucose fail to coaggregate.[18] The interaction appears to be mediated by dextran, which is produced by the streptococci only when grown with sucrose. Although it may be the exception, the possibility does exist that culture medium may be an important variable in determining coaggregation properties.

Cell Concentration of Coaggregating Species. Two separate aspects of cell concentration are important variables in coaggregation assays (Fig. 1): (1) the ratio, or relative concentration, of cells of the different strains used in the assay,[7,10,12] for example, a 10 : 1 ratio of species A to species B; and (2) the total concentration of cells used in the assay,[13,19] for example, 10^8 versus 10^9 cells per milliliter.

Studies examining the relative concentration of bacterial cells[10,12,14] have revealed that maximal levels of coaggregation are achieved with a cell ratio of approximately 1 : 1 (species A : species B). Excess of either cell type (e.g., ratio of 10 : 1 or 1 : 10) results in diminished levels of measurable coaggregation (see Fig. 1A). In the latter case, adherence between individual cells may occur[24]; however, the excess of one cell type prohibits the formation of aggregates large enough for detection in assay systems measuring coaggregation of bacterial cells in suspension. Of note, in studies designed to examine potential inhibitors of coaggregation, the sensitivity of the assay to inhibition depends on the relative concentration of the coaggregating cells.

Coaggregation is further dependent on the total concentration of cells, with a dose-dependent relationship between cell concentration and per-

[23] R. Borinski and S. C. Holt, *Infect. Immun.* **58,** 2770 (1990).

[24] P. E. Kolenbrander and R. N. Andersen, *Appl. Environ. Microbiol.* **54,** 1046 (1988).

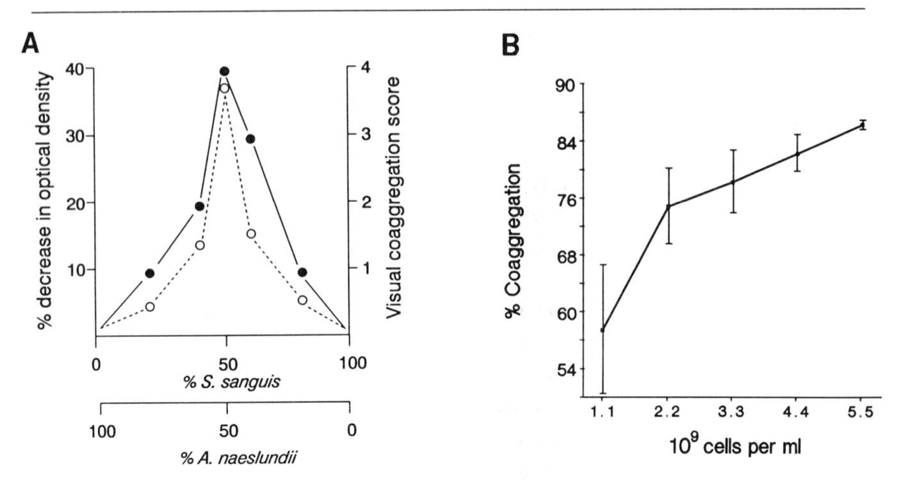

FIG. 1. Effect of cell concentration on coaggregation. (A) Variation in the extent of coaggregation with alterations in the relative proportion of cells of *Actinomyces naeslundii* strain 1 and *Streptococcus sanguis* strain 34. Coaggregation was scored using a semiquantitative visual assay (●) and a quantitative spectrophotometric assay (○). (Adapted from Gibbons and Nygaard[19] with permission from the authors and publisher.) (B) Variation in the extent of coaggregation with the total concentration of cells. *Fusobacterium nucleatum* T18 and *Porphyromonas gingivalis* T22 were suspended in equal concentrations (ratio of 1 : 1) in the quantitative microcoaggregation assay. The standard deviations of triplicate determinations of the percentage coaggregation are indicated by the horizontal lines. (Reproduced from Kinder and Holt[13] with permission from the authors and publisher.)

centage coaggregation (Fig. 1B). A cell concentration in the range of 10^9 is typically used in assays of coaggregation of bacterial cells in suspension.[9,10,13,25]

Incubation Conditions. Coaggregation reactions have been observed in a wide variety of buffers, including the commonly used phosphate-buffered saline (PBS) and Tris buffers.[10] The pH of the buffer systems used has essentially no effect on coaggregation over the range pH 5 to 10,[9,10,13] and typically a neutral to slightly basic pH is used.[13,26] Most studies of coaggregation have used incubations in aerobic environments[9,10] and comparisons of anaerobic versus aerobic incubation demonstrated no differences.[12] Temperature has been found to affect the rate of coaggregation, with retarded kinetics at lower temperatures.[12,13]

[25] P. E. Kolenbrander and R. N. Andersen, *J. Bacteriol.* **168,** 851 (1986).
[26] J. O. Cisar, P. E. Kolenbrander, and F. C. McIntire, *Infect. Immun.* **24,** 742 (1979).

Semiquantitative Assays of Bacterial Coaggregation in Suspension: Macrocoaggregation Assays

Basic Methodology

The basic methodology behind the macrocoaggregation assay system[13] was originally described by Gibbons and Nygaard.[19] Bacterial cells are harvested, washed, and resuspended in the reaction buffer [coaggregation buffer (CB): 1.0 m*M* Tris–HCl, 0.1 m*M* $MgCl_2$, 0.1 m*M* $CaCl_2$, 0.15 *M* NaCl; or phosphate-buffered saline].[13,26] The cells may be used immediately[13] or stored in buffer with 0.02% azide at 4°.[9,10,26] For use in the assay, cells are resuspended to a standard cell concentration, typically in the range of 10^9 cells/ml. This is most easily accomplished using a turbidity measurement with a spectrophotometer, based on previous determinations of the relationship between optical density and microscopic or viable cell counts. The standardized suspensions of bacterial cells are then combined in a glass test tube with an equal volume of another bacterial suspension or with an equal volume of buffer as a control for autoaggregation. A final volume of 1 to 2 ml is generally adequate for visualization of the reaction. The reaction mixtures are vortexed for 10 sec, then incubated at a specific temperature for a predetermined time (i.e., typically from a few minutes up to 2 hr). Examples of the macrocoaggregation assay are illustrated in Fig. 2.

Measurement of Extent of Coaggregation

Determination of the extent of coaggregation for the semiquantitative assay is based on a visual assessment of aggregation (see Fig. 2), with the assignment of a score ranging from 0 (no coaggregation) to +4 (maximal coaggregation). Two different versions of the scoring system are presented in Table I.[26a] In the first system (A), three variables are used to determine the score: the formation, precipitation, and rate of precipitation of aggregates in the reaction mixtures. In the second system (B) the time element has been eliminated, as this evaluation becomes cumbersome when a large number of samples are examined. Thus, the scores are based on the formation and precipitation of aggregates at a specific time point.

Scoring system B further advocates the use of control incubations in which cells of the individual strains are combined with buffer to evaluate the extent of autoaggregation. This control has not been routinely used in coaggregation studies and may be of particular significance when exam-

[26a] P. E. Kolenbrander, R. N. Andersen, and L. V. Holdeman, *Infect. Immun.* **48,** 741 (1985).

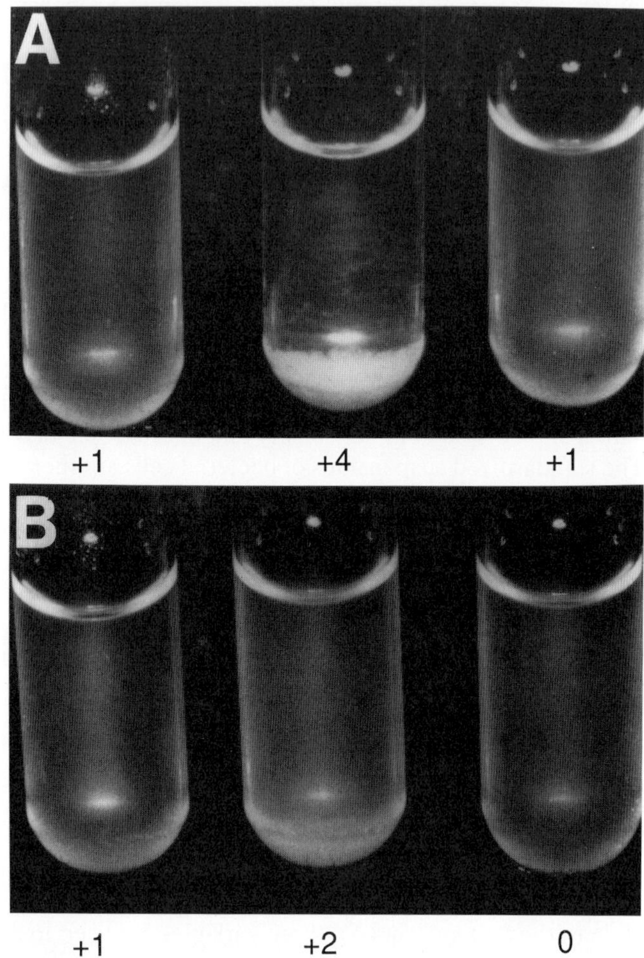

FIG. 2. Semiquantitative macrocoaggregation assay. The extent of coaggregation as determined by the visual scoring system B of Kinder and Holt.[13] The interaction between *Fusobacterium nucleatum* T18 and *Porphyromonas gingivalis* T22 (center tube, A) resulted in a +4 reaction, whereas control incubations of *F. nucleatum* T18 (left tube, A) and *P. gingivalis* T22 (right tube, A) resulted in +1 reactions. In comparison, the interaction between *P. gingivalis* T22 and *Actinomyces viscosus* T30 resulted in a +2 reaction (center tube, B) and +1 and 0 reactions for the *P. gingivalis* T22 (left tube, B) and *A. viscosus* T30 (right tube, B) control incubations, respectively.

TABLE I
SEMIQUANTITATIVE COAGGREGATION SCORING SYSTEMS

Score	Scoring system A[a]	Scoring system B[b]
0	No change in turbidity; no evidence of coaggregates	Evenly turbid supernatant; no evidence of aggregates
+1	Turbid supernatant with finely dispersed coaggregates	Evenly turbid supernatant; precipitation of aggregates evident
+2	Definite coaggregates visible, but they do not precipitate immediately	Turbid supernatant with aggregates; precipitation of aggregates evident
+3	Slightly turbid supernatant remaining with formation of large precipitating coaggregates	Turbid supernatant with aggregates; precipitation of aggregates evident
+4	Clear supernatant and large coaggregates formed, but they precipitate immediately	Clear supernatant without turbidity or aggregates; precipitation of aggregates evident

[a] From Kolenbrander *et al.*[26a]
[b] From Kinder and Holt.[13]

ining the interactions of bacteria that normally aggregate, for example, *Streptococcus* species. The level of aggregation of the reaction mixture containing both species is interpreted as autoaggregation plus coaggregation, and significant coaggregation is thus evident when the reaction between the different species exceeds that of the reaction of the controls, at a given time point.

Uses of Semiquantitative Assays

Semiquantitative assays are rapid and allow for the examination of a large number of reactions, making them the ideal system to screen for coaggregating species. They have also been used for an initial determination of the effect of potential inhibitory substances, such as sugars or antibody preparations[10,27,28] and for the isolation of naturally occurring coaggregation-deficient mutant strains.[12,29]

Quantitative Assays of Bacterial Cells in Suspension

Basic Methodology

The basic methodology is essentially the same as for the semiquantitative assays described above. Additional modifications of the assay

[27] P. E. Kolenbrander and R. N. Andersen, *Infect. Immun.* **57,** 3204 (1989).
[28] E. I. Weiss, J. London, P. E. Kolenbrander, R. N. Andersen, C. Fischler, and R. P. Siraganian, *Infect. Immun.* **56,** 219 (1988).
[29] P. E. Kolenbrander, *Infect. Immun.* **37,** 1200 (1982).

systems depend on the method of quantitation used and are discussed below.

Microcoaggregation Assay

This assay, as described by Kinder and Holt[13] (Fig. 3), is based on modifications of the original procedure described by Kolenbrander and Andersen.[25] The methodology involves the metabolic radiolabeling of one of the two bacterial strains to be examined. The specific radionucleotide may vary depending on its uptake by different bacterial species, but tritiated labels (e.g., [^{3}H]thymidine, [^{3}H]adenine) are preferred based on the relative safety of the reagent. The labeling is accomplished by incorporating the radionucleotide into the liquid growth medium at the time of inoculation,[13] at concentrations of 2 to 10 μCi/ml. The extent of labeling varies between species, and specific activity in the range of 10^4 to 10^5 bacterial cells/cpm is sufficient for the purpose of quantitation. The labeled cells are harvested and prepared as described for the semiquantitative assay. The radiolabeled cells (0.2 ml) are combined with similarly prepared

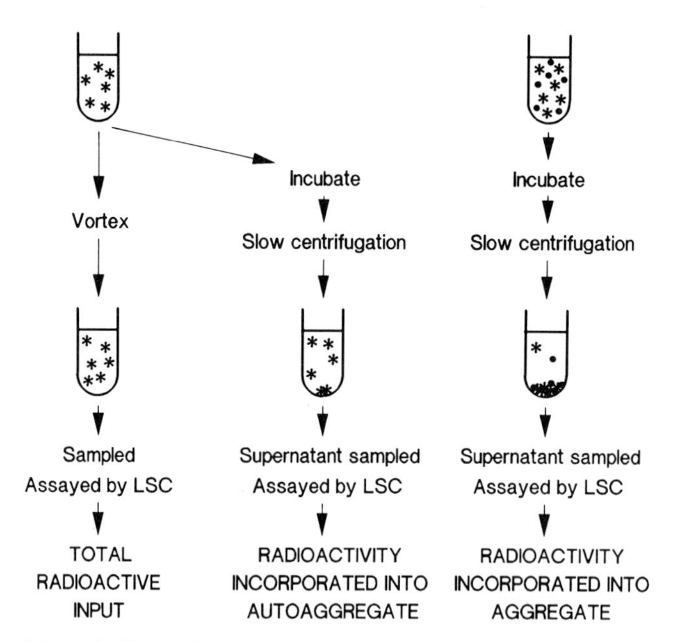

FIG. 3. Schematic illustration of the quantitative microcoaggregation assay. (*) Radiolabeled bacterial strain; (●) unlabeled bacterial strain. LSC, liquid scintillation counting. See text for details of the method. (Reproduced from Kinder and Holt[13] with permission from the authors and publisher.)

unlabeled cells (0.2 ml) of the other species to determine the level of aggregation or with buffer to determine the total radioactive input and the level of autoaggregation of the radiolabeled cells (Fig. 3). The cells are combined in Tween 20-coated microfuge tubes with a final volume of 0.4 ml and are incubated at 37° for 30 min. The microfuge tubes are coated by immersing the tubes in 0.05% (v/v) Tween 20 and then drying overnight at 55°. The use of coated microfuge tubes has been found to improve significantly the recovery of radiolabeled cells, probably as a result of a reduction in the absorption of bacterial cells to the tube itself.[12] After incubation, the reaction mixtures are subject to a slow-speed centrifugation to delineate the aggregated from nonaggregated cells. Centrifugation protocols that have been used for this purpose vary, ranging from 7 *g* for 2 min[26,30] to 86 *g* for 1 min.[13] The appropriate *g* force varies with different species and should be determined by examining the precipitation of individual cells at comparable concentrations over a range of *g* forces and time periods using a variable speed microcentrifuge (Eppendorf Micro Centrifuge 5415; Brinkmann Instruments, Inc., Westbury, NY). A 0.1-ml sample of the supernatant is removed from each reaction mixture and assayed by liquid scintillation counting. Equations (1)–(3) are used to determine the percentage coaggregation[13]:

$$\%\ \text{autoaggregation} = [1 - (\text{cpm autoaggregate supernatant/cpm radioactive input})] \times 100 \quad (1)$$

$$\%\ \text{aggregation} = [1 - (\text{cpm aggregate supernatant/cpm radioactive input})] \times 100 \quad (2)$$

$$\%\ \text{coaggregation} = [(\%\ \text{aggregation} - \%\ \text{autoaggregation})/(1 - \%\ \text{autoaggregation})] \times 100 \quad (3)$$

All determinations are carried out in triplicate to assess variability, and the data are expressed as the mean and standard deviation. The results of the microcoaggregation assays correlate well with results of the semiquantitative macrocoaggregation assay,[12,13,25] and are highly reproducible with some variation observed as a function of bacterial batches.[12,13] As a consequence of the reproducibility of the data, the difference between statistical significance and biological significance must be carefully considered in the interpretation of these results.

This assay system is also useful in characterizing the effect of biological fluids such as saliva or serum, sugars, chelating agents, antibody preparations, and isolated bacterial cell wall components on coaggregation.[9,10,12,13,18,26,30] In addition, bacterial cells may be pretreated with heat or enzymes that modify protein or carbohydrate structures and then exam-

[30] F. C. McIntire, A. E. Vatter, J. Baros, and J. Arnold, *Infect. Immun.* **21,** 978 (1978).

ined for coaggregation properties. This approach is useful in characterizing the bacterial cell surface molecules involved in adherence.[13,16,27] Data from this type of experiment are used to determine either the percentage inhibition[13,30] [Eq. (4)] or the relative percentage coaggregation[13] [Eq. (5)].

$$\% \text{ inhibition} = ([\% \text{ coaggregation in the absence of inhibitor} - \% \text{ coaggregation in the presence of inhibitor}]/\% \text{ coaggregation in the absence of inhibitor}) \times 100 \quad (4)$$

$$\text{relative } \% \text{ coaggregation} = (\% \text{ coaggregation using treated bacterial cells}/\% \text{ coaggregation using untreated bacterial cells}) \times 100 \quad (5)$$

Filter-Retention Assay

The retention of coaggregates on filters is another means of generating quantitative data representing the extent of coaggregation. This method, described by Lancy and co-workers,[14,15] has been used to examine the interaction of species with an elongated cell morphology, such as *F. nucleatum* and *Bacteroides matruchotii,* with streptococci. The streptococci are metabolically radiolabeled with [^{3}H]thymidine, and the reaction mixtures are prepared and combined in a manner similar to the assays described above. After incubation on a rocking platform for 1 hr, the mixtures are passed through a 5-μm filter (Nucleopore Corp., Pleasanton, CA), and aggregates retained are quantitated by liquid scintillation counting of the recovered filters. The data are expressed as the number of streptococci bound, and this is determined based on the specific activity of the radiolabeled cells.

Spectrophotometric Assay

The spectrophotometric assay first described by McIntire and co-workers[30] is similar in principle to the microcoaggregation assay described above. An advantage of the spectrophotometric assay is that it does not require radiolabeling of the bacterial cells; a disadvantage is that it does require larger volumes of cells to carry out the quantitative measurements. Bacterial cells prepared as described above are combined in 10 × 75-mm culture tubes with a total volume of 1 ml. The reaction mixtures include the individual strains alone as controls for autoaggregation in addition to the combination of the two strains being examined for coaggregation. The mixtures are vortexed for 10 sec and incubated for a defined period ranging from 30 min to 2 hr. The reaction mixtures are centrifuged at low speed (ca. 7 g for 2 min); then 0.6 ml of supernatant is carefully pipetted off and analyzed by spectrophotometry to determine the A_{650}. The level of coaggregation for species X and Y is determined according to the equation[30]

$$\% \text{ coaggregation} = [([\{A_{650}\,\text{X} + A_{650}\,\text{Y}\}/2] - A_{650}\,(\text{X} + \text{Y}))/\{A_{650}\,\text{X} + A_{650}\,\text{Y}\}/2] \times 100 \qquad (6)$$

When potentially inhibiting substances are examined in this assay system, the extent of inhibition is calculated according to Eq. (4) as described for the microcoaggregation assay.

Solid-State Coaggregation Assays

Principle

Another approach to study coaggregation, or interbacterial adherence, is to examine the adherence of one "test" bacterium to a second "base" bacterium that has previously been fixed to a surface. This approach models the *in vivo* situation of the adherence of bacterial cells of one species to bacterial cells of another species already resident on a surface. These assays (Fig. 4) involve the following common steps: (1) fixing of

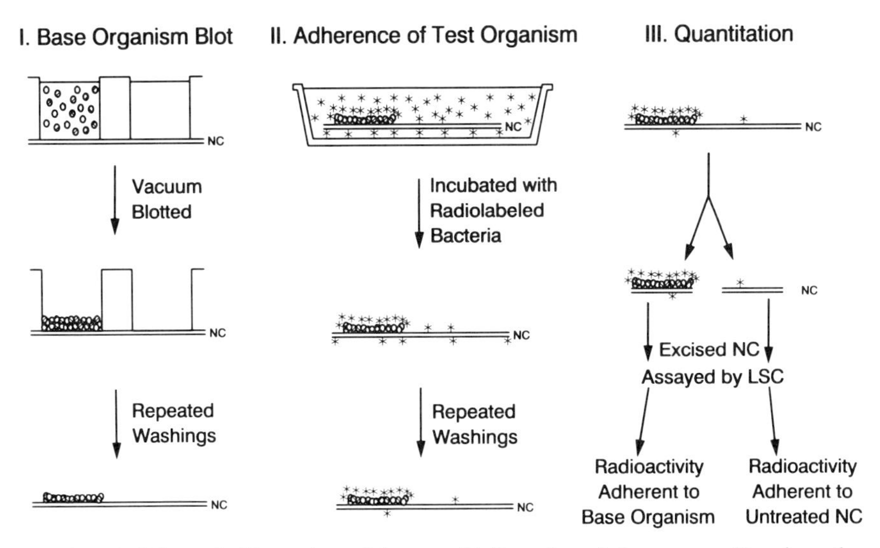

FIG. 4. Schematic illustration of the quantitative nitrocellulose assay. Based on the nitrocellulose assay described by Lamont and Rosan,[31] the steps of development of the base organism blot (I), adherence of the test organism to the base organism blot (II), and quantitation (III) are outlined. (*) Radiolabeled bacterial strain; ⊕, unlabeled bacterial strain. NC, nitrocellulose; LSC, liquid scintillation counting.

the base organism to the substrate, (2) adherence of the test organism to the immobilized base organism, and (3) detection of the adherence of the test organism. The solid substrates that have been used include nitrocellulose,[31] a biological adhesive coating on plastic or enamel,[32] and beads of agarose or saliva-coated hydroxyapatite.[33–35] The principles of these assay systems are very similar and two systems, the nitrocellulose assay and the biological adhesive assay, are reviewed below as examples.

Nitrocellulose Assay

The nitrocellulose assay, described by Lamont and Rosan,[31] involves the immobilization of cells of the base organism on nitrocellulose, followed by examination of the adherence of cells of the test organism to the nitrocellulose-bound cells.

Initial studies are carried out to assess the binding of the base organism to nitrocellulose. The bacterial strains are metabolically radiolabeled by incorporating 10 μCi [^{3}H]thymidine/ml into the growth medium, to achieve a specific activity in the range of 10^4 to 10^5 cells/cpm. Bacterial cells are harvested, then washed twice and resuspended in KCl buffer (5 m*M* KCl, 2 m*M* K_2PO_4, 1 m*M* $CaCl_2$, pH 6.0) to varying concentrations based on previously determined correlations between spectrophotometric readings and bacterial cell counts. Volumes of 50 μl of the bacterial cell suspensions are placed in individual wells of a dot-blot apparatus (Bio-Rad Laboratories, Richmond, CA) with a 0.45-μm nitrocellulose membrane (Schleicher & Schuell Co., Keene, NH), and a vacuum is applied to draw off the buffer and concentrate the bacterial cells on the nitrocellulose surface. The nitrocellulose, with bound cells, is removed from the apparatus and washed three times (KCl-T: KCl buffer with 0.1% Tween 20, 15 min per wash) to remove loosely bound cells. The areas of the nitrocellulose exposed to the bacterial cells are excised and the radioactivity is determined by liquid scintillation counting. Calculation of the number of bacteria bound is based on the specific activity of the bacterial cell preparation. The maximum number of base organisms bound is determined and this level is then used to prepare base organism blots with nonradiolabeled cells for examination of the adherence of the test organism. The base

[31] R. J. Lamont and B. Rosan, *Infect. Immun.* **58,** 1738 (1990).

[32] W. F. Liljemark, C. G. Bloomquist, M. C. Coulter, L. J. Fenner, R. J. Skopek, and C. F. Schachtele, *J. Dent. Res.* **67,** 1455 (1988).

[33] S. Schwarz, R. P. Ellen, and D. Grove, *Infect. Immun.* **55,** 2391 (1987).

[34] J. E. Ciardi, G. F. A. McCray, P. E. Kolenbrander, and A. Lau, *Infect. Immun.* **55,** 1441 (1987).

[35] M. W. Stinson, K. Safulko, and M. J. Levine, *Infect. Immun.* **59,** 102 (1991).

organism blots are used immediately after preparation or may be stored at 4° in a humid chamber for a time period documented to have no effect on adherence.

To examine the adherence of the test organism, bacterial cells are metabolically radiolabeled and prepared as described above. The test bacterial cells are resuspended in KCl-T to the desired cell concentration. Three milliliters of the test organism is added to the base organism blot and rotated at 37° for 2.5 hr. The blot is then washed four times (KCl-T, 15 min each), the experimental area of the blot is excised, and the associated radioactivity is determined by liquid scintillation counting. The data are expressed as the number of bacterial cells bound, based on the specific activity of the bacterial cell preparation. Controls for nonspecific binding consist of examination of an equivalent area of nitrocellulose lacking the bound base organism that was exposed to the test organism during the assay. As an alternative initial screening procedure, the test organisms may be radiolabeled with ^{32}P and visualized by autoradiography.

Biological Adhesive Assay

The biological adhesive CellTak (BioPolymers, Inc., Farmington, CT) has been used on plastic tissue culture plates to promote the attachment of a continuous layer of bacterial cells. This surface of base organisms is then used to examine the subsequent adherence of a second test organism. The procedure of Liljemark and co-workers[32] to prepare a continuous bacterial surface on tissue culture plates for use in an adherence assay is described below.

The coating of tissue culture dishes (35 × 100 mm, Falcon No. 3001) with 250 mg (w/v) of CellTak diluted in 5% acetic acid is carried out according to the manufacturer's recommendations. Bacterial cells (2 ml at 10^{10} cells/ml) suspended in buffer (50 m*M* KCl, 1 m*M* K_2PO_4, 1 m*M* $CaCl_2$, 1 m*M* $MgCl_2$, pH 6.0) are added to the plates, which are then centrifuged (30 min, 1200 *g*, 37°). The plates are washed twice and the coating procedure is repeated a second time. An additional wash is performed with the plates rotating (50 rpm) with 2 ml of buffer at 37°, followed by rinsing with buffer. Scanning electron microscopy is used to assess the uniformity of coating of the tissue culture wells with bacteria.

The adherence of a second bacterial strain is then examined. The test organism is metabolically radiolabeled with [^{3}H]thymidine (10 μCi/ml in the culture medium), washed, and resuspended in buffer at the desired cell concentration (ranging from 10^7 to 10^9 cells/ml), and 2 ml is added to the wells of the tissue culture plate. The plates are gently rocked for 1.5 hr at 37°, the solution is decanted, and the plates are washed twice with buffer and then allowed to dry. Several pieces of the tissue culture dish with adherent bacteria are removed from the plates with a metal punch and are counted in a liquid scintillation counter. The number of bacteria adhering is calculated based on the specific activity of the bacterial cell preparation.

Similar procedures can be used to coat surfaces of materials that may then be placed in an *in vivo* environment and used to assess adherence. For example, enamel chips have been coated with a base organism, then placed in orthodontic appliances *in vivo*.[32]

Sensitivity and Specificity of Coaggregation Assay Systems

The sensitivity of the solid-state assays of coaggregation are generally greater than that of the assays evaluating coaggregation of bacteria in suspension. The macro- and microcoaggregation assay systems are able to detect the adherence of approximately 10^9 bacterial cells,[9,10,13] whereas the filter-retention assay as well as the solid-state assays are able to detect the adherence of approximately 10^7 bacterial cells.[31] Despite the increased sensitivity, a significant drawback of the solid-state assays is that they are more time consuming than the assays of coaggregation of bacteria in suspension. In either case, the specificity of adherence is an important property to delineate, and this is accomplished by demonstrating the adherence of certain strains but the clear lack of adherence of other strains in the same assay system.[9,10,13]

Summary

Bacterial coaggregation, or interbacterial adherence, is one mechanism involved in the development of bacterial biofilms that are found on surfaces in nature. Assays used to measure coaggregation rely on the interaction of bacterial cells in suspension or attachment of one species to a second species that has been fixed to a solid substrate. Both semiquantitative and quantitative assays are described. These methods have also been used to determine the nature of the adherence and molecules involved in mediating the interaction, to characterize potential inhibitors, to isolate the bacterial adhesins and receptors, and to isolate adherence-deficient mutant strains. Each of the assay systems offers different advantages, with significant variations in sensitivity. Selection of a particular assay system should depend on the goals of the study to be performed.

[19] Purification of Somatic Pili

By GARY K. SCHOOLNIK

Introduction

Pili are nonflagellar, proteinaceous appendages of the bacterial outer membrane that are elaborated by a wide variety of gram-negative species.[1] F pili are involved in the conjugal transfer of genetic material between bacteria; their purification and characterization is beyond the scope of this article. Somatic pili mediate a multiplicity of other functions which, in the most general sense, enable bacteria to colonize host surfaces or to occupy inanimate environmental habitats. Depending on the bacterial species and pilus type, somatic pili have been implicated as effectors of the following functions: specific, i.e., ligand–receptor based, adherence of bacteria to host cell surfaces; nonspecific adsorptive binding of bacteria to abiotic surfaces; translocation of attached bacteria across surfaces by twitching motility; resistance to phagocytosis; transformation competence; and the formation of interbacterial linkages within colonies. The purification of pili has been carried out to determine the structural basis for these functional attributes, to identify antigenic determinants, and to study pili as vaccines.

All pili comprise a repeating polypeptide subunit, termed *pilin,* which polymerizes to form the pilus rod, fiber, or filament. Thus, pili are macromolecular assemblies of pilins and because each filament is composed of an estimated 1000 or more of these subunits, pili are usually visible at magnifications as low as 20,000 when negatively stained and examined by transmission electron microscopy. Members of the type IV pilus family (see below) seem to be homopolymers composed entirely of an identical pilin within which resides the important functional domains and antigenic determinants of the entire filament. In contrast, *Escherichia coli* type 1 and pyelonephritis-associated pili (PAPs) are heteropolymers composed principally of their respective pilins, but containing in addition relatively small numbers of specialized accessory proteins. For each of these *E. coli* pilus types there is a lectinlike accessory protein that specifically binds the corresponding carbohydrate receptor of the pilus filament.

The pilus polymer is stabilized by strong, noncovalent hydrophobic interactions which normally remain intact during most of the purification methods described below. As a consequence, preparations of purified

[1] C. C. Brinton, Jr., *Nature* (*London*) **183,** 782 (1959).

pilus filaments not only contain a large concentration of pilin, but may also contain small concentrations of accessory pilus proteins, depending on the pilus type. Dissociation of the pilus polymer can be accomplished through the use of detergents or denaturants, yielding pilin monomers, pilin oligomers, and accessory proteins, if present. This article considers only the preparation of intact pilus filaments; the preparation of pilus subunits, including accessory proteins, is discussed in [20].

Pili are classified by taxonomic, functional, structural, and serologic criteria. However classified, they are remarkably diverse organelles, reflecting the influence of different pilin amino acid sequences and accessory proteins on their ultrastructure, function, and antigenicity. At the same time, their filamentous morphology and polymeric nature constitute a common structural plan that has led to purification strategies that are similar for many of the pilus types. Accordingly, general principles of pili purification are discussed in the next section. Chemical differences between pilus types have led to the development of specialized purification schemes; these are described in the last section of the article for the pathogenically significant *E. coli* pilus types and for one example of the type IV pilus type (the pili of *Neisseria gonorrhoeae*).

Preparative Methodologies: Growth-Dependent Pilus Expression and Pilus Purification Strategies

Pilus expression is usually modulated *in vivo* and *in vitro*. For example, many pilus-expressing bacteria exhibit phase variation of some kind in which some bacteria within a population of bacteria are phenotypically piliated (P+), whereas others are nonpiliated (P−). At rates that depend on the bacterial species and strain and the conditions of growth, P+ bacteria spawn P− progeny and P− bacteria in turn give rise to P+ daughter cells. Phase variation is usually a back-and-forth phenomenon, but for some species mutations in pilus expressing genes occur or pilus-expressing plasmids are lost yielding variants of the original strain that are irreversibly P−. Phase variation also refers to the oscillating expression of two different pilus types by the same strain; usually both pilus types are never simultaneously expressed by the same bacterium although occasionally a P− state is achieved when neither pilus type is expressed. Although a description of the genetic mechanisms underlying phase variation is beyond the scope of this article, the phenomenon needs to be recognized and controlled to obtain high yields of pure pili that are homogeneous; for this reason, most of the detailed purification protocols discussed in the last section of the article include growth conditions for optimal *in vitro* pilus expression. Phase variation and its significance for the purification of pili is discussed in greater detail below.

In general, nonselective passage of bacteria *in vitro* leads to an increasing percentage of P− organisms either as a result of phase variation and the selective outgrowth of the P− variants on the medium used or through the loss of a pilus-encoding plasmid. This problem can be ameliorated through the use of aliquots of freshly isolated P+ organisms that have been stored at −70°.

P+ organisms often can be recovered from a P− population of organisms—providing they have retained the capacity to express pili—by passage of the bacteria through an appropriate animal model where the P+ phenotype confers a selective growth advantage. Similarly, P+ organisms can be obtained through the serial passage of bacteria that adhere via pili to tissue culture cell monolayers. The power of both of these methods comes from two independent effects: first, the probable induction of pilus-expressing genes by exposure of the bacteria to an environment containing physicochemical signals that lead to transcriptional activation of pili genes; and second, selection of P+ bacteria as a consequence of a pilus-mediated growth advantage. For hemagglutinating pilus types, P+ variants can be obtained by incubating red cells with a mixed population of P+ and P− bacteria, washing the red cells free of the nonadhering P− variants, and then inoculating the red cells together with their attendant bacteria onto a nutrient agar.

P+ and P− variants can often be distinguished by colonial morphology and the P+ colonies visually identified and cloned by selective, serial daily passage. This has been extraordinarily useful in studies of type 1 *E. coli* pili and gonococcal pili. Colonies composed of P+ bacteria are frequently smaller, have sharp borders, and pit the agar; when these colonies are viewed at low magnification through a dissecting microscope the edge of a P+ colony may exhibit twitching motility.

Most pilus types are optimally expressed at a particular temperature, and for some, specific components of the growth medium significantly increase or inhibit pili expression. For example, the expression of *E. coli* type 1 pili is usually maximal for organisms cultivated at 41° in a static liquid broth where the piliated variants form a surface pellicle at the air–liquid interface. In contrast, expression of type 1 pili is reduced when the same *E. coli* strain is grown at 37° on a solid medium; however, growth at 37° on a solid medium favors the expression of PAP pili by uropathogenic *E. coli* strains. Because effects of this kind seem to modulate the expression of pili by a great variety of bacterial species and strains, it is important to perform baseline physiological studies to identify optimal conditions of growth for pilus expression and to initiate the purification process with

P+ variants of the organism. Pilus expression is most easily assessed through transmission electron microscopy using negatively stained specimens. Alternatively, if a pilus-specific antiserum is available, an immunoassay can be used for this purpose. In some instances, pilus expression can be monitored functionally by performing hemagglutination assays if it is known that the pili are a hemagglutinin and if they are the only hemagglutinin of the organism.

In contrast to the phase variation phenomenon described above, some bacterial species simultaneously express two or more pilus types. This may impede the purification of a homogeneous preparation of pili as all pilus types are detached from the surface of the organism by the methods described below. As a result, additional purification steps must be used to separate biochemically distinct members of the organism's repertoire of pilus types. This difficulty can be circumvented if growth conditions can be identified that greatly favor the expression of only one pilus type. An alternative strategy entails the molecular cloning of each of the constituent pilus gene clusters, their separate expression in nonpiliated host strains, and the use of these clones for the purification of the expressed pilus filaments.

Overview of Pili Purification Methods

Detachment of Pili from Bacterial Surface

Unlike integral proteins of the outer membrane, pili are readily detached from the bacterial surface through the use of a mechanical blender. Alternatively, some pilus types can be detached by heating a suspension of bacteria (60° for 30 min) or by extracting the pili with hypertonic concentrations of NaCl. The objective of these methods is the quantitative separation of morphologically, functionally, and serologically intact pili from the outer membrane without the release of other components of the bacterial envelope from which the pili will need to be purified. Biochemical analysis of crude pilus extracts prepared by mechanical agitation, heating, or salt extraction, however, always reveals several contaminating structures including lipopolysaccharides, flagella, outer membranes, and occasionally capsular material. As a consequence, further purification steps are required.

Cyclic Pilus Solubilization and Crystallization as Purification Strategy

Brinton[2] was the first to describe a very general pilus purification method based on the capacity to cause pili to form paracrystalline aggre-

[2] C. C. Brinton, Jr., *Trans. N.Y. Acad. Sci.* [2] **27,** 1003 (1965).

gates under one set of conditions and their solubilization into separate, nonaggregated pilus filaments under a different set of conditions. In this context, solubilization does not refer to disassembly of the pilus filament into its constituent subunits, but rather to the retention of individual pilus filaments in the supernatant during a centrifugation experiment. Stepwise purification is achieved through centrifugation and alternate cycles of pilus aggregation and solubilization. Under one condition, pilus aggregates are collected by centrifugation, leaving soluble contaminants in the supernatant which is then discarded; under a second condition, the aggregates are solubilized, insoluble contaminants are removed by centrifugation, and the pellet is discarded and the soluble pilus filaments in the supernatant are retained. This purification technique can yield highly pure preparations of intact pilus filaments after several cycles of aggregation and solubilization without the need for additional purification steps. Additional methods may, however, be required particularly if more than one pilus type is present in the initial crude extract.

The identification of conditions that cause the solubilization or aggregation of pilus filaments is an empirical process: A crude pilus extract is examined visually for the presence of streaming birefringence or optical clarity indicating, respectively, the presence of aggregated or solubilized pili. Alternatively, aggregates of pili can be visualized by phase-contrast microscopy. Finally, centrifugation experiments can be conducted to determine under which conditions the pili are either in the supernatant (indicating that the pilus filaments are soluble) or in the pellet (indicating that the pilus filaments have formed aggregates). In general, pilus filaments aggregate at pH values close to their isoelectric point, in buffers of high ionic strength and in buffers containing divalent cations. In contrast, pilus aggregates are usually solubilized at pH values more alkaline than their isoelectric point and in buffers of low ionic strength that lack calcium and magnesium salts.

Criteria for Purity of Pilus Preparations

Generally accepted purity criteria for amino acid sequencing and analysis and for most immunochemical and functional experiments are as follows: The pilus preparation migrates as a single pilin species when analyzed by sodium dodecyl sulfate–polyacrylamide gel electrophoresis (SDS–PAGE); membranous contaminants are absent on negative staining and examination by transmission electron microscopy; and a single N-terminal residue is released after Edman degradation. Isopycnic and analytical ultracentrifugation and ultraviolet absorption spectroscopy have also been employed for this purpose. Finally, the empirically determined amino acid composition of a newly purified batch of pili can be compared with a previously obtained, accepted standard.

Most preparations of pili that have been purified using the solubilization–aggregation technique described above contain lipopolysaccharide that is tightly, but noncovalently associated with the pilus filament through hydrophobic interactions. Pilus accessory proteins may also be present, depending on the pilus type, as integral constituents of the heteropolymeric fiber. These may be detected by SDS–PAGE as faintly staining bands when compared with the stoichiometric major pilin band. If a particular application requires the removal of lipopolysaccharide, loosely associated nonpilus contaminants, or accessory proteins, greater purity can be achieved by depolymerization of the pilus filament and purification of the resulting pilin monomers; in some cases these spontaneously reassemble into pilus-like filaments on removal of denaturants or detergents.

Detailed Methods of Pili Purification

Purification of E. coli Type 1 Pili

The method of Salit and Gotschlich[3] solubilizes pili in a low-ionic-strength Tris buffer, aggregates pili in an acetate buffer close to the isoelectric point of type 1 pilin, and then carries out cyclic solubilization and aggregation of pili in the Tris buffer and in 10% (v/v) saturated ammonium sulfate, respectively; homogeneity is then achieved by isopycnic centrifugation. Specifically, piliated (P+) colonial variants are inoculated into minimal glucose medium (2000 ml in a 2800-ml Fernbach flask, Belco, Vineland, NJ) and incubated at 41° at 60 rpm in a gyratory shaker for 36–48 hr. Alternatively, P+ colonies are inoculated onto nutrient agar and grown for 16–18 hr at 41°. The bacteria are collected from the liquid medium by centrifugation (10,000 *g* at 4° for 20 min) or from agar using a bent glass rod, and the bacteria are then suspended in ice-cold 0.05 *M* Tris–HCl, pH 7.8. The bacterial suspension is sheared in a Sorvall Omnimixer (highest setting for 2 min), and the depiliated bacteria are removed by centrifugation (2000 *g* for 20 min); residual bacteria are removed from the supernatant by a second centrifugation (10,000 *g* for 30 min). The supernatant is collected and dialyzed against 0.1 *M* acetate buffer, pH 3.9, leading to the aggregation of the pilus filaments; these are harvested by centrifugation (2000 *g* for 20 min). The resulting pellet is washed in the same acetate buffer, and the pilus aggregates are then suspended in and solubilized by the initial 0.05 *M* Tris, pH 7.8, buffer. Saturated ammonium sulfate is then added dropwise to this optically clear

[3] I. E. Salit and E. C. Gotschlich, *J. Exp. Med.* **146,** 1169 (1977).

solution to a final volume of 10%, resulting in the aggregation of pilus filaments and the appearance of streaming birefringence. The pilus aggregates are collected by centrifugation (4000 *g* for 15 min), and the pellet is resuspended in the original Tris buffer. Additional cycles of aggregation and solubilization are performed until SDS–PAGE analysis of the preparation shows only pilin and a 38,000-Da major outer membrane protein. A homogeneous preparation of pili is obtained by subjecting this material to isopycnic centrifugation in cesium chloride, which yields a single band of pili with a density of 1.29 g/cm^3. Approximately 35 mg of pili is obtained from 100 g (wet weight) of bacteria.

The method of Brinton *et al.*[4] uses alternate cycles of solubilization in phosphate-buffered saline (PBS) and aggregation in $MgCl_2$ followed by removal of an outer membrane protein contaminant using ethylenediaminetetraacetic acid (EDTA) and a nonionic detergent. Specifically, the piliated bacteria are harvested into PBS (4 m*M* sodium phosphate, 0.85% NaCl, pH 7.2) and sheared as described above. The depiliated bacteria are removed by centrifugation and the pili aggregated by the addition of $MgCl_2$ to 0.1 *M* and collected by centrifugation (30,000 *g* for 60 min). The pellet is then solubilized in PBS and insoluble contaminants are removed by centrifugation. Three to four alternating cycles of solubilization in PBS and aggregation in $MgCl_2$ resulted in a pilus preparation that was 99% pure; a contaminating outer membrane protein was removed by suspending the aggregated pili in 5 m*M* EDTA–10 m*M* Tris, pH 7.2, followed by the addition of Triton X-100 to 0.2%; the pili are then precipitated from this solution by the addition of saturated ammonium sulfate to a volume of 5%; finally the purified, aggregated pili are collected by centrifugation (30,000 *g* for 60 min).

Purification of PAP Pili from Uropathogenic E. coli

Korhonen *et al.*[5] describe a method that uses a low-ionic-strength Tris buffer for pilus solubilization, ammonium sulfate for the precipitation of aggregated pili from solution, and then sodium deoxycholate (DOC) to separate hydrophobic outer membrane components from pilus filaments. This is followed by the use of 6 *M* urea to depolymerize flagella (but not pili), thus allowing the separation of flagellin monomers from pilus filaments by column chromatography. Specifically, the bacteria are grown in a manner that enhances PAP pilus production and suppresses type 1 pili expression (Luria agar, 37° for 18 hr), the bacteria harvested into 10 m*M* Tris buffer, pH 7.5, the pili detached by mechanical shearing, and

[4] M. S. Hansen, J. Hempel, and C. C. Brinton, Jr., *J. Bacteriol.* **170,** 3350 (1988).

[5] T. K. Korhonen, E. Nurmiaho, H. Ranta, and C. Svanborg-Eden, *Infect. Immun.* **27,** 569 (1980).

the depiliated bacteria removed by centrifugation. The pili are then precipitated by the addition of crystalline ammonium sulfate to the supernatant to 50% saturation and the aggregated pili collected by centrifugation (10,000 *g* for 1 hr). The pellet is suspended in the original Tris buffer, ammonium sulfate removed by dialysis against the Tris buffer, and DOC then added to the dialyzate to a concentration of 0.5%; dialysis is continued for 48 hr against the Tris buffer containing 0.5% (w/v) DOC. Exposure of the ammonium sulfate-precipitated pili to DOC was found to disaggregate pili into individual filaments (but not into pilin subunits) and from outer membrane vesicles. DOC-insoluble contaminants (principally outer membrane fragments) are removed by centrifugation(10,000 *g* for 10 min) and the pellet is discarded. The supernatant containing the DOC-solubilized pili is concentrated by ultrafiltration using an ×50 Amicon (Danvers, MA) membrane and the pili are obtained in a band with a density of 1.10 to 1.15 g/cm^3 after ultracentrifugation in a discontinuous sucrose gradient. Material of this density is dialyzed against the original Tris buffer, concentrated DOC is added to a concentration of 0.5%, and the sucrose gradient ultracentrifugation step is repeated. The pilus-containing band is again dialyzed against Tris buffer and crystalline urea is added to a concentration of 6 *M*; 2 hr later this solution is subjected to column chromatography in the urea buffer using Sepharose 4B. The pilus polymer is stable in 6 *M* urea and thus elutes in the void volume of the column free of flagellin monomers, which elute later. Finally, the urea is removed by dialysis against water. The estimated yield of pili from this procedure is 4 to 6 mg from 35 g (wet weight) of bacteria.

Purification of Colonizing Factor Antigen I from Enterotoxigenic E. coli

Colonizing factor antigen I (CFA I) is a plasmid-encoded pilus adhesin of certain enterotoxigenic *E. coli* (ETEC) serotypes. It can be prepared according to the method of Evans *et al.*[6] from appropriate ETEC strains that are grown under conditions that suppress type I pilus expression using nonmotile variants that lack flagellae, thus simplifying CFA I pili purification. Bacteria are grown on CFA agar medium at 37° for 48 hr, and harvested into 0.1 *M* sodium phosphate buffer, pH 7.2, and the CFA I pili are detached by homogenization in a Waring blender. The depiliated bacteria are removed by centrifugation (12,000 *g* for 20 min), the supernatant is collected and filtered through a 0.65-μm membrane Millipore (Bedford, MA) filter, and the filtrate is then subjected once again to the same

[6] D. G. Evans, D. J. Evans, Jr., S. Clegg, and J. A. Pauley, *Infect. Immun.* **5**, 738 (1979).

centrifugation and filtration steps. Solid ammonium sulfate is added to this filtrate to a saturation of 20% and the precipitate removed by centrifugation (20,000 *g* for 20 min) and discarded. Solid ammonium sulfate is then added to the supernatant to a final saturation of 40% and the resulting precipitate collected by centrifugation and found to contain CFA I; the supernatant of this second ammonium sulfate precipitation is discarded. The pellet of the second ammonium sulfate precipitation is dissolved in 0.05 *M* sodium phosphate buffer, pH 7.2, and then dialyzed extensively against the same buffer. Final purification is achieved using ion-exchange chromatography: The dialyzed sample is applied to a DEAD-Sephadex A-50 column that has been equilibrated with the 0.05 *M* sodium phosphate buffer described above. CFA I elutes in the 0.05 *M* sodium phosphate buffer in the void volume peak, whereas the non-pilus-contaminating proteins elute in a buffer containing 1.0 *M* NaCl. The final yield of CFA I protein is approximately 4 mg/g (wet weight) of bacteria.

Purification of E. coli Surface Antigens 1 and 3 of Colonization Factor Antigen II from Enterotoxigenic E. coli

Unlike CFA I which is composed of a single protein species, CFA II ETEC strains express combinations of three adhesin proteins which collectively encode the CFA II antigen: coli surface antigen 1 (CS1), coli surface antigen 2 (CS2), and coli surface antigen 3 (CS3). CS1 and CS2 are typical 6-nm-diameter rigid pilus filaments, whereas CS3 is a thinner (2-nm), flexible fibrillar structure. Because CFA II-expressing ETEC strains usually produce CS3 together with CS1 or CS2, Levine *et al.*[7] developed a method to purify separately CS1 and CS3 from the same strain. Their method depends on the fact that the CS1 or CS2 pilus structures exist on the bacterial surface as morphologically distinct appendages that are physically separated from the fibrillar CS3 structure. Specifically, a CFA II-positive ETEC strain is grown on CFA agar at 37° for 16–18 hr, and the bacteria are then harvested, suspended in PBS, pH 7.2, and subjected to mechanical shearing in a Sorvall Omnimixer (Dupont, New Brunswick, NJ). Depiliated bacteria are removed by centrifugation (8000 *g* for 15 min at 4°) and membrane fragments are removed by a second centrifugation (45,000 *g* 2 hr at 4°). The supernatant is then subjected to ultracentrifugation to obtain the CS1 and CS3 proteins (190,000 *g* for 2 hr at 4°); the pellet, which contains both proteins, is dissolved in PBS and subjected to isopycnic cesium chloride centrifugation using a self-generating gradient of density 1.29 g/cm^3. After ultracentrifugation (160,000 *g* for 18 hr) the

[7] M. M. Levine, P. Ristaino, G. Marley, C. Smyth, S. Knutton, E. Boedeker, R. Black, C. Young, M. L. Clements, C. Cheney, and R. Patnaik, *Infect. Immun.* **44,** 409 (1984).

CS1 and CS3 proteins are found in separate, but closely juxtaposed bands; thus repeated cycles of cesium gradient centrifugation are required to separate the two proteins fully. The yield of CS1 is 150 μg per 4 g wet weight of harvested bacteria.

Purification of CS3 can be facilitated by using a strain that produces only CS3, i.e., not CS1. Unlike the procedure described for the purification of CS1 from strains producing both CS1 and CS3, the filtrate is subjected to sequential 20 and 40% saturated ammonium sulfate precipitation. The 40% ammonium sulfate precipitate is retained, dialyzed against sodium phosphate buffer (0.1 *M*, pH 7.0), and loaded onto a Sepharose CL-6B column equilibrated with 6 *M* guanidine hydrochloride in the same sodium phosphate buffer. Fractions containing fibriller subunits are pooled and loaded onto a Sephacryl S-200 column (equilibrated with the same 6 *M* guanidine/sodium phosphate buffer) to remove high-molecular-weight contaminants. Fractions containing 14.5 and 15.5 kDa subunits are pooled; each band reacts with a CS3 antiserum, suggesting that these polypeptides are antigenically and possibly structurally related and that the CS3 polymer might be composed of each.

Purification of K99 Antigen of Enterotoxigenic E. coli Calf Strains

According to the method of de Graaf *et al.*,[8] the K99 pilus antigen can be purified to homogeneity by heat extraction of the protein in a phosphate–urea buffer, Sepharose CL-4B chromatography, and solubilization of this material in a DOC buffer as described by Korhonen *et al.*[5] for the purification of the PAP pili of uropathogenic *E. coli,* as described above. Specifically, K99-expressing bacteria are grown in minimal medium until late log phase, the bacteria harvested by centrifugation, and the pili extracted (i.e., detached) by heating the dispersed bacterial pellet at 60° for 20 min in 50 m*M* phosphate buffer, pH 7.2, containing 2 *M* urea. The bacteria are removed by centrifugation (30,000 *g* for 15 min) and the pili precipitated from the supernatant by the addition of ammonium sulfate to 60% saturation. After centrifugation, the pellet is suspended in the phosphate–urea buffer against which it is dialyzed for 16 hr at 4° to remove the ammonium sulfate. The dialyzate is loaded onto a Sepharose CL-4B column equilibrated in the phosphate–urea buffer; the pili, which elute in the void volume, are collected and dialyzed against 50 m*M* phosphate buffer, pH 7.5, and the dialyzate is then brought up to 0.5% (w/v) with sodium DOC. After exhaustive dialysis against the DOC–phosphate buffer, the DOC-insoluble material is removed by centrifugation. The dialyzate is found to contain pure K99 pilus subunits which migrate with an apparent molecular mass of 18,500 Da when analyzed by SDS–PAGE.

[8] F. de Graaf, P. Klemm, and W. Gaastra, *Infect. Immun.* **33,** 877 (1980).

Purification of 987 Pilus of Enterotoxigenic E. coli Piglet Strains

The method of Issacson and Richter,[9] adapted from the method of Brinton,[2] employs sequential cycles of pilus aggregation and solubilization for the purification of this pilus type. Trypticase soy broth is inoculated with a colony containing 987 P+ phase variant organisms and the bacteria are grown for 18 hr with aeration. The bacteria are collected by centrifugation (10,000 *g* for 30 min), the pellet is suspended in 0.01 *M* morpholinopropanesulfonic acid buffer (MOPS), pH 7.2, and the pili are detached using a Sorvall Omnimixer. After removal of the bacteria by centrifugation (10,000 *g* for 30 min), the pili are precipitated from the supernatant by decreasing the pH to 3.9 by the addition of acetic acid; after 30 min the pili are collected by centrifugation and the pellet is dissolved in the MOPS buffer. Insoluble nonpilus contaminants are then removed by centrifugation. The pili are reprecipitated from the supernatant by the addition of $MgCl_2$ (an equal volume of 0.09 *M* MOPS buffer, pH 7.2, containing 0.2 *M* $MgCl_2$ and 1.7% NaCl is added to the supernatant). The pili are again collected by centrifugation. After five additional cycles of solubilization (in the original MOPS buffer) and precipitation (in the $MgCl_2$-containing buffer) the pili are judged to be free of contaminants.

Purification of Gonococcal Pili

Brinton *et al.*[10] have proposed two methods for gonococcal pilus purification; most investigators have used variants of these methods not only for gonococcal pili purification,[11] but for the purification of other type IV pilus types including the somatic pili of *Moraxella bovis, Pseudomonas aeruginosa,* and *Vibrio cholerae* and the bundle-forming pili of enteropathogenic *E. coli.* The gonococcal protocols follow. Colonies of P+ phase variant bacteria are propagated on GC typing agar by serial daily passage. For pilus production, a suspension of these colonies in GC liquid medium is evenly inoculated onto the entire surface of GC agar plates and the plates are then incubated at 36.5° in 5% (v/v) CO_2 for 18 hr. The bacteria are harvested into 0.05 *M* Tris-buffered saline, pH 8.0, and the organisms then collected by centrifugation (13,000 *g* for 30 min). The pellet is resuspended in 0.15 *M* ethanolamine buffer, pH 10.5, and the pili are mechani-

[9] R. E. Isaacson and P. Richter, *J. Bacteriol.* **146,** 784 (1981).

[10] C. C. Brinton, Jr., J. Bryan, J. A. Dillon, N. Guerina, L. J. Jacobson, A. Labik, S. Lee, A. Levine, S. Lim, J. McMichael, S. Polen, K. Rogers, A. C. C. To, and S. C. M. To, *in* "Immunobiology of *Neisseria gonorrhoeae*" (G. F. Brooks, E. C. Gotschlich, K. K. Holmes, W. D. Sawyer, and F. E. Young, eds.), p. 159. Am. Soc. Microbiol., Washington, DC, 1978.

[11] G. K. Schoolnik, R. Fernandez, and E. C. Gotschlich, *J. Exp. Med.* **159,** 1351 (1984).

cally detached in a Sorvall Omnimixer. The bacteria are collected by centrifugation and discarded; the supernatant is dialyzed against 0.05 *M* Tris-buffered saline, pH 8.0, resulting in the aggregation of pilus filaments and the appearance of streaming birefringence. The pilus aggregates are collected by centrifugation (13,000 *g* for 60 min) and the pellet is resuspended in the same ethanolamine buffer, resulting in the solubilization of the pilus filaments. Insoluble contaminants are removed by centrifugation (23,000 *g* for 60 min) and the supernatant once again is dialyzed against the Tris–saline buffer. Successive cycles of solubilization and aggregation led to a pure preparation of pilus filaments. Alternatively, pili are precipitated from the high-pH ethanolamine buffer by the addition of ammonium sulfate to 10% saturation. Because the same gonococcal strain can produce multiple pilus types which have amino acid substitutions, deletions, or additions in variable domains of the subunit, it may be necessary to use somewhat different conditions of pH and ionic strength for their purification. This is ordinarily accomplished on an empirical basis.

[20] Genetic, Biochemical, and Structural Studies of Biogenesis of Adhesive Pili in Bacteria

By Meta J. Kuehn, Françoise Jacob-Dubuisson, Karen Dodson, Lynn Slonim, Robert Striker, and Scott J. Hultgren

Introduction

Bacterial adherence is an important early step in infection and results from the specific binding of a bacterial protein, called an adhesin, to a defined eukaryotic cell receptor. In many cases, the bacterial adhesins are associated with rodlike structures that radiate from the surface of bacteria, called pili. In general, pili are heteropolymeric organelles composed of approximately a thousand copies of a major pilin subunit and several minor pilins. Although the adhesive characteristics of pili can be conferred by the major pilin subunit, more often the adhesin is a minor pilus protein that is either joined to the pilus at the distal tip or intercalated at distinct points along the rod.[1,2]

An investigation of pilus biogenesis ultimately leads to basic questions concerning the postsecretional folding, targeting, and assembly of pro-

[1] S. J. Hultgren, S. N. Abraham, and S. Normark, *Annu. Rev. Microbiol.* **45,** 383 (1991).
[2] F. K. de Graaf, *Curr. Top. Microbiol. Immunol.* **151,** 29 (1990).

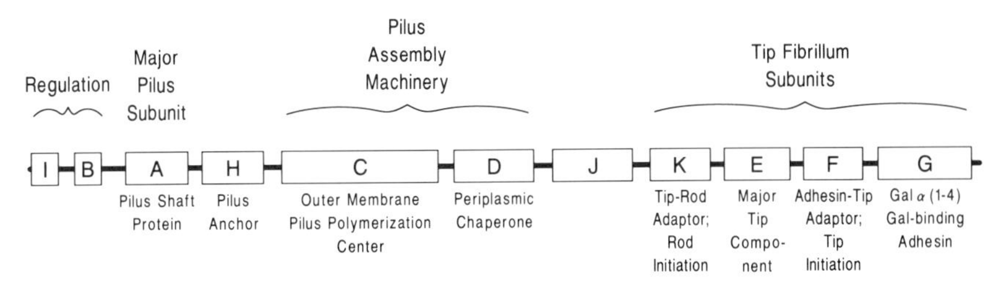

FIG. 1. The *pap* gene cluster. The proposed functions of the gene products are indicated and discussed in the text.

teins. For example, after the secretion of pilus subunits across the cytoplasmic membrane, what controls the folding of each subunit type into domains that can serve as assembly modules for building up adhesive pili at outer membrane assembly sites? Research in this area has unveiled two new classes of proteins located in the periplasmic space and outer membrane of gram-negative bacteria. These proteins function by guiding protein subunits along productive folding and assembly pathways and by converting soluble subunits into architecturally distinct surface fibers called pili.

In this article we discuss how the molecular details of pilus biogenesis, including the roles of pilus subunits and accessory proteins, have been studied using a powerful blend of genetics, biochemistry, carbohydrate chemistry, X-ray crystallography, and high-resolution electron microscopy techniques. We focus on the P-pilus system as a model for a detailed analysis of postsecretional assembly.

P Pilus Model System

P pili, encoded by the *pap* operon (Fig. 1), are found on uropathogenic *Escherichia coli* isolates,[3] and bind specifically to the Galα(1–4)Gal moiety via the adhesin, PapG.[4] The P pilus is a composite fiber consisting of a long rigid pilus rod composed of the major pilin subunit, PapA,[5] and a thin flexible fibrillum extending from the distal tip of the rod, composed

[3] G. Kallenius, R. Molby, S. B. Svenson, I. Helin, H. Hultberg, B. Cedergren, and J. Winberg, *Lancet* **2,** 1369 (1981).

[4] B. Lund, F. Lindberg, B. I. Marklund, and S. Normark, *Proc. Natl. Acad. Sci. U.S.A.* **84,** 5898 (1987).

[5] M. Baga, S. Normark, J. Hardy, P. O'Hanley, D. Lark, O. Olsson, G. Schoolnik, and S. Falkow, *J. Bacteriol.* **157,** 330 (1984).

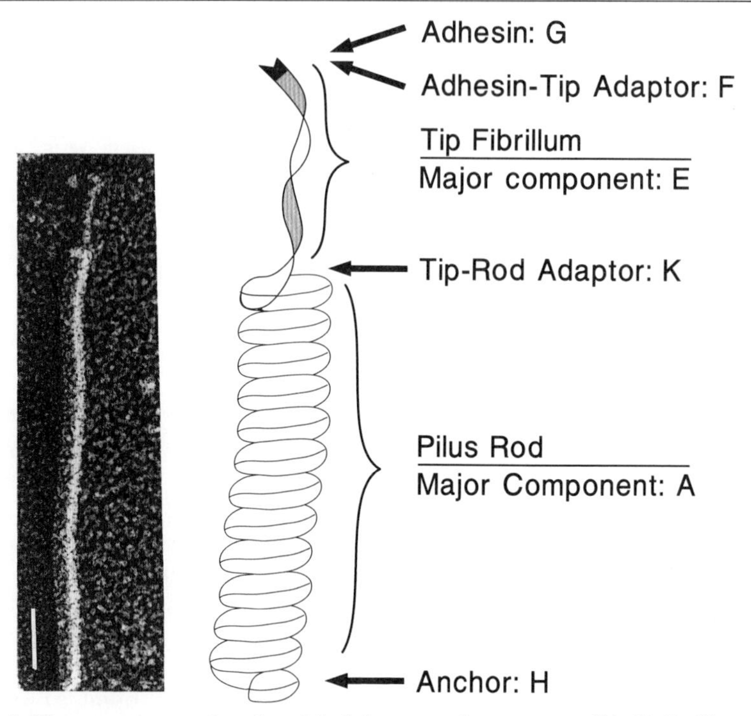

FIG. 2. Electron micrograph and model of the composite structure of P pili. P pili were purified and examined by high-resolution freeze-etch electron microscopy.[6] The functions of the minor pilin subunits were identified by analyzing subunit-deficient pili from bacteria with genetic insertions, by complementation of the minor subunit genes, and by biochemical analysis of purified tip fibrillae.[6,41] Bar: 0.025 μm.

of minor pilin subunits PapE, PapK, PapF, and the PapG adhesin (Fig. 2).[6] The tip fibrillar structure has been proposed to allow the PapG adhesin the maximal steric freedom needed to bind to digalactoside-containing receptors on uroepithelial cells. The pili are anchored to the cells via a minor pilin protein, PapH.[7] The assembly of P pili requires the action of two assembly proteins: a periplasmic chaperone, PapD, and an outer membrane protein, PapC.[8]

Many pilus systems have been described to date, and although there are differences in the specifics of each system, their assembly mechanisms appear to be similar, with the exception of type IV[1] which is not discussed here. Each system contains a major pilin and minor pilins. In addition, pilus assembly in every system requires a periplasmic chaperone and a

[6] M. J. Kuehn, J. Heuser, S. Normark, and S. J. Hultgren, *Nature (London)* **356,** 252 (1992).
[7] M. Baga, M. Norgren, and S. Normark, *Cell (Cambridge, Mass.)* **49,** 241 (1987).
[8] M. Norgren, M. Baga, J. M. Tennent, and S. Normark, *Mol. Microbiol.* **1,** 169 (1987).

large outer membrane protein, which are related to each other in structure and function.

Identification of Genes Encoding Accessory and Minor Pilin Products

Clearly, a map of the pilus gene cluster is an invaluable aid in the analysis of pilus assembly and accessory proteins. Pilus genes are usually found clustered in a typical arrangement, exemplified by the P-pilus operon (Fig. 1).[9] The relative positions of the genes and functions of the gene products can be obtained using transposon or *Xho*I linker insertions[10,11] or genetic deletions[12] to inactivate a gene. The mutant gene cluster is then assayed for the ability to produce functional pili. In addition, minicell or maxicell analyses have often been used to determine the effect of mutations on the expression of particular gene products.[11–13] Minicells obtained from *Escherichia coli* strains that have *minA* and *minB* mutations are small cell-like vesicles containing no chromosome but all necessary factors for transcription and translation. Incubation of minicells carrying a plasmid in a medium containing radiolabeled amino acids allows detection of the expression of the genes encoded by the plasmid, as no chromosomally encoded proteins are produced. Comparison of the proteins expressed from particular subclones of an operon, with the proteins produced by the plasmid vector alone, allows the identification of the proteins encoded by the cloned DNA fragment. In addition, insertion mutants can be compared with the wild-type clone to identify the gene products that are lacking. Pulse–chase experiments are used to identify which proteins are processed. Minicell analysis allows physical mapping of the genes in the operon.

Minicell Analysis

This procedure is based on the work of Thompson and Achtman.[14]

Media

10× BSG: 1.45 *M* NaCl, 0.02 *M* KH_2PO_4, 0.04 *M* Na_2HPO_4, 0.1% gelatin. Sterilize by autoclaving.

[9] J. Tennent, S. Hultgren, B.-I. Marklund, K. Forsman, M. Goransson, B.-E. Uhlin, and S. Normark, *in* "The Bacteria" (B. H. Iglewski and V. C. Clark, eds.), Vol. 11, p. 79. Academic Press, San Diego, 1990.

[10] F. P. Lindberg, B. Lund, and S. Normark, *EMBO J.* **3,** 1167 (1984).

[11] M. Norgren, S. Normark, D. Lark, P. O'Hanley, G. Schoolnik, S. Falkow, C. Svanborg-Eden, M. Baga, and B. E. Uhlin, *EMBO J.* **3,** 1159 (1984).

[12] F. R. Mooi, A. Wijfjes, and F. K. de Graaf, *J. Bacteriol.* **154,** 41 (1983).

[13] F. Lindberg, L. Westman, and S. Normark, *Proc. Natl. Acad. Sci. U.S.A.* **82,** 4620 (1985).

[14] R. Thompson and M. Achtman, *Mol. Gen. Genet.* **165,** 295 (1978).

2× Minicell Minimum Medium (M9M): Mix 40 ml 10× M9 medium,[15] 4 ml 20% glucose, 60 μl 5 mg/ml vitamin B_1, 1.4 ml 4% proline, 1.4 ml 4% leucine, 1.4 ml 4% threonine. Bring to 200 ml with distilled water and sterilize by filtration.

Labeling medium: For 5 ml, mix 2.5 ml 2× M9M (without glucose), 0.5 ml 50% glycerol, 0.5 ml 10× labeling medium (Difco, Detroit, MI), and 1.5 ml H_2O. Sterilize by filtration.

Procedure

On the day before,

1. Inoculate 5 ml LB + selecting antibiotic with a minicell strain (ORN103, AA10) carrying the plasmid of interest. Dilute 1000-fold into 500 ml fresh LB medium + antibiotic and grow in an orbital shaker at 37° overnight.
2. Prepare sucrose gradients in one SS-34 tube and two Corex glass tubes by adding 22% sucrose in 1× BSG. Freeze the tubes for at least 3 hr and let thaw standing overnight at 4° to form the gradient.

On the following day,

1. Centrifuge the overnight culture to pellet the cells.
2. Resuspend the pellet in 2 ml BSG. Pass the resuspended cells through a Pasteur pipette several times to dissociate the minicells from the mother cells.
3. Layer the cell slurry on the SS34 gradient using a Pasteur pipette. Centrifuge in a swing-out bucket rotor (JS-13) at 4000 *g* for 20 min at 4°.
4. Remove the upper two-thirds of the minicell band to a clean SS34 tube and add 1× BSG to a final volume of 5 ml.
5. Pellet the cells in a JA20 rotor at 12,000 *g* for 10 min at 4°.
6. Resuspend the cells in 2 ml 1× BSG and layer them on the first Corex gradient. Centrifuge the tubes in the SW rotor at 4000 *g* for 20 min at 4°.
7. Repeat the previous two steps in the second Corex tube gradient.
8. Remove the minicell band and bring to 11 ml total volume with 1× BSG in a SS34 tube.
9. Measure A_{600} of 1 ml of the minicells. Pellet the remaining 10 ml in a JA20 rotor at 12,000 *g* for 10 min at 4° and resuspend the minicells to A_{600} = 1.0 in the labeling medium.
10. Preincubate 0.5 ml of the minicells in an Eppendorf tube at 37° without shaking for 20 min.
11. Add 80–90 μCi[^{35}S]methionine and label for 10 min at 37°.

[15] J. Sambrook, E. F. Fritsch, and T. Maniatis, "Molecular Cloning: A Laboratory Manual," 2nd ed. Cold Spring Harbor Lab., Cold Spring Harbor, NY, 1989.

12. Chase by adding 0.5 ml LB supplemented with 100 μg/ml unlabeled methionine. Incubate for 10 min at 37°.

13. Spin down the labeled cells. Wash once with 1× BSG and resuspend in 50 μl sodium dodecyl sulfate–polyacrylamide gel electrophoresis (SDS–PAGE) loading buffer.

14. Load 5–10 μl of the sample and run the gel. Fix the gel for 15 min in 45% methanol, 9% acetic acid (v/v); stain it for 30 min at 50° in 45% methanol, 9% acetic acid, 0.1% Coomassie blue; and destain it at 50°. Enhance with 1 *M* sodium salicylate or another commercial enhancer. Dry the gel at 60° for 2 hr and autoradiograph it at −70°.

Analyses of Adhesion

There are several prerequisites for analysis of pilus-associated adhesins and assembly proteins in a pilus system. It is generally best to work with the cloned pilus genes under regulated control, such as the *tac, lac,* or *ara* promoters, as natural pilin promoters are often under phase variation and subject to environmental control signals. A direct assay for piliation such as hemagglutination (HA), adherence to eukaryotic cells, and electron microscopy are invaluable in the analysis of phenotypes of mutant alleles. Knowledge of the specific factor to which the adhesin binds is a tremendous advantage, particularly in the purification of the adhesin. Adhesin-binding receptors have been identified by testing the ability of specific glycosidic enzymes to inhibit HA, by testing agglutination of erythrocytes deficient in certain types of blood group antigens, and by overlaying bacteria on glycolipids using thin-layer chromatography.[16,17]

Hemagglutination assays (see [16]) measure the ability of a suspension of bacteria to cross link (agglutinate) red blood cells. Multiple adhesins on the bacterial cell surface binding to multiple adhesin receptors on the red blood cell surface cause hemagglutination to occur. When these criteria are met, hemagglutination assays provide a quick and easy method for determining the relative abundance of properly presented adhesins on the surface of the bacteria.

Hemagglutination Assay

This procedure is adapted from Hultgren *et al.*[18]

[16] G. C. Hansson, K.-A. Karlsson, G. Larson, N. Stromberg, and J. Thurin, *Anal. Biochem.* **146,** 158 (1985).

[17] K. A. Karlsson and N. Stromberg, this series, Vol. 138, p. 220.

[18] S. J. Hultgren, J. L. Duncan, A. J. Schaeffer, and F. K. Amundsen, *Mol. Microbiol.* **4,** 1311 (1990).

Phosphate-Buffered Saline. Phosphate-buffered saline (PBS) comprises 120 m*M* NaCl, 2.7 m*M* KCl, 10 m*M* phosphate buffer salts, pH 7.4 (Sigma, St. Louis, MO).

Bacterial Suspension

1. Suspend bacteria (fresh, overnight culture) in PBS to $A_{540} = 1.0$.
2. Spin down 1 ml of the cell solution in a microcentrifuge for 1 min.
3. Resuspend cell pellet in 100 μl PBS.

Red Blood Cell Suspension

1. Wash blood in PBS by repeated mixing and centrifugation until supernatant is clear (no lysed cells).
2. Suspend red blood cells (RBCs) in PBS to $A_{640} = 1.9$.

Hemagglutination Assay

1. In a microtiter plate with pointed bottom (Costar, Cambridge, MA), dispense 25 μl PBS/well.
2. Add 25 μl bacterial suspension in first well. Mix. Make serial dilutions by taking 25 μl from the first well, transferring to the next well, mixing, taking 25 μl from the second well, mixing, and so on.
3. Add 25 μl RBC suspension in all wells.
4. Mix plate gently by tapping side of plate. Cover and place at 4° for 1–12 hr.
5. Read titer by determining the maximum dilution of bacteria that yields 50% hemagglutination. In the absence of hemagglutination the red blood cells are able to settle into a dotlike pool at the bottom of the well. In the presence of hemagglutination the red blood cells and bacteria form a diffuse sheet of clumped cells over the entire well.

Identification of Adhesin

There are several known strategies by which pili present adhesins to eukaryotic cells, represented by different pilus architectures. The P pilus, for example, localizes the adhesin at the distal end of a flexible fiber composed of minor pilin subunits linked end to end to the pilus rod (Fig. 2).[6] This strategy allows the adhesin steric freedom to recognize host cell receptors, but still places the adhesin sufficiently far from the interfering negative charge of the cell. The type 1 pilus intercalates the adhesin and minor pilin proteins at distinct intervals within the pilus structure and at the tip.[19] This strategy apparently ensures that the pili are adhesive even

[19] S. Ponniah, R. O. Endres, D. L. Hasty, and S. N. Abraham, *J. Bacteriol.* **173,** 4195 (1991).

if the stalks are broken, as they tend to break where the adhesin is intercalated; however, the FimH adhesin only binds the receptor when it is exposed at the pilus tip.[19] The K88 and K99 pili differ in that they are much thinner and the major subunit of the pilus is also the adhesin.[2] The overall adhesiveness and flexibility of these fibers may be an advantage in bacterial colonization.

To verify that a gene product is the adhesin, the putative adhesin can be purified from the tip of pili,[20,21] or from periplasmic extracts as a preassembly complex with the chaperone.[22] These techniques are necessary to demonstrate that the one protein alone is responsible for the binding activity in the absence of other pilus subunits. For P pili, PapG was also identified as the actual adhesin using transcomplementation experiments between two related gene clusters [*pap* and *prs* (pap-related sequences)] showing different binding specificities. Only PapG could complement the *prsG*$^-$ mutant to give PapG-specific adherence properties.[4]

Pilins

Identification of Major Pilin

The major pilin subunit is readily identified as the major band obtained on SDS–PAGE of purified pili. Pili often are removed from bacteria by shearing the cells in a blender or by heating the cells, and the pili are purified from the supernatant of these preparations by precipitation and/or sucrose gradient centrifugation[21,23,24] (see also [19]). Inactivation of the gene encoding the major subunit abolishes pilus formation; however, if the major subunit is not the adhesin, the adhesive properties of the cells are not necessarily abolished.[18,25]

Identification of Minor Pilins

Virtually all pilus gene clusters described so far contain several genes encoding pilinlike proteins distinct from the major subunit. They range in

[20] T. Moch, H. Hoschutzky, J. Hacker, K. D. Kroncke, and K. Jann, *Proc. Natl. Acad. Sci. U.S.A.* **84,** 3462 (1987).

[21] H. Hoschutzky, F. Lottspeich, and K. Jann, *Infect. Immun.* **57,** 76 (1989).

[22] S. J. Hultgren, F. Lindberg, G. Magnusson, J. Kihlberg, J. M. Tennent, and S. Normark, *Proc. Natl. Acad. Sci. U.S.A.* **86,** 4357 (1989).

[23] Y. Eshdat, F. J. Siverblatt, and N. Sharon, *J. Bacteriol.* **148,** 308 (1981).

[24] T. K. Korhonen, E. Nurmiaho, H. Ranta, and C. Svanborg-Eden, *Infect. Immun.* **27,** 569 (1980).

[25] B. E. Uhlin, M. Norgren, M. Baga, and S. Normark, *Proc. Natl. Acad. Sci. U.S.A.* **82,** 1800 (1985).

size from 14 to 30 kDa and have sequence homology among themselves and with the major subunit, mainly at their N and C termini.[26] They are referred to as "minor" pilins because of their low abundance in the pilus compared with the major subunit which forms the bulk of the structure. Nevertheless, they are indispensable for the proper assembly of the pilus. Expression of the minor pilins has been detected by minicell analysis using various subclones of the operon. Alternatively, insertions of the T7 promoter at different positions along the gene cluster and induction of transcription by the T7 RNA polymerase *in vivo* constitute a very sensitive technique.[27] TnPhoA fusions have also been used to detect the products of these genes.[27] Minor pilins are located at the tip, along intervals of the fimbriae, or at their base, and have been proposed to be involved in adhesion, adhesin presentation, modulation of the pilus length, anchoring of the pilus onto the outer membrane, and initiation of pilus biogenesis (see below).

Location of Minor Components in Pilus Using Electron Microscopy

The minor pilus components often remain undetected by SDS–PAGE of purified pili because of their low abundance. Radioiodination of pili prior to SDS–PAGE allows detection of some minor pilins[28]; however, identification and localization of the minor subunits in the pilus structure have been achieved mostly by the use of immunogold electron microscopy. Monospecific antiserum specific for a minor pilin is obtained by adsorbing antiserum raised against purified pili with recombinant bacteria carrying mutant operons having a linker insertion in the specific minor pilin gene.[29] Alternatively, a specific antiserum can be obtained using a fusion protein with the subunit of interest,[30] or antibodies can be raised against peptides corresponding to amino acid sequences of the pilin.[31]

In enteropathogenic *E. coli* strains with K88 or K99 pili, the major subunit that forms the bulk of the fibrillae is also the adhesin. Immunoelectron microscopy has shown that other minor pilins are also components of the structure and are located at the tip or laterally associated with the

[26] S. Normark, M. Baga, M. Goransson, F. P. Lindberg, B. Lund, M. Norgren, and B.-E. Uhlin, *in* "Microbial Lectins and Agglutinins: Properties and Biological Activity" (D. Mirelman, ed.), p. 113. Wiley (Interscience), New York, 1983.

[27] D. M. Schifferli, E. H. Beachey, and R. K. Taylor, *J. Bacteriol.* **173,** 1230 (1991).

[28] F. Lindberg, B. Lund, and S. Normark, *Proc. Natl. Acad. Sci. U.S.A.* **83,** 1981 (1986).

[29] F. Lindberg, B. Lund, L. Johansson, and S. Normark, *Nature (London)* **328,** 84 (1987).

[30] B. L. Simons, P. T. J. Willemsen, D. Bakker, B. Roosendaal, F. K. de Graaf, and B. Oudega, *Mol. Microbiol.* **4,** 2041 (1990).

[31] S. N. Abraham, J. D. Goguen, D. Sun, P. Klemm, and E. H. Beachey, *J. Bacteriol.* **169,** 5530 (1987).

shaft (FaeC,[32] FanF[33]). In other pilus systems where the adhesin is a minor subunit, it has been shown to be located at the tip only or at the tip and along the fiber, often in a complex with other minor subunits.

In P pili of $F7_1$, $F7_2$, F11 and F13 serotypes, the adhesin was found complexed with the E subunit by immunocytochemical double labeling and located at the pilus tip only, whereas in the F9 serotype these complexes were located along the pilus shaft as well.[33] These adhesins were only loosely associated with the rod, as they could be detached using a detergent without breaking the pili. In contrast, FimH in complex with FimF and/or FimG is integrated into the pilus structure in type 1 pili.[34] Freeze–thaw experiments on whole pili demonstrated that the intercalation of these complexes makes pili more fragile and that breakage exposes new adhesive sites formed by adhesins along the pili.[19]

In the *pap* system, immunogold electron microscopy specifically labeled PapE, PapF, and PapG at the tip of the pilus.[20] A high-resolution electron microscopy technique was employed to analyze the fine-structural details of the pilus architecture and revealed a composite structure with a helical rod and a thin linear fiber joined end to end (Fig. 2).[6] Various linker insertion mutants were used to assess more precisely the location of the minor pilin subunits. PapE was found to be the major component of the thin-tip fibrillum, as inactivation of the gene resulted in the disappearance of the fibrillum. In contrast, significantly longer tip fibrillae were observed in *papF1* and *papK1* mutant strains, suggesting a role for these subunits in regulating the length of that structure. The PapG adhesin was localized to the distal end of the tip fibrillum using immunogold EM.

Roles of Minor Pilins

The functions of the minor subunits have been investigated by deletions, mutations, or frameshift inactivations in the operon, and complementations by providing one or more genes *in trans*. The phenotype of the mutant strains (adhesive properties, level of piliation, morphology, and composition of the pili) was then analyzed using adherence assays, electron microscopy (EM), enzyme-linked immunosorbent assay (ELISA), radioimmunoassay (RIA), Western blotting, and immunogold labeling.

[32] B. Oudega, M. de Graaf, L. de Boer, D. Bakker, C. E. M. Vader, F. R. Mooi, and F. K. de Graaf, *Mol. Microbiol.* **3,** 645 (1989).

[33] N. Riegman, H. Hoschutzky, I. van Die, W. Hoekstra, K. Jann, and H. Bergmans, *Mol. Microbiol.* **4,** 1193 (1990).

[34] K. A. Krogfelt and P. Klemm, *Microb. Pathog.* **4,** 231 (1988).

The SfaS adhesin in S pili is responsible for the initiation of pilus formation together with SfaH,[35] as is the case for FstG/FstF in $F7_2$ pili.[36] More often, one or more minor pilins distinct from the adhesin are devoted to initiation of polymerization. Some pilus systems require more than one initiator protein and these may have nonoverlapping functions, in which case the inactivation of only one of them renders cells defective in pilus biogenesis, such as FaeC and FaeF (K88[11,32]) or FanG, FanH, and FanF (K99[30,37]). FanF also is responsible for elongation of the pilus via continuous reinitiation of pilus polymerization.[30] In type 1 pili, the minor subunit FimH was shown to be the adhesin.[31] In addition inactivation of the *fimF* gene significantly decreased piliation and inactivation of the *fimG* gene resulted in very long pili.[38] These results suggested that the minor subunits FimF and FimG are the initiator and the terminator of pilus polymerization, respectively. In 987P pili, two different proteins regulate the length of the pilus.[27] Thus in many cases it appears that the same subunit assumes several roles, affecting initiation, elongation, and presentation of the adhesin. Other pilinlike proteins have been proposed to act mainly as adaptors between the adhesin and the rest of the structure, such as SfaG in S pili.[35] Finally, anchoring the pilus to the bacterial cell may be performed by a specific minor subunit, such as MrkF in type 3 pili of *Klebsiella pneumoniae*[39] or PapH in the P pili.[7]

All the general ideas stated above are best exemplified by the studies performed on the *pap* system, briefly summarized below.

Incorporation of PapH in the rod was demonstrated to terminate the growth of the pilus fiber and mediate its anchoring onto the surface of the cell.[7] PapH could not be detected in pilus preparations because it was present in very low abundance compared with the structural subunit or because it remained cell associated during the pilus purification procedure. Insertional inactivation of *papH* did not decrease piliation, but did result in a large proportion of pili free in the culture medium. Modification of the PapA/PapH ratio by overproducing one of the two genes *in trans* with the wild-type operon showed that the pilus length was determined by the stoichiometric relationship between the two.

[35] T. Schmoll, H. Hoschutzky, J. Mortschhaueser, F. Lottspeich, K. Jann, and J. Hacker, *Mol. Microbiol.* **3,** 1735 (1989).

[36] N. Riegman, I. van Die, J. Leunissen, W. Hoekstra, and H. Bergmans, *Mol. Microbiol.* **2,** 73 (1988).

[37] B. Roosendaal, A. A. C. Jacobs, P. Rathman, C. Sondermeyer, F. Stegehuis, B. Oudega, and F. K. de Graaf, *Mol. Microbiol.* **1,** 111 (1987).

[38] P. Russel and P. Orndorff, *J. Bacteriol.* **174,** 5923 (1992).

[39] B. Allen, G. Gerlach, and S. Clegg, *J. Bacteriol.* **173,** 916 (1991).

The function and location of the four proteins that constitute the tip fibrillum were investigated using high-resolution EM, genetics, biochemistry, and carbohydrate chemistry. Tip fibrillae were expressed in the absence of the major pilin PapA in an inducible system comprising genes *papCDJKEFG*. Tip fibrillae were purified by Galα(1–4)Gal affinity chromatography using the specific adhesion properties of PapG. All four minor subunits, PapG, F, E, and K, were found to be associated to form these structures.[40] Tip fibrillae from various inactivation mutants (*papE1, papF1, papK1*) were also obtained to assess the role of each protein in the assembly process. PapG was not polymerized to the distal end of the tips in the absence of PapF, yet tip structures were assembled as shown by the purification of PapE oligomers using sucrose gradient centrifugation. In the absence of PapE, no fibers were formed and only PapF coeluted with the adhesin, whereas in the absence of PapK, fibrillae composed of PapG, PapF, and PapE were obtained. These results suggested that PapF is an adaptor between the adhesin and the PapE oligomer, and PapK joins the tip fibrillum to the pilus rod.[40] Thus, the assembly of the tip fibrillum appears to be strictly ordered and this order is defined by the interactions between complementary surfaces from different pilin subunits.

The presence of the tip fibrillum was shown to be required for the polymerization of the pilus shaft, as no pili were assembled in the absence of all four tip proteins.[40] In the absence of the four tip fibrillar proteins, *trans*-complementation with *papK* could restore pilus assembly, suggesting that only PapK could initiate the formation of the pilus rod. In addition, the incorporation of PapK was found to terminate growth of the tip fibrillum, consistent with its adaptor function to join the tip fibrillum to the pilus rod.

Cells carrying the *pap* operon in which the *papF* gene was inactivated were shown to be nonadhesive and less fimbriated.[10,29] No such effects were detected in the absence of PapK. But the simultaneous inactivation of *papF* and *papK* resulted in a virtual lack of piliation that could be complemented fully by expressing PapF *in trans* (which restored full piliation and hemagglutination) or partly by providing PapK (which could restore an intermediate level of piliation but no hemagglutination).[40] These results demonstrated the absolute requirement for PapF or PapK as initiators of pilus assembly.

Despite some variations regarding the specificity of the adhesin, the precise role of the minor subunits, and their location in the structure, common features emerge as to how gram-negative pathogens assemble

[40] F. Jacob-Dubuisson, J. Heuser, K. Dodson, and S. Hultgren, *EMBO J.* **12,** 837 (1993).

composite surface fibers to present adhesins to specific eukaryotic receptors.

Periplasmic Pilus Chaperones

Function of Periplasmic Pilus Chaperones

A periplasmic chaperone has been found in every well-characterized pilus system in gram-negative bacteria, with the exception of type IV pili.[1,41,42] The periplasmic pilus chaperone is generally 22–29 kDa, has an isoelectric point greater than 9.0, and can be identified by its sequence homology to the chaperone family.[41] Inactivation of the chaperone gene results in no piliation and, often, degradation of pilus subunits in the periplasm.[11,22,43] Periplasmic pilus chaperones have been found to be essential to guide pilus subunits along a biologically productive pathway. In the absence of the chaperone, unfolded pilus proteins emerge from the cytoplasmic membrane; aggregate, probably because of exposed hydrophobic surfaces; and eventually are proteolytically degraded. The binding of the chaperone prevents aggregation of the subunits and may assist in their correct folding. Chaperone–subunit complexes are then targeted to outer membrane assembly sites.

Purification and Analysis of Chaperones

A strong or inducible promoter (such as *tac*) can be used to overproduce the chaperone gene product in large quantities without harming the cells. The chaperone and chaperone–pilus subunit preassembly complexes accumulate in the periplasmic space in the absence of the outer membrane usher protein. The chaperones have been purified from periplasmic extracts using common techniques.[44,45]

Periplasm Preparation

This procedure is adapted from Hultgren *et al.*[22]

1. Grow cell culture with induction of promoter, if appropriate.
2. Pellet cells at 10,000 *g* for 10 min at 4°, and weigh the cell pellet.

[41] A. Holmgren, M. Kuehn, C.-I. Branden, and S. Hultgren, *EMBO J.* **11,** 1617 (1992).
[42] F. Jacob-Dubuisson, M. Kuehn, and S. Hultgren, *Trends Microbiol.* **1,** 50 (1993).
[43] F. R. Mooi, C. Wouters, A. Wijfjes, and F. K. de Graaf, *J. Bacteriol.* **150,** 512 (1982).
[44] F. P. Lindberg, J. M. Tennent, S. J. Hultgren, B. Lund, and S. Normark, *J. Bacteriol.* **171,** 6052 (1989).
[45] D. Bakker, C. E. M. Vader, B. Roosendaal, F. R. Mooi, B. Oudega, and F. K. de Graaf, *Mol. Microbiol.* **5,** 875 (1991).

3. Resuspend cells at 4° in 4 ml of 20 m*M* Tris, pH 8.0, 20% sucrose per gram of cells.
4. Add 200 μl 0.1 *M* ethylediaminetetraacetic acid (EDTA), pH 8.0, and 40 μl 15 mg/ml lysozyme per gram of cells.
5. Incubate on ice for 40 min.
6. Add 160 μl 0.5 *M* $MgCl_2$ per gram of cells.
7. Pellet spheroplasts at 15,000 *g* for 20 min at 4°.
8. Supernatant contains the periplasmic contents. Store at −20°.

Most periplasmic pilus chaperones sequenced or purified to date have either experimentally or theoretically derived isoelectric points of greater than 9.0.[41,44,45] The PapD chaperone was purified from the periplasm using cation-exchange chromatography,[44] and the FaeE chaperone has been purified using isoelectric focusing.[45] Further purification was achieved using a hydrophobic interaction column (CAA-HIC, Beckman, or Phenyl Superose, Pharmacia, Piscataway, NJ) with a 1 to 0 *M* ammonium sulfate gradient in 50 m*M* phosphate, pH 7.0.[46]

Cation-Exchange Chromatography of Periplasmic Pilus Chaperones

This procedure is adapted from Lindberg *et al.*[44]

1. Precipitate periplasm with 30% ammonium sulfate, spin at 16,000 *g* (30 min, 4°), and save the supernatant.
2. Precipitate supernatant with 70% ammonium sulfate and spin again. Resuspend pellet in 20 m*M* KMES, pH 6.5.
3. Dialyze 30–70% fraction with 20 m*M* KMES, pH 6.5.
4. Spin down debris at 22,000 *g* for 30 min at 4°.
5. Filter supernatant with 0.22-μm filter and apply to Mono S cation-exchange column (Pharmacia).
6. Elute from Mono S column with a 0 to 0.15 *M* KCl gradient at a rate of 25 m*M*/10 ml.
7. Monitor fractions at 280 nm and analyze peak fractions using SDS–PAGE and Western blotting.

Three-Dimensional Structure of Periplasmic Chaperone Proteins

More than twelve pilus chaperones from various pilus systems in various bacterial species have been found to be members of a highly related periplasmic chaperone family of which PapD is the prototype member.[41,42]

[46] A. Holmgren, C.-I. Branden, F. Lindberg, and J. M. Tennent, *J. Mol. Biol.* **203,** 279 (1988).

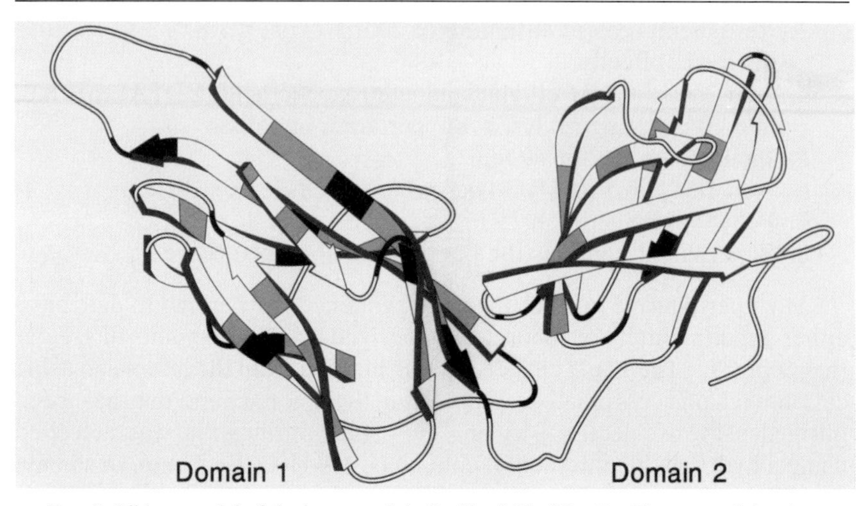

FIG. 3. Ribbon model of the immunoglobulin-like fold of PapD. Alignment of the chaperone sequences superimposed on the crystal structure of PapD revealed the positions of invariant residues (black boxes) and conserved residues (shaded boxes).[42] Pilus subunit interactions involve conserved and invariant residues in the cleft of the molecule.[56]

The crystal structure of PapD was solved by Holmgren and Branden.[47] PapD was found to have an immunoglobulin-like fold, consisting of two antiparallel β barrels connected by a hinge region (Fig. 3). According to analysis of sequence homology in relation to structural features, the entire family of periplasmic pilus chaperones are predicted to share the same immunoglobulin-like fold as PapD. Novel chaperone sequences were aligned to the PapD sequence without introducing gaps in the β strands of the known PapD structural framework. The consensus pilus chaperone deduced from this alignment had 12 invariable and 36 highly conserved residues.[42] The amino-terminal domains of the proteins were more homologous to each other than the carboxyl-terminal domains. Many of the conserved and invariant residues occurred at sites critical to the fold of the protein; however, several were surface exposed and conserved for no obvious structural reason.[41]

Complementation experiments have been used to determine whether there is a similar binding surface used by all of these chaperones and whether the chaperones recognize similar elements in related and unrelated pilus subunits. Several pilus systems appear to have interchangeable parts. For example, the *E. coli* F41, CS31A, and K88 assembly gene products (outer membrane protein and periplasmic chaperone) were able

[47] A. Holmgren and C. I. Branden, *Nature* (*London*) **342,** 248 (1989).

to assemble each other's respective major subunits and functionally express these on the surface of the cells.[48] The *E. coli pap* assembly system (*pap* operon with a deletion of the adhesin gene) was able to incorporate the *Klebsiella pneumoniae* adhesin, MrkD, into the P pilus so that it could mediate MrkD-specific hemagglutination, whereas the *E. coli* type 1 system could not.[49] PapD from the *E. coli pap* system was able to complement the *E. coli* type 1 pilus operon in which the chaperone gene *fimC* had been inactivated, resulting in the production of adhesive type 1 pili[50]; however, the *E. coli* chaperones FanE and FaeE from the K99 and K99 pilus systems, respectively, were unable to complement each other.[45] Further, the DaaE adhesin from the F1845 pilus system and the DraA from the Dr hemagglutinin can be assembled in the Dr and F1845 pilus structures, respectively.[51] The ability to interchange the assembly proteins between pilus systems suggests that general mechanisms govern pilus biogenesis in gram-negative bacteria.

Further complementation analysis of subunits and assembly systems provides a testing ground for theories of evolutionary divergence; however, complementation experiments need to be carefully monitored to detect whether the subunit is stabilized in the periplasm, whether the subunit is polymerized into a pilus, and whether the subunit exists in the polymerized pilus in a functional conformation. Initiator and adaptor proteins are also critical to pilus expression and may influence where and when a subunit is assembled.

Periplasmic Preassembly Complexes

Use of Adhesin Specificity to Purify Adhesin–Chaperone Complex

Insertional inactivation and minicell analysis have demonstrated that periplasmic pilus chaperones are generally required for the stabilization of subunits in the periplasm.[10,12,45,50] The chaperone associates with pilus proteins in periplasmic preassembly complexes, which can then be isolated from periplasmic extracts. These periplasmic chaperone–subunit complexes can be identified using immunoprecipitation with antisubunit antisera. The most well-characterized preassembly complex, PapD–PapG, was affinity purified with Galα(1-4)Gal-Sepharose using the ability of the PapG adhesin to bind to the immobilized receptor.

[48] M. J. Korth, J. M. Apostol, Jr., and S. L. Moseley, *Infect. Immun.* **60,** 2500 (1992).

[49] G.-F. Gerlach, S. Clegg, and B. L. Allen, *J. Bacteriol.* **171,** 1262 (1989).

[50] C. H. Jones, J. Pinkner, A. Nicholes, L. Slonim, S. Abraham, and S. Hultgren, *Proc. Natl. Acad. Sci. U.S.A.* **90,** 8397 (1993).

[51] T. N. Swanson, S. S. Bilge, B. Nowicki, and S. L. Moseley, *Infect. Immun.* **59,** 261 (1991).

Affinity Purification of Preassembly PapD–PapG Complex

This purification is adapted from Hultgren *et al.*[22]

Materials

The preparation of the affinity matrix is described in Dahmen *et al.*[52] Briefly, 2-(2-methoxycarbomethlythiol)-ethyl-4-*O*-α-D-galactopyranosyl-(1 → 4)-β-D-galactopyranoside is coupled to aminated Sepharose beads.

A soluble high-affinity analog of the receptor is required for the elution step.

Procedure

1. Grow cells carrying a plasmid coding for the chaperone and the adhesin. Induce for expression if necessary.
2. Harvest and prepare periplasmic extracts from the cells.
3. Bring the periplasmic fraction to 50% ammonium sulfate saturation.
4. Centrifuge at 15,000 *g* for 20 min to pellet the precipitated proteins.
5. Redissolve the pellet in a small volume of PBS.
6. Apply the concentrated periplasmic fraction to a slurry of Galα(1–4)Gal-Sepharose.
7. Rock overnight at 4°.
8. Pellet the Sepharose beads by gently centrifuging.
9. Extensively wash the beads using PBS/1 *M* NaCl to remove nonspecifically bound proteins.
10. Elute the adhesin complex by rocking the beads in 50 m*M* ethyl-β-galabioside in PBS/1 *M* NaCl followed by gentle centrifugation.
11. Dialyze the adhesin complex-containing supernatant extensively against PBS.

Purification of Major Subunit–Chaperone Complex

Most pilus subunit–chaperone complexes do not have a receptor binding specificity to exploit for purification; however, monoclonal antibodies raised against the subunit or the chaperone can be used to purify the complex from periplasmic extracts. Immunoaffinity techniques such as immunoprecipitation using antibodies specific to either the chaperone or a subunit are useful to demonstrate the association of that pilus subunit with the chaperone in periplasmic extracts. However, it might be difficult to elute chaperone–subunit complexes bound to antibodies without dissociating the complexes. The purification of several chaperone–subunit com-

[52] J. Dahmen, T. Frejd, and G. Magnusson, *Carbohydr. Res.* **137,** 219 (1983).

plexes in the *pap* pilus system was performed using various chromatographic steps including ion exchange and hydrophobic interaction columns.[53] Great care must be taken during the purification, since some of these complexes are very prone to self-aggregation and precipitation due probably to interactive surfaces on the subunits. In fact, even ion exchange chromatography was disruptive so only anion exchange resin was used to bind out periplasmic contaminants as an initial fractionation step before the hydrophobic interaction chromatography. Due to the cationic character of PapD and its homologs, even chaperone–subunit complexes do not bind the anion exchange resin but most *E. coli* periplasmic proteins do at pH 6 or higher.[53] PapD–PapA complexes were purified by hydrophobic interaction chromatography, which separated them from most other periplasmic proteins. The PapD–PapA complex was retained longer on the column than the PapD chaperone alone, indicating that the complex was more hydrophobic than the chaperone alone.[53] The purified PapD–PapA complex provides a self-aggregating protein in a nonaggregated state and will be used to study principles of polymer assembly. Interestingly, actin and microtubules direct their own aggregation into a filamentous structure.[54] Whether the P-pilus structure is a product of self-aggregation of the major subunit or whether the structure of the fiber is influenced by the periplasmic chaperone and the outer membrane usher can now be addressed. Finally, purification and subsequent crystallization of the complex will provide detailed structural information which will aid in the development of vaccines.

Analysis of Periplasmic Preassembly Complexes

Affinity purification of the chaperone–adhesin periplasmic complex has allowed this to be the first preassembly complex to undergo biochemical analysis. The PapD–PapG preassembly complex has been studied using various *in vitro* assays to discern the folding state of the proteins in the complex, the composition of the complex, and the function of PapD in forming this interaction. Purified PapD–PapG complex was analyzed under both native and denaturing conditions. The complex was first run on native and isoelectric focusing gels under nondenaturing conditions. The band(s) corresponding to the complex was then excised, boiled in SDS–PAGE sample buffer, and analyzed by SDS–PAGE and Western blotting. This type of gel analysis has shown that the PapD–PapG chaperone–adhesin complex contains an equimolar ratio of chaperone to ad-

[53] R. Striker, F. Jacob-Dubuisson, C. Frieden, and S. Hultgren, *J. Biol. Chem.* (1994), in press.

[54] C. Frieden, *Annu. Rev. Biophys. Biophys. Chem.* **14,** 189 (1985).

hesin.[55] This was confirmed by amino acid analysis of the acid-hydrolyzed purified complex. Isoelectric focusing and native gel analysis of radiolabeled K88 periplasm from minicells indicated that FaeE also forms an equimolar complex with FaeG.[12]

Circular dichroism spectra and SDS–PAGE analysis of the PapD–PapG complex in the presence and absence of a reducing agent have indicated that PapG in the PapD–PapG complex has a nativelike structure. The circular dichroism spectrum of PapD–PapG subtracted from the spectrum of PapD revealed that PapG has secondary structure while in the complex, and SDS–PAGE analysis revealed that the disulfide bridges within PapG were intact.[55] The ability of the complex to bind to the receptor provided further evidence of the nativelike structure of the adhesin within the complex.[55] Although the chaperone presumably binds to nascently translocated, unfolded subunits as they emerge into the periplasm, the subunit seems to be in a nativelike state in the complex. Whether the chaperone actually assists in the folding of the subunits, or binds to the subunits immediately after their folding to prevent their premature aggregation is unknown.

Mapping Subunit Binding Site of Chaperone

The three-dimensional structure of PapD as well as the alignment of the entire family of chaperone sequences showing the conserved and invariant residues within the family can be used to the fullest advantage to construct site-directed mutations to study interactive surfaces of the chaperone. Site-directed mutations of PapD were critical in determining the active site of the PapD chaperone.[56] The expression of pili assembled by wild-type and mutant PapD proteins was synchronized using an inducible genetic system to compare accurately their respective activities. Mutant and wild-type *papD* genes were cloned downstream of the inducible *tac* promoter. These plasmids were then used to complement strains carrying a plasmid that contains the entire *pap* gene cluster with a linker insertion in *papD*. In the absence of the inducer, isopropylthiogalactoside (IPTG), the chaperone was not expressed, therefore the bacteria did not assemble pili and the pilus subunits were degraded in the periplasm. The amount of PapD in the periplasm and pili on the bacterial cell surface was directly proportional to the amount of IPTG added to cultures. Hemagglutination assays, quantitation of surface-localized pili, electron microscopy, and analysis of the stability of pilus subunits in the periplasm were then

[55] M. J. Kuehn, S. Normark, and S. J. Hultgren, *Proc. Natl. Acad. Sci. U.S.A.* **88,** 10586 (1991).

[56] L. Slonim, J. Pinkner, C. I. Branden, and S. J. Hultgren, *EMBO J.* **11,** 4747 (1992).

used to study the effects of point mutations on the mechanism of action of PapD.

Site-directed mutations were made in the cleft of PapD at residues corresponding to the invariant and conserved residues of the periplasmic chaperone family (see Fig. 3). Mutations at the invariant residue arginine 8 (R8) and lysine 112 (K112) abolished PapD activity completely, rendering the bacteria HA negative and devoid of surface-localized pili.[56,58] Although it was localized to the periplasm, the mutant chaperone could not bind to most of the pilus subunits in the periplasm, resulting in degradation of the pilus subunits. R8 is located on a bulge in the first β strand of PapD, with its side chain pointing straight into the empty cleft. R8 and K112 do not interact with other side chains, indicating that they have no apparent structural function. Therefore, R8 and K112 seem to be required to mediate pilus subunit binding.

A point mutation in methionine-172 (M172), another residue in the cleft of PapD, abolished piliation when the mutant PapD was expressed in small amounts.[56] It was determined that under these conditions the interaction between PapD and PapF, an initiator/adaptor protein, was abolished. M172 protrudes into the cleft from the second domain of PapD and does not appear to be involved in protein structure. This highly conserved residue in the periplasmic chaperone family also seems to play a role in chaperone function. Point mutations made in a residue at the lip of the cleft of PapD, corresponding to a position of variable residues in the chaperone family, did not abolish the function of PapD. These types of biochemical analyses of the mutant PapD chaperones have revealed that PapD may recognize different pilus subunits with different affinities.

Conserved Subunit Motif Recognized by Pilus Chaperones

Both *in vitro* evidence and *in vivo* evidence suggest that the chaperone recognition site on the subunit appears to be, at least in part, composed of the carboxyl terminus of the subunit protein. The carboxyl termini of pilus protein subunits have been found to be very hydrophobic and to contain two highly conserved residues, the penultimate tyrosine and a glycine 14 amino acids from the C terminus.[26,57] A genetic deletion corresponding to the 13 amino acids at the C terminus of PapG was found to abolish association with PapD, and resulted in proteolytic cleavage of

[57] B. L. Simons, P. Rathman, C. R. Malij, B. Oudega, and F. K. de Graaf, *FEMS Microbiol. Lett.* **67,** 107 (1990).

[58] M. Kuehn, D. Ogg, J. Kihlberg, L. Slonim, K. Flemmer, T. Bergfors, and S. Hultgren, *Science* **262,** 1234 (1993).

PapG.[22] *In vitro* biochemical assays testing the ability of PapD to bind to synthetic peptides corresponding to the carboxyl terminus of PapG also indicate that PapD has a high affinity for this region.[58] Mutations made at the penultimate tyrosine residues of FanC, the K99 major subunit, abolished the piliation and the HA phenotype of the cells and caused instability of the mutant FanC in the periplasm, indicating that this site was important for chaperone binding.[57] These results indicate that the hydrophobic, conserved carboxyl terminus of pilin subunits may form a specific motif recognized by periplasmic chaperones.

Outer Membrane Usher Proteins

Identification and Comparisons of Pilus Outer Membrane Proteins

In addition to encoding a periplasmic chaperone protein, all pilus gene clusters contain a gene encoding a larger outer membrane protein (greater than 80 kDa). In the case of proteins whose genes have been sequenced, FaeD,[59] FanD,[60] FimD,[61] MrkC,[40] F17C,[62] SfaF,[63] and an 84-kDa *Salmonella* protein,[64] each share approximately 28% identity and 49% similarity with the outer membrane protein, PapC.[8] Further, all of these proteins share high contents of glycine and serine and have conserved cysteine residues. The similarities of these proteins suggest that they each perform analogous functions in their respective pilus systems.

Deletions or insertion mutations of several outer membrane usher proteins, PapC,[8] FaeD,[12] and FimD,[65] have shown that they are necessary for production of their respective pili. In addition, it was shown that when PapC was overproduced, the number of pili also increased.[8] In the absence of PapC, the pilin subunits have been shown to build up in the periplasm in complexes with the chaperone, PapD. This was also shown for FaeD. These studies proposed that outer membrane proteins such as PapC are necessary to remove pilus subunit proteins from the chaperone complexes, in a process referred to as uncapping, and to incorporate the subunits into the pilus.

[59] F. R. Mooi, I. Claasen, D. Baker, H. Kuipers, and F. K. de Graaf, *Nucleic Acids Res.* **14,** 2443 (1986).

[60] B. Roosendaal and F. K. de Graaf, *Nucleic Acids Res.* **17,** 1263 (1989).

[61] P. Klemm and G. Christiansen, *Mol. Gen. Genet.* **220,** 334 (1990).

[62] P. Lintermans, Thesis, Rijksunivsersiteit Ghent, Belgium (1990).

[63] T. Schmoll, J. Morschhauser, M. Ott, B. Ludwig, I. van Die, and J. Hacker, *Microb. Pathog.* **9,** 331 (1990).

[64] C. R. Rioux, M. J. Friedrich, and R. J. Kadner, *J. Bacteriol.* **172,** 6217 (1990).

[65] P. E. Orndorff and S. Falkow, *J. Bacteriol.* **159,** 736 (1984).

Outer Membrane Preparation

This procedure is adapted from Achtman *et al.*[66]

1. Harvest cells from liquid culture by centrifugation at 8000 *g* for 15 min.
2. Resuspend cells in 10 m*M* Tris, pH 8.0.
3. Break cells open by sonication or French press.
4. Centrifuge at 8000 *g* for 15 min to pellet unbroken cells.
5. Carefully transfer supernatant to clean tube.
6. Centrifuge supernatant at 48,000 *g* for 60 min to pellet membranes.
7. Resuspend pellet in sterile H_2O.
8. Freeze membranes at −70°.
9. Thaw. Add 8 vol of 1.67% Sarkosyl, 11.1 m*M* Tris, pH 8.0, and incubate at room temperature for 20 min to solubilize the inner membrane.
10. Centrifuge at 48,000 *g* for 90 min to pellet outer membranes. Remove supernatant (inner membrane). Resuspend outer membranes in appropriate buffer.

PapC as a Targeting/Ushering Protein

Recently it was shown that periplasmic chaperone–subunit complexes were able to recognize and bind to PapC in *in vitro* assays.[67] Extracts containing PapC were run on SDS–PAGE and electroblotted to PVDF paper. The PVDF-bound PapC was then exposed to periplasmic extracts containing the PapD chaperone and subunits proteins, and binding was detected by secondary immunoblotting with antisera to the pap proteins. Outer membrane extracts with or without PapC were also bound to ELISA plates overnight followed by sequential incubations with PapD-subunit preparations and anti-PapD antisera. With these methods it was demonstrated that the different chaperone–subunit complexes had different affinities for PapC. The PapD–PapG complexes appeared to have the greatest affinity for PapC, whereas PapD–PapA complexes had little to no binding affinity for PapC. This suggested that *in vivo,* subunits are targeted to the outer membrane by virtue of their ability to bind to PapC.[67] Further, it suggested an ushering function for PapC, in which differential affinities for PapC might be one mechanism by which the subunits are able to recognize PapC in an ordered fashion and thus be incorporated into the final pilus structure in a specific order. In the current model (Fig. 4),

[66] M. Achtman, A. Mercer, B. Kusecek, A. Pohl, M. Heuzenroeder, W. Aaronson, A. Sutter, and R. P. Silver, *Infect. Immun.* **39,** 315 (1983).

[67] K. Dodson, F. Jacob-Dubuisson, R. Striker, and S. J. Hultgren, *Proc. Natl. Acad. Sci. U.S.A.* 3670 (1993).

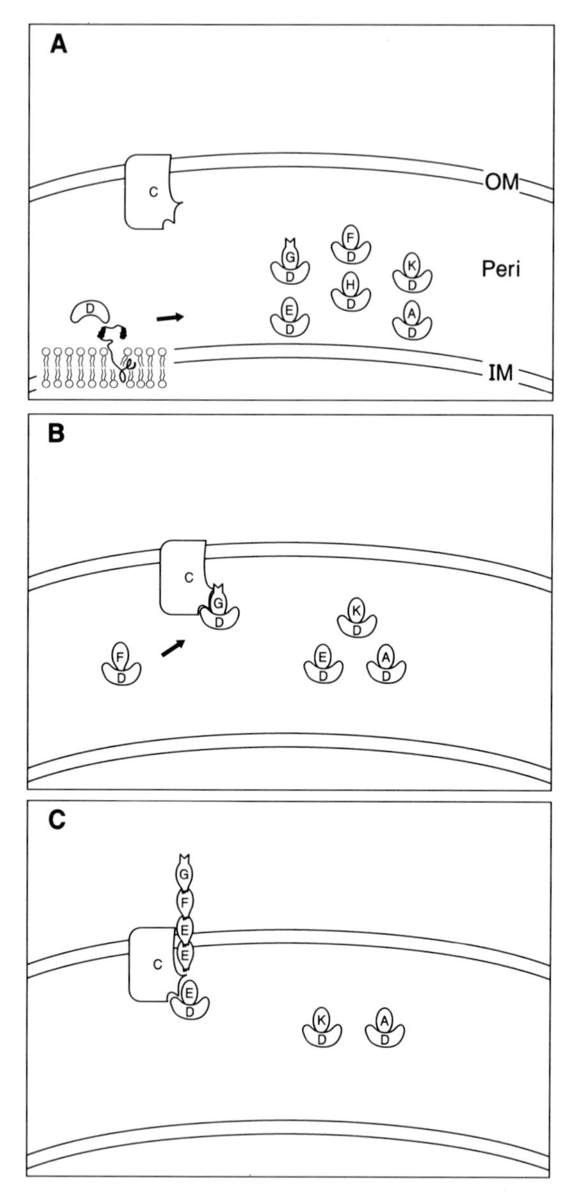

FIG. 4. Model of P-pilus assembly. Subunits and assembly proteins are indicated by letters. (A) PapD caps interactive surfaces on each subunit type after they have translocated across the inner membrane. (B) Initiation of tip assembly occurs when PapF–PapD complexes interact with PapD–PapG in the assembly site. (C) Polymerization of PapE results in the tip fibrillum structure. (D) PapK terminates tip fibrillum growth and initiates pilus rod polymerization. (E) Polymerization of PapA forms the bulk of the pilus rod. See text for details. OM, Outer membrane; Peri, periplasm; IM, inner membrane.

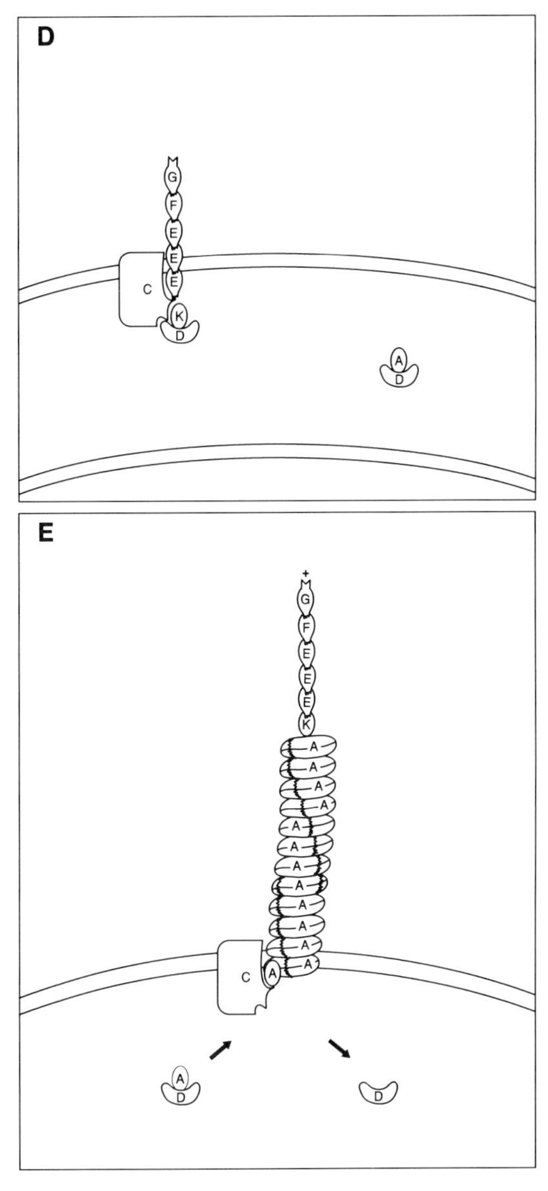

FIG. 4. (*continued*)

PapD–PapG complexes are the first to recognize the PapC assembly sites due to their high relative binding affinity for PapC. Next, PapD–PapF complexes bind to PapC, and PapG is subsequently incorporated at the tip of the P pilus. PapG is then joined to PapE via the PapF adaptor protein. Incorporation of PapK caps the growth of the tip fibrillum fiber and triggers the targeting and uncapping of PapD–PapA complexes to the assembly site, thus initiating the polymerization of the pilus rod at the end of the tip fibrillum.

Conclusions

Pathogenic organisms appear to have an abundant repertoire of adhesive organelles, probably only a fraction of which have been identified. The different architectures of pili represent a variety of strategies that these pathogens have evolved to adhere effectively to host tissues. Although the minor and major pilus subunits have a diversity of function, interestingly, they are incorporated into adhesive structures using common mechanisms of assembly. The ongoing efforts to identify, isolate, and analyze products of pilin gene clusters will undoubtedly lead to the discovery of additional, highly related members of the immunoglobulin-like chaperone family and the outer membrane usher family, as well as the discovery of a seemingly diverse group of minor pilin subunits. The sequence comparisons of the pilus chaperone family have already highlighted highly conserved amino acid residues which may be critical to chaperone function. Site-directed mutations in these residues have indicated that residues within the cleft are critical to the ability of the chaperone to bind pilus subunits. The addition of members to both chaperone and usher families will lead to determining which features are strictly universal and which features are specific to the system. The emerging understanding of the molecular details of periplasmic chaperone-assisted assembly, folding, targeting, and ordered assembly of the composite pilus fiber at outer membrane sites has developed into a complex, exciting new perspective on the secretion and assembly of extracellular structures.

Acknowledgments

This work was supported by grants to S.J.H. from the Lucille P. Markey Charitable Trust, Washington University/Monsanto Biomedical Research Contract, National Institutes of Health (Support Grant 1RO1 AI-29549), Institutional Biomedical Research (Support Grant 2S07 RR-5389), and the American Cancer Society (Grant IN-36). M.K. and L.S. received support from NIH Training Grant AI-07172. F.J.-D. was the recipient of a Long Term Post Doctoral EMBO Fellowship (1991–1993), and K.D. of a Markey Fellowship. R.S. is supported by Medical Scientist Training Program NIH Grant 5T32 GM-07200.

[21] Isolation and Identification of Eukaryotic Receptors Promoting Bacterial Internalization

By GUY TRAN VAN NHIEU and RALPH R. ISBERG

Introduction

Many pathogenic bacteria can be internalized by normally nonphagocytic host cells.[1,2] In some cases, this process may be essential for the replication of microorganisms such as *Chlamydia trachomatis*, a bacterium responsible for the most common sexually transmitted diseases.[3] In the case of other pathogens such as *Salmonella* and *Yersinia* species, internalization by host cells is probably a prerequisite for the establishment of the infection and enables bacteria to disseminate and colonize various body compartments.[4] In many instances, as in the case of piliated bacteria, interactions between bacterial surface components and cellular receptors result in simple attachment of the bacterium to the surface of the host cell.[5-9] As binding of the bacterium to cellular receptors is also the primary event leading to the internalization of the bacterium, the discrimination between surface adhesion and internalization appears to be determined by (1) the selection of specific receptors by bacterial ligands and (2) the physical nature of the interaction with the receptor. Identifying such receptors and characterizing the interaction between the host receptor and the bacterial ligand constitute an important first step toward investigating the molecular strategies used by these pathogens. We describe in this article the affinity chromatography technique used to identify members of the integrin family of cell adhesion molecules as cellular receptors for invasin,[10] a *Yersinia pseudotuberculosis* surface protein that allows bacterial internalization,[11] as well as assays allowing the study of the interaction of invasin with the purified $\alpha_5\beta_1$-integrin.

[1] S. Formal, T. L. Hale, and P. J. Sansonetti, *Rev. Infect. Dis.* **5,** S702 (1983).
[2] J. W. Moulder, *Microbiol. Rev.* **49,** 298 (1985).
[3] R. L. Hodinka, C. H. Davis, J. Choong, and P. B. Wyrick, *Infect. Immun.* **56,** 1456 (1988).
[4] T. Une, *Microbiol. Immunol.* **21,** 205 (1977).
[5] J. Swanson, *Rev. Infect. Dis.* **4,** S678 (1983).
[6] A. Urisu, J. L. Cowell, and C. R. Manclark, *Infect. Immun.* **52,** 695 (1986).
[7] B. Lund, B. I. Marklund, N. Stromberg, F. Lindberg, K. A. Karlsson, and S. Normark, *Mol. Microbiol.* **2,** 255 (1988).
[8] K. K. Lee, P. Doig, R. T. Irvin, W. Paranchych, and R. S. Hodges, *Mol. Microbiol.* **3,** 1493 (1989).
[9] R. R. Isberg, *Science* **252,** 934 (1991).
[10] R. R. Isberg and J. M. Leong, *Cell (Cambridge, Mass.)* **60,** 861 (1990).
[11] R. R. Isberg, D. L. Voorhis, and S. Falkow, *Cell (Cambridge, Mass.)* **50,** 769 (1987).

Isolation of Invasin Receptor from Mammalian Cells

Affinity Matrix Preparation

Use of Maltose-Binding Protein–Invasin Hybrid Protein. The limiting step in the affinity matrix preparation is the amount of purified bacterial ligand available. In the case of invasin, the use of strains that overexpress hybrid proteins consisting of the cell binding domain of invasin fused to the *Escherichia coli* maltose-binding protein (MBP) overcomes this problem.[12] By use of commercially available vectors (New England Biolabs, Beverly, MA) the carboxy-terminal domain of invasin is fused to MBP, and large amounts of the resulting hybrid protein, which still retains cell binding activity, are isolated in a single-step procedure via amylose resin affinity chromatography.[12] The full-length hybrid protein is subsequently purified from any degradation products that arise during isolation by precipitation of the hybrid protein with 40% ammonium sulfate.[13] The resulting product is then crosslinked to Affi-Gel 10 beads (Bio-Rad, Richmond, CA) for preparation of the affinity matrix.

Coupling to Affi-Gel 10

Reagents

100 m*M* HEPES, pH 8.0 (coupling buffer)
1 *M* $CaCl_2$

The purified hybrid protein MBP–Inv479 containing the cell binding domain of invasin is dialyzed against coupling buffer. In a typical coupling procedure, 10 ml of protein is coupled with 5 ml of Affi-Gel 10 (Bio-Rad) at a final protein concentration of 2 to 3 mg/ml for 4 hr at 4° with gentle shaking, following the manufacturer's instructions. After 4 hr, $CaCl_2$ is added to the coupling buffer at a final concentration of 80 m*M* and the coupling is allowed to continue overnight at 4°. The addition of $CaCl_2$ to the coupling buffer increases the efficiency of coupling of the MBP–Inv479 hybrid protein to the Affi-Gel 10 beads. The calculated isoelectric point of the hybrid protein is quite acidic (p*I* 6.5), and in theory, Affi-Gel 15 beads should be more adequate for coupling of such acidic proteins. Practically, however, for reasons that are not clear, this procedure seems to be the most efficient for the MBP–Inv hybrids. Under these conditions, the efficiency of coupling varies from 50 to 75% as monitored by absorbance at 280 nm after hydrolysis of interfering absorbing material with 50 m*M*

[12] J. M. Leong, R. S. Fournier, and R. R. Isberg, *EMBO J.* **9,** 1979 (1990).
[13] G. Tran Van Nhieu and R. R. Isberg, unpublished (1990).

HCl. The succinimidyl reactive groups on the beads are subsequently blocked by adding ethanolamine (100 m*M* final concentration) and incubated at 4° for another 2 hr. For invasin affinity chromatography, the beads are equilibrated with 10 vol of 25 m*M* HEPES buffer, pH 7.0, 150 m*M* NaCl, 1 m*M* $MgCl_2$, 0.5 m*M* $CaCl_2$. If not used immediately, invasin-coupled beads can be resuspended in 0.2% (w/v) sodium azide, 100 m*M* Tris, pH 8.0, and stored at 4° for at least 6 months without significant loss of activity.

To differentiate specific from nonspecific binding, a control column in which native MBP is coupled to Affi-Gel 10 should be prepared.

Preparation of Cellular Extract

For identification of the mammalian surface protein binding to invasin, it has proved convenient to work with a relatively small amount of cellular extract, allowing the use of a correspondingly small amount of invasin–agarose and shortening the length of the chromatography procedure.[10] Surface labeling of cellular receptors can be readily achieved by biotinylation using the impermeant reagent NHS-LC-biotin (Pierce, Rockford, IL), and nanogram amounts of the purified labeled receptors can be visualized after Western blot by probing with streptavidin and standard immunodetection methods. To obtain larger amounts for further characterization of the interaction between invasin and its integrin receptors, a two-step purification procedure involving isolation of glycoproteins by lectin affinity chromatography is required to limit the level of contamination due to increasing nonspecific binding to the invasin resin.[14]

Small-Scale Purification from Tissue Culture Cell Extract. A schematic representation of the surface labeling with biotin and the invasin affinity chromatography procedure is given on Fig. 1A.[15]

BIOTINYLATION OF CELL SURFACE RECEPTORS

Reagents

- NHS-LC-biotin (Pierce) freshly made at 20 mg/ml in dimethyl sulfoxide (DMSO)
- Phosphate-buffered saline (PBS: 20 m*M* PO_4, pH 7.0, 150 m*M* NaCl) + 2 m*M* ethylenediaminetetraacetic acid (EDTA)
- PBS–Ca/Mg (PBS, 1 m*M* $MgCl_2$, 0.5 m*M* $CaCl_2$)
- Confluent cultured mammalian cells, freshly split (1 : 2) 18 hr previously (10^8 to 10^9 cells).

[14] G. Tran Van Nhieu and R. R. Isberg, *J. Biol. Chem.* **266,** 24367 (1991).

[15] R. R. Isberg, *Mol. Biol. Med.* **7,** 73 (1990).

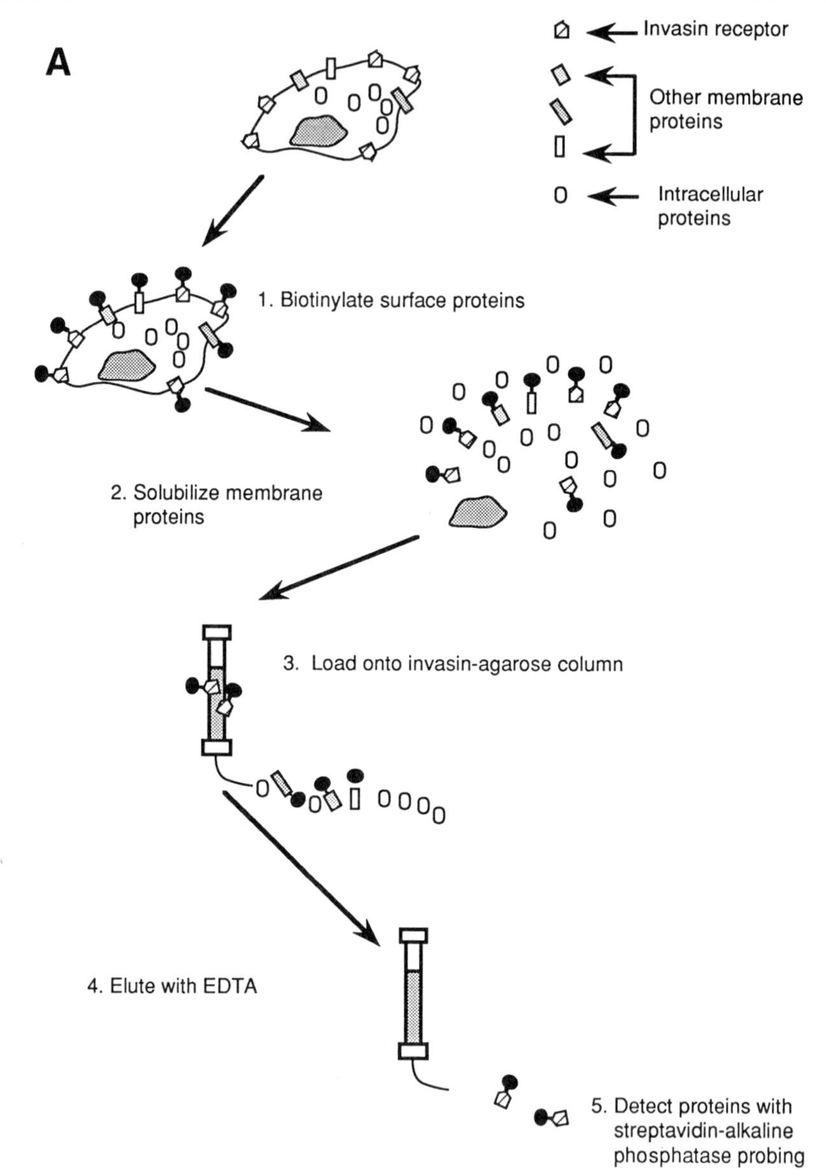

FIG. 1. (A) Identification of cellular receptors for invasin. Cultured cells were surface-labeled using NHS-LC-biotin, proteins were extracted in octylglucoside-containing buffer and loaded onto an invasin–agarose column (see text). Invasin receptors were eluted with EDTA and analyzed by SDS–PAGE, followed by probing with streptavidin–alkaline phosphatase (From Isberg[15]). (B) SDS–PAGE analysis of invasin receptors from HEp-2 cells obtained after invasin affinity chromatography. Fractions eluted from the invasin–agarose

(*continued*)

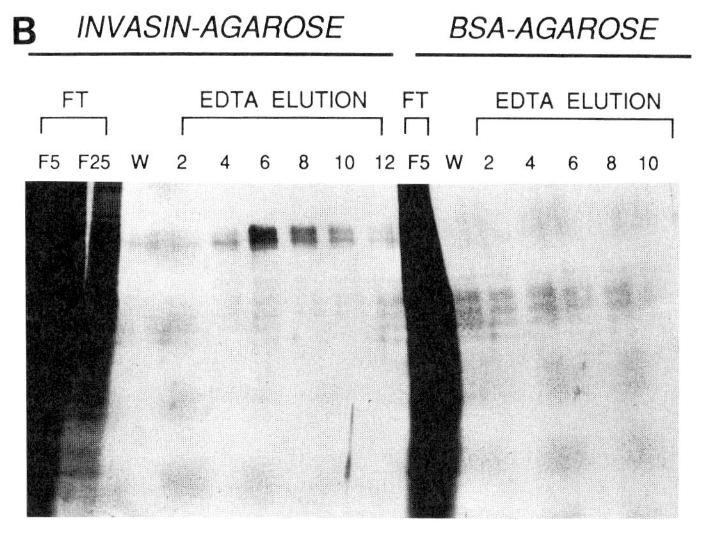

column (invasin agarose) or a bovine serum albumin–agarose column (BSA–agarose) were fractionated on a 7.5% polyacrylamide–SDS gel, and transferred to an Immobilon-P filter. The protein bands were visualized by probing with streptavidin–alkaline phosphatase (see text). The invasin receptors are characteristic of β_1-chain integrins; in the case of HEp-2 cells, two species are observed with approximate molecular masses of 115 and 140 kDa (invasin agarose, lanes 4–10) corresponding to the β_1 chain and the α_3 chain, respectively.

1. Lift off adherent cells from plastic surface with 10 ml of PBS containing 2 m*M* EDTA per 175-mm tissue culture flask. In the case of nonadherent cells, wash the cells once with PBS, 2 m*M* EDTA before proceeding in the same following fashion.
2. Wash cells twice, by centrifugation for 10 min at 1000 rpm (approximately 550 *g*), followed by resuspension in 10 ml PBS–Ca/Mg per 10^7 cells.
3. Resuspend cells at a density of 10^6 to 10^7 cells/ml in PBS–Ca/Mg.
4. Add NHS-LC-biotin to the cell suspension to a final concentration of 200 μg/ml; invert rapidly to mix.
5. Incubate cells at 22° for 60 min, occasionally inverting to resuspend cells.
6. Add a 1 : 20 dilution of 1 *M* ethanolamine, pH 7.0 (50 m*M* final concentration) to stop the labeling reaction. Incubate for 30 min at 22°, inverting occasionally.
7. Wash the cells twice with PBS–Ca/Mg and immediately proceed to the extraction procedure. Alternatively, cell pellets with the supernatants decanted may be flash frozen and stored at −80° for up to 4 months.

PREPARATION OF CELL EXTRACT AND INVASIN AFFINITY CHROMATOGRAPHY

Reagents

Extraction buffer: PBS–Ca/Mg containing 200 m*M* octyl-β-D-glucopyranoside (Pierce), 1 m*M* phenylmethylsulfonyl fluoride (PMSF), 1 μg/ml pepstatin, and 2.5 μg/ml leupeptin

Washing buffer: PBS–Ca/Mg, 0.1% Nonidet P-40, 1 m*M* PMSF

Elution buffer: PBS–Ca/Mg, 10 m*M* EDTA, 0.1% Nonidet P-40, 1 m*M* PMSF

1 ml of invasin–Affi-Gel 10 per extract obtained from 10^8 to 10^9 cells

All reagents are kept at 4° before extraction, and all steps are carried out at 4° unless otherwise stated.

1. Pour 1 ml of resin into a disposable column (Bio-Rad, Catalog No. 737-0704), and wash with 10 column volumes of wash buffer followed by 1 column volume of extraction buffer.
2. Resuspend the labeled cells in a volume of extraction buffer equal to the weight of the cell pellet, typically about 500 μl/10^8 cells. Let stand on ice for 60 min.
3. Pellet the cell debris at 170,000 *g* for 45 min.
4. Load the cleared extract over a 2-hr period onto the invasin–agarose column.
5. Wash the column with 10 column volumes of washing buffer.
6. Elute the bound receptors with elution buffer,[16] and collect 400-μl fractions in tubes containing 8 ml of 1 *M* $MgCl_2$. Mix immediately by gentle swirling.
7. The pH of each fraction should be checked and neutralized with 1 *M* Tris–HCl, pH 8.0, if necessary.
8. Analyze 20 μl of each fraction by sodium dodecyl sulfate–polyacrylamide gel electrophoresis (SDS–PAGE) under reducing or nonreducing conditions.[10]
9. Transfer proteins to Immobilon filters,[17] and probe with streptavidin–alkaline phosphatase (Zymed Laboratories, South San Francisco, CA) followed by standard chromogenic detection methods.[10]

As shown in Fig. 1B, the protein species observed after invasin affinity chromatography of an extract made from HEp-2 cells followed by strepta-

[16] The use of EDTA as an eluting agent is based on the fact that invasin binding to integrins, as for other extracellular matrix proteins (i.e., fibronectin), requires the presence of traces of divalent cations. β_1-Chain integrins contain several binding sites for divalent cations such as Ca^{2+}, and it is assumed that cation binding is required for the receptor to assume its proper active conformation.

[17] H. Towbin, T. Staeheli, and J. Gordon, *Proc. Natl. Acad. Sci. U.S.A.* **76,** 4350 (1979).

vidin probing show features characteristic of β_1-chain integrins.[10] Members of this receptor family are heterodimers that differ in electrophoretic migration depending on the reducing conditions used.[18] Numerous antibodies specific for different integrin receptors have been isolated,[19,20] allowing immunological identification of the integrin receptors that bind invasin.[10]

Affinity chromatography analysis of extracts from different cell lines demonstrated that four β_1-chain integrins ($\alpha_3\beta_1$, $\alpha_4\beta_1$, $\alpha_5\beta_1$, and $\alpha_6\beta_1$) act as receptors for invasin.[10] In the case of the α_6 subunit, identification of the receptor involved the purification of receptor from approximately 10 g of platelets, followed by N-terminal microsequencing of protein species transferred to Immobilon-P.[10]

Large-Scale Isolation from Human Placenta. A convenient source of receptors is human placenta. About 500 μg of integrin receptor can be isolated from 200 g of placenta by invasin affinity chromatography,[14] using a procedure similar to the one described for the isolation of fibronectin receptors.[21]

PREPARATION OF HUMAN PLACENTA EXTRACT AND AFFINITY CHROMATOGRAPHY

Reagents

Invasin affinity chromatography: same buffers as above.

Lectin affinity chromatography:

Washing buffer: PBS–Ca/Mg, 0.1% Nonidet P-40, 1 m*M* PMSF

Elution buffer II: PBS–Ca/Mg 5% *N*-acetylglucosamine, 0.1% Nonidet P-40, 1 m*M* PMSF

Wheat germ agglutinin (Pharmacia)

Fifty-gram aliquots of placenta are stored frozen at −80° and thawed immediately before use.

1. Homogenize placental tissue with no added buffer in a Waring blender three times with 30-sec pulses. The Waring blender should be chilled on ice before extraction, and the homogenization of placenta should be performed as rapidly as possible.
2. Add ice-cold extraction buffer (50 ml/50 g of placenta) the homogenized placenta, and transfer the suspension to a 250-ml centrifuge bottle.
3. Let stand 60 min on ice.
4. Centrifuge at 7000 rpm for 10 min.

[18] R. O. Hynes, *Cell (Cambridge, Mass.)* **48,** 549 (1987).

[19] S. M. Albelda and C. A. Buck, *FASEB J.* **4,** 2868 (1990).

[20] M. E. Hemler, *Annu. Rev. Immunol.* **8,** 365 (1990).

[21] R. Pytela, M. D. Pierschbacher, S. Argraves, S. Suzuki, and E. Ruoslahti, this series, Vol. 114, p. 475.

5. Transfer supernatant into ultracentrifugation tubes, and proceed as above (step 3–8), using 10 ml of invasin resin per 50 g of placenta.

WHEAT-GERM AGGLUTININ AFFINITY CHROMATOGRAPHY. When working with a large amount of placental extract, analysis of the protein eluted from the invasin–agarose column reveals that several protein species are usually observed by SDS–PAGE, probably resulting from nonspecific interactions with the chromatography matrix. Integrins can be subsequently purified from nonglycoproteins via lectin affinity chromatography.[14,21]

1. Pool the fractions from the invasin affinity chromatography containing the peak of eluted proteins.
2. Dialyze immediately against washing buffer. (The presence of EDTA in the extract can inactivate the lectin resin.)
3. Load onto a 5-ml wheat germ agglutinin–agarose column previously equilibrated with washing buffer at a flow rate of about 5 ml/hr.
4. Wash the column with 10 column volumes of washing buffer.
5. Elute with elution buffer II, collecting 400-μl fractions.
6. Analyze by SDS–PAGE as above.

Figure 2 shows the result of SDS–PAGE analysis of the fractions eluted from the wheat germ agglutinin column, following invasin affinity chromatography. The main invasin receptor species isolated from placenta is the $\alpha_5\beta_1$-integrin. The $\alpha_5\beta_1$-integrin isolated after this procedure is stable for at least 2 months at 4°. For long-term storage, the receptor preparation is divided into aliquots and stored at −80°. Repeated freezing and thawing result in rapid loss of receptor binding activity.

Analysis of Invasin Binding to Isolated $\alpha_5\beta_1$-Integrin Receptor Immobilized on Filter

An important step in the study of the interaction between invasin and integrin receptors is the development of an assay allowing the direct visualization of invasin binding to the $\alpha_5\beta_1$-integrin. The feasibility of such an assay depends on the stability of the receptors under electrophoresis conditions: The $\alpha_5\beta_1$-integrin is inactivated during standard SDS–PAGE, but retains invasin binding activity when electrophoresis is performed under nondissociating conditions. The demonstration of direct binding on a filter assay has also allowed the development of a microtiter assay to quantitate *in vitro* binding and inhibition studies using very small amounts of proteins.[14]

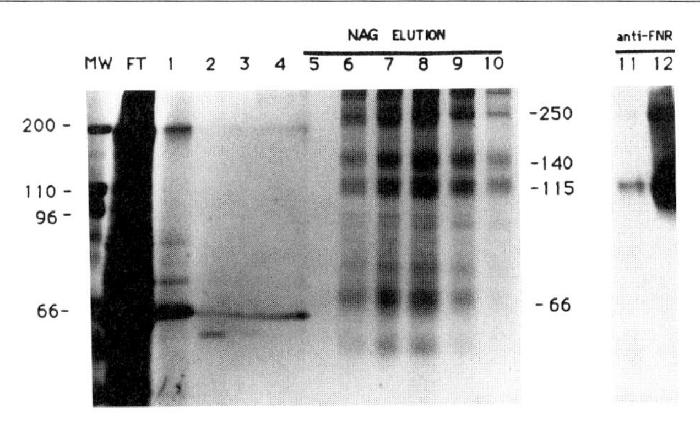

FIG. 2. Isolation of invasin receptors from human placenta. Receptor fractions were obtained by affinity chromatography on invasin–agarose and on wheat germ agglutinin–agarose (see text). The samples were analyzed on an SDS gel containing 7.5% polyacrylamide, under nonreducing conditions. Protein bands were visualized by Coomassie blue staining. Lane MW, molecular weight standards ($\times 10^{-3}$); lane FT, cell extract; lanes 1–5, wash fractions; lanes 6–10, fractions eluted from invasin–agarose column with buffer containing 5% *N*-acetyl-D-glucosamine. The receptor appears as 250-, 140-, and 115-kDa bands in lanes 6–10. Lanes 11 and 12; Western blot of lanes 5 and 6 with rabbit anti-FNR antiserum. (From Tran Van Nhieu and Isberg.[14])

Native Gel Filter Assay

Reagents

Running buffer: 192 m*M* glycine, 25 m*M* Tris, pH 8.4
- 2× loading sample buffer: 62 m*M* Tris, pH 6.8, 10% glycerol, 0.005% bromphenol blue
- Separating gel, for 10 ml: 1.7 ml of acrylamide–bisacrylamide (30 : 08); 2.5 ml of 1.5 *M* Tris, pH 8.8; 5.75 ml of H_2O; 20 μl 10% ammonium persulfate; 20 μl TEMED
- Stacking gel, for 10 ml: 1.25 ml of acrylamide–bisacrylamide (30 : 08); 2.5 ml of 0.5 *M* Tris, pH 6.8; 6.3 ml of H_2O; 20 μl 10% ammonium persulfate; 20 μl TEMED
- Immobilon-P filter membrane (Millipore, Bedford, MA).
- Running buffer containing 20% methanol (transfer buffer)
- PBS—Ca/Mg containing 1% bovine serum albumin (BSA)

1. Mix about 5 ng of receptor preparation with an equal volume of 2× loading sample buffer.
2. Immediately load on 5% polyacrylamide gel, and run at constant current of 25 mA.
3. Transfer to Immobilon-P membrane for 2 hr at 80 V at 4°. (We found that traces of SDS, i.e., 0.001%, in the transfer buffer increases

the efficiency of transfer from the native gel to the Immobilon-P membrane.)

4. Block the filter with PBS, 1% BSA for 2 hr at 22°.
5. Probe with MBP–Inv479 at 1 μg/ml (corresponding to approximately twice the K_d) in PBS–Ca/Mg, 1% BSA for 2 hr at 22°, using about 0.2 ml of probing solution per square centimeter of filter.
6. Wash three times with PBS–Ca/Mg.
7. Reveal invasin binding by probing with anti-MBP antiserum, and anti-rabbit IgG–alkaline phosphatase, followed by standard chromogenic detection methods.[10]

Immunodepletion experiments on the receptor preparation prior to electrophoresis allow the identification of $\alpha_5\beta_1$ as the main invasin receptor isolated from placenta[14] (Fig. 3).

Dot-Blot Filter Assay

The use of a dot-blot apparatus allows high concentrations of receptor to be immobilized on a filter membrane on a single 100-μl dot correspond-

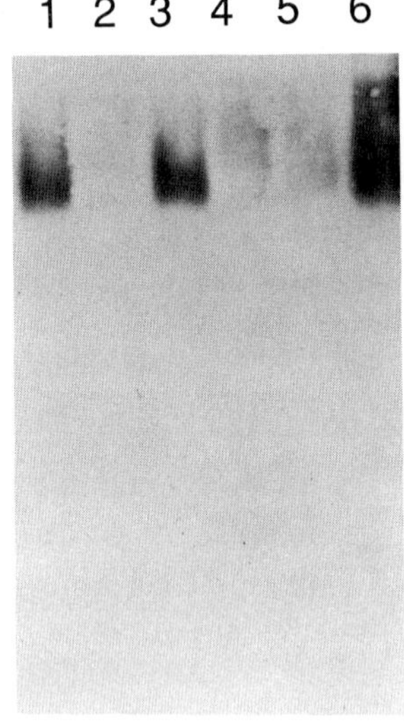

FIG. 3. Identification of placental invasin receptors on a filter transfer assay. Aliquots of the placental receptor preparation were depleted with no antibody, protein A beads (lane 1); anti-FNR antiserum, protein A beads (lane 2); no antibody, anti-rat beads (lane 3); rat MAb anti-α_5, anti-rat beads (lane 4); rat MAb anti-β_1, anti-rat beads (lane 5); no antibody, no beads (lane 6). Immunodepleted extracts were submitted to electrophoresis on a native gel containing 5% polyacrylamide. An Immobilon-P filter replica of the gel was then probed with MBP–Inv479, followed by immune detection (see text), to detect invasin receptor activity. (From Tran Van Nhieu and Isberg.[14])

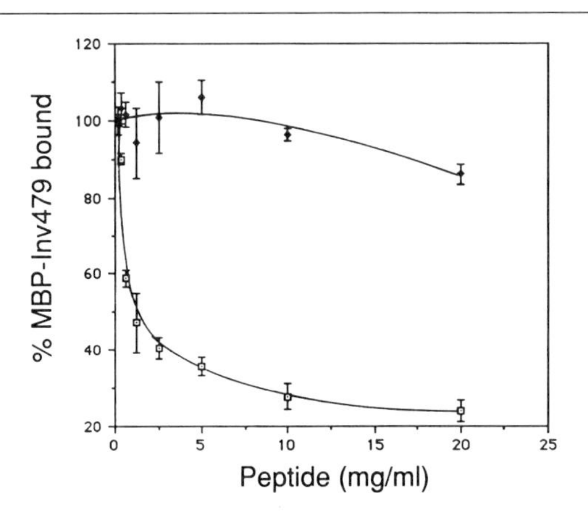

FIG. 4. The GRGDSP hexapeptide inhibits [^{125}I]MBP–Inv479 binding to filter-immobilized $\alpha_5\beta_1$. Samples containing about 50 ng of filter-immobilized invasin receptor preparation were incubated in 100 μl 0.4% BSA–PBS–Mg/Ca containing [^{125}I]MBP-Inv479 (2 × 10^6 cpm/μg, 2 μg/ml) in the presence of increasing concentrations of either GRGDSP (□) or GRGESP (◆). After incubation for 3 hr at 22°, the filters were washed with PBS, and the bound radioactivity was determined. Each point represents the mean of three determinations. (From Tran Van Nhieu and Isberg.[14])

ing to a well of a 96-well apparatus. The samples prepared this way yield a high specific binding activity and can be incubated with ligands in small volumes to quantitate binding.

Materials and Buffers

96-well dot-blot apparatus (Bio-Rad No. 170-6545)
PBS–Ca/Mg, 1% BSA
PBS–Ca/Mg, 0.1% Nonidet P-40
Iodinated invasin (2 × 10^6 cpm/mg)[14]
Receptor preparation (500 ng/ml in PBS–Ca/Mg, 0.1% NP-40, 1 m*M* PMSF)
Immobilon-P filter (Millipore)

Each step is carried out at room temperature.

1. Spot about 5 ng of the purified receptor preparation in 100 μl of PBS–Ca/Mg, 0.1% Nonidet P-40 on an Immobilon-P membrane using a 96-well dot-blot apparatus connected to a laboratory vacuum aspirator.
2. Rinse each well with 200 μl of PBS–Ca/Mg, 0.1% Nonidet P-40.
3. Remove the filter membrane from the dot-blot apparatus and block in PBS–Ca/Mg, 1% BSA for 2 hr at room temperature.
4. Cut individual dots and transfer to 96-well microtiter plate.
5. Incubate with ^{125}I-labeled invasin[14] in PBS–Ca/Mg, 1% BSA for 2 hr at 22°.

6. Wash the wells three times with 0.1% Nonidet P-40, PBS–Ca/Mg.
7. Count the filter spots for γ irradiation.

For inhibition experiments, various concentrations of inhibitors are added during the incubation with the labeled tracer protein (i.e., [^{125}I]MBP–Inv479). Figure 4 shows inhibition of invasin binding to the $\alpha_5\beta_1$-integrin by the fibronectin RGD peptide, consistent with the notion that invasin and fibronectin interact with the same site on the $\alpha_5\beta_1$ receptor.[14]

Conclusion

The use of affinity chromatography has allowed the identification of several β_1-chain integrins as receptors for invasin.[10] It is likely that these β_1-chain integrins identified as receptors for invasin, and normally involved in cell attachment to the extracellular matrix, are directly involved in the process of internalization of the bacteria as it appears that high-affinity binding to integrins (i.e., $\alpha_5\beta_1$ or $\alpha_3\beta_1$) is sufficient to promote efficient bacterial internalization.[13] The development of the assays described in this article has allowed us to study the interaction between invasin and the $\alpha_5\beta_1$-integrin receptor isolated from placenta. These studies have shown that invasin binds to the receptor with a much higher affinity than does fibronectin,[14] an important property in discriminating between bacterial internalization and simple extracellular adherence to human epithelial cells.[9]

[22] Identification of Fibronectin as a Receptor for Bacterial Cytoadherence

By JOHN F. ALDERETE, ROSSANA ARROYO, and MICHAEL W. LEHKER

Introduction

The idea that organisms are coated with soluble, host-derived serum proteins may be an obvious one. After all, it is difficult to envision a situation where a microbial pathogen will not, at any time after entry into the host environment and regardless of the site of infection, be in contact, indeed continuously bathed, with host fluids as complex as serum. In support of this view, the literature is replete with reports of specific host proteins associating with microbial surfaces, and many of these examples have been discussed in numerous reviews.[1,2] From the beginning, though, studies on the associations between bacterial pathogens and certain host proteins likely resulted from serendipitous observations. Such was the case involving fibronectin and *Staphylococcus aureus*.[3] The same could be said of perhaps many other examples of bacterial pathogens and fibronectin associations, and whether or not a bacterial pathogen bound a particular host protein was the result of a rather random screening of a library of commercially available proteins. Not surprisingly, then, there are few reports in the literature where the uniqueness and/or nature of the relationship between the host and a bacterial pathogen has drawn attention to whether a host serum or cellular component might be recognized and bound by the organism.

The significance of the early report on the *S. aureus*–fibronectin association[3] is underscored by the explosion in the number of articles that have appeared showing interactions by pathogenic bacteria,[4–8] protozoa,[9–11] yeast,[12] virus,[13,14] and parasites[2] with extracellular matrix (ECM) proteins. The consequences to the host of being infected by organisms interacting with host serum proteins in general and ECM proteins in particular are numerous, apart from the obvious feature of host parasitism through enhanced cytoadherence. For example, masking of the microbial surface might interfere with antigen presentation or prevent recognition of the foreign organism or antigens leading to an overall immune evasion strategy by the pathogen. Host proteins might impair further the more specialized functions of immune cells whose job is ligand recognition for phagocytosis. Autoimmune reactions might be envisioned, possibly the result of an

[1] M. Höök and L. M. Switalski, *in* "Biology of Extracellular Matrix" (D. Mosher, ed.), Vol. 5, p. 295. Academic Press, San Diego, 1989.

[2] D. J. Wyler, *Rev. Infect. Dis.* **9,** Suppl., S391 (1987).

[3] P. Kuusela, *Nature (London)* **276,** 718 (1978).

[4] K. M. Peterson, J. B. Baseman, and J. F. Alderete, *J. Exp. Med.* **157,** 1958 (1983).

[5] L. M. Switalski, P. Speziale, M. Höök, T. Wadström, and R. Timpl, *J. Biol. Chem.* **259,** 3734 (1984).

[6] S. B. Baloda, A. Faris, G. Fröman, and T. Wadström, *FEMS Microbiol. Lett.* **28,** 1 (1985).

[7] C. Abon-zeid, T. Garbe, R. Lathigra, H. G. Wiker, M. Harboe, G. A. W. Rook, and D. B. Young, *Infect. Immun.* **59,** 2712 (1991).

[8] M. Haapasalo, U. Singh, B. C. McBride, and V. Uitto, *Infect. Immun.* **59,** 4230 (1991).

[9] D. J. Wyler, J. P. Sypek, and J. A. McDonald, *Infect. Immun.* **49,** 305 (1985).

[10] S. T. Pottratz, J. Paulsrud, J. S. Smith, and W. J. Martin, II, *J. Clin. Invest.* **88,** 403 (1991).

[11] M. A. Onaissi, D. Afchain, A. Capron, and J. A. Grimaud, *Nature (London)* **308,** 380 (1984).

[12] A. Kalo, E. Segal, E. Sahar, and D. Dayon, *J. Infect. Dis.* **157,** 1253 (1988).

[13] K. Wang, R. J. Kuhn, E. G. Strauss, S. On, and J. H. Strauss, *J. Virol.* **66,** 4992 (1992).

[14] I. Julkunen, A. Hautanen, and J. Keski-oja, *Infect. Immun.* **40,** 876 (1983).

altered conformation of the host protein and elicitation of autoreactive antibody, and contribute to disease pathogenesis.[15]

From this vantage point, might it be possible to develop a strategy by which investigators could (1) examine the general association between organisms and host serum or cellular components; (2) identify the protein or proteins preferentially enriched onto that microbial surface (as compared with those proteins just loosely associated with the surface); and (3) determine the role that the acquired host protein plays in the biology of the host–parasite interrelationship?

An experimental approach as outlined here would enable the examination of the extent to which microorganisms associate themselves with many serum proteins and might allow for the questions posed above to be answered. Of course, the nature of the host protein–bacteria interactions, whether a loose association of serum proteins with the microorganism (in essence coating or masking the bacterial surface) versus one of a high-affinity, receptor-mediated binding, would be examined in step 1. The identification of the preferentially bound serum protein(s) (step 2) would employ its own substrategy, such that from this information the investigator could then plan experiments to test experimentally the contribution of the host protein to the biology of the host–bacteria relationship (step 3). Considering the complex composition of serum, evidence for a preferential enrichment by the bacterial surface for specific serum proteins would itself be informative, as it is unlikely that receptor-mediated acquisition of a few proteins from serum would occur without relevance.

Therefore, the examination of the interaction between host proteins and microorganisms appears to be central to many questions involving molecular aspects of pathogenesis, from establishment of infection to survival within the adverse environment of the host, through nutrient acquisition and immune evasion, to possible autoimmune manifestations. These types of issues became prominent after the report showing the specific fibronectin binding by *S. aureus,* and it was at this time that the approach mentioned above was attempted for the syphilis spirochete, *Treponema pallidum*. What follows, then, is a description of the approach and brief relevant methodology, except that it is put into a historical context, thereby showing the stepwise nature of the investigation that led to the identication of fibronectin and the highly defined RGD sequence as a receptor for host cytoadherence by this spirochete. Another technique, called the ligand assay and developed by Baseman and Hayes,[16] was used in separate but parallel studies. The importance of this technique is also

[15] R. E. Baughn, *Rev. Infect. Dis.* **9,** Suppl., S372 (1987).

[16] J. B. Baseman and E. C. Hayes, *J. Exp. Med.* **151,** 573 (1980).

addressed here because it initiated the investigation for the fibronectin-binding proteins of *T. pallidum* organisms, which were later verified. Thus, the host protein binding studies and the ligand assay both came together to aid in our understanding of the molecular nature of host parasitism by this infectious agent. Finally, the importance of these approaches has since been affirmed and realized by showing their usefulness in other models of microbial pathogens, including a protozoan parasite, *Trichomonas vaginalis*.

Identification of Fibronectin as a Host Cell Receptor

Loosely Associated Serum Proteins

In 1963 a suggestion was made that the syphilis spirochete was surrounded by a protective, host-derived surface coat.[17] The indication that serum proteins were contaminating the organisms was reinforced in observations made from stained patterns of total proteins after electrophoresis.[18] Prominent bands disappeared during the various washings of the spirochetes in phosphate-buffered saline (PBS). Even more intriguing was the reacquisition of the protein bands by incubation of the washed bacteria with rabbit serum. It is noteworthy that *T. pallidum* was isolated from extractions of infected rabbit testes, as *in vitro* cultivation was and remains a major impediment to the study of this pathogen. Nonetheless, it was fortuitous that these *in vivo*-derived organisms were being examined immediately after isolation, as a feature of the host protein interaction was the rather loose and reversible association of serum proteins with the spirochete. The loss of the loosely associated serum protein was confirmed by comparison of the patterns of bands from stained gels with fluorograms of intrinsically labeled proteins.

Agglutination Assay to Demonstrate Avid Binding of Serum Proteins

At the time, the protein A-bearing *S. aureus* was being used in radioimmunoprecipitation assays (RIPAs),[19] and these fixed bacteria were now adapted in agglutination assays to determine the avid association of serum proteins to the surface of *T. pallidum*. After extensive washing, the freshly extracted organisms were radiolabeled for a brief period before being incubated with commercially available antiserum to specific serum proteins. Binding of antibody to proteins not removed by the extensive wash-

[17] S. Christiansen, *Lancet* **1,** 423 (1963).
[18] J. F. Alderete and J. B. Baseman, *Infect. Immun.* **26,** 1048 (1979).
[19] S. W. Kessler, *J. Immunol.* **117,** 1482 (1976).

ing of the spirochetes with PBS was then monitored by an agglutination assay involving the addition of formaldehyde-fixed *S. aureus*. The staphylococcus–antibody–spirochete complex was stable, highly specific, and readily precipitable and separable from free treponemes by low-speed centrifugation, which allowed for quantitation of radioactivity of the pellet. Agglutination could be abolished only by enzymatic (trypsin) treatment of the treponemes for release of the avidly bound host proteins. As a control, it was important to show the absence of immunoreactivity of the antisera with treponemal proteins, and this was demonstrated by the RIPA.

This agglutination assay was again exploited several years later to show the general feature of avidly associated host serum proteins with *Trichomonas vaginalis,* a sexually transmitted protozoan parasite.[20] Here, too, ultrastructural work showed the appearance of a host-derived "coat" which was lost during *in vitro* cultivation. In this case, *T. vaginalis* organisms grown in a complex medium supplemented with heat-inactivated human serum, a requirement for growth and multiplication of trichomonads, were washed numerous times to remove loosely bound serum proteins and added to a suspension of *S. aureus* that was first pretreated with specific antibodies to individual host proteins. The bacteria–antibody–trichomonad agglutination was extensive, and the complex readily separated itself from *S. aureus* alone or unbound trichomonads.

Identification of Specific Plasma Proteins Bound onto the Bacterial Surface

Preparation of Plasma and Fibronectin for Binding Studies

Plasma proteins were iodinated using lactoperoxidase as described earlier.[4] Radiolabeled proteins were dialyzed for 3 days with numerous changes of PBS at 4° prior to incubation of iodinated plasma with live treponemes. Fractionation of the plasma containing about 2.2 g of protein was performed by cold ethanol precipitation procedures,[4] which fractionated plasma into eight defined Cohn fractions listed in Table I. Fibronectin (Fn) was purified on gelatin–agarose and examined by sodium dodecyl sulfate–polyacrylamide gel electrophoresis (SDS–PAGE)[21] to ensure the use of only purified Fn fractions for important acquisition experiments. In this case, Fn was obtained from normal human plasma or fraction I + III-3 (Table I).

[20] K. M. Peterson and J. F. Alderete, *Infect. Immun.* **37,** 755 (1982).
[21] D. D. Thomas, J. B. Baseman, and J. F. Alderete, *J. Exp. Med.* **161,** 514 (1985).

TABLE I
COMPARATIVE ACQUISITION OF ^{125}I-RADIOLABELED PLASMA PROTEIN PREPARATIONS BY *T. pallidum* AND *T. phagedenis* BIOTYPE REITER[a]

Sample number	Protein preparation added to treponemes[b]	Composition[c]	cpm avidly bound (% of total cpm added) *T. pallidum*	*T. phagedenis*
1	Normal human plasma	—	1,900 (0.7)	Same as control[d]
2	I + III-3	Plasminogen, fibronectin, fibrinogen	3,190,000 (7.6)	Same as control[d]
3	II	γ-Globulins	20,800 (0.4)	Same as control[d]
4	III-0	β-Lipoproteins, euglobulins, ceruloplasmin	34,600 (1.4)	Same as control[d]
5	III-1,2	Prothrombin, isoagglutinins	18,000 (0.6)	Same as control[d]
6	IV-1	α-Lipoproteins	19,500 (0.5)	Same as control[d]
7	IV-6,7	β-Metal binding protein, α_2-mucoprotein, choline esterase, α_2-glycoprotein	13,700 (0.3)	Same as control[d]
8	V	Albumin	6,700 (0.5)	Same as control[d]
9	VI	α_1-Glycoprotein, small proteins and peptides	1,411,000 (0.4)	Same as control[d]
10	Fibronectin	—	5,430,000 (1.3)	Same as control[d]

[a] Reproduced from Peterson *et al.*,[4] by copyright permission of the Rockefeller University Press.
[b] 50 μl of radioiodinated plasma, specific Cohn fractions, or purified fibronectin was added to 100 μl containing 5×10^9 organisms and incubated at 37° for 30 min as described under Materials and Methods.
[c] Specific activities for individual protein preparations (cpm/ng protein): (1) 2700, (2) 2800, (3) 900, (4) 800, (5) 520, (6) 700, (7) 750, (8) 500, (9) 5600, (10) 4500. Purified fibronectin was prepared from fraction I + III-3; Cohn fractions were obtained as described[34] from normal human plasma.
[d] Tubes without organisms but handled identically to those with organisms were used to determine the level of nonspecific binding of radioactivity. Values never exceeded 1% the level detected for *T. pallidum* acquisition. *T. phagedenis* yielded cpm values equivalent to those of control tubes.

Avid Binding of Iodinated Plasma Proteins to Freshly Purified Treponemes

The procedure demonstrating the specific association of fibronectin and other plasma proteins with the syphilis spirochete is detailed in the original publication.[4] A brief description of the basic protocol is as follows: The reactions are carried out in siliconized 1.5-ml microfuge tubes, always pretreated with 1% horse serum to reduce nonspecific binding. After the tubes are washed, a small volume (usually 100 μl) of a suspension containing 5×10^9 freshly harvested treponemes is mixed with 50-μl volumes of iodinated plasma or plasma fractions. The final volume is immediately adjusted to 300 μl with PBS. Tubes are then placed in a 37° incubator and shaken gently about every 5 min to ensure uniform distribution of the organisms. At designated times the treponemes are pelleted at 17,000 *g* and washed twice before finally resuspending in cold PBS and transferring to another microfuge tube. An aliquot of 100 μl is precipitated with trichloroacetic acid (TCA) for preparation of proteins for SDS–PAGE autoradiography.

Figure 1 (part I) illustrates the dramatic, avid binding of several proteins from iodinated plasma onto the surface of *T. pallidum* organisms (lane A). That this represented an enrichment of just a few proteins was readily demonstrated by comparing two-dimensional protein patterns of autoradiograms of the total plasma proteins with those that were bound. A hallmark of this and numerous other experiments was the absence of binding of significant amounts of any plasma protein by the nonpathogenic oral spirochete, *Treponema phagedenis* (lane B). At this point, evaluation of certain biochemical parameters like time, temperature, pH, and saturation kinetics as well as competition experiments provided initial evidence for the ligand–receptor type of interaction between the bacterial surface and the host proteins, something reaffirmed later in studies involving particular proteins and the microorganism.

Establishing the Identity of the Acquired Fn

The use of Cohn fractions[4] seemed a logical next step in attempts to identify the proteins specifically and avidly bound to the treponemal surface. These fractions have been well characterized and have been traditionally used as sources of enriched and practically purified proteins, such as albumin and immunoglobulins. Table I shows that Sample 2 gave levels of binding greater than those seen for the other fractions. Again, the control avirulent Reiter spirochete did not bind proteins of the various fractions. The next step, then, was to examine which of the three enriched proteins in this fraction (Fn, plasminogen, or fibrinogen) might represent the predominantly bound material. It was apparent that Fn purified from this fraction or directly from serum was indeed readily and avidly bound by *T. pallidum*. Proof of specific Fn acquisition was then demonstrated (Fig. 1, part II). In this case incubation of live, freshly extracted organisms with either plasma (lane C), fraction I + III-3 (lane E), or purified Fn (lane G), but not with Fn-free plasma or other fractions without Fn (lane F), readily bound Fn, as evidenced from immunoblots with anti-Fn serum.

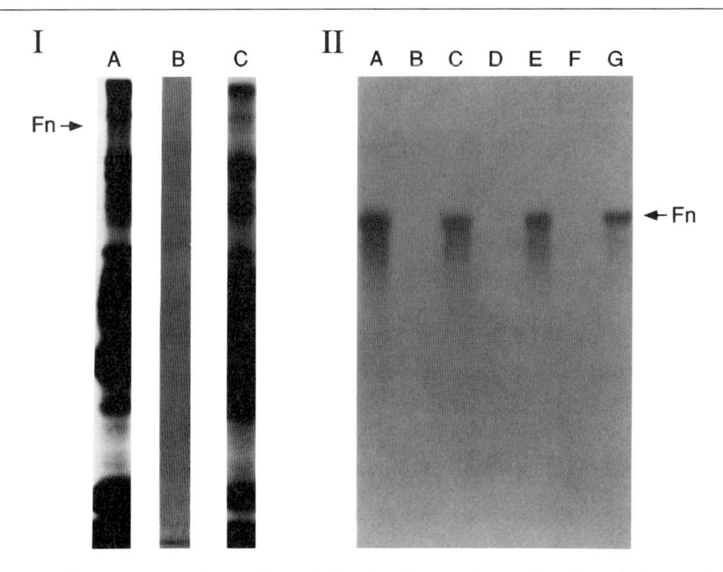

FIG. 1. Left: Representative sodium dodecyl sulfate–polyacrylamide gel electrophoresis/autoradiography of total proteins after incubation of *T. pallidum* with radioiodinated normal human plasma. Lane A represents ^{125}I-labeled plasma proteins avidly bound to *T. pallidum*. Lane B shows the lack of plasma protein acquisition by the avirulent spirochete, *T. phagedenis* biotype Reiter, handled similarly. Lane C is a profile of ^{125}I-labeled normal human plasma used in these acquisition assays. The location of fibronectin (Fn) was established by electrophoresis of ^{125}I-labeled purified human Fn under similar conditions. Right: Electrophoretic transfer and immunodetection of unlabeled fibronectin acquired by motile *T. pallidum* incubated with normal human plasma (lane C), Cohn fraction I + III-3 (lane E), and purified fibronectin (lane G). Lane A shows the immunodetection of purified fibronectin alone. Lane B demonstrates the lack of reactivity of normal goat serum and ^{125}I-labeled protein A with fibronectin alone or with blotted *T. pallidum* proteins (lane D). Lane F shows the lack of detection using antifibronectin antibody and ^{125}I-labeled protein A of *T. pallidum* incubated with plasma depleted of fibronectin. (Reproduced from Peterson *et al.*,[4] by copyright permission of the Rockefeller University Press.)

Fibronectin Involvement in the Cytoadherence Property of T. pallidum

Three reasons motivated the testing of the possibility that Fn was involved in treponemal attachment to host cells. Numerous investigators described the important role in pathogenesis of *T. pallidum* adherence to host cells,[22,23] and the presence of a specialized tip structure on these treponemes was reported as the functional organelle mediating host cell

[22] N. S. Hayes, K. E. Muse, A. M. Collier, and J. B. Baseman, *Infect. Immun.* **17,** 174 (1977).
[23] T. J. Fitzgerald, J. N. Miller, and J. A. Sykes, *Infect. Immun.* **11,** 1133 (1975).

surface parasitism.[22] The second development was evident in the flourish of information regarding the structure–function properties of Fn and other ECM proteins residing on mammalian cells.[24–29] Finally, there was already debate at this time regarding the role of Fn in cytoadherence by bacterial pathogens, including *S. aureus*. Thus, it seemed logical to test for the possibility that Fn was a receptor for treponemal attachment to host cells, which could be performed readily using cell monolayer cultures.

Fn, a dimeric glycoprotein, was immobilized onto glass coverslips to visualize the binding by highly motile syphilis spirochetes via the typical polarized fashion involving the tip structure.[22] As presented in Fig. 2B, only Fn allowed for the tip-oriented adherence as visualized by dark-field microscopy. Subsequent experiments were performed with radiolabeled treponemes to confirm the observations quantitatively. Increasing numbers of organisms gave correspondingly elevated levels of binding as determined by the associated radioactivity. Also, titration of Fn adsorbed onto the coverslips gave corresponding changes in the levels of bound organisms, and only treatment of Fn-coated coverslips with anti-Fn antibodies impaired the treponemal association with immobilized Fn.

Parallel approaches were used to show that Fn on the host cell surface was the target for recognition. Fn was found to be exposed and accessible on the epithelial cells used before to demonstrate cytoadherence.[22] The inhibition of attachment was demonstrated by pretreatment of the host cell surfaces with specific anti-Fn antibodies. Specificity was shown by the lack of inhibition with control antibodies (antialbumin, for example) or with antibodies to other ECM proteins (antilaminin or anticollagen). Dose–response curves showed levels of inhibition as a function of antibody titer. Other parameters that were tested supported the notion of specificity of the spirochete–Fn associations.

Targeting of Fibronectin Cell-Binding Domain and RGD Sequence by Syphilis Spirochete

During these early studies, a wealth of information appeared on the biochemistry of the Fn monomers, so much so that techniques were avail-

[24] M. D. Pierschbacher, E. Ruoslahti, J. Sundelin, P. Lind, and P. A. Peterson, *J. Biol. Chem.* **257,** 9593 (1982).

[25] M. D. Pierschbacher, E. G. Hayman, and E. Ruoslahti, *Cell (Cambridge, Mass.)* **26,** 259 (1981).

[26] M. D. Pierschbacher and E. Ruoslahti, *Nature (London)* **309,** 30 (1984).

[27] K. M. Yamada and K. Olden, *Nature (London)* **275,** 179 (1978).

[28] E. Pearlstein, L. I. Gold, and A. Garcia-Pardo, *Mol. Cell. Biochem.* **29,** 103 (1980).

[29] E. Ruoslahti, E. G. Hayman, E. Engvall, W. C. Cothran, and W. T. Butler, *J. Biol. Chem.* **256,** 7277 (1981).

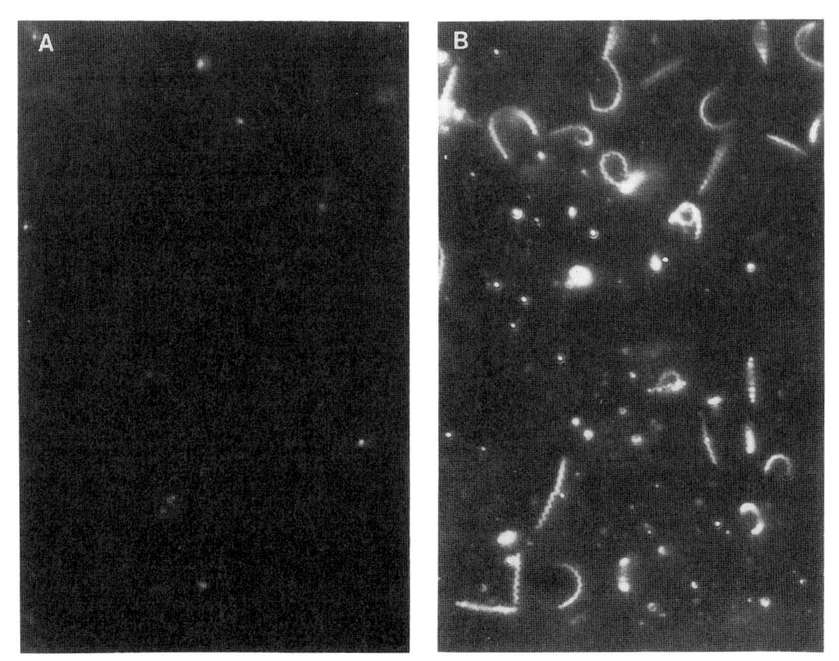

FIG. 2. A representative dark-field view of tip-mediated adherence of freshly harvested *T. pallidum* incubated with fibronectin-coated coverslips (B), compared with the lack of attachment of treponemes from the same extraction to untreated or albumin-coated coverslips (A). (Reproduced from Peterson *et al.*,[4] by copyright permission of the Rockefeller University Press.)

able for purifying the corresponding cell, heparin, and gelatin binding domains.[28,29] In addition, it is noteworthy that monoclonal antibodies (mAbs) directed to amino-terminal, to carboxy-terminal, and to the more central regions of the Fn molecule soon became available,[25] and clearly, these reagents defined the precise domain on Fn being recognized by the organism. Therefore, definition of the structural and functional domain of Fn involved in the tip-oriented treponemal parasitism of host cells was both a natural extension of the earlier work and a requirement for a more molecular understanding of cytoadherence.

The cell and heparin binding domains of purified human Fn were generated through proteolytic digestion of Fn according to the procedure established by Hayashi and Yamada.[30] By use of established protocols, the

[30] M. Hayashi and K. M. Yamada, *J. Biol. Chem.* **258,** 3332 (1983).

TABLE II
INHIBITION OF SYPHILIS SPIROCHETE CYTOADHERENCE[a] BY Fn CELL-BINDING DOMAIN SYNTHETIC PEPTIDES[b]

Expt.	Treatment reagent	μg/ml	Radiolabel recovered[c]	
			HT1080	HEp-2
1	DMEM	—	24,718 ± 2,116 (100)	26,455 ± 3,218 (100)
	GRGDSPC	50	13,077 ± 1,919 (53)	13,761 ± 2,119 (52)
	GRADSPC	50	22,114 ± 1,602 (90)	24,884 ± 2,273 (94)
	GKGDSPC	50	23,325 ± 1,881 (94)	27,043 ± 2,881 (102)
2	DMEM	—	20,565 ± 861 (100)	19,561 ± 580 (100)
	GRGDSPC	25	14,017 ± 1,050 (68)	12,271 ± 261 (63)
	GRGDSPC	50	8,019 ± 164 (39)	8,265 ± 111 (42)
	GRGDSPC	250	7,749 ± 128 (38)	7,712 ± 138 (39)
	GRGDSPC	500	7,318 ± 96 (36)	7,392 ± 165 (38)
	GRGDSPC	750	7,386 ± 186 (36)	7,315 ± 161 (37)

[a] Cytoadherence assays were performed as previously described.[3] Radiolabeled treponemes (7×10^7 cells/reaction volume) were incubated with the indicated synthetic peptide or medium for 30 min at 34° before addition to cultured cell monolayers.

[b] Reproduced from Thomas *et al.*,[33] by copyright permission of the Rockefeller University Press.

[c] Each value represents the mean cpm ± SD of three separate determinations. Numbers in parentheses give data as percentages of control.

gelatin binding domain resulted from thermolysin digestion of Fn.[31,32] The use of these domains in treponemal binding studies has been detailed.[21]

Freshly harvested organisms recognized and bound to the cell binding domain of Fn, results that were further reinforced by the efficient inhibition of attachment to Fn-coated coverslips by specific mAbs. This stage in the studies of Fn targeting was concluded by competition experiments with heptapeptides, reagents that were becoming available following the discovery of the RGD sequence involved in the receptor-mediated binding of Fn by mammalian cells.[26] Only the GRGDSPC peptide, but not three other peptides altered within the RGD sequence, gave concentration-dependent inhibition of cell binding domain acquisition by purified spirochetes. Even more exciting was the quantitative demonstration of a concentration-dependent inhibition of host cytoadherence by *T. pallidum*. Table II shows the inhibitory effect on cell attachment of pretreatment of ^{35}S-labeled organisms with the RGD heptapeptide. Only GRGDSPC, but not the

[31] B. A. Bernard, S. K. Akiyama, S. A. Newton, K. M. Yamada, and K. Olden, *J. Biol. Chem.* **259,** 9899 (1984).

[32] K. Sekiguchi and S. Hakomori, *J. Biol. Chem.* **258,** 3967 (1983).

control peptides, diminished parasitism of two epithelial cell types that have been used by investigators in treponemal binding studies.[33]

This example illustrated the extent of cross-fertilization among scientific disciplines. The cell biology and role of Fn in cell growth and differentiation, the biochemistry of the structure–function properties of Fn regions and defined peptides, the technology for generating mAbs to the various Fn domains, and the *in vitro* manipulation of the syphilis spirochete all came together simultaneously.

Identification of the Treponemal Fn-Binding (Adhesin) Proteins

Ligand Assay

HEp-2 epithelial cells, used as the *in vitro* cell culture model system to study the host cytoadherence property of *T. pallidum*, are resuspended in PBS and washed well before addition of formaldehyde. This is done using 3×10^6 cells/ml placed in a 25-ml siliconized glass flask to which is added 1% (final concentration) formaldehyde. Fixation occurs at 22° for 1 hr, followed by extensive washing in PBS and resuspension in a 150 m*M* NaCl–5 m*M* ethylenediaminetetraacetic acid (EDTA)–50 m*M* Tris (NET) buffer containing phenylmethylsulfonyl fluoride (PMSF), 0.1% SDS, and 1% Triton X-100 as detailed elsewhere.[16] These conditions are optimal for showing the cellular integrity of the HEp-2 cells for the duration of the experiment.

To these fixed host cells is added a detergent extract of ^{35}S-labeled *T. pallidum* organisms. These bacteria are solubilized by taking a suspension containing 5×10^8 treponemes that is already at room temperature (RT) and adding 100 μl of 1% SDS in NET buffer. After 15 min at RT, ovalbumin (10 mg/ml in NET buffer) is added (to bind excess SDS) followed by addition of 100 μl of 10% Triton X-100. This mixture is heated to 37° for 15 min and insoluble material removed by centrifugation. The supernatant is then diluted to a 1-ml final volume, and one-half of this extract is used for incubating with 3×10^5 formaldehyde-fixed HEp-2 cells under conditions described in the first adaptation of this procedure.[16] After incubation at 34° for 1 hr and mixing at 15-min intervals, HEp-2 cells are extensively washed four times in NET–0.05% Triton X-100. The addition of Triton X-100 facilitates the removal of nonspecifically bound treponemal extract components from the fixed cell surfaces. Finally, 1% SDS is added and the pelleted cells are resuspended and stirred vigorously for removal of the avidly bound, radiolabeled treponemal proteins. The supernatant

[33] D. D. Thomas, J. B. Baseman, and J. F. Alderete, *J. Exp. Med.* **162,** 1715 (1985).

is electrophoresed and specific *T. pallidum* proteins are visualized by fluorography.

The employment of this procedure for identifying possible epithelial cell-binding proteins of the syphilis spirochete preceded the studies involving host protein acquisition as described earlier. This ligand assay identified three proteins as the putative adhesins. Two concerns were paramount during this assay. First, the fixation of host cells might destroy the surface-exposed binding sites (receptors) and thereby not allow for accommodation of the bacterial counterparts, the adhesin proteins, if indeed they existed. Second and equally important, the solubilization of the parasite, especially using harsh conditions, might destroy the receptor recognition function of the adhesins.

In the event that this ligand assay might have followed the demonstration of Fn as the receptor, then experiments with anti-Fn antibodies would likely have been performed to show immunoreactivity and, therefore, epitope integrity of Fn on the fixed host cells. In addition, inhibition experiments with anti-Fn antibody labeling of the fixed host cells would have been performed to show possibly diminution of adhesin protein recognition and binding.

Fibronectin-Affinity Purification of Treponemal Proteins

Knowledge of Fn as the likely receptor mediating host parasitism by the live organisms and of putative adhesins paved the way for examining the interaction of treponemal proteins by Fn-affinity chromatography.[4]

About 2×10^{10} ^{35}S-labeled spirochetes, washed well to remove contaminating host material from the bacterial surface, are suspended in 1% Zwittergent 3-12 (Calbiochem-Behring Corp., La Jolla, CA) detergent.[4] The labeled organisms are gently homogenized, and insoluble material is removed by centrifugation at 100,000 *g* for 30 min. The soluble treponemal extract is then diluted further with PBS until the concentration of Zwittergent 3-12 is 0.05%, at which time the extract is passed over a 2×10-cm Fn-affinity Sepharose column at the rate of 1 ml/cm^2/hr. Extensive washing of the column is carried out sequentially with PBS, 2 *M* potassium bromide, and 10% SDS. After the initial washing with PBS, the other treatments of the column did not release bound treponemal proteins, illustrating the difficulty in recovering Fn-associated molecules from an affinity column. This material was finally displaced by boiling the beads for 3 min in electrophoresis dissolving buffer. The radioactively labeled proteins released from the column were analyzed and fluorograms revealed three bands, which corresponded to proteins absent or diminished from the total proteins of extract passed over the column.[4] Furthermore, the three eluted proteins had electrophoretic mobilities and characteristics similar to those detected years earlier by the ligand assay.[16]

Discussion on the Usefulness of the Strategies to Yield Other Important Information and to Work on Other Microbial Pathogens

Nutrient Acquisition by Syphilis Spirochetes and Other Microorganisms

The host protein binding strategy contributed tremendously to knowledge about other aspects of the biology of the host–parasite interrelationship of the syphilis spirochete. The acquisition of host iron-binding proteins (lactoferrin and transferrin)[33] and of lipoproteins[34] showed the nutritional dependency of this organism on its host. This information is important as it may help clarify previously unknown issues regarding survival of the organisms within various host sites and tissues, as well as aid in the delineation of metabolic deficiencies that make the organisms dependent on the host for survival. This information may also explain the inability to cultivate the organism *in vitro*, an experimental problem that must be overcome.

Because these experiments demonstrated the value of the information gained from specific host protein acquisition studies, an attempt was made to examine the usefulness of the strategy on a sexually transmitted *Trichomonas vaginalis* parasite.[20] In the case of this protozoan, some of the avidly associated serum proteins included fibronectin, lipoproteins, iron-binding proteins, α_2-macroglobulin, α_1-antitrypsin, and immunoglobulins. The consequences of binding these and possibly other host proteins onto the parasite surface are numerous and are being dissected. For example, these organisms do not have the biosynthetic capacity for lipid or sterol biosynthesis, so early on the receptor-mediated binding of apoprotein CIII was found to be a mechanism by which the trichomonads obtained lipids, fatty acids, and cholesterol from binding to lipoproteins containing apoprotein CIII.[35] *T. vaginalis* has a very high iron requirement for optimal metabolism and energy generation, and it is now appreciated that a repertoire of parasite receptors for iron-binding and iron-containing proteins of the host satisfy the parasite's demands for iron.[35–37] The trichomonads have also been shown to produce up to 25 distinct cysteine proteinases,[38]

[34] J. F. Alderete and J. B. Baseman, *Genitourin. Med.* **65,** 177 (1989).

[35] K. M. Peterson and J. F. Alderete, *J. Exp. Med.* **160,** 1261 (1984).

[36] K. M. Peterson and J. F. Alderete, *J. Exp. Med.* **160,** 398 (1984).

[37] M. W. Lehker, T. H. Chang, D. C. Dailey, and J. F. Alderete, *J. Exp. Med.* **171,** 2165 (1990).

[38] K. A. Neale and J. F. Alderete, *Infect. Immun.* **58,** 157 (1990).

so that the coating of the surface with host-derived proteinase inhibitors (α_2-macroglobulin and α_1-antitrypsin) may confer resistance against the parasite's own degradative enzymes. In fact, the inhibitory capacity of α_1-antitrypsin, while on the surface of the parasite, was demonstrated.[39]

Ligand Assay for Detection of Microbial Adhesins

The ligand assay, although seemingly straightforward, can be used on other microbial models, but only after empirically testing and optimizing conditions. Problems, as mentioned below and in the next section, can preclude obtaining information about adhesin candidates.

Incubation of glutaraldehyde-fixed HeLa or vaginal epithelial cells with a deoxycholate extract of radiolabeled *T. vaginalis* has shown the existence of four distinct adhesin proteins.[40,41] It is noteworthy that Zwittergent 3-12, used for making treponemal extracts and also used for RIPA of *T. vaginalis,*[42] failed to solubilize the epithelial cell-binding proteins of this protozoan. This selective solubilization highlights the need to test adequately multiple detergent systems for optimal results from the ligand assay. That these proteins are adhesins has been proven through fulfillment of criteria summarized by Beachey.[43]

Problems Demonstrating Extracellular Matrix Proteins as Host Receptors and Existence of Adhesins by the Ligand Assay

Although the host protein binding strategy and the ligand assay proved worthwhile for the examples involving *T. pallidum* and the protozoan *T. vaginalis,* this last section highlights a few examples of problems that might be encountered.

One of the constant hazards in experiments is the *in vitro* cultivation of microbial pathogens. The growth medium likely will not be reflective of the microenvironments of the host during infection. It is reasonable to presume that these environments result in adaptive responses of the invading organisms,[44] which allow for expression of certain virulence factors, possibly adhesins. A recent report showed that laminin and collagen were

[39] K. M. Peterson and J. F. Alderete, *Infect. Immun.* **40,** 640 (1983).
[40] J. F. Alderete and G. E. Garza, *Infect. Immun.* **56,** 28 (1988).
[41] R. Arroyo, J. Engbring, and J. F. Alderete, *Mol. Microbiol.* **6,** 853 (1992).
[42] J. F. Alderete, *Infect. Immun.* **39,** 1041 (1983).
[43] E. H. Beachey, "Molecular Mechanisms of Microbial Adhesion," p. 1. Springer-Verlag, New York, 1989.
[44] C. A. Lee and S. Falkow, *Proc. Natl. Acad. Sci. U.S.A.* **87,** 4304 (1990).

recognized by *Streptococcus gordonii*.[45] This work was significant in that it demonstrated that induction of the laminin-binding protein of *S. gordonii* occurred under positive control. The upregulation of laminin-binding proteins required exposure of the bacteria to laminin.

Equally noteworthy was that the levels of cytoadherence and amounts of adhesins of *T. vaginalis* were regulated by the concentration of iron in the growth medium.[46] Thus, batch cultures of organisms may not allow for expression of ECM-binding receptors, and in fact, it was exceedingly difficult to detect the trichomonad adhesins by the ligand assay in laboratory isolates grown *in vitro* for extended periods, conditions that favor loss of expression of the adhesins. Thus, recent work indicating the ECM proteins may represent host cell receptors which bind the *T. vaginalis* adhesins requires a priori knowledge of the environmental cues which favor expression of cytoadherence of ECM protein targeting. In both cases, the lack of basic knowledge regarding expression of the adhesins, which bind ECM proteins, would preclude the ability to identify them by the ligand assay.

Last, detergent extracts must include inhibitors of proteinases, as pathogenic human trichomonads and undoubtedly numerous other pathogens possess proteinases which may degrade ECM proteins[47] or, alternatively, within minutes after solubilization, degrade adhesins detectable by the ligand assay.[41] This means that basic research into the existence and type of proteinases produced by microbial pathogens is necessary for the analyses presented here. In both scenarios the presence of active microbial proteinases during the assays may result in inaccurate observations or results.

[45] P. Sommer, C. Gleyzal, S. Guerret, J. Etienne, and J. Grimaud, *Infect. Immun.* **60,** 360 (1992).

[46] M. W. Lehker, R. Arroyo, and J. F. Alderete, *J. Exp. Med.* **174,** 311 (1991).

[47] M. Wikström and A. Linde, *Infect. Immun.* **51,** 707 (1986).

[23] Interactions of Bacteria with Leukocyte Integrins

By EVA ROZDZINSKI and ELAINE TUOMANEN

Introduction

Integrins are a family of cell surface glycoproteins which are known to mediate cell–cell and cell–extracellular matrix adhesion in eukaryotic systems. They consist of an α chain and a β chain which both take part in forming a ligand recognition site. At least 11 different α and 6 β subunits are known which pair with each other in various combinations.[1,2] Leukocytes express several β_2-, β_1-, and α_v-integrins as well as an $\alpha_4\beta_p$-integrin.

Bacteria have been shown to interact with only one family of leukocyte integrins, the β_2-integrins, which includes CR3 ($\alpha_M\beta_2$, Mac-1, CD11b/CD18), LFA-1 ($\alpha_L\beta_2$, CD11a/CD18), and p150/95 ($\alpha_X\beta_2$, CR4, CD11c/CD18). This family of integrins is restricted to leukocytes and participates in phagocytosis of complement-coated particles and in leukocyte transmigration from blood into tissues. The β_2-integrins have been highly targeted by bacteria probably because they mediate uptake into the phagocyte without an oxidative burst.[3]

CR3 is expressed on neutrophils, monocytes, and macrophages and binds characteristically to proteins containing the Arg–Gly–Asp (RGD) triplet, e.g., complement 3bi (C3bi) and presumably a receptor on endothelia.[4] In addition, CR3 has distinct non-RGD binding sites for endotoxin[5,6] and factor X of the coagulation cascade.[7,8] LFA-1 is found on neutrophils, monocytes, macrophages, and B and T lymphocytes and binds the surface proteins ICAM-1 and ICAM-2 on endothelia and tumor cells. This interaction does not involve recognition of RGD. p150/95 is present on neutrophils, monocytes, and macrophages but its natural ligand is not well characterized. An interaction with fibrinogen has been described.[9]

Recent work suggests that bacteria co-opt the integrin network to adhere to and enter into human cells.[10] Three strategies are used by bacteria to infiltrate this system and thereby promote intracellular survival

[1] E. Ruoslahti, *J. Clin. Invest.* **87,** 1 (1991).

[2] T. A. Springer, *Nature (London)* **346,** 425 (1990).

[3] S. D. Wright and S. C. Silverstein, *J. Exp. Med.* **158,** 2016 (1983).

[4] S. D. Wright, P. A. Reddy, M. T. C. Jong, and B. W. Erickson, *Proc. Natl. Acad. Sci. U.S.A.* **84,** 1965 (1987).

[5] S. D. Wright and M. T. C. Jong, *J. Exp. Med.* **164,** 1876 (1986).

[6] S. D. Wright, S. M. Levin, M. T. C. Jong, Z. Chad, and L. G. Kabbash, *J. Exp. Med.* **169,** 175 (1989).

[7] D. C. Altieri and T. S. Edgington, *J. Biol. Chem.* **263,** 7007 (1988).

[8] D. C. Altieri, O. R. Etingin, D. S. Fair, T. K. Brunck, J. E. Geltosky, D. P. Hajjar, and T. S. Edgington, *Science* **254,** 1200 (1991).

[9] J. D. Loike, B. Sodeik, L. Cao, S. Leucona, J. I. Weitz, P. A. Detmers, S. D. Wright, and S. C. Silverstein, *Proc. Natl. Acad. Sci. U.S.A.* **88,** 1044 (1991).

[10] A. I. M. Hoepelman and E. I. Tuomanen, *Infect. Immun.* **60,** 1729 (1992).

TABLE I
BACTERIA THAT RECOGNIZE CD18 INTEGRINS ON LEUKOCYTES

Strategy	Example	Refs.
Mimickry of RGD	Filamentous hemagglutinin of *Bordetella pertussis*	Relman *et al.*[11]
Ancillary binding site recognition (non-RGD)	Type I fimbriae of *Escherichia coli*	Sharon *et al.*[11a]
	Endotoxin	
Masking with C3bi	*Legionella pneumophila*	Payne and Horwitz[11b]
	Mycobacterium tuberculosis	Schlesinger *et al.*[12]

(Table I)[11,11a,b,12]: (1) mimickry—the bacterium expresses a protein containing an RGD triplet or a sequence so like RGD as to bind to the RGD recognition site on the integrin; (2) ancillary binding site recognition—the bacterium binds to a non-RGD recognition site, for instance, to a carbohydrate on the glycosylated integrin; (3) masking—the bacterium adsorbs a natural ligand for the integrin onto its surface leading to entry into the host along natural pathways. Once the bacterium is attached to the host cell, factors like the activation of the receptor, the ligand used by the bacterium, the signal transduction pathway from the integrin to the cytoskeleton of the host cell, and the presence of cytokines determine the fate of the pathogen.

We describe a sequence of experiments designed to distinguish the following: (1) Does a bacterium bind to leukocyte β_2-integrins? (2) Which of the three mechanisms of binding is used: mimickry, ancillary binding site recognition, or masking? (3) What is the fate of the bound bacterium?

Our laboratory has used this experimental paradigm to investigate the interaction of *Bordetella pertussis* with macrophages and we use these studies as examples. These assays build on methods published previously by Wright in this series.[13] The reader is referred to these studies for additional details.

[11] D. Relman, E. Tuomanen, S. Falkow, D. T. Golenbock, K. Saukkonen, and S. D. Wright, *Cell (Cambridge, Mass.)* **61,** 1375 (1990).

[11a] N. Sharon, A. Gbarah, C. G. Gahmberg, I. Ofek, and U. Jacobi, *Infect. Immun.* **59,** 4524 (1991).

[11b] N. R. Payne and M. A. Horwitz, *J. Exp. Med.* **166,** 1377 (1987).

[12] L. S. Schlesinger, C. G. Bellinger-Kawahara, N. R. Payne, and M. Horwitz, *J. Immunol.* **144,** 2771 (1990).

[13] S. D. Wright, this series, Vol. 132, p. 204.

Methods

Interaction of Bacterium with β_2-Integrins on Leukocytes

The interaction of a bacterium with a leukocyte β_2-integrin is suspected when bacterial adherence is decreased in the presence of anti-CD18 antibody in what is known as a downmodulation experiment. This technique, illustrated in Fig. 1, takes advantage of the ability of integrins to move in the plane of the membrane. As a consequence of this mobility, CD18 receptors can be sequestered beneath leukocytes when these cells spread on surfaces coated with an anti-CD18 antibody, for instance, monoclonal antibody (MAb) IB4.[14] As the CD18 integrins are trapped under the leukocyte, the apical surface becomes relatively depleted of receptors.[15,16] If the subsequent binding of bacteria to the apical surface is markedly decreased, one can conclude that the adherence involves β_2-integrins. If the adherence of bacteria is only partially blocked, bacterial interactions with non-β_2 receptors (e.g., carbohydrates or non-β_2 proteins) may account for residual binding.

Procedure

Preparation of Macrophages, C3bi Erythrocytes, and Fluorescein-Labeled Bacteria

Isolation and Culture of Mononuclear Phagocytes. Macrophages express integrins on their surfaces constitutively, whereas polymorphonuclear leukocytes require activation to do so. Thus macrophages are the preferred cell in the following assays. Techniques for isolating both cell types are described in detail in this series.[13] Human monocytes are purified using centrifugation on Percoll gradients. The cells are then cultured in suspension using Teflon beakers.[16] Over 4 to 10 days, the monocytes mature into macrophages. If neutrophils are used, they can be purified by neutrophil isolation medium (NIM) as described by the manufacturer (Cardinal Associates Inc., Santa Fe, NM). Once harvested, both macrophages and neutrophils should be kept on ice until seeded onto Terasaki plates (Robbins Scientific, Sunnyvale, CA) for the adherence assay.

[14] S. D. Wright, P. E. Rao, W. C. van Voorhis, L. S. Craigmyle, K. Iida, M. A. Talle, E. F. Westberg, G. Goldstein, and S. C. Silverstein, *Proc. Natl. Acad. Sci. U.S.A.* **80,** 5699 (1983).

[15] J. Michl, M. Pieczonka, J. C. Unkeless, G. I. Bell, and S. C. Silverstein, *J. Exp. Med.* **157,** 2121 (1983).

[16] S. D. Wright and S. C. Silverstein, *J. Exp. Med.* **156,** 1149 (1982).

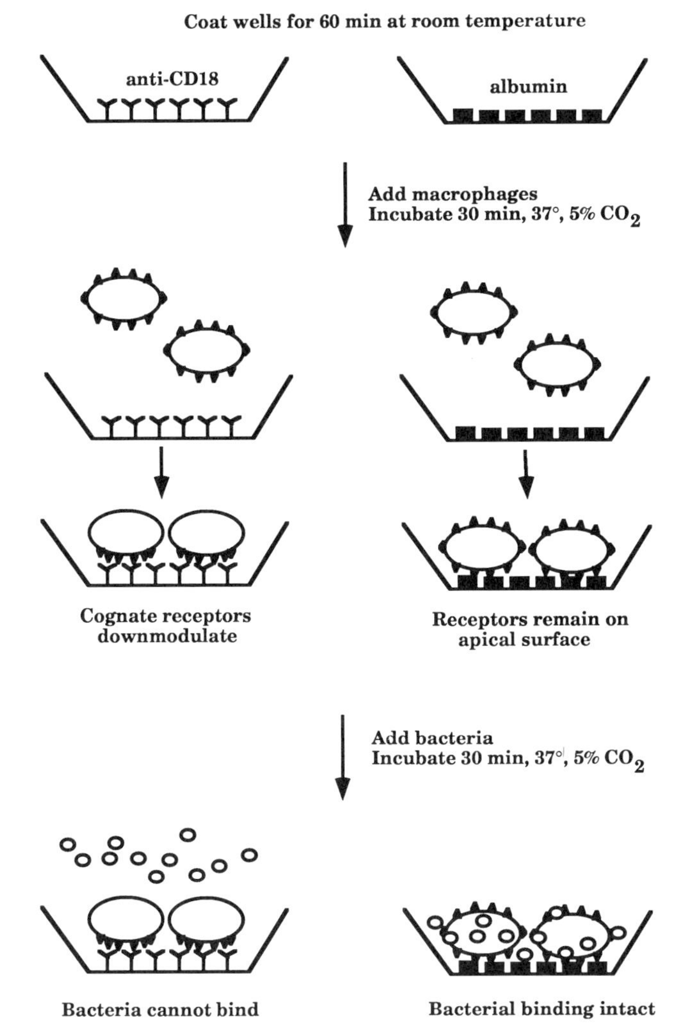

FIG. 1. Flowsheet: Downmodulation assay to detect interaction of leukocyte integrins with bacteria.

Preparation of C3bi-Coated Erythrocytes. This is a simplification of the method as per S.D.[13]

1. Wash erythrocytes: Wash sheep erythrocytes (E) in Alsever's solution (Kroy Medical Inc., Stillwater, MN) three times in equal volumes of cold (4°) $DGVB^{2+}$ buffer (2.5 m*M* Veronal buffer, pH 7.5, 75 m*M* NaCl,

2.5% dextrose, 0.05% gelatin, 0.15 mM $CaCl_2$, 0.5 mM $MgCl_2$) and suspend to a final concentration of 10^8 E/ml.

2. Coat E with IgM: E are coated with rabbit IgM antibodies to sheep erythrocytes (Diamedix Corp., Miami, FL): Incubate 1 ml E(10^9/ml) with 20 μl anti-E IgM in the presence of 1 ml ethylene diaminetetraacetic acid (EDTA)/GVB^{2-} (5 mM Veronal buffer, pH 7.5, 140 mM NaCl, 10 mM EDTA, 0.1% gelatin) at 37° for 30 min, then on ice for another 30 min. Wash the sheep erythrocytes coated with IgM antibodies (E^{IgM}) three times in $DGVB^{2+}$ and resuspend to 10^9/ml.

3. Load E^{IgM} with C3bi: Combine 500 μl of E^{IgM} (10^9/ml), 250 μl $DGVB^{2+}$, and 25 μl human C5-deficient serum (Sigma Chemical Co., St. Louis, MO). Incubate the mixture at 37° for 60 min, with gentle agitation after the first 30 min. This procedure activates the classical complement pathway. Add 500 μl EDTA/GVB^{2-} to stop the reaction, and incubate the cell suspension on ice for 10 min. Finally, wash the C3bi-coated sheep erythrocytes (E^{C3bi}) four times in cold $DGVB^{2+}$ and resuspend in 5 ml $DGVB^{2+}$ to yield a final concentration of 10^8/ml. Erythrocytes are stable for 1–2 weeks when stored on wet ice at 4°. For prevention of bacterial growth 100 U penicillin and 100 μg streptomycin can be added per milliliter cell suspension.

Fluorescent Labeling of Bacteria. Bacteria are labeled with fluorescein isothiocyanate (FITC, Sigma Chemical Co., St. Louis, MO) by a modification of the method of Hed.[17] Bacteria (lifted from agar with an inoculating loop or pelleted from liquid media) are suspended in 1 ml of Dulbecco's phosphate-buffered saline with Ca^{2+} and Mg^{2+} (DPBS 1×, Whittaker Bioproducts, Walkersville, MD) and centrifuged at room temperature. The pellet is resuspended in 0.5 ml FITC solution (1 mg FITC per 1 ml 50 mM sodium carbonate, 100 mM NaCl; allow settling of particulate FITC, use supernatant only). The mixture is wrapped in aluminum foil to prevent bleaching by light and incubated at room temperature for 20 min. Bacteria are then washed three times with DPBS containing 3 mM glucose, 0.5 mg/ml human serum albumin (HSA, New York Blood Center Inc., New York), and 0.2 unit/ml aprotinin (Sigma Chemical Co., St. Louis, MO). The bacterial concentration is adjusted to 2×10^8 colony-forming units (cfu)/ml (for *E. coli* or *B. pertussis* this density corresponds to an OD at 620 nm of 0.1). Bacteria can be stored at 4° in the dark for up to 2 weeks if the density of bacteria is rechecked before use.

[17] J. Hed, *FEMS Microbiol. Lett.* **1,** 357 (1977).

Similar techniques using alternative labels such as auramine–rhodamine and acridine orange are described elsewhere.[12,18]

Downmodulation Assay to Detect Integrin–Bacteria Interactions

Coating of Terasaki Plates with Antibodies. Terasaki tissue culture plates (flat bottom, Miles Laboratories, Inc., Naperville, IL) are coated by adding 5 μl of monoclonal antibody IB4 (10–50 μg/ml, Merck Sharpe & Dohme Inc., Rahway, NJ).[14] Two controls are appropriate: (1) antibody against a nonintegrin leukocyte receptor (50 μg/ml), such as W6/32 against HLA,[19] and (2) human serum albumin (1 mg/ml DPBS). After incubation for at least 60 min at room temperature, the plates are washed with DPBS by flooding and aspirating the overlying fluid.[14] The distribution of antibodies and albumin in a Terasaki plate is shown in Fig. 2.

Bacterial Adherence Assay

1. Plate macrophages: Because adhesion mediated by β_2-integrins requires divalent cations and warm temperatures,[16,20] experiments are performed in media supplemented with Ca^{2+} and Mg^{2+}, and cells are incubated at 37° in 5% CO_2.

To each well coated with albumin, IB4 or W6/32, add 5 μl of macrophages (0.5×10^6 cells/ml) suspended in DPBS, containing 3 mM glucose, 0.5 mg/ml HSA, and 0.2 unit/ml aprotinin. Cover the plate and allow the macrophages to spread at 37° in 5% CO_2 for 45 min. Wash the plates three times by flooding the plate with DPBS. To prevent damage to the cells, carefully aspirate the residual fluid in each well after the last wash by placing the tip of a Pasteur pipette connected to vacuum onto the edge of the wells. Proceed rapidly to the next step to prevent drying of cells.

2. Add bacteria or E^{C3bi}: Add 5 μl fluorescein-labeled bacteria (2×10^8/ml) or E^{C3bi} to appropriate wells (see Fig. 2), cover the plate, and incubate for 30 min at 37° in 5% CO_2. Wash off nonadherent bacteria by flooding the plate three times with DPBS. Fix cells by flooding the plate with 2.5% (v/v) glutaraldehyde in DPBS for 3 min. Plates may be stored in fixative at 4° in the dark until ready to be counted.

3. Quantitate adherence (see Fig. 2): Before counting, discard glutaraldehyde and flood the plate with DPBS. Determine the attachment of

[18] P. Rainard, *J. Immunol. Methods* **94,** 113 (1986).

[19] C. J. Barnstable, W. F. Bodmer, G. Brown, G. Galfre, C. Milstein, A. F. Williams, and A. Ziegler, *Cell (Cambridge, Mass.)* **14,** 9 (1978).

[20] S. Shaw, G. F. Ginther Luce, R. Quinones, R. E. Gress, T. A. Springer, and M. E. Sanders, *Nature (London)* **323,** 262 (1986).

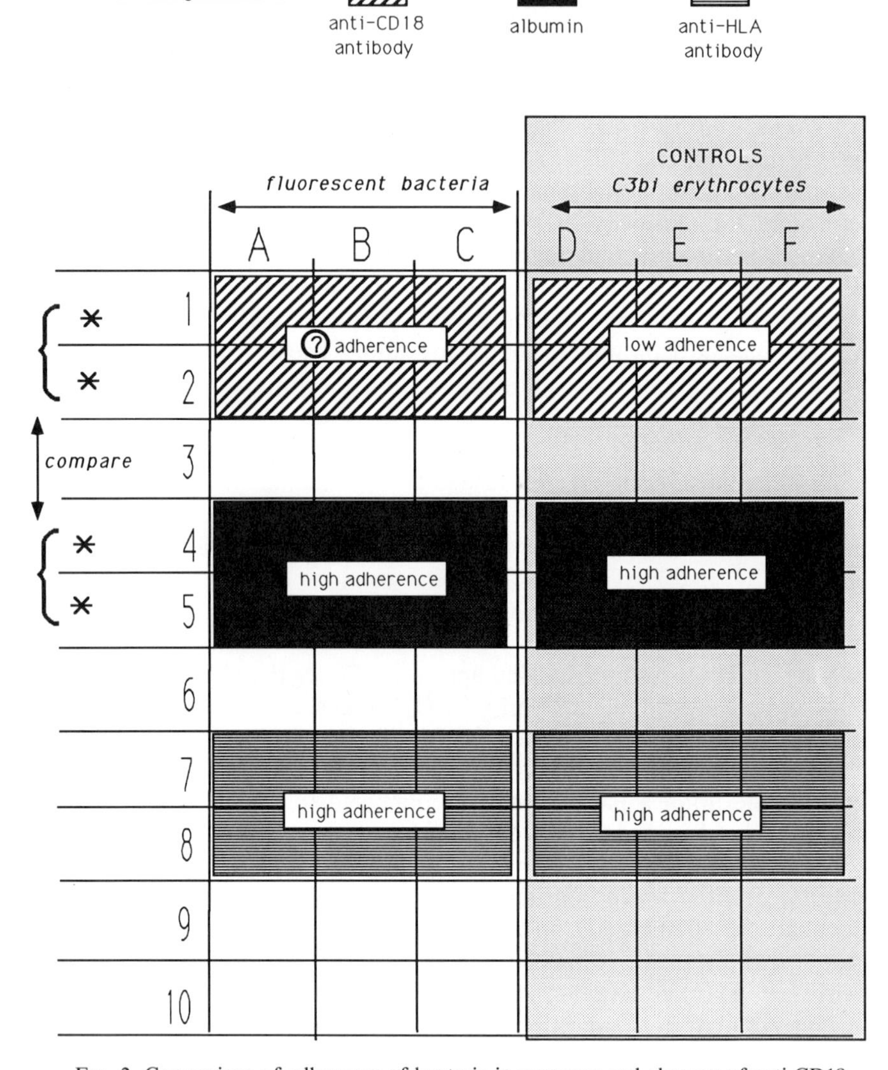

FIG. 2. Comparison of adherence of bacteria in presence and absence of anti-CD18.

bacteria, defined as attachment index, by calculating the mean number of bacteria on 100 macrophages viewed under an inverted fluorescence microscope (Nikon). At least 200 macrophages should be evaluated in duplicate wells. Note that adherence to individual macrophages within a field will vary from zero to numerous. Macrophages with no bound bacteria are counted as well as those with bound bacteria.

The key comparison is the attachment index for macrophages plated on albumin versus those plated on antibody IB4. If β_2-integrins are active in the bacterial adherence process the attachment index in anti-CD18-coated wells is significantly lower than in albumin- and anti-HLA-coated wells.

4. Use negative control: Coat Terasaki wells with antibodies against antigens that do not affect binding to integrins (e.g., W6/32 against HLA[19]).

5. Use positive control to show that β_2-integrins were actually down-modulated by the IB4 antibody: Add 5 μl of E^{C3bi} (0.5×10^8/ml $DGVB^{2+}$), which are specific ligands for CD18 receptors, instead of bacteria as above. E^{C3bi} should adhere to macrophages plated on albumin or W6/32 but not to macrophages plated on IB4. As IgM-coated erythrocytes are not recognized by phagocytes, they can be used as a negative control in the rosetting assay. The attachment index is calculated by counting the number of adherent erythrocytes per 100 macrophages.

Identification of β_2-Integrin That Mediates Adherence of Bacterium on Leukocyte

To examine which β_2-integrin is involved in the bacterial adherence, a variation of the bacterial adherence assay is performed using monoclonal antibodies specific to one of the three β_2-integrins. The Terasaki plates are coated with 30–50 μg/ml monoclonal antibodies, such as TS1/22 against α_L = CD11a of LFA-1 (Dr. T. A. Springer, Dana-Farber Cancer Institute, Boston, MA[21]), OKM1 or OKM10 against α_M = CD11b of CR3 (Dr. G. Goldstein, Ortho Pharmaceuticals, Raritan, NJ),[14] or Leu-M5 against α_x = CD11c of p150/95 (Dr. Louis Lanier, Becton, Dickinson & Co., Mountain View, CA).[22] The assay is performed as described above. The integrin targeted by the bacteria is identified by the antibody that blocks the adherence of the microorganism. The bacteria may bind to one, two, or three members of the β_2 family. If partial inhibition is observed, combinations of antibodies may summate to the level of inhibition seen with IB4. Alternatively, if any one of the three integrins is sufficient for binding, then each anti-CD11 will not inhibit individually, but a combination of all CD11 antibodies will inhibit as effectively as anti-CD18.

[21] F. Sanchez-Madrid, A. M. Krensky, C. F. Ware, E. Robbins, J. L. Strominger, S. J. Burakoff, and T. A. Springer, *Proc. Natl. Acad. Sci. U.S.A.* **79,** 7489 (1982).

[22] L. L. Lanier, M. A. Arnaout, R. Schwarting, N. L. Warner, and G. D. Ross, *Eur. J. Immunol.* **15,** 713 (1985).

Bacterial Surface Protein Operating as Ligand for Integrin

Three criteria can be used to suggest that a candidate bacterial protein serves as the adhesive ligand for an integrin.

1. Bacterial mutants that do not express the putative adhesin should have greatly decreased ability to adhere to macrophages spread on HSA in comparison to the wild type. Two additional tests can confirm this result.
2. Soluble purified putative adhesin should compete with bacteria for binding to macrophages. Bacterial adherence to macrophages plated on HSA is compared in the presence and absence of high concentrations of purified protein or peptide. For this, 4 μl of bacteria is added together with 4 μl of either the purified protein (1–5 mg/ml DPBS) or buffer alone to macrophages spread on HSA-coated Terasaki plates. The purified protein competes with the bacterial protein for the integrin and therefore reduces the bacterial binding to the macrophage in comparison to buffer alone.[11]
3. Coating a surface with the putative adhesin should downmodulate the macrophage integrin. Another method to prove the relevance of a bacterial protein to the adherence process uses Terasaki plates coated with the purified protein (50 μg/ml). If the integrin on the surface of the macrophage binds directly to the protein, it will downmodulate to the bottom of the cell, depleting the apical surface of receptor. Subsequent addition of bacteria will result in a low adherence index.

These three criteria suggest but do not prove a direct interaction between a bacterial protein and a eukaryotic integrin. In particular, it has been noted that RGD-containing peptides can nonspecifically inhibit interactions that do not involve RGD or RGD-like sequences. If the above-described tests suggest involvement of a particular domain of the adhesin, for instance an RGD region, then experimental avenues outlined in the next section can be used to show definitively direct binding. Ultimately, purified integrins and purified adhesin should be shown to interact *in vitro* in a cell free system.

Mechanism of Binding of Bacteria to Integrins

The previous assays define the ligands involved on the side of the bacteria and on the side of the eukaryotic cell. To determine the mechanisms of the interaction, three possibilities must be distinguished[10]: (1) direct binding of the bacterial adhesin to the RGD recognition site on the CD18 integrin (mimickry); or (2) direct binding of the bacterial adhesin to non-RGD recognition sites, e.g., carbohydrates, on the CD18 integrin (ancillary binding site recognition); or (3) binding of the bacterial surface protein to a serum protein, e.g., C3bi, which then acts as a bridge to the CD18 integrin (masking). The following examples illustrate assays that distinguish these possibilities using CD11b/CD18 as an example.

Direct Binding of RGD Site

This mechanism is most obvious when the bacterial adhesin contains an RGD triplet in its sequence. The most direct indication of the involvement of the triplet (or another region acting like an RGD triplet) in integrin recognition is the loss of adherence consequent to a site-directed mutation in the triplet itself (e.g., RGD → RAD). For example, the binding of filamentous hemagglutinin (FHA) of *Bordetella pertussis* to CR3 involves an RGD sequence. Bacterial mutants in which the RGD sequence is changed into RAD fail to adhere.[11] Loss of adherence indicates that the RGD participates in binding, mimicking natural recognition of RGD ligands by integrins.

Further support for RGD–integrin recognition can be obtained by competition assays using RGD-containing peptides (1–5 mg/ml). To support the specificity of inhibition by a peptide, an altered peptide, e.g., in which the glycine of the RGD sequence is substituted by alanine, should cause no inhibition of the binding of the wild-type or C3bi-coated erythocytes to macrophages.[11] This type of assay must, however, be interpreted with caution as RGD peptides inhibit non-RGD-dependent interactions and non-RGD peptides inhibit RGD-dependent interactions. Thus, conclusions from these types of competition experiments are presumptive.

Recognition of Ancillary Non-RGD Binding Site

To examine whether a bacterial adhesin binds to carbohydrate determinants on leukocytes, bacterial adherence to macrophages plated on HSA-coated Terasaki plates is measured in the presence of isotonic buffers containing up to 50 m*M* soluble sialylated or nonsialylated carbohydrates (e.g., sialic acid, fucose, mannose, glucose, galactose, *N*-acetylglucosamine). Decreased adherence indicates a carbohydrate-dependent recognition event. If the carbohydrate determinant is on an integrin, then the ability of carbohydrates to inhibit adherence will be lost when macrophages are plated on anti-CD18.[23] Conversely, if the carbohydrate is independent of the integrin, then plating macrophages on anti-CD18 will add to the inhibition by soluble carbohydrates. It should be kept in mind that

[23] J. Van't Wout, W. N. Burnette, V. L. Mar, E. Rozdzinski, S. D. Wright, and E. I. Tuomanen, *Infect. Immun.* **60,** 3303 (1992).

bacteria are able to bind to CR3 by LPS (endotoxin). This recognition occurs at an ancillary non-RGD binding site.[5]

Masking of Bacterial Adhesin by Serum Proteins

CR3 binds C3bi-coated particles, leading to phagocytosis. Pathogens, such as *Legionella* and *Mycobacteria*, capture C3bi on their surfaces and thereby bind to CR3. To determine whether bacteria use a C3 degradation product as a bridge to bind to CR3, bacteria are incubated for 20 min at 37° in 50% fresh or heat-inactivated nonimmune serum, respectively. Then the bacteria are washed twice vigorously with phosphate-buffered saline without Ca^{2+} or Mg^{2+} (PBS, Whittaker Bioproducts, Walkersville, MD) before they are used in the adherence assay.[12] If the bacteria incubated in fresh serum exhibit a markedly higher adherence to macrophages in comparison to those incubated in heat-inactivated serum, this is strong evidence that opsonization with complement is an important factor for adherence.

This hypothesis can be further tested by incubation of macrophages with bacteria in the presence of heterologous C3-depleted serum or C3-depleted serum to which C3 has been added back.[12] The fixation of C3 on the bacterial surface can be confirmed by a whole-cell enzyme-linked immunosorbent assay (ELISA) specific for human C3 which is described elsewhere.[12]

To differentiate if a serum-dependent adherence effect involves binding to CR3 a downmodulation experiment is carried out in which macrophages are plated on an anti-CD11b/CD18 antibody. Binding of the serum-treated bacteria is abrogated if CR3 is a prominent receptor.

Interaction between Bacterial Recognition Site and Integrin May Promote Engulfment and/or Killing of Bacterium

Human macrophages are purified and cultured as described above. At days 4–6 they are harvested and, while they are in suspension, incubated with bacteria in medium RPMI 1640 (Sigma Chemical Co., St. Louis, MO) at a relative density of 1 macrophage : 100 bacteria for 30 min on a tumbler at 37°. The suspension is centrifuged at 250 *g* (10 min, 37°) to sediment the macrophages; the nonadherent, extracellular bacteria in the supernatant are discarded. The location and viability of bacteria associated with the macrophages are determined in separate aliquots as follows.

To determine the number of intra- versus extracellular bacteria, the bacteria–macrophage mixture is incubated at 37° in 5% (v/v) CO_2 for 4–6 hr. Samples of the mixture (50–100 μl) are taken at hourly intervals and divided equally onto two slides to air-dry. One slide of each pair is fixed divided equally onto two slides to air-dry. One slide of each pair is fixed in each fixative: (1) 3.3% neutral buffered formalin (Sigma Chemical Co.) or (2) 100% methanol. Cells fixed in formalin exclude subsequent fluorescent antibody from the intracellular bacteria, thereby providing a measure of the extracellular bound population. In contrast, methanol fixation allows the fluorescent antibody to stain intra- and extracellular bacteria.[24] To count the bacteria per macrophage, the samples are overlaid with a fluorescent antibody to the bacteria commonly diluted to 1 : 40 in PBS and incubated for 30 min at 37° in a humidified chamber. The difference in the number of bacteria in the formalin versus methanol preparations is a measure of the number of intracellular bacteria. If, for example, on the formalin-fixed slide 30 fluorescent bacteria per 100 macrophages are seen and on the methanol-fixed slide 100 fluorescent bacteria per 100 macrophages are seen, then 70 microorganisms have been ingested by the macrophages. Compare test results between wild-type and mutant bacteria that do not express the adhesin.

To determine the viability of intracellular and extracellular bacteria, the gentamicin killing assay of S. Falkow and co-workers is used (see [39]).[25] Cells with adherent bacteria are incubated at 37° in RPMI 1640 supplemented with gentamicin (50 μg/ml) to kill extracellular bacteria. To determine the number of viable intracellular bacteria, samples are taken at 30-min intervals over 6 hr. The macrophages are separated from the gentamicin by centrifugation for 5 min at 10,000 *g* at 4°. The macrophages are resuspended in 0.1% (v/v) Triton X-100 to lyse the macrophages but not kill the bacteria. The total viable bacterial count is titered by plating on agar and determining colony-forming units. Colony-forming units at the time of addition of gentamicin (time 0) will be high, indicating the total extra- and intracellular bacterial population. After 60 min, gentamicin-induced killing of the extracellular bacteria will be significant and the colony-forming units will be expected to decline. If the number of colony-forming units reaches a plateau after 2 hr, then the concentration of bacteria at the plateau level can be considered viable and intracellular.

[24] G. J. Noel, D. M. Mosser, and P. J. Edelson, *J. Clin. Invest.* **85,** 208 (1990).

[25] R. R. Isberg, D. L. Voorhis, and S. Falkow, *Cell (Cambridge, Mass.)* **50,** 769 (1987).

[24] Solid-Phase Binding of Microorganisms to Glycolipids and Phospholipids

By Carolyn D. Deal and Howard C. Krivan

Introduction

Human and other eukaryotic cells are sugar coated, that is, the surface of the cells are covered with a layer of carbohydrate, much of which exists in the form of oligosaccharides linked covalently to lipids and proteins embedded in the membrane. These glycolipids and glycoproteins form a network of outer surface projections and contain particular carbohydrate sequences which encode very precise binding domains recognized by a variety of microorganisms that may come into contact with the host cell. Cell surface carbohydrate molecules that "function" by binding microorganisms have been called adhesion receptors.

The binding of viruses, bacteria, toxins, and antibodies to glycoconjugates immobilized on solid supports has been described by a number of investigators.[1–4] The technique has emerged as a important tool to identify putative receptors for microbial adhesion to eukaryotic target tissues. Specificity and tissue tropism of various pathogens have been studied by analyzing the binding of these microbes to isolated neutral and acidic glycolipids, including gangliosides, and phospholipids.[5–9] This method identifies carbohydrate binding specificities for a particular microbe; however, the carbohydrate moiety may be naturally located on both glycosphingolipids and glycoproteins on the eukaryotic cell surface. The fact that glycolipids contain one carbohydrate moiety per molecule rather than

[1] J. L. Magnani, D. F. Smith, and V. Ginsburg, *Anal. Biochem.* **109,** 399 (1980).

[2] G. C. Hansson, K. A. Karlsson, G. Larson, N. Stromberg, and J. Thurin, *Anal. Biochem.* **146,** 158 (1985).

[3] K. A. Karlsson and N. Stromberg, this series, Vol. 138, p. 220.

[4] H. C. Krivan, V. Ginsburg, and D. D. Roberts, *Arch. Biochem. Biophys.* **260,** 493 (1988).

[5] C. D. Deal and H. C. Krivan, *J. Biol. Chem.* **265,** 12774 (1990).

[6] N. Stromberg, C. Deal, G. Nyberg, S. Normark, M. So, and K. A. Karlsson, *Proc. Natl. Acad. Sci. USA* **85,** 4902 (1988).

[7] H. C. Krivan, B. Nilsson, C. A. Lingwood, and H. Ryu, *Biochem. Biophys. Res. Commun.* **175,** 1082 (1991).

[8] R. Lindstedt, G. Larson, P. Falk, U. Jodal, H. Leffler, and C. Svanborg, *Infect. Immun.* **59,** 1086 (1991).

[9] C. A. Lingwood, M. Cheng, H. C. Krivan, and D. Woods, *Biochem. Biophys. Res. Commun.* **175,** 1076 (1991).

the potentially multiple and heterogeneous carbohydrate structures of glycoproteins simplifies this initial analysis.

The binding specificity identified may define one of the initial events necessary for microbial adherence to host cells and may initiate subsequent events such as invasion. Alternatively, several binding specificities may contribute synergistically to an organism's ability to bind to eukaryotic cells. Dissecting individual binding epitopes/sequences may eventually enable us to reconstruct a series of binding events necessary for microbial pathogenesis. This knowledge may lead to new ways of intervening in this complex biological process, such as developing carbohydrate drugs to inhibit adhesion and infection or producing vaccines that contain the corresponding microbial adhesin protein.

Thin-Layer Chromatography Overlay Assay

Preparation of Glycolipids

Purified gangliosides, neutral glycolipids, and phospholipids used as standards or to be analyzed in this assay can be purchased from a variety of commercial sources. Alternatively, extracts can be prepared from relevant tissue types. Total lipids can be prepared from various tissues or tissue culture cells by extraction with chloroform/methanol/water (4 : 8 : 3).[8,10] The total lipid extract can be further separated into neutral and acidic fractions by anion-exchange chromatography.[11,12] Stock solutions (typically 1–10 mg/ml) in chloroform/methanol (1 : 1) are stored at −20°.

Thin-Layer Chromatography

Thin-layer chromatography (TLC) of lipids is typically carried out on aluminum-backed silica gel high-performance plates (Merck, Darmstadt, Germany), although plastic-backed plates can also be used. By use of a 10-μl Hamilton syringe, lipid standards, or mixtures thereof, are applied to plates in designated lanes in the range of 0.5 to 5 μg per lipid species. The amount of lipid needed and applied from tissue extracts can greatly vary. For example, human trachea acidic glycolipids from 100 mg wet weight of tissue were applied to visualize receptor-active lipids for *Myco-*

[10] L. Svennerholm and P. Fredman, *Biochim. Biophys. Acta* **617,** 97 (1980).

[11] H. C. Krivan, D. D. Roberts, and V. Ginsburg, *Proc. Natal. Acad. Sci. USA* **85,** 6156 (1988).

[12] K. A. Karlsson, this series, Vol. 138, p. 212.

plasma,[13] whereas total lipids from 5 mg wet weight of HeLa cells were sufficient for chlamydial binding.[7] The plates are developed with a suitable solvent system such as chloroform/methanol/water (60 : 35 : 8) or chloroform/methanol/0.25% KCl in water (50 : 40 : 10) until the solvent front is approximately 5 mm from the top of the plate. After chromatography the plates are air-dried, and separated glycolipids are visualized by spraying one of the TLC plates with orcinol (0.5% in dilute sulfuric acid) or cupric acetate (30 g Cu acetate/500 ml distilled H_2O, 55 ml 85% phosphoric acid, and distilled H_2O to 1000 ml).

Solid-Phase Binding Assay

Plates to be used for microbial binding are dipped for 1 min in 0.1% polyisobutylmethacrylate in hexane or diethyl ether. After the plates are air-dried, they are sprayed with Tris-buffered saline (TBS) and then immersed for 2 hr in 2% bovine serum albumin in the same buffer (TBS–BSA) to block nonspecific binding sites. After excess buffer is drained from the plates, the plates are placed horizontally on a pedestal (i.e., a few microscope slides in a petri dish) and overlaid for 2 hr with 60 μl of bacteria/cm^2 of surface of the TLC plate. The bacteria are suspended to a concentration of approximately 10^7 to 10^8 bacteria/ml in a suitable buffer such as Hanks' balanced salt solution (HBSS) or TBS–BSA. After incubation, the plates are gently washed four times with cold buffer (TBS or phosphate-buffered saline) to remove unbound bacteria.

Detection of Bacteria Binding

Bacterial binding can be detected by several methods. The first is radiolabeling, either by metabolic labeling or by iodination of nonlabeled bacteria. Conditions for metabolic labeling are determined by the growth requirements of the specific bacteria. Various isotopes such as ^{35}S, ^{32}P, ^{3}H, and ^{14}C can be added to the growth medium. Iodination of nonlabeled bacteria can be performed as described using Na^{125}I[3–5] with the lodo-Gen method.[14] Briefly, bacteria are placed in a 10 $\times$ 75-mm test tube coated with 100 μg of Iodogen and reacted with 0.5 to 1 mCi of Na^{125}I at 4° for 5–10 min. The reaction is terminated by removing the cells and washing three times in buffer. Typically the cells are resuspended to a level of 2.0 $\times$ 10^6 cpm/ml. After the 2-hr binding assay is completed, the thin-layer plate is washed as described above and autoradiographed (XAR-5 X-ray film, Eastman Kodak).

[13] H. C. Krivan, L. D. Olson, M. F. Barile, V. Ginsburg, and D. D. Roberts, *J. Biol. Chem.* **264,** 9283 (1988).

[14] P. J. Fraker and J. C. Speck, *Biochem. Biophys. Res. Commun.* **80,** 849 (1978).

Alternatively, bacteria can be detected by immunochemical methods using monoclonal or polyclonal antibodies specific for the bacteria. Following bacterial binding, plates are incubated for 1 hr with primary antibody. Plates are washed five times with buffer and incubated for 1 hr with either radiolabeled or enzyme-linked secondary antibody. Following five washes with buffer, plates with radiolabeled secondary antibody are dried and autoradiographed (XAR-5 X-ray film, Eastman Kodak). The method of development for plates treated with enzyme-linked secondary antibody is determined by the enzyme. For example, for bacteria detected with alkaline phosphatase-labeled secondary antibodies, the plates can be immersed in TBS containing 2 mg of fast red and 1 mg of naphthol phosphate per plate[15] and bound bacteria are visualized by the appearance of red bands.

To identify receptor-active lipids, autoradiograms or enzyme-developed plates are compared with identical thin-layer chromatographic patterns of lipids visualized by chemical detection,[16] such as orcinol. An example is shown in Fig. 1, where nonpiliated *Pseudomonas aeruginosa* labeled with ^{125}I binds to ganglio-series (lanes 2) and lacto-series (lanes 3) glycolipids. Another example is shown in Fig. 2, where ^{125}I-labeled *Neisseria gonorrhoeae* is shown to bind strongly to paragloboside (lanes 3) and lacto-*N*-triaosylceramide (lanes 4).

Solid-Phase Binding in Microtiter Plates

Analysis of binding of microorganisms to purified lipids adsorbed on microtiter plates (polyvinyl chloride plates) allows for quantification of binding and further definition of specificity by comparison of the relative avidities of different receptors. Purified glycolipids or phospholipids are serially diluted in 25 μl of methanol containing 0.1 μg each of the auxiliary lipids cholesterol and phosphatidylcholine. We have found that use of auxiliary lipids reduces background noise and enhances reproducibility. Dilutions may range from 0.005 to 10 μg of glycolipid per well. The lipids are dried by evaporation at room temperature (typically 1–2 hr). The wells are then filled with 1–2% BSA in a suitable buffer (such as TBS) to block unbound sites. After 1 hr the wells are emptied and rinsed with TBS. Bacteria (25–50 μl) are dispensed in each well and incubated for 2 hr. The wells are washed five times with cold phosphate-buffered saline using a multichannel pipet or Pasteur pipet to remove unbound bacteria. Auto-

[15] N. R. Baker, V. Minor, C. Deal, M. S. Shahrabadi, D. A. Simpson, and D. E. Woods, *Infect. Immun.* **59,** 2859 (1991).

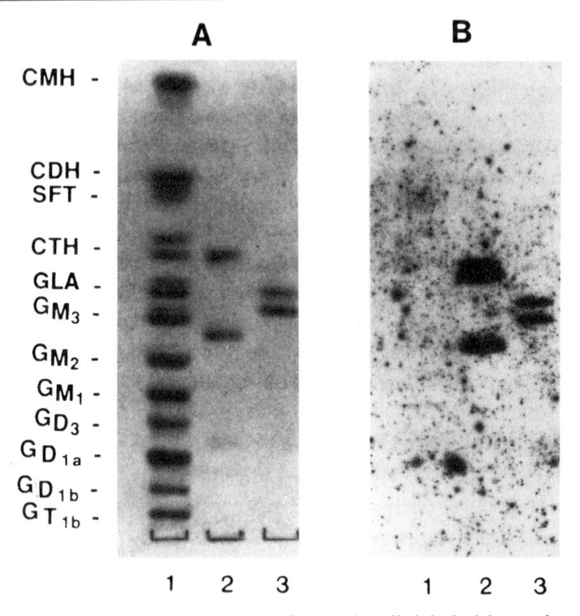

FIG. 1. Comparison of autoradiogram of bound radiolabeled bacteria with chemically detected thin-layer chromatographic patterns of glycolipids. (A) Glycolipids detected by orcinol reagent. (B) Autoradiogram of a chromatogram overlaid with ^{125}I-labeled *Pseudomonas aeruginosa*. Lanes 1, 1 μg each of the standard glycolipids, galactosylceramide (CMH), lactosylceramide doublet (CDH), sulfatide (SFT), trihexosylceramide (CTH), globoside (GLA), and gangliosides G_{M3}, G_{M2}, G_{M1}, G_{D3}, G_{D1a}, G_{D1b}, and G_{T1b}; lanes 2, 1 μg each of asialo-G_{M1} and asialo-G_{M2}; lanes 3, 2 μg of paragloboside (doublet).[16]

matic suction plate washes are not recommended as the shear force is too high. Bound bacteria are then quantified.

Binding of bacteria to glycolipid-coated microtiter plates is detected by radioisotopic labeling and scintillation counting or enzyme-linked antibody detection. If metabolically labeled bacteria or isotopically labeled secondary antibodies are used, wells can be cut out and counted in a scintillation counter. If immunochemical methods are used, plates can be incubated with primary and secondary antibodies, developed, and measured in an enzyme-linked immunosorbent assay (ELISA) plate reader. Curves are normally approximated from triplicate determinations. Binding to cholesterol- and phosphatidylcholine-coated plates or to plates coated with nonreactive glycolipids determines levels of nonspecific binding. Normal control levels are usually less than 1% of the total specific binding. This assay quantitates binding of bacteria to a specific glycoconjugate or lipid. Fig-

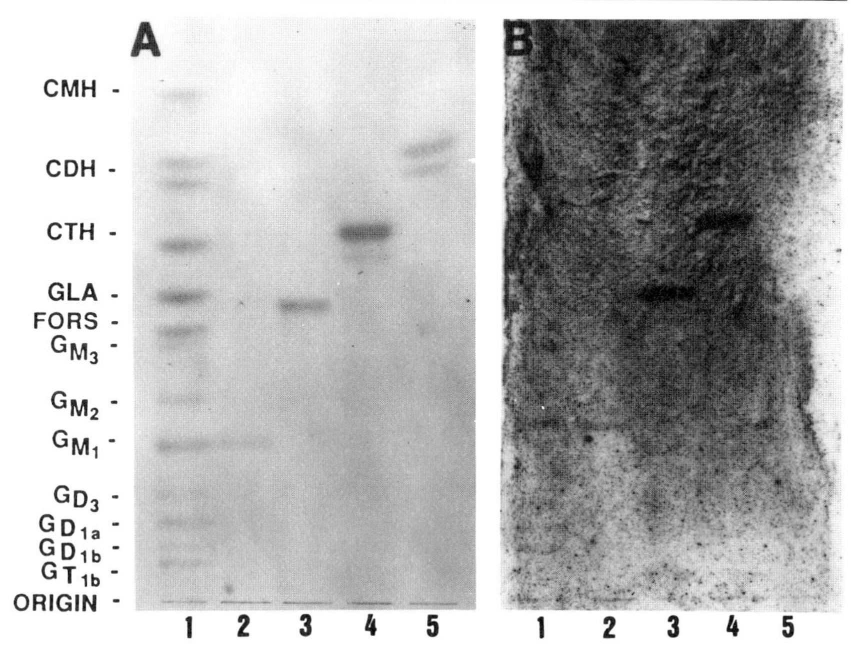

FIG. 2. Binding of bacteria to glycolipids separated on a thin-layer chromatogram. Glycolipids were separated by chromatography in chloroform/methanol/0.25% KCl in water (5 : 4 : 1). (A) Sprayed with orcinol to identify the glycolipids. (B) Overlaid with ^{125}I-labeled *Neisseria gonorrhoeae* and autoradiographed on XAR-5 X-ray film. Lanes 1, glycolipid standards consisting of 1 μg each of galactosylceramide (CMH), lactosylceramide doublet (CDH), trihexosylceramide (CTH), globoside (GLA), Forssman glycolipid (FORS), and gangliosides G_{M3}, G_{M2}, G_{M1}, G_{D1a}, G_{D1b}, and G_{T1b}; lanes 2, sialylparagloboside; lanes 3, paragloboside; lanes 4, lacto-*N*-triaosylceramide; lanes 5, lactosylceramide.[5]

ure 3 illustrates a typical binding curve in which gonococci bind avidly to asialo-G_{M2}.

Comments

The binding of microorganisms to separated glycolipids is a powerful tool for identifying binding specificities; however, as in many assays caution must be exercised in the interpretation of results. Binding of bacteria to bands in complex mixtures can be initially evaluated by comparing the comigration of the lipid band with standards; however, careful analysis is cautioned in that bands might overlap in migration. For example, a band contained in a tissue extract may comigrate with the standard but

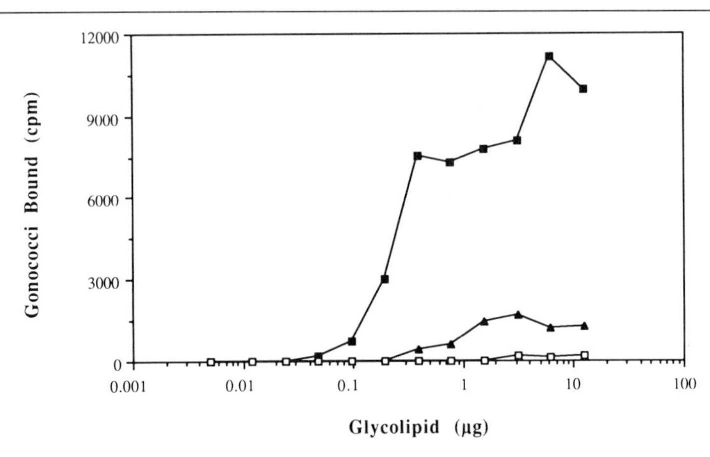

FIG. 3. Microtiter well binding assay of bacteria to immobilized glycolipids. Glycolipid standards were evaporated in flat-bottom wells of polyvinyl chloride microtiter plates. After blocking with 1% BSA/HBSS, the wells were incubated for 2 hr with 25 μl of ^{125}I-labeled *Neisseria gonorrhoeae* (approximately 10^5 cpm). The wells were washed, cut from the plate, and counted in a scintillation counter. Curves are indicated for asialo-G_{M2} (■), paragloboside (▲), and lactosylceramide (□).[5]

be a different structure. Identity of the unknown band must be confirmed either immunochemically by antibody specific for the glycolipid or biochemically by isolation and chemical/structural characterization.

Interpretation of biological significance is influenced by whether or not the bound compound is present on relevant target cells for the particular bacteria. For example, Krivan *et al.* reported that *Chlamydia trachomatis* binds to phosphatidylethanolamine.[7] A biological role for this phospholipid is suggested because phosphatidylethanolamine is present in HeLa and McCoy cells which are required for growing chlamydiae.

Another issue is whether or not the identified compound is accessible on the cell surface. Binding results from thin-layer chromatography may or may not predict bacterial binding to intact cells. Lindstedt *et al.* report differences between bacterial binding to glycolipids on TLC plates and to intact cells for uropathogenic *E. coli.*[8] Conversely, binding of bacteria to eukaryotic cells cannot predict the presence or absence of receptor-active glycolipids on those cells.

Finally, the carbohydrate binding specificity observed may reflect a specificity found on both glycolipids and glycoproteins. Several of the same terminal carbohydrate sequences are present on both glycolipids

[16] H. C. Krivan, L. Plosila, L. Zhang, V. Holt, and M. Kyogashima, "Microbial Adhesion and Invasion," p. 1, Springer-Verlag, Berlin/New York, 1991.

and glycoproteins in the cell membrane. Therefore, the assay described is an excellent tool to identify binding specificities. Once these specificities are known, further work can be done to examine and evaluate the nature of the specificities and their possible relevance and role in biological processes.

[25] Characterization of Microbial Host Receptors

By PER FALK, THOMAS BORÉN, and STAFFAN NORMARK

Introduction

Bacteria present on epithelial cell surfaces have a restricted range of hosts and tissues (i.e., cell lineages) that they are able to colonize. This is referred to as *tropism* and is dictated by a complex set of host- and microbe-derived factors, most of which are unknown to date. It is thought that competition between bacterial species for space and nutrients selects for bacteria able to colonize specific niches. Biochemical parameters, including pH, antimicrobial peptides (e.g., defensins),[1,2] and growth-promoting substances (e.g., Bifidus factor),[3] are also some potential modulating factors.[4]

Attachment is a prerequisite for bacteria to colonize epithelial surfaces, and prokaryotes express surface-associated adhesion molecules, i.e., adhesins,[5,6] with often fine-tuned specificities for eukaryotic cell surface protein or carbohydrate structures. If bacteria are unable to adhere to the epithelial cells, they will be rapidly removed by the local nonspecific host defense mechanisms, i.e., peristalsis, ciliary activity, fluid convection, and turnover of the epithelial cell populations and the mucus layer. The biological impact of this initial step of colonization and the notion that novel approaches such as antiadhesive compounds (soluble receptor analogs) might be a rational alternative drug design for antimicrobial therapy

[1] M. E. Selsted, S. I. Miller, A. H. Henschen, and A. J. Ouellette, *J. Cell Biol.* **118,** 929 (1992).

[2] K. Lal, R. P. Santarpia, III, L. Xu, F. Mansurri, and J. J. Pollock, *Oral Microbiol. Immunol.* **7,** 44 (1992).

[3] R. Lambert and F. Zilliken, *Biochim. Biophys. Acta* **110,** 544 (1965).

[4] D. C. Savage, *Annu. Rev. Microbiol.* **31,** 107 (1977).

[5] G. W. Jones and R. E. Isaacson, *CRC Crit. Rev. Microbiol.* **10,** 229 (1983).

[6] N. Sharon, *in* "The Lectins; Properties, Functions, and Applications in Biology and Medicine" (I. E. Liener, N. Sharon, and I. J. Goldstein, eds.), p. 494. Academic Press, Orlando, FL., 1986.

have nourished an increasing interest in the structures on the bacterial and eukaryotic cell surfaces involved in these interactions. Soluble receptor analogs could be "tailormade" to inhibit competitively pathogenic bacteria without interfering with the indigenous flora, circumventing the negative effects of broad-spectrum antibiotics that nonspecifically affect the microbiota.[7] The latter can cause serious superinfections, particulary common in the immunocompromised host.[8] In addition, because the composition of potential adhesion molecules expressed on epithelial cell surfaces varies among species[9] and cell lineages[10,11] and as a function of the developmental/differentiation stage,[12,13] tropism could conceivably be partly determined by receptor distribution.

In this article we present a strategy for identifying and characterizing eukaryotic cell surface and soluble body fluid molecules involved in specific interactions with microbes. The topic is illustrated by the authors' data as well as by examples from the literature. Figure 1 summarizes the outlined approaches in flowchart form.

Species- and Cell Lineage-Specific Distribution of Receptor Molecules

Epidemiological observations generally provide the first clues to the elucidation of a specific host receptor for a bacterial species. Commensal bacteria are isolated from specific habitats in their hosts or set of hosts. Pathogens are associated with a characteristic spectrum of host symptoms depending on which tissues they colonize. For instance, *Neisseria gonorrhoeae* is known to cause sexually transmitted genital disease and sometimes serious infections of the eye.[14] *Haemophilus influenzae* is a common cause of focal respiratory tract disease and meningitis,[15] whereas *Mycoplasma pneumoniae* produces primarily lower respiratory tract infections.[16] Commensal bacteria like *Escherichia coli* sometimes express virulence-promoting factors, such as P fimbriae, associated with urinary tract

[7] S. P. Boriello and H. E. Larson, *J. Antimicrob. Chemother.* **7,** Suppl. A, 53 (1981).

[8] P. J. Tutschka, *Pediatr. Infect. Dis. J.* **7,** Suppl. 22 (1988).

[9] G. C. Hansson, *Adv. Exp. Med. Biol.* **228,** 465 (1988).

[10] S. Björk, M. E. Breimer, G. C. Hansson, K.-A. Karlsson, and H. Leffler, *J. Biol. Chem.* **262,** 6758 (1987).

[11] Y. Umesaki K. Takamizawa, and M. Ohara, *Biochim. Biophys. Acta* **1001,** 157 (1989).

[12] T. Muramatsu, *J. Cell. Biochem.* **36,** 1 (1988).

[13] E. A. Muchmore, N. M. Varki, M. Fukuda, and A. Varki, *FASEB J.* **1,** 229 (1987).

[14] T. C. Quinn, and W. Cates, Jr., *in* "Sexually Transmitted Diseases" (T. C. Quinn, ed.), p. 1. Raven Press, New York, 1992.

[15] D. C. Turk, *J. Med. Microbiol.* **18,** 1 (1984).

[16] A. M. Collier and W. A. Clyde, Jr., *Infect. Immun.* **3,** 694 (1971).

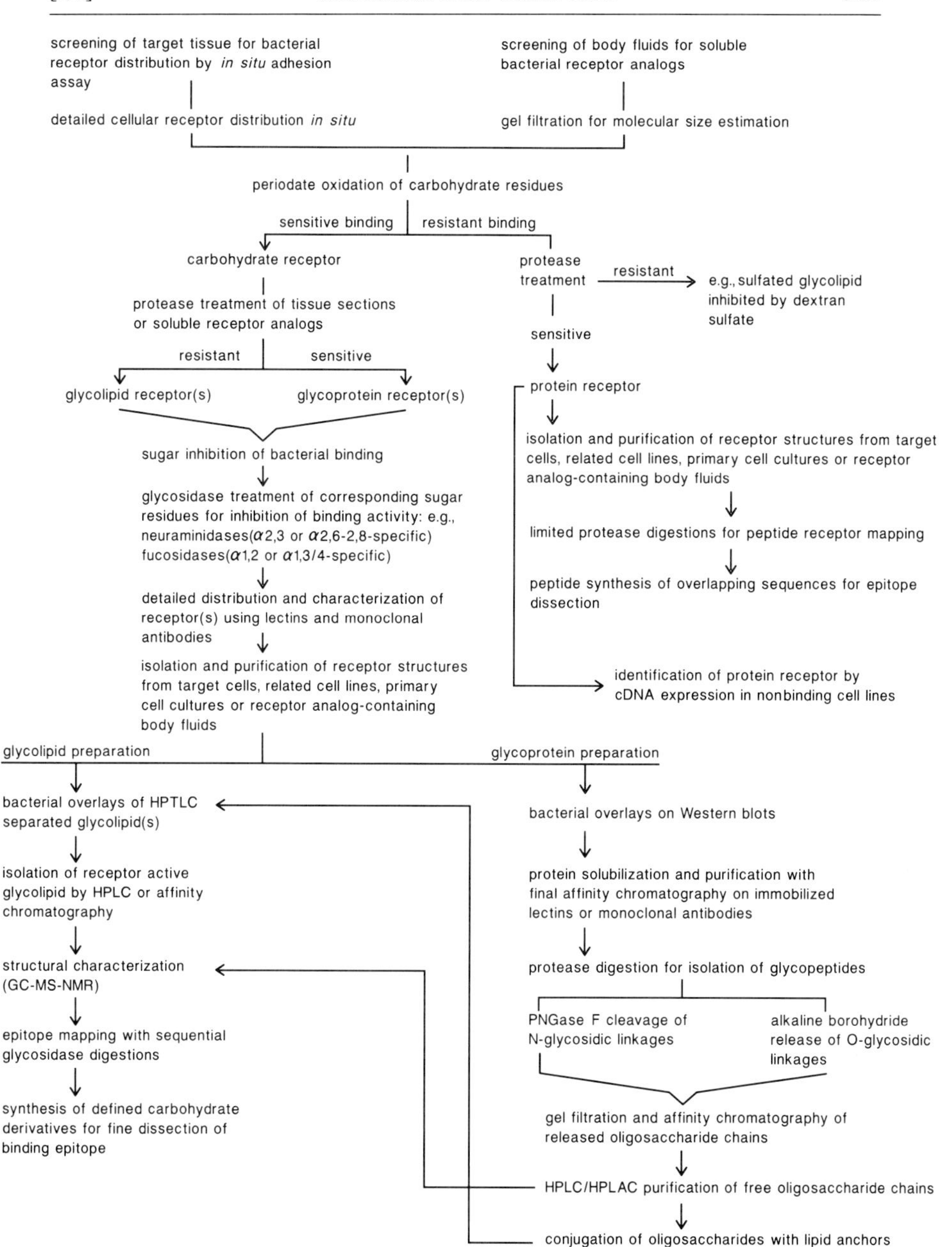

FIG. 1. Approaches to characterization, purification, and identification of bacterial host receptors.

infections (UTIs),[17] and S fimbriae, causing neonatal UTIs and meningitis.[18] *Actinomyces naeslundii/viscosus* are implicated in cemental caries[19,20] and periodontal disease.[21] *Helicobacter pylori,* a recently defined bacterial species,[22] is associated with dyspeptic syndrome[23,24] and gastric adenocarcinoma.[25]

The host and tissue tropism of pathogenic microorganisms, as illustrated by the clinical features of the infection, restricts the search for receptor molecules to a specific set of tissues. This has been suggested for the *E. coli pap/prs* system in which three different classes of adhesins have been defined based on their different affinities for globotriaosylceramide, globotetraosylceramide (globoside), and globopentaosylceramide (the Forssman antigen), respectively.[26] These three adhesin specificities affect the variation in target cell range (tropism) among bacteria expressing the different adhesins. This is probably due to minor conformational shifts induced in the receptor molecules inserted in the plasma membrane, as a consequence of increased sugar chain length.[27]

In Situ Screening of Host Receptor Distribution

In situ screening of histological sections could reveal the cell-specific distribution of bacterial receptors in target tissues using histochemical[28] and immunohistochemical[29] staining, *in situ* DNA hybridization,[30] or elec-

[17] H. Leffler and C. Svanborg-Edén, *FEMS Microbiol. Lett.* **8,** 127 (1980).

[18] T. K. Korhonen, M. V. Valtonen, J. Parkkinen, V. Väisänen-Rehn, J. Finne, F. Ørskov, I. Ørskov, S. B. Svensson, and P. H. Mäkelä, *Infect. Immun.* **48,** 486 (1985).

[19] R. P. Ellen, D. W. Banting, and E. D. Fillery, *J. Dent. Res.* **64,** 1377 (1985).

[20] B. Nyvad and M. Killians, *Infect. Immun.* **58,** 1628 (1990).

[21] B. C. Williams, R. M. Pantelone, and J. S. Sherries, *J. Periodontal Res.* **11,** 1 (1976).

[22] C. S. Goodwin, J. A. Armstrong, T. Chilvers, M. Peters, M. D. Collins, L. Sly, W. McConnel, and W. E. S. Harper, *Int. J. Syst. Bacteriol.* **39,** 397 (1989).

[23] J. I. Wyatt and M. F. Dixon, *J. Pathol.* **154,** 113 (1988).

[24] D. Y. Graham, *Gastroenterology* **96,** 615 (1989).

[25] A. Nomura, G. N. Stemmermann, P.-H. Chyou, I. Kato, G. I. Perez-Perez, and M. J. Blaser, *N. Engl. J. Med.* **325,** 1132 (1991).

[26] N. Strömberg, B.-I. Marklund, B. Lund, D. Ilver, A. Hamers, W. Gaastra, K.-A. Karlsson, and S. Normark, *EMBO J.* **9,** 2001 (1990).

[27] N. Strömberg, P. G. Nyholm, I. Pascher, and S. Normark, *Proc. Natal. Acad. Sci. U.S.A.* **88,** 9340 (1991).

[28] J. I. Wyatt, B. J. Rathbone, G. M. Sobala, T. Shallcross, R. V. Heatley, A. T. R. Axon, and M. F. Dixon, *J. Clin. Pathol.* **43,** 981 (1990).

[29] L. P. Andersen, S. Holck, and C. O. Povlsen, *Acta Pathol. Microbiol. Immunol. Scand.* **96,** 559 (1988).

[30] F. M. van der Berg, H. Zijlmans, W. Langenberg, E. Rauws, and M. Schipper, *J. Clin. Pathol.* **42,** 995 (1989).

tron microscopy,[31] as has been described for *H. pylori*. Sections of potential target tissues could also be overlaid with labeled bacteria or adhesins in an *in situ* overlay assay. This approach has been used to characterize the binding patterns of fluorescein isothiocyanate (FITC)-labeled S-fimbriated *E. coli* to cryosectioned human kidney.[32] We have used formalin-fixed tissues to investigate the cell-specific distribution of receptor(s) for *H. pylori*,[33] an organism considered to have an exclusive niche in the human gastric mucosa.[34] Bacteria were surface-labeled with FITC according to Korhonen *et al.*[32] Blood agar-grown *H. pylori* were washed with 0.1 *M* carbonate buffer, pH 9. FITC was added and the suspension was incubated for 1 hr, at room temperature in the dark. Bacteria were washed three times in phosphate-buffered saline (PBS)/0.05% Tween 20. Standard immunocytochemical procedures[35] were used with minor modifications. Tissues were formalin-fixed, embedded in paraffin, cut into 5-μm-thick sections, and affixed to microscope glass slides. Prior to overlay with bacteria, sections were deparaffinized and blocked for 30 min at room temperature by incubation in PBS/0.2% bovine serum albumin (BSA)/0.05% Tween 20. Bacteria were overlayed for 1 hr, followed by six washes ($\times$ 5 min) in PBS prior to inspection.

Host and Tissue Receptor Distribution for *Helicobacter pylori*

FITC-labeled *H. pylori* were overlaid on sections of human esophagus, stomach, jejunum, proximal colon, kidney, brain, endometrium, and cervix uteri. Four of five tested strains, i.e., NCTC 11637 and 11638,[36] WV-229,[37] and P-466 (clinical isolate kindly provided by Dr. T. U. Westblom, St. Louis Medical Center, St. Louis, MO) were found to attach exclusively to gastrointestinal epithelium. Binding was found primarily to the gastric mucosa and confined to the surface mucous cell population in the pit region of the gastric unit, whereas no binding was found to the cell populations in

[31] S. J. Hessey, J. Spencer, J. I. Wyatt, G. Sobala, B. J. Rathbone, A. T. R. Axon, and M. F. Dixon, *Gut* **31,** 134 (1990).

[32] T. K. Korhonen, J. Parkkinen, J. Hacker, J. Finne, A. Pere, M. Rehn, and H. Holthöfer, *Infect. Immun.* **54,** 322 (1986).

[33] P. Falk, K. A. Roth, T. Borén, T. U. Westblom, J. I. Gordon, and S. Normark, *Proc. Natl. Acad. Sci. U.S.A.* **90,** 2035 (1993).

[34] S. L. Hazell, A. Lee, L. Brady, and W. Henessy, *J. Infect. Dis.* **153,** 658 (1986).

[35] D. A. Sweetser, E. H. Birkenmeier, P. C. Hoppe, D. W. McKeel, and J. I. Gordon, *Gen. Dev.* **2,** 1318 (1988).

[36] B. J. Marshall, *in* "*Campylobacter pylori* in Gastritis and Peptic Ulcer Disease" (M. J. Blaser, ed.), p. 7. Igaku-Shoin, Tokyo, 1989.

[37] T. U. Westblom, E. Madan, M. A. Subik, D. E. Duriex, and B. R. Midkiff, *Scand. J. Gastroenterol.* **27,** 249 (1992).

the neck and gland regions (parietal cells, chief cells, mucous neck/gland cells).[38] Mucous neck cells situated in the upper portions of the glandular segments of gastric units were completely devoid of bacteria, indicating that the bacteria were able to distinguish between two differentiated mucus-producing cell lineages in the stomach.[39] Strong binding was also found to the epithelial cell lineages along the entire villus (enterocytes, goblet cells and enteroendocrine cells),[40] but not in the crypts of small intestine. Only weak attachment was detected to the epithelial cuffs in the colon. An interesting finding was the attachment to esophageal submucosal glandular structures and glandular ducts, whereas the squamous epithelium of the esophagus was negative. This might reveal a "silent site" of colonization for *H. pylori*. The binding pattern to human stomach and intestine was compared with binding to gastrointestinal mucosa from other species (P. Falk and T. Borén, unpublished results). Strain WV-229 bound to the gastric pit regions of Rhesus and Green monkey stomach mucosa. Both strains WV-229 and P-466 attached to the gastric pit regions of rat stomach, to forestomach, consisting of a squamous epithelium comparable to human esophagus, and to Brunner's glands. In contrast, no adhesion was detected to Rhesus monkey, Green monkey, or rat small intestine and colon, to dog gastric mucosa, or to any part of mouse gastrointestinal epithelium. All of these attachment patterns were strain specific, and interestingly, one of the strains, MO19,[41] did not attach to any of the tissues tested. These results indicate a complexity in *H. pylori*–host interactions with variations in host-receptor distribution, both within and between species, as well as in strain-specific expression of bacterial adhesin types.

Biochemical Nature of Receptor Molecules

Once the cellular distribution of receptor epitopes in the target tissues has been established, the next step is to characterize the cellular receptor for the bacterial adhesin(s). The first well-characterized carbohydrate epitope to be established as a microbial receptor was the ganglioside G_{M1}, used by cholera toxin for attachment to enterocytes.[42] Since then, an overwhelming number of reports have established the ability of microbial

[38] E. R. Lee, J. Trasler, S. Dwivedi, and C. P. LeBlond, *Am. J. Anat.* **164,** 187 (1982).

[39] H. Ota, T. Katsuyama, K. Ishii, J. Nakayama, T. Shiozawa, and Y. Tsukahara, *Histochem. J.* **23,** 22 (1991).

[40] J. I. Gordon, *J. Cell Biol.* **108,** 1187 (1989).

[41] T. U. Westblom, J. S. Barthel, T. U. Kosunen, and E. D. Everett, *Scand. J. Infect. Dis.* **21,** 311 (1989).

[42] J. Holmgren, I. Lönnroth, J.-E. Månsson, and L. Svennerholm, *Proc. Natl. Acad. Sci. U.S.A.* **72,** 2520 (1975).

ligands to attach to carbohydrates. Bacterial binding to specific carbohydrate epitopes includes, among others, P-fimbriated *E. coli* binding Galα1,4Gal sequences[43]; S-fimbriated *E. coli* binding NeuAcα2,3Galβ1[44]; *N. gonorrhoeae*[45] and several pulmonary pathogens,[46] including *H. influenzae*, binding GalNAcβ1,4Gal sequences, e.g., asialo-G_{M1}; *M. pneumoniae* binding sialylated glycoproteins[47] and sulfated glycolipids[48]; and *A. naeslundii* binding to GalNAcβ-containing glycoconjugates.[49] In addition, a number of bacteria have been reported to bind to host protein receptors, e.g., *Salmonella typhimurium* binding to the EGF receptor,[50] curli-expressing *E. coli* binding fibronectin,[51] and type 1 fimbriated *A. viscosus* binding salivary PRP proteins.[52] Here, however, we focus on carbohydrate receptors since host microbial protein–protein interactions are covered in other articles of this volume.

To characterize partially the carbohydrate epitope(s) mediating the cell lineage-specific attachment of *H. pylori* to human gastric surface mucous cells, we initially set out to characterize the composition of complex carbohydrates on the target cells *in situ*. Previous work had shown that *H. pylori* agglutinates erythrocytes[53] and binds to mouse adrenal gland (Y1) cells[54] in a sialic acid-dependent fashion. Furthermore, specific binding of *H. pylori* to acid glycosphingolipids extracted from human gastric mucosa, such as the ganglioside G_{M3} and sulfatide,[55] has been reported. We probed formalin-fixed sections of human gastric mucosa

[43] K. Bock, M. E. Breimer, A. Brignole, G. C. Hansson, K.-A. Karlsson, G. Larson, H. Leffler, B. E. Samuelsson, N. Strömberg, C. Svanborg-Edén, and J. Thurin, *J. Biol. Chem.* **260,** 8545 (1985).

[44] J. Parkkinen, J. Finne, M. Achtman, V. Väisänen, and T. K. Korhonen, *Biochem. Biophys. Res. Commun.* **111,** 456 (1983).

[45] N. Strömberg, C. Deal, G. Nyberg, S. Normark, M. So, and K.-A. Karlsson, *Proc. Natl. Acad. Sci. U.S.A.* **85,** 4902 (1988).

[46] H. C. Krivan, D. C. Roberts, and V. Ginsburg, *Proc. Natl. Acad. Sci. U.S.A.* **85,** 6157 (1988).

[47] D. D. Roberts, L. D. Olson, M. F. Barile, V. Ginsburg, and H. C. Krivan, *J. Biol. Chem.* **264,** 9289 (1980).

[48] H. C. Krivan, L. D. Olson, M. F. Barile, V. Ginsburg, and D. D. Roberts, *J. Biol. Chem.* **264,** 9283 (1989).

[49] N. Strömberg and K.-A. Karlsson, *J. Biol. Chem.* **265,** 11251 (1990).

[50] J. E. Galán, J. Pace, and M. J. Hayman, *Nature (London)* **351,** 588 (1992).

[51] A. Olsén, A. Jonsson, and S. Normark, *Nature (London)* **338,** 652 (1989).

[52] R. J. Gibbons, D. I. Hay, J. O. Cisar, and W. B. Clark, *Infect. Immun.* **56,** 2990 (1988).

[53] D. G. Evans, D. J. Evans, Jr., and D. Y. Graham, *Infect. Immun.* **56,** 2896 (1988).

[54] D. G. Evans, D. J. Evans, Jr., J. J. Moulds, and D. Y. Graham, *Infect. Immun.* **57,** 2272 (1989).

[55] T. Saitoh, H. Natomi, W. Zhao, K. Okuzumi, K. Sugano, M. Iwamori, and Y. Nagai, *FEBS Lett.* **282,** 385 (1991).

with the *N*-acetyl/glycoloylneuraminic acid (NeuAc/Gc)-specific lectins *Limax flavus* agglutinin (LFA), with a broad specificity for NeuAc/Gc,[56] *Maackia amurensis* agglutinin (MAA), recognizing NeuAcα2,3Galβ1, 4GlcNAc,[57] and *Sambucus nigra* agglutinin (SNA), recognizing NeuAcα2,6Gal/GalNAc,[58] as well as the cholera toxin B subunit, recognizing internal NeuAcα2,3Gal residues.[59] The LFA, MAA, and SNA lectins did not recognize any cell lineages in the human gastric epithelium, but showed cell lineage-specific staining in mouse and dog gastrointestinal epithelium, verifying that these lectins could serve as tools for detecting species-related differences in glycosylation patterns (P. Falk and T. Borén, unpublished results). The staining pattern of cholera toxin B subunit, the only NeuAc-specific marker that stained human gastric mucosa *in situ,* was distinctly separate from the binding pattern of all four *H. pylori* strains attaching to gastric surface mucous cells, as cholera toxin exclusively stained cells in the isthmus/neck region. Furthermore, we looked for potential receptor analogs/inhibitors among sialylated glycoconjugates. Thus, *H. pylori* strains were preincubated with the following sialic acid-containing glycoconjugates prior to overlay on sections of human gastric epithelium: fetuin (100 μg/ml), glycophorin (100 μg/ml), NeuAcα2,3- and α2,6-lactose (5 μg/ml). Asialofetuin (100 μg/ml) and asialoglycophorin (100 μg/ml) served as controls. None of these compounds resulted in any inhibition in bacterial adhesion.

Sialic acid-independent *H. pylori* adhesion to HeLa cells has, however, been shown,[60] and *campylobacter spp*. binding to an intestinal epithelial cell line (INT407) was partially inhibited by monovalent fucose.[61] This probably reflects the presence of multiple binding modes among different *H. pylori* strains. We therefore investigated the distribution of fucosylated glycoconjugates in the gastric mucosa, using immunocytochemical staining with the α-*L*-fucose-specific lectin *Ulex europaeus* type 1 (UEA1)[62] and monoclonal antibodies directed to the blood group antigens A, B, H, Le^a, and Le^b.[63] The cellular distribution of fucosylated sugar epitopes was completely coexpressed with the receptor(s) for *H. pylori,* and the

[56] R. L. Miller, this series, Vol. 138, p. 527.

[57] R. N. Knibbs, I. J. Goldstein, R. M. Ratcliffe, and N. Shibuya, *J. Biol. Chem.* **266,** 83 (1991).

[58] N. Shibuya, I. J. Goldstein, E. J. M. van Damme, and W. J. Peumans, *J. Biol. Chem.* **263,** 728 (1988).

[59] C.-L. Schengrund and N. Ringler, *J. Biol. Chem.* **264,** 13233 (1989).

[60] J.-L. Fauchére and M. J. Blaser, *Microb. Pathog.* **9,** 424 (1990).

[61] M. Cinco, E. Banfi, E. Ruaro, D. Crevatin, and D. Crotti, *FEMS Microbiol. Lett.* **21,** 347 (1984).

[62] S. Sugii and E. A. Kabat, *Carbohydr. Res.* **99,** 99 (1982).

[63] H. Clausen and S.-I. Hakomari, *Vox Sang.* **56,** 1 (1989).

bacterial binding was significantly reduced when the tissue was preincubated with the monoclonal antibodies.

To discriminate between protein- and carbohydrate-mediated binding, selective chemical hydrolysis of carbohydrate epitopes can be achieved by metaperiodate oxidation under mild acidic conditions.[64] This treatment cleaves carbon–carbon linkages carrying vicinal hydroxyl groups while leaving peptides intact. Sections of human stomach were incubated with 10 m*M* periodate, pH 4.5, for 1 hr at room temperature. The oxidation completely abolished the binding of *H. pylori* and UEA1. In a control experiment using rat gastric mucosa, an antiserum to intrinsic factor (IF) was used to monitor the integrity of peptide antigens in the tissue, and UEA1 lectin was used as carbohydrate marker in the same section. The staining intensity of IF was retained, whereas UEA1 staining was completely abolished by the periodate treatment. By performing the periodate reaction under less oxidating conditions (pH 5.5, 10 min, on ice), selective cleavage of the unsubstituted side chain (carbons 7–9) of sialic acid can be achieved.[65] This was evident *in situ* by the complete loss of binding of the SNA lectin to dog gastric mucosa. No effect on the ability of *H. pylori* or of UEA1 to bind surface mucous cells was seen when the latter treatment was applied to sections of human gastric mucosa.

Tissue sections were also incubated with proteinase K, which resulted in a drastic decrease in the number of bacteria adhering to the surface mucous cells. These results argue that the receptor epitope contains an essential carbohydrate component, probably presented on a glycoprotein, and terminal nonsubstituted sialic acid is not a part of the binding epitope.

Characterization of Carbohydrate Receptors

Once the biochemical nature of a receptor activity has been determined, a directed search for the receptor molecules can begin. One approach is to work with solubilized material as described by Magnani *et al.*[66] for the characterization of epitopes binding monoclonal antibodies. This procedure involves a separation of the receptor-active material into lipophilic and protein components, i.e., organic solvent- and water-soluble material, respectively. The extraction steps are followed by screening for receptor activity, using immunological detection with labeled antibodies/bacteria/bacterial adhesins on thin-layer plates (for glycolipids) or in micro-

[64] M. P. Woodward, W. W. Young, Jr., and R. A. Bloodgood, *J. Immunol. Methods* **78,** 143 (1985).

[65] A. E. Manzi, A. Dell, P. Azadi, and A. Varki, *J. Biol. Chem.* **265,** 8094 (1990).

[66] J. L. Magnani, S. L. Spitalnik, and V. Ginsburg, this series, Vol. 138, p. 195.

titer wells (for protein components). Solubilization of membrane-integrated glycoproteins under nondenaturating conditions can be performed with detergents,[67,68] providing more conformationally preserved glycoproteins as compared with extraction procedures using organic solvents. This also facilitates further handling of the protein fraction because of the reduced need for extensive resolubilization. Invasive bacteria often adhere to proteins that span the membrane bilayer and are associated with the cytoskeleton.[69,70] Solubilization of protein receptors anchored to cytoskeletal structures can result in low recovery. A combination of detergent, high ion strength, and ATP has proven efficient in solubilizing such receptor complexes.[70]

Glycolipids are the glycosides of *N*-acylsphingosine (ceramide) and act as carriers of complex carbohydrates in eukaryotic plasma membranes,[71] where they could potentially serve as anchoring molecules for microbial ligands. Many studies of specific carbohydrate receptor–bacterial ligand interactions used glycolipids and a thin-layer chromatogram immunostaining (overlay) technique,[72] where the binding of isotope-labeled bacteria can be compared to antibody[73] and/or lectin[74] immunostainings and chemical stainings. Anisaldehyde stains glycolipids in general,[75] whereas resorcinol is specific for sialic acid-containing glycolipids (gangliosides).[76] The glycolipids are isolated from target tissues, e.g., P-fimbriated *E. coli* binding globoside extracted from human kidney.[43] Panels of purified glycolipids from different sources can also be employed.[77] The latter approach has been used to identify potential receptor candidates, as well as to deduce minor differences in receptor motifs where the use of large panels of glycolipids might be powerful.[78] A microscale method for purification of total glycolipid fractions has been recently described,[79] allowing identifi-

[67] C. Tanford and J. A. Reynolds, *Biochim. Biophys. Acta* **457,** 133 (1976).

[68] D. Lichtenberg, R. J. Robson, and E. A. Dennis, *Biochim. Biophys. Acta* **737,** 285 (1983).

[69] S. Falkow, *Cell* (*Cambridge, Mass.*) **65,** 1099 (1991).

[70] P. Rafiee, H. Leffler, J. C. Byrd, F. J. Cassels, E. C. Boedecker, and Y. S. Kim, *J. Cell Biol.* **115,** 1021 (1991).

[71] S.-i. Hakomori, *in* "Sphingolipid Biochemistry" (J. N. Kanfer, and S.-i. Hakomori, eds.) p. 1. Plenum, New York, 1983.

[72] G. C. Hansson, K.-A. Karlsson, G. Larsson, N. Strömberg, and J. Thurin, *Anal. Biochem.* **146,** 158 (1985).

[73] M. Brockhaus, J. Magnani, M. Blaszczyk, Z. Steplewski, H. Koprowski, K.-A. Karlsson, G. Larson, and V. Ginsburg, *J. Biol. Chem.* **256,** 13223 (1981).

[74] B. V. Torres, D. K. McCrumb, and D. F. Smith, *Arch. Biochem. Biophys.* **262,** 1 (1988).

[75] E. Stahl, "Dünnschichtschromatographie," p. 817. Springer-Verlag, Berlin, 1967.

[76] L. Svennerholm, *J. Neurochem.* **10,** 613 (1963).

[77] K.-A. Karlsson, this series, Vol. 138, p. 212.

[78] K.-A. Karlsson, *Annu. Rev. Biochem.* **58,** 309 (1989).

[79] P. Falk, L. C. Hoskins, R. Lindstedt, C. Svanborg, and G. Larson, *Arch. Biochem. Biophys.* **290,** 312 (1991).

cation of glycolipid receptors from minute quantities of receptor-active material. Further purification of glycolipids can be achieved by lectin affinity chromatography[80] and high-performance liquid chromatography (HPLC).[81] In analogy with the thin-layer chromatogram binding assay for glycolipids, a Western blot overlay approach can be used to detect binding of bacteria to glycoproteins (see below).

The final step in receptor characterization involves the elucidation of the molecular structure of the carbohydrate. Complete structural characterization of an oligosaccharide includes determination of glycoside residue number and composition, the linkage positions and anomeric configurations (α or β) of the glycosidic bonds, absolute configurations (D or L form), and ring forms (pyranose or furanose) of the consistuent sugars, and finally the sequence of the glycosyl residues. In mammalian systems, all sugar residues isolated hitherto except fructose are in the pyranose ring form, and all except fucose are in the D form. The absolute configuration can be determined with gas chromatography of derivatized sugars.[82] Identification of sugar residues and estimation of their molar ratios can be performed with gas chromatography after hydrolysis of the carbohydrate molecule.[83] By acid degradation of permethylated samples, reduction, and finally acetylation, the binding positions can be elucidated with gas chromatography[84] of the resulting alditol acetates. The rapid development and refinement of mass spectrometry (MS) techniques[85] have considerably potentiated structural carbohydate characterization. The most powerful technique for structural characterization of oligosaccharides is nuclear magnetic resonance spectroscopy (NMR). NMR provides information on glycosidic composition, sequence, and anomerity as well as position of glycosidic linkages.[86] But even though the entire sequence of a receptor structure is known, it does not allow final identification of the actual bacterial binding receptor epitope. This could be worked out by performing sequential degradations with defined exo- and endoglycosidases[87,88] and assaying for loss of receptor activity.[70] Receptor epitope mapping using synthesized oligosaccharides[89] of overlapping fragments of the sugar chain

[80] D. F. Smith and V. B. Torres, this series, Vol. 179, p. 30.

[81] R. Kannagi, K. Watanabe, and S.-I. Hakomori, this series, Vol. 138, p. 3.

[82] K. Leontien, B. Lindberg, and B. Lönngren, *Carbohydr. Res.* **62,** 359 (1978).

[83] S. J. Rickert and C. C. Sweely, *J. Chromatogr.* **147,** 317 (1973).

[84] K. Stellner, H. Saito, and S.-I. Hakomori, *Arch. Biochem. Biophys.* **155,** 464 (1973).

[85] R. A. Laine, this series, Vol. 193, p. 539.

[86] J. Holgersson, P.-Å. Jovall, B. E. Samuelsson, and M. E. Breimer, *J. Biochem.* (*Tokyo*) **108,** 766 (1990).

[87] A. Kobata, *Anal. Biochem.* **100,** 1 (1979).

[88] C. J. Edge, T. W. Rademacher, M. R. Warmold, R. B. Parekh, T. D. Butters, D. R. Wing, and R. A. Dwek, *Proc. Natl. Acad. Sci. U.S.A.* **89,** 6338 (1992).

[89] H. M. Flowers, this series, Vol. 138, p. 359.

and/or with substitutions in defined epitope positions[90] can be used to reveal subtle details of the bacterial binding site.

Receptor Purification and Identification

When outlining strategies for the detailed molecular nature of a bacterial receptor, it is important to take into consideration whether the primary goal is to identify the biological cell surface receptor in the target tissue in order to understand subsequent events such as invasion and cellular cytokine responses.[91] This requires knowledge of the exact composition and location of the receptor structure, and the receptor must consequently be isolated and purified from the specific bacterial target tissue or from established cell lines/primary cell cultures derived from that tissue. On the other hand, if the primary goal is to identify an efficient receptor analog that inhibits bacterial binding in different assay systems or for subsequent clinical applications, alternative approaches might be considered as are discussed below.

An established method for initial receptor characterization employs red blood cells and hemagglutination (HA) reactions to define the nature of the binding component on the erythrocyte membrane. Based on HA/HA inhibition data it is possible to define further receptor-active molecules. Classic examples are the finding that D-mannose prevents HA mediated by type 1 fimbriated *E. coli*[92] and the demonstration that *Vibrio cholerae* sialidase (receptor-destroying enzyme) eliminates the binding of influenza virus.[93] To define whether bacteria discriminate between glycoprotein and glycolipid receptors or bind to the peptide core of a membrane protein, protease-treated erythrocytes can be very useful. S-fimbriated *E. coli* binds terminal α2,3-linked sialic acid residues (NeuAcα2,3Gal) exclusively on integrated membrane glycoproteins such as glycophorin A. Limited trypsin treatment of human erythrocytes therefore inhibited the bacterial binding completely.[94] HA by clinical uropathogenic and meningitis isolates of S-fimbriated *E. coli* differs in sensitivity to limited chymotrypsin treatment of erythrocytes. As trypsin and chymotrypsin have different cleavage sites on glycophorin A,[95] the two S-fimbria types may

[90] K. Bock, T. Frejd, J. Kihlberg, and G. Magnusson, *Carbohydr. Res.* **176,** 253 (1988).

[91] A. I. M. Hoepleman and E. I. Toumanen, *Infect. Immun.* **60,** 1729 (1992).

[92] J. P. Duguid and D. C. Old, *in* "Bacterial Adherence" (E. H. Beachey, ed.), Recept. Recognition, Ser. B, Vol. 6, p. 185. Chapman & Hall, London, 1980.

[93] F. M. Burnet and J. D. Stone, *Aust. J. Exp. Med. Sci.* **25,** 227 (1947).

[94] J. Parkkinen, G. N. Rogers, T. Korhonen, W. Dahr, and J. Finne, *Infect. Immun.* **54,** 37 (1986).

[95] J. K. Dzandu, M. E. Deh, and G. E. Wise, *Biochem. Biophys. Res. Commun.* **126,** 50 (1985).

recognize different clusters of O-linked sialyl-oligosaccharides on glycophorin A, possibly indicting minor differences in receptor recognition, resulting in different tissue tropism. This is also illustrated by different HA titers for human and bovine erythrocytes.[96] Similarly, *M. pneumoniae* also recognizes NeuAcα2,3Gal exclusively in glycoproteins,[47] whereas another adhesin is specific for sulfated glycolipids.[48] The less discriminating *A. naeslundii* ATCC 12104 (Rockville, MD) binds the Galβ/GalNAcβ residues in both glycoproteins and glycolipids.[97] Protease treatment of human erythrocytes enhanced hemagglutination by *A. naeslundii*, probably because receptor density was higher in the glycolipid layer as compared with the glycoprotein layer. The reverse was seen with chicken erythrocytes, where hemagglutination was abolished by protease treatment. This probably reflects the receptor distribution in chicken erythrocytes, where *A. naeslundii* receptors are confined exclusively to the glycoproteins. Topographic variations in receptor distribution on erythrocyte surfaces from different species were also verified with overlays of ^{35}S-labeled *A. naeslundii* to glycolipids separated on high-performance thin-layer chromatography (HPTLC) plates, where binding was detected to human glycolipids, whereas no binding was seen to chicken glycolipids.[97] Binding of P-fimbriated *E. coli* recognizing Galα1,4Gal-containing structures appears to be restricted to the glycolipid layer. Therefore, protease treatment of host cells has no inhibitory effect on P-fimbriated *E. coli*,[96] which could make P-fimbriated *E. coli* suitable as a control for the integrity of the glycolipid receptors after protease digestions. In addition, red blood cells from different species with defined glycolipid compositions have been used to identify glycolipid receptors in natural membranes,[26] where the accessibility of the carbohydrate part of the receptor structure is partly governed by the relation to the hydrophobic lipid membrane.[98] This approach allows dissection of tissue tropism involving bacterial adhesins with minor amino acid differences.[27]

The "red cell" approach uses integrated membrane components as receptor substrates and is a valuable method for initial studies because it also allows receptor binding studies of bacteria with adhesins of low affinity that might need a high receptor density to bind. For bacteria binding with higher affinity, or lower valency, binding experiments to soluble oligosaccharides or glycoconjugates can be a feasible way to identify a receptor. The human body contains several fluids with secreted

[96] G. Nyberg, Thesis (No. 333), Umeå University (1992).

[97] N. Strömberg and T. Borén, *Infect. Immun.* **60**, 3268 (1992).

[98] P. G. Nyholm, I. Pascher, and S. Sundell, *Chem. Phys. Lipids* **52**, 1 (1990).

and soluble receptor substances, e.g., breast milk (colostrum),[99–101] saliva,[102–104] tears,[105,106] urine,[107] lung secretions,[108] gastrointestinal mucus,[109] and meconium.[110–112] These secretions can be incubated with bacteria to investigate binding inhibitory activity. Initial characterization by gel fractionation[113] makes it possible to define an approximate molecular size of the inhibitory substance (glycoprotein versus oligosaccharide). This information provides some directions for further purification approaches such as combined size fractional/ion exchange,[114] and affinity chromatography[115] for glycoproteins, whereas a combination of affinity chromatography[115] followed by HPLC,[116] HPTLC[117] and/or paper chromatography[118] might be better suited for oligosaccharides.

The bacterial binding properties of salivary proteins separated by gel filtration can be analyzed by adsorbing these proteins onto hydroxyapatite beads.[119,120] Radiolabeled bacteria are then incubated with the beads, and after several washes the binding is measured by liquid scintillation. Recep-

[99] P. A. Montgomery, S. Patton, G. E. Huston, and R. V. Josephson, *Comp. Biochem. Physiol.* **86,** 635 (1987).

[100] A. Kobata, K. Yamashita, and Y. Tachibana, this series, Vol. 50, p. 216.

[101] D. F. Smith, D. A. Zopf, and V. Ginsburg, this series, Vol. 50, p. 221.

[102] R. J. Gibbons, and D. Y. Hay, *Infect. Immun.* **56,** 439 (1988).

[103] B. L. Gillece-Castro, A. Prakobphol, A. L. Burlingame, H. Leffler, and S. J. Fisher, *J. Biol. Chem.* **266,** 17358 (1991).

[104] N. Strömberg, T. Borén, A. Carlén, and J. Olsson, *Infect. Immun.* **60,** 3278 (1992).

[105] D. V. Seal, J. I. McGill, I. A. Mackie, G. M. Liakos, T. Jacobs, and J. Golding, *Br. J. Ophthalmol.* **70,** 122 (1986).

[106] R. J. Boukes, A. Boonstra, A. C. Breeboat, D. Reits, E. Glasius, L. Luyendyk, and A. Kijlstra, *Doc. Ophthalmol.* **67,** 105 (1987).

[107] A. Lundblad, this series, Vol. 50, p. 226.

[108] M. Callaghan Rose, this series, Vol. 179, p. 3.

[109] D. K. Podolsky, *J. Biol. Chem.* **260,** 8262 (1985).

[110] K.-A. Karlsson and G. Larson, *J. Biol. Chem.* **254,** 9311 (1979).

[111] M.-C. Herlant-Peers, J. Montreuil, G. Strecker, L. Dorland, H. van Halbeek, G. A. Veldink, and J. F. G. Vliegenhart, *Eur. J. Biochem.* **117,** 291 (1981).

[112] C. Capon, Y. Leroy, J.-M. Wieruszeski, G. Ricart, G. Strecker, J. Montreuil, and B. Fournet, *Eur. J. Biochem.* **182,** 139 (1989).

[113] K. Yamashita, T. Mizuochi, and A. Kobata, this series, Vol. 83, p. 105.

[114] J. Muenzer, C. Bildstein, M. Gleason, and D. M. Carlson, *J. Biol. Chem.* **254,** 5623 (1979).

[115] H. Lis and N. Sharon, *in* "The Lectins: Properties, Functions, and Applications in Biology and Medicine" (I. E. Liener, N. Sharon, and I. J. Goldstein, eds.), p. 293. Academic Press, Orlando, FL, 1986.

[116] M. Fukuda, M. Lauffenberger, H. Sasaki, M. E. Rogers, and A. Dell, *J. Biol. Chem.* **262,** 11952 (1987).

[117] J. L. Magnani, this series, Vol. 138, p. 208.

[118] A. Kobata, this series, Vol. 28, p. 262.

[119] L. Gahnberg, J. Olsson, B. Krasse, and A. Carlén, *Infect. Immun.* **37,** 401 (1982).

[120] R. J. Gibbons and D. I. Hay, *Infect. Immun.* **56,** 439 (1988).

tor-active fractions can then be further purified by ion-exchange or lectin affinity chromatography and reprobed in the hydroxyapatite binding assay. For example, parotid saliva has been screened for the distribution of soluble receptors for *A. naeslundii*. The saliva proteins were size-separated on Sephacryl S-200 (200 × 5 cm, Pharmacia-LKB Biotechnology, Sollentuna, Sweden) and assayed for bacterial receptor activity. *A. naeslundii* ATCC 12104 was shown to have an affinity for parotid secretory IgA (S-IgA), as well as for the lower-molecular-weight acidic proline-rich proteins. The binding could be abolished by preincubating the bacteria with 0.2 mg/ml of GalNAcβ1,3Galα-*O*-ethyl.[104] This interaction was further investigated by Western blot overlays of *A. naeslundii* to sodium dodecyl sulfate–polyacrylamide gel electrophoresis (SDS–PAGE)-separated total parotid saliva proteins and human colostrum S-IgA. The neoglycoprotein $(GalNAc\beta1,3Gal\alpha O)_{25}$–BSA (BioCarb, Lund, Sweden) containing 25 synthesized carbohydrate chains covalently linked to BSA molecules served as positive binding control. Strong binding could be shown to the heavy chain (α) of S-IgA and the interaction was inhibited by 2.0 mg/ml lactose (Fig. 2), indicating binding to the Galβ residues.

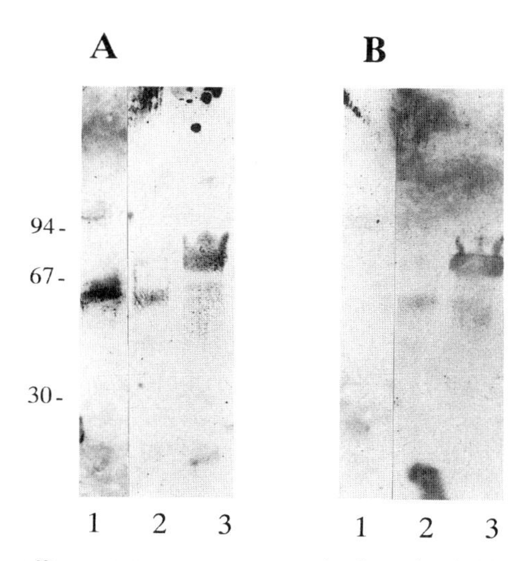

FIG. 2. Binding of ^{35}S-labeled *Actinomyces naeslundii* strain 12104 to total parotid saliva proteins and colostrum S-IgA on Western blots. The results of bacterial binding to the fractions in the absence (A) and presence (B) of lactose at a concentration of 2.0 mg/ml are shown. Lanes 1–3 contain 10 μl of total parotid saliva (lane 1), 5 μg of colostrum S-IgA (lane 2), and 5 μg of GalNAcβ1,3Galα–BSA (lane 3). Molecular weights ($\times 10^{-3}$) are indicated on the left-hand side.

This binding also correlated to the peanut agglutinin (PNA)[121] binding exclusively to α-chain components.[122] The ease and speed with which soluble receptors are obtained from fluids such as saliva and colostrum reduce the harmful effects of released cellular proteases and glycosidases and allow preparation of large quantities of material. These can later be used as a natural source for purifying biologically active material for clinical applications, which is important given that oligosaccharide synthesis of complex structures in preparative scale still is laborious. The possibility that a natural soluble structure might in fact be an optimal receptor analog for competitive inhibition of bacterial adherence is intriguing. A receptor isolated from glycolipids could be optimal for solid-phase interactions where high affinity might not be crucial because of high receptor density on the surface. On the other hand, a soluble receptor analog with a higher affinity, to compensate for reduced valency, could for this reason be the structure of choice for clinical receptor analog evaluations. The reason bacteria use a membrane-integrated receptor of slightly lower affinity could be to allow for small variations in carbohydrate epitope presentation, based on minor differences in conformation, to direct bacteria competitively to the specific ecological niche where the host–bacterial receptor–adhesin interactions will be optimal, i.e., result in tissue tropism.

Characterization and Identification of Receptor Analogs for *Helicobacter pylori*

We investigated the inhibitory activity of human colostrum S-IgA (Cappel Organon Technica, West Chester, PA), a molecule carrying a highly variable set of N- and O-linked oligosaccharides,[123,124] on the adhesion of *H. pylori* to human gastric surface mucous cells. S-IgA was recently shown to aggregate type 1-fimbriated *E. coli*,[125] and has been suggested to mediate both immunological and "nonimmunological" mucosal protection.[125–127] The effects of S-IgA were compared with those of human serum IgA (Cappel). Interestingly, S-IgA completely blocked the cell lineage-specific

[121] M. J. Swamy, D. Gupta, S. K. Mahanta, and A. Surolia, *Carbohydr. Res.* **213,** 59 (1991).

[122] T. Borén , D. Landys, and N. Strömberg, in preparation.

[123] A. Pierce-Crétel, M. Pamblanco, G. Strecker, J. Montreuil, and G. Spik, *Eur. J. Biochem.* **114,** 169 (1981).

[124] A. Pierce-Crétel, J.-P. Decottignies, J.-M. Wieruszeski, G. Strecker, J. Montreuil, and G. Spik, *Eur. J. Biochem.* **182,** 457 (1989).

[125] A. E. Wold, J. Mestecky, M. Tomana, A. Kobata, H. Ohbayashi, T. Endo, and C. Svanborg-Edén, *Infect. Immun.* **58,** 3073 (1990).

[126] I. Ofek and N. Sharon, *Infect. Immun.* **56,** 539 (1988).

[127] J. C. Davin, J. Senterre, and P. R. Mahieu, *Biol. Neonate* **59,** 121 (1991).

attachment of *H. pylori* P-466 and WV-229 in concentrations of 15 μg/ml, whereas serum IgA did not exhibit any inhibition at all in concentrations up to 100 μg/ml. The following series of experiments were performed to characterize the receptor epitope(s) on the S-IgA molecules. S-IgA was submitted to metaperiodate oxidation in 10 m*M* periodate, pH 4.5, for 30 min at room temperature. S-IgA was incubated with 100 mU of bovine kidney α-L-fucosidase or *Vibrio cholerae* neuraminidase (both from Boehringer-Mannheim, Mannheim, Germany). Glycosidases were inactivated by incubation at 85°. Untreated S-IgA, incubated at 85°, was included as control. The results obtained suggested that the inhibitory activity of S-IgA was not mediated by a classic immunoglobulin recognition event, involving the hypervariable region of the Fab fragments, or by a bacterial Fc receptor: (1) The inhibitory activity was not affected by preincubation at 85° for 30 min; (2) inhibition was abolished by metaperiodate oxidation; and (3) α-L-fucosidase digestion (but not sialidase digestion) nearly abolished the blocking activity. The carbohydrate-mediated inhibitory activity of colostrum S-IgA is of biological interest, as this could be a natural scavenging mechanism[127] preventing colonization of *H. pylori*. These data suggest that the cell surface receptor epitope contains a vital fucose residue, also present on the S-IgA molecule, and support the data suggesting that the surface mucous cell-specific attachment of *H. pylori* is independent of nonsubstituted terminal sialic acid.

To delineate the S-IgA receptor epitope for *H. pylori* we are using fucose-recognizing tools such as lectins and monoclonal antibodies in addition to enzymes, i.e., glycosidases with specificity for different fucose linkages (α-L-fucosidases), and chemical defucosylation.[128]

Several lectins have been described to have a defined specificity for terminal α-L-fucose such as the blood group H(O) antigen, e.g., the α1,2-linked fucose-recognizing lectins *Ulex europaeus* type 1 (UEA1)[62] (E-Y-Lab, San Mateo, CA) and *Anguilla anguilla* agglutinin (AAA)[129] (E-Y-Lab). The *Lotus tetragonolobus* agglutinin (LTA)[62] (E-Y-Lab) seems to be slightly less inhibited by α1,6-linked fucose residues in N-linked glycoproteins[130] and by α1,3-linked fucose[131] as compared with UEA1 and AAA. In contrast, the *Aleuria aurantia* agglutinin [orange peel fungus agglutinin (OPA), Boehringer-Mannheim][132] is specific for α1,6-linked fucose, but

[128] T. Borén, P. Falk, K. A. Roth, G. Larson, and S. Normark, *Science* **262,** 1892 (1993).

[129] V. Horejsi, M. Ticha, and J. Kocourek, *Biochim. Biophys. Acta* **499,** 290 (1977).

[130] H. Debray, D. Decout, G. Strecker, G. Spik, and J. Montreuil, *Eur. J. Biochim.* **117,** 41 (1981).

[131] M. E. A. Pereira, E. C. Kisailus, F. Gruezo, and E. A. Kabat, *Arch. Biochem. Biophys.* **185,** 108 (1978).

[132] H. Debray and J. Montreuil, *Carbohydr. Res.* **185,** 15 (1989).

does not recognize α1,2- or α1,3-linked fucose in glycoproteins. These properties make these lectins useful for characterizing fine differences in fucosylation patterns.

Monoclonal antibodies with specificity for the Lewis (Le) antigens[63] (Accurate Corp., Westbury, NY) are useful markers for detection of the fucosylated antigens Le^a, Le^b, Le^x, and Le^y, indicating the composition of the core chain (type 1 or 2) as well as the presence of mono- or difucosylated terminals in the glycoconjugates.[63] Additional fucose-specific markers could include organisms such as *Candida albicans* adhering to terminal Fucα1,2Galβ residues[133] and *Vibrio cholerae*[134] used as probes to define fucose-containing carbohydrate epitopes and receptor distribution.

Glycosidase from beef kidney (Boehringer-Mannheim) with a broad specificity for α-L-linked fucose[135] is useful, as loss of binding activity after fucosidase treatment of a receptor-active glycoconjugate indicates a vital role for fucose residue(s) in the interaction. Further dissection of the epitope using fucosidases from *Arthrobacter oxidans* F1 (Takara Biochemical Inc., Berkley, CA) and *Clostridium perfringens*[136] with specificity for α1,2-linked fucose or fucosidases from *Streptomyces* sp. 142 (Takara Biochemical Inc.) and almond emulsion[137,138] with specificity for α1,3/4-linked fucose, in combination with the monoclonal antibodies and lectins, could reveal structural requirements for the interaction.

Chemical defucosylation is possible by mild acid hydrolysis with 0.1 *M* trichloroacetic acid (TCA) at 100° for 1 h.[139] This treatment desialylates glycoproteins and removes most fucose residues resistant to fucosidases. Separate removal of terminal sialic acid with neuraminidase (sialidase) from *V. cholerae* recognizing α2,3-, α2,6- and α2,8-linked terminal sialic acid is used to evaluate the involvement of sialic acid residues in the fucosidase-sensitive *H. pylori* interaction.

Human milk oligosaccharides contain a very large number of complex structures, many of which are fucosylated.[100] Human colostrum S-IgA is highly glycosylated, reflecting this complexity.[140–142] The O-glycosidic-

[133] D. Brassart, A. Woltz, M. Golliard, and J.-R. Neeser, *Infect. Immun.* **59,** 1605 (1991).

[134] J. Sanchez, and G. Jonson, *Acta Pathol. Microbiol. Immunol. Scand.* **98,** 353 (1990).

[135] D. J. Opheim, and O. Touster, this series, Vol. 50, p. 505.

[136] D. Aminoff, this series, Vol. 28, p. 763.

[137] A. Kobata, this series, Vol. 83, p. 625.

[138] P. Scudder, D. C. A. Neville, T. D. Butters, G. W. J. Fleet, R. A. Dwek, T. W. Rademacher, and G. S. Jacob, *J. Biol. Chem.* **265,** 16472 (1990).

[139] J.-R. Neeser, A. Chambaz, M. Golliard, H. Link-Amster, V. Fryder, and E. Kolodziejczyk, *Infect. Immun.* **57,** 3727 (1989).

[140] J. Baenziger and S. Kornfeld, *J. Biol. Chem.* **249,** 7260 (1974).

[141] A. Pierce-Crétel, M. Pamblanco, G. Strecker, J. Montreuil, G. Spik, L. Dorland, H. van Halbeek, and J. F. G. Vliegenthart, *Eur. J. Biochem.* **125,** 383 (1982).

[142] A. Pierce-Crétel, H. Debray, J. Montreuil, and G. Spik, *Eur. J. Biochem.* **139,** 337 (1984).

linked glycans are located at the hinge region of the α chains,[123,124] and the N-glycosidic-linked glycans are present on the α chains,[140] the J chain,[143] and the secretory component.[144,145] In contrast, serum IgA is comparatively less glycosylated.[146]

Our efforts to delineate the receptor analog for *H. pylori* have focused on the distribution of fucose residues in the S-IgA molecule. Western blots of S-IgA and serum IgA probed with panels of lectins and monoclonal antibodies identified common structures such as terminal α-linked fucose (UEA1 and anti-blood group H antibody), α1,6-linked fucose (OPA), and Galβ1,3-linked carbohydrate chains (PNA). Carbohydrate epitopes specific for S-IgA were identified with the anti-Le^b antibody and the LTA lectin, both binding exclusively to the secretory component. This indicates that the fucosidase-sensitive binding epitope for the *H. pylori* interaction with gastric epithelium might also be carried in the secretory component and is dependent on additional fucose residues besides the α1,2-linked fucose terminal. Glycoprotein molecules carry different types of carbohydrate chains, linked either via GlcNAcβ1-*N*-Asn (N-linked)[147] or GalNAcα1-*O*-Ser/Thr (O-linked)[148] anchors to the peptide core. Consequently, receptor-positive glycoproteins will have to be further dissected, as opposed to the glycolipids which have only one sugar chain per molecule. To correlate the binding epitope to the N-linked glycans, as the O-linked glycans are located exclusively at the hinge region of the α chains,[123,124] the S-IgA can be digested by the *N*-glycosidic-specific enzyme *N*-glycosidase F (PNGase F) from *Flavobacterium meningosepticum*.[149] Endoglycosidase F from *F. meningosepticum* is restricted to the biantennary hybrid and complex oligosaccarides in N-linked glycoproteins, and it does not cleave tri- and tetraantennary complex structures.[150] This provides information about the structural complexity of the *H. pylori* binding epitope in S-IgA.

Overlay of the bacteria on Western blots[151,152] of S-IgA and IgA will identify the receptor-active polypeptide, and this subunit component can

[143] J. U. Baenziger, *J. Biol. Chem.* **254,** 4063 (1979).

[144] S. Purkayastha, C. V. N. Rao, and M. E. Lamm, *J. Biol. Chem.* **254,** 6583 (1979).

[145] A. Mizoguchi, T. Mizuochi, and A. Kobata, *J. Biol. Chem.* **257,** 9612 (1982).

[146] J. Descamps, M. Monsigny, and J. Montreuil, *C. R. Hebd, Seances Acad. Sci., Ser. D* **266,** 1775 (1968).

[147] R. Kornfeld and S. Kornfeld, *Annu. Rev. Biochem.* **54,** 631 (1985).

[148] N. Jentoft, *Trends Biochem. Sci.* **15,** 291 (1990).

[149] T. H. Plummer, Jr., J. H. Elder, S. Alexander, A. W. Phelan, and A. L. Tarentino, *J. Biol. Chem.* **259,** 10700 (1984).

[150] A. L. Tarentino, C. M. Gomez, and T. H. Plummer, Jr., *Biochemistry* **24,** 4665 (1985).

[151] J. M. Gershoni, M. Lapidot, N. Zakai, and A. Loyter, *Biochim. Biophys. Acta* **856,** 19 (1986).

[152] A. Prakobphol, P. A. Murray, and S. J. Fisher, *Anal. Biochem.* **164,** 5 (1987).

then be purified after cleavage of disulfide bonds under reducing conditions to release the different subunits [H(α)/L chains, secretory component, and J chain]. Separation of polypeptides can be achieved by gel filtration and affinity chromatography using monoclonal antibodies against the different subunits or polypeptide-specific lectins such as the LTA lectin recognizing the secretory component and the α-chain-specific PNA and Jacalin[153] (IgA_1-specific) lectins.

Receptor-active polypeptides are then subjected to extensive proteolytic digestion to degrade the peptide core to short glycopeptides.[154,155] The glycopeptides can then be group-separated on affinity columns,[156] and the receptor properties of the purified glycopeptides can be assayed. N-linked carbohydrates are released by PNGase F treatment.[150] O-linked oligosaccharides are attached to the peptide core in clusters[157] and are converted to free oligosaccarides by mild alkaline borohydride treatment.[158] Subsequent gel filtration separates the resulting generally small O-linked glycosidic chains from the N-linked oligosaccharides.[159] HPLC-purified[116]- and/or high-performance liquid affinity chromatography (HPLAC)-purified[160] receptor active-oligosaccharides are then subjected to detailed structural characterization using combinations of GC,[83] MS,[85] NMR,[86] and sequential glycosidase digestions[87,88] as described above. Conjugation of oligosaccharides with lipid anchors, e.g., phosphatidylethanolamine-dipalmitoyl,[161] allows further separation on HPTLC and characterization as above.

Potential Diagnostic and Therapeutic Applications

Receptor glycopeptides[162] or oligosaccharides[163] can be conjugated to carrier proteins such as albumin. These neoglycoproteins can be labeled with fluorochromes (FITC, TRITC), enzymes (peroxidase, alkaline phosphatase), or immunological tags (biotin, digoxigenin) and used as probes

[153] H. Ahmed and B. P. Chatterjee, *J. Biol. Chem.* **264,** 9365 (1989).
[154] J. Finne and T. Krusius, this series, Vol. 83, p. 269.
[155] M. Fukuda, this series, Vol. 179, p. 17.
[156] R. K. Merkle and R. D. Cummings, this series, Vol. 138, p. 232.
[157] I. Funakosi and I. Yamashina, *J. Biol. Chem.* **257,** 3782 (1982).
[158] D. M. Carlsson, *J. Biol. Chem.* **243,** 616 (1968).
[159] J. Finne and T. Krusius, *Eur. J. Biochem.* **102,** 583 (1979).
[160] D. Zopf, S. Ohlson, J. Dakour, W. Wand, and A. Lundblad, this series, Vol. 179, p. 55.
[161] T. Mizouchi, R. W. Loveless, A. M. Lawson, W. Chai, P. J. Lachman, R. A. Childs, S. Thiel, and T. Feizi, *J. Biol. Chem.* **264,** 13854 (1989).
[162] Y. C. Lee and R. T. Lee, this series, Vol. 179, p. 253.
[163] P. W. Tang, H. C. Gooi, M. Hardy, Y. C. Lee, and T. Feizi, *Biochem. Biophys. Res. Commun.* **132,** 474 (1985).

in diagnostic and clinical screening procedures. The genetically conserved bacterial adhesins might be more consistently identified than the capsular polysaccharides and fimbria subunit proteins that are characterized by high antigenic drift.[164] Receptor identification could also allow highly specific drug delivery systems in antimicrobial therapy, with antibiotics coupled to receptor-active molecules, promoting specific adherence to the pathogenic bacteria. Antibiotics could also be linked to purified (bacterial) adhesins, priming the antimicrobial substances for the specific niches of the pathogenic bacteria.

Functional Aspects of Microbial Receptors

Characterization of a microbial receptor involves establishment of a functional correlation between the presence of a certain cell surface molecule (the putative receptor), attachment, and biological events triggered by the microbial interaction. An elegant approach reported by Paulson and Rogers[165] demonstrated that influenza virus hemagglutination, abolished after neuraminidase treatment of the cell surfaces, was restored by resialylation of the erythrocytes.

The rapidly expanding field of transgenic techniques offers new, powerful tools for studying molecular microbial pathogenesis *in vivo*. The poliovirus protein receptor has recently been expressed in transgenic mice,[166,167] providing an animal model for poliomyelitis. This illustrates the potential for creating animal models for additional human-specific pathogens, such as *N. gonorrhoeae*. Advances in glycosyltransferase characterization, including cloning of the genes coding for the enzymes responsible for synthesis of major blood group antigens,[168,169] could allow for the introduction of novel glycosylation patterns in transgenic animals. Carbohydrate synthesis is, however, a sequential process depending on multiple enzymes acting in fine-tuned coordinace.[170] Therefore, introducing new carbohydrate epitopes into animals could be considerably more unpredictable. One prerequisite is that the receptor precursor, which also will have to

[164] A.-B. Jonsson, J. Pfeifer, and S. Normark, *Proc. Natl. Acad. Sci. U.S.A.* **89,** 3204 (1992).

[165] J. C. Paulson and G. N. Rogers, this series, Vol. 138, p. 162.

[166] R. Ren, F. Costantini, E. J. Gorgacz, J. J. Lee, and V. R. Racaniello, *Cell (Cambridge, Mass.)* **63,** 353 (1990).

[167] R. Ren and V. R. Racaniello, *J. Virol.* **66,** 296 (1992).

[168] F. Yamamoto, H. Clausen, T. White, J. Marken, and S.-I. Hakomori, *Nature (London)* **345,** 229 (1990).

[169] J. F. Kukowska-Latallo, R. D. Larsen, R. P. Nair, and J. B. Lowe, *Gen. Dev.* **4,** 1288 (1990).

[170] J. C. Paulson and K. J. Colley, *J. Biol. Chem.* **264,** 17615 (1989).

qualify as an acceptor molecule for the glycosyltransferase, is present in the cells expressing the transgene. When such technical problems have been solved, these approaches will be most rewarding for both the scientific and therapeutic disciplines as they enable direct evaluation of the biological impact of complex carbohydrates on microbial pathogenesis.

Acknowledgments

This work was supported by grants from Symbicom, Umeå, Sweden (S.N.) and supported by the Swedish Institute, the Swedish Medical Research Council, the Swedish Society for Medical Sciences (P.F. and T.B.). T.B. is supported by a Public Health Service Fogarty International Research Fellowship (nr. TWO4669-02, number 1430653611Al), and the Swedish Society of Medicine. P.F. is the recipient of postdoctoral fellowships from the William Keck Foundation.

Section IV

Invasion and Intracellular Survival in Eukaryotic Cells

[26] Culture and Isolation of *Chlamydia trachomatis*

By JULIUS SCHACHTER and PRISCILLA B. WYRICK

Introduction

The genus *Chlamydia* contains a diverse group of microorganisms, although only three species (*Chlamydia trachomatis, Chlamydia psittaci,* and *Chlamydia pneumoniae*) are recognized. It is thus impossible to present a single cell line and set of conditions that provide for optimal growth of all these organisms. There are differences in susceptibility of cell lines, incubation temperatures, etc., within a single species. The only known host system that supports growth of all *Chlamydia* is the yolk sac of the embryonated hen's egg. This may still be useful to some workers for production of high yields of fastidious *Chlamydia* isolates. The procedures set forth below include a monolayer system which can be used for isolation of *Chlamydia* or growth of the organism for laboratory purposes. That procedure can use surface areas ranging from single wells in 96-well plates to tissue culture flasks. For those *Chlamydia* that are cytopathic and highly efficient in cell-to-cell transmission, a suspension cell method is described. Finally, a recently developed microcarrier bead procedure that generates high yields is also presented. Chlamydial elementary and reticulate bodies may be purified from tissue culture harvests by use of differential renografin density gradients and cushions.

Growth and Culture

Because *Chlamydia* are obligate intracellular parasites, it is necessary to supply living host cells to support their growth.[1] The introduction of tissue culture isolation procedures made chlamydial isolation clinically relevant. The organism can be recovered from patients in 48 to 72 hr, a period consistent with many other bacteriological procedures. A number of cell lines have been used and a variety of treatments, physical or chemical, have been employed to increase the susceptibility of these cells to chlamydial infection.[2]

Modification of the cell surface charge by pretreatment of the monolayers with diethylaminoethyl (DEAE)-dextran will enhance attachment by non-lymphogranuloma venereum (LGV) *C. trachomatis* strains and

[1] J. W. Moulder, *Microbiol. Rev.* **49,** 298 (1985).
[2] C. C. Kuo, S.-P. Wang, and J. T. Grayston, *Infect. Immun.* **8,** 74 (1973).

some *C. psittaci* strains.[3] With many chlamydial strains of different species, the single most important step in enhancing infection is centrifugation of the inoculum onto the tissue culture monolayer.[4] With the exception of LGV biovar, the *C. trachomatis* strains are not efficient at attaching to and infecting cells *in vitro*. Indeed, a convenient method of differentiating LGV from the trachoma biovar strains is to measure the enhancement of infectivity achieved with centrifugation, for it is minimal with the LGV strains and greater than 10^2 for the non-LGV strains. The centrifugal forces used (ca. 2500 *g*) are inadequate to sediment the chlamydial particles, and the effects may be on the cultured cell membranes.

The most commonly used procedure involves treatment of cells with cycloheximide after the inoculation. Cycloheximide inhibits protein synthesis of the host mammalian cell, providing the chlamydiae with a nutritional advantage. The optimal concentration should be determined in each laboratory. Usually ca. 1 μg/ml provides maximal inclusion counts. Hormone levels may affect chlamydial growth and metabolism. Estradiol enhances the growth of *C. psittaci* and *C. trachomatis* in HeLa cells.[5,6]

Incubation Temperatures

The optimal temperature for incubating chlamydial cultures varies depending on the specific strains being tested. For example, a trachoma biovar has optimal growth at approximately 35°, whereas the LGV biovar *C. trachomatis* grows better at 37°. With *C. psittaci,* different isolates have shown optimal yields at temperatures ranging between 35 and 39°. Thus, the choice of any single temperature must be a compromise. We tend to use 35° where *C. trachomatis* may be involved, but change with other known materials.

Timing Passage

In suspension cultures or in monolayers with rapidly growing isolates capable of cell-to-cell transfer, it is possible to take advantage of the amplifying effects of the initial bursts of infection and maintain the cultures for longer than one or two cycles of chlamydial replication to obtain maximum infectivity. For other instances, as described below in the spinner cultures, the best results are obtained with a high initial infection and a synchronous cycle. Taking advantage of spontaneous cell-to-cell

[3] C. C. Kuo *et al., J. Infect. Dis.* **125,** 665 (1972).

[4] E. Weiss and H. R. Dressler, *Proc. Soc. Exp. Biol. Med.* **103,** 691 (1960).

[5] S. K. Bose and P. C. Goswami, *Infect. Immun.* **53,** 646 (1986).

[6] E. B. Moses *et al., Curr. Microbiol.* **11,** 265 (1984).

infection obviously reduces the need for blind passages and can be labor saving. Thus, there are real advantages to inoculating host cells and incubating for 5–7 days before examination. With isolates not capable of efficient cell-to-cell transfer, where centrifugation may be required to increase infectivity, the optimal time for passage depends on the organism being studied and its developmental life cycle. As most chlamydial species have a cycle in cell culture systems of 48–72 hr, the optimal time appears to be around 72–96 hr for passage as there are always late-maturing inclusions and very large inclusions toward the end of the cycle.

Monitoring or Quantifying Infection

For the most part, detecting chlamydial infection in cell culture systems depends on microscopy to visualize characteristic intracellular inclusions. With experience it is possible to recognize them in unstained living cells. Quantification of infection can be expressed in terms of inclusion-forming units (IFU), either as a percentage of cells infected or per microscopy field. Plaque assays have been described for cytopathic strains.[7]

Control of Bacterial Contamination

Because *Chlamydia* are bacteria, the selection of antibiotics to prevent other bacterial contamination is restricted. Broad-spectrum antibiotics such as tetracyclines, macrolides, and penicillin must be excluded. Aminoglycosides and fungicides are the mainstays. The chlamydial specimens should be refrigerated if they can be processed within 48 hr of collection. If not, they should be frozen at $-70°$.

Biosafety Considerations

The trachoma biovar of *C. trachomatis* is not considered to be a particularly dangerous pathogen to handle in the laboratory. There have been a number of laboratory infections, usually manifested as follicular conjunctivitis. The LGV biovar, however, is a more invasive organism and severe cases of pneumonia have occurred when research workers were exposed to aerosols created by laboratory procedures such as sonication.[8] *C. psittaci* also causes invasive disease and must be considered as a potentially dangerous organism to handle in the laboratory. For many years it was a major cause of laboratory infections. These usually resulted from exposure to aerosols. The stability of the organism is an additional

[7] J. Banks *et al.*, *Infect. Immun.* **1,** 259 (1970).
[8] D. I. Bernstein *et al.*, *N. Engl. J. Med.* **311,** 1543 (1984).

potential problem; the organism should not be handled in laboratories without appropriate containment facilities. *C. pneumoniae* has also caused laboratory infections (after centrifuge accidents), but there is less experience with this organism.

Primary Isolation

All known chlamydiae grow in the yolk sac of the embryonated hen's egg. With centrifugation of the inoculum, it appears that all chlamydiae (with some variability) grow in tissue culture; psittacosis and LGV agents are capable of serial growth in tissue culture without centrifugation. A number of different cell lines have been used to support the growth of chlamydiae. It does not appear that any single cell line has a marked superiority to others, as successful studies have been performed using monkey kidney cells, HeLa cells, L cells and McCoy cells, among others. There is little experience with *C. pneumoniae*. This organism grows poorly, but HL cells have been recommended,[9] and HEp-2 and Vero cells are reasonably susceptible.

The most common technique involves inoculation of clinical specimens onto cycloheximide-treated McCoy cells[10] by centrifugation at approximately 3000 rpm for 1 hr, incubation of monolayers for 48 to 72 hr, and then staining and microscopic examination for inclusions.

Propagation of McCoy Cells

The cells are grown in Eagle's minimum essential medium (MEM) in Earle's salts, containing 10% heat-inactivated fetal calf serum and 2 n*M* L-glutamine, adjusted to pH 7.8 with sodium bicarbonate using a 250-ml tissue culture flask, and incubated at 35° in 5% CO_2. After removal of spent medium and washing of the cultures with phosphate-buffered saline (PBS), the monolayer is trypsinized and the cells are resuspended in growth medium and counted using a hemocytometer. One-milliliter volumes containing at least 50,000 cells/ml are distributed into shell vials and incubated at 37° in a CO_2 incubator with foil caps or a regular incubator if vials are sealed for 3 days. Vials not used for isolation after 3 days of incubation can be kept at room temperature for 3 more days before being discarded.

[9] L. Cles and W. E. Stamm, *J. Clin. Microbiol.* **28,** 938 (1990).
[10] T. Ripa and P.-A. Mardh, *J. Clin. Microbiol.* **6,** 328 (1977).

First-Passage Isolation

The clinical specimen tubes containing swab/cytobrush should be shaken (Vortex) with glass beads for 2 min prior to inoculation. This is safer and more convenient than sonication, although the latter may increase sensitivity. Tissues or aspirates are processed as swab samples except that the material is ground with a mortar and pestle and inoculated at several dilutions, e.g., 1 : 2, 1 : 10, and 1 : 50. Collection medium (growth medium with 100 μg/ml vancocin, 10 μg/ml gentamicin, 10 U/ml mycostatin, 3 μmol/ml glucose, adjusted to pH 7.2–7.4 with sodium bicarbonate) is used throughout the processing of specimens that may contain a heavy growth of bacteria, e.g., rectal samples. Approximately 1.0 ml of isolation medium (same as collection medium but with 50 μg/ml vancocin) is added to the specimen tube. After mixing, each vial (washed once with PBS) is inoculated with 1.0 ml of sample (two vials per specimen) and centrifuged at 3000 rpm for 60 min at room temperature, and the infected monolayers are further incubated in isolation medium containing 1.0 μg/ml cycloheximide for 3 days at 35° in 5% CO_2. One of the two vials for each specimen is stained.

Second Passage

Four days after first passage, the one remaining vial of each specimen is rubber-stoppered and vortexed to detach the cells. Isolation medium (1.0 ml) is added and two fresh vials (prepared for inoculation as described above) are inoculated. After centrifugation and incubation (see above), one vial of each specimen is fixed and stained. If no inclusions are found on reading the first- and second-passage coverslips, then the specimen is considered negative for *C. trachomatis*.

Modifications for Isolation of Chlamydia psittaci

For *C. psittaci* isolation attempts, it may be convenient to lengthen the incubation period to 5 to 10 days before examining the coverslips for inclusions. These organisms do not usually require mechanical assistance for cell-to-cell infection.[11]

Use of Plates Instead of Vials

For laboratories processing large numbers of specimens, it may be convenient to use flat-bottomed 96-well microtiter plates rather than

[11] J. Schachter, N. Sugg, and M. Sung, *J. Infect. Dis.* **137,** 44 (1978).

vials.[12] Cells are planted onto coverslips or directly onto the plates. Processing and incubation are as above, but microscopy is modified to use either long working objectives or inverted microscopes. This procedure is less sensitive than the shell vial technique but offers considerable savings in terms of reagents and time, and may be suitable for settings where mostly symptomatic patients are being screened. These patients usually yield higher numbers of chlamydiae and this minimizes the impact of the decreased sensitivity of the test. Plates with 24 or 48 wells are more sensitive than the 96-well plates.

Staining

Iodine is used for *C. trachomatis* to detect the glycogen-positive inclusions although fluorescent antibody staining may allow earlier detection of the inclusion. Use of fluorescein-conjugated monoclonal antibodies represents the most sensitive method for detecting *C. trachomatis* inclusions in cell culture.[13] The procedure requires more attention to staining than the iodine technique and is more costly. For *C. psittaci* and *C. pneumoniae* the inclusions can be demonstrated with genus-specific monoclonal antibodies or by the Giemsa stain. The manufacturer's instructions should be used for the fluorescent antibody procedures.

Iodine Staining

Aspirate medium, wash coverslip with 1.0 ml of PBS, and fix the infected cells with methanol for 10 min. Following air drying of the coverslip in the vial, stain it with Gram's iodine for 5 min. Remove the stained coverslip from the vial, invert as a wet mount preparation, and examine microscopically. The McCoy cells stain a pale yellow which contrasts with the medium red-brown to dark brown staining of the *C. trachomatis* intracytoplasmic inclusions. For a more permanent record of positive cultures, the coverslip may be returned to the vial for Giemsa staining.

Giemsa Stain (Permanent Stain for Positive Coverslips)

Aspirate iodine from coverslip, decolorize with methanol for 10 min to decolorize, and allow the coverslips to air dry. Add 1.0 ml of dilute Giemsa stain to each vial for 30 min. Following staining, quickly dip the

[12] B. L. Yoder *et al.*, *J. Clin. Microbiol.* **13,** 1036 (1981).
[13] W. E. Stamm *et al.*, *J. Clin. Microbiol.* **17,** 666 (1983).

coverslip in absolute ethanol (twice). Mount coverslip in permamount and examine microscopically.

Freezing

Chlamydiae can be maintained viable by storage at −70°. The coverslip should have a minimum of 200 inclusions on examination. Infected cultures in the remaining vials are removed as described for second passage, inactivated fetal calf serum (50%) is added as a cryoprotectant, and the samples are stored in sealed ampules.

Stationary Cultures

This technique is best suited for those chlamydiae that are capable of unassisted infection of host cells; however, some success has been achieved even with the slow-growing trachoma biovar. Serial passage of the isolates in the relatively susceptible HeLa-229 cell line has resulted in better growth.[14]

Infection

Monolayers are pretreated with 30 μg/ml DEAE-dextran in Hanks' buffered salt solution (HBSS), pH 7.0 (10 ml per 150-cm^2 flask), for 30 min at room temperature. After removal of the solution, the chlamydial stock inoculum is thawed quickly (50°), diluted to the desired concentration with sterile sucrose-phosphate glutamate (SPG), and added (3 ml per 150-cm^2 flask) to the cell monolayers. The monolayers are tilted gently every 15 min during the 2-hr adsorption period at room temperature. The monolayers are washed twice with prewarmed HBSS (37°) and the infected cultures replenished with prewarmed MEM-10 (MEM with 10% fetal bovine serum/1% L-glutamine) containing 1 μg/ml cycloheximide (optional). Incubation is at 35° for the desired period.

Harvest

Decant MEM-10 and add sterile glass beads until they cover half of the bottom of the flask when standing on end. Add at least 10 ml of precooled HBSS (4°) to each flask, and tighten the cap securely. Rock the glass beads over the monolayer until all the cells are dislodged and stream down the flask wall when the flask is stood on end. Agitate violently

[14] T. R. Croy, C.-C. Kuo, and S.-P. Wang, *J. Clin. Microbiol.* **1**, 434 (1975).

(back and forth) in the on-end position for at least 1 min to break open all cells. Withdraw this suspension and transfer to a precooled centrifuge bottle. Then wash each flask with another 10 ml of precooled HBSS and transfer washes to the bottle. Keep all suspensions cold! Sonicate suspensions 6 × 5 sec at 100 W (Inter-probe, Braun). Pellet cell debris at 1000 rpm (IEC) for 10 min and transfer the supernatants to a suitably sized polycarbonate centrifuge tube/bottle. Centrifuge at 36,400 *g* in a refrigerated fixed-angle rotor for 30 min. Discard the supernatants and resuspend the pellets in cold SPG using a No. 17 cannula and an appropriately sized disposable syringe. The procedure can be stopped here by immediately placing the suspensions at −70°. If *Chlamydia* are to be passed directly, use resuspended inoculum to infect fresh monolayers as already described. It is recommended to always test the inoculum for sterility with a thioglycolate tube and blood agar slant for each strain harvested.

Suspension Cultures

To obtain large quantities of infectious elementary bodies (EBs) for freezer stocks and metabolically active reticulate bodies (RBs) for experimental analyses, suspension cultures are available and are more cost effective than propagating the organisms in monolayer cultures in flasks or roller bottles. Certain strains of *C. psittaci* and the LGV biovar grow well in the L929 cell line adapted for suspension culture by Tamura and Higashi[15]; however, members of the *C. trachomatis* non-LGV biovar do not grow, or grow poorly, in L929 cells. Recently, the microcarrier bead culture technology was exploited for use with McCoy cells so that the attributes of suspension culture could be applied to the *C. trachomatis* biovar.[16] The latter method was described using a genital *C. trachomatis* isolate (serovar E), but the technology should be readily accessible for growing the ocular serovars (A, B, Ba, and C) because they, too, are routinely cultured in McCoy cells.

The reason for the high yields of infectious EBs from the microcarrier bead culture system is believed to be twofold. Growth of the anchorage-dependent McCoy cells on a porous matrix permits the cells to feed naturally from their basolateral domain. As such, the nutritional status of the cells is greater than when the cells are grown to confluency on plastic surfaces. A healthier host cell, in turn, appears to be important for optimal

[15] A. Tamura and N. Higashi, *Virology* **20,** 596 (1967).
[16] J. E. Tam *et al., Biotechniques* **13,** 374 (1992).

chlamydial development. The result appears to be more efficient conversion of noninfectious RBs into infectious EBs.

L929 Suspension Cells

Infection

Resuspend L929 cells (2×10^5 cells/ml) in prewarmed suspension medium (MEM-10) and pool in a siliconized 125-ml Erlenmeyer flask with a screw-cap top (final volume 35–50 ml). Add a desired amount of inoculum, and place the sealed flask in a rotary shaker water bath for 1 hr at 35° with moderate/low agitation to allow the chlamydiae to absorb. Sediment the infected suspension, and resuspend in prewarmed (35°) suspension medium (final volume 1000 ml), and incubate. After a few minutes, remove an aliquot containing 2×10^5 cells, and plant in a 1-dram shell vial containing prewarmed MEM-10 (final volume 1 ml). This is used to determine the rate of infection.

Harvest and Purification

Centrifuge (1000 rpm) infected suspension cultures for 10 min, resuspend the infected cell pellets in HBSS (150 ml), and sonicate (e.g., 6×5 sec) to release EBs.[17] Remove cell debris by slow-speed centrifugation (e.g., 1000 rpm) for 10 min, and transfer the supernatants into polycarbonate centrifuge bottles for high-speed centrifugation (36,400 *g*) for 90 min. Resuspend the pellet(s) containing EBs and subcellular components in HBSS (30 ml), layer (5 ml per gradient) on discontinuous Renografin gradients (40, 44, and 54%), and centrifuge for 1 hr (18,000 rpm in SW 27). Collect the EB band above 54% renografin, wash it, and resuspend in SPG.

McCoy Cells on Microcarrier Beads

Seeding the Cytodex Beads

Cytodex 3 Microcarrier beads (Pharmacia; distributed by Sigma) are coated with denatured collagen and supplied as a dry powder. A stock of 1 g beads in 50 ml PBS is prepared, autoclaved, and stored in a silicon-coated bottle at 4°. The bead concentration in the stock can be estimated by making three 1 : 10 serial dilutions and then counting the number of beads in several 50-μl drops by light microscopy.

Add 5×10^4 to 1×10^5 sterile beads/ml (~15–20 ml stock, or 2.5 mg/ml Cytodex beads) to 100 ml Eagle's MEM containing glutamine (2 m*M*)

[17] H. D. Caldwell *et al.*, *Infect. Immun.* **31,** 1161 (1981).

and 10% heat-inactivated fetal bovine serum. Place the suspension in a silicon-coated Pyrex (125-ml capacity) slow-speed stirring flask (Corning) and let it equilibrate a few hours or overnight (optional). With too many beads, the mechanical force of the paddle blades hitting the beads or the beads bumping together shears the McCoy cells from the bead surface. Harvest a near-confluent monolayer of McCoy cells from a 75-cm^2 flask ($5–6 \times 10^6$ cells) in 10 ml medium and add to the beads. The ratio of cells per bead should be ~50 to 100 : 1. Place the flask on a magnetic stir plate set at the lowest possible speed sufficient to permit rotation of the paddle blades to keep the beads in suspension (Fig. 1A). Incubate the flask at 37° for 2–4 days until the McCoy cells have formed a confluent monolayer on the beads. Each day, examine by light microscopy a drop of the suspension to monitor the progress of McCoy cell monolayer development (Figs. 2A, B, E, F). Note that there will always be a few beads devoid of McCoy cells. Epithelial cells, such as HeLa and HEC-1B cells, can also be grown on the beads, but it is more difficult to control the evenness of the monolayer with these cells; there is a tendency for the epithelial cells to form clumps, or "organoids," on the beads (Fig. 2C), which eventually break off.

Infecting the Bead Cultures

Remove the culture flask from the stir plate and allow the beads to settle to the bottom of the flask (5–10 min; Fig. 1B). Discard most of the spent culture medium (~110 ml), leaving about 15 ml to keep the cells and beads moist. Add 50–100 ml sterile, prewarmed PBS to the flask, gently rotate it to wash the bead cultures, and discard the wash fluid (Fig. 1C). Rapidly thaw a stock of *C. trachomatis* EBs, sonicate in a water bath sonicator for several minutes to break up any EB aggregates, and dilute the stock in 2× SPG to a concentration that will give a 90% infection in 5×10^7 to 1×10^8 McCoy cells (one roller bottle equivalent). Add this inoculum to the washed bead culture slurry and bring the volume to a total of 25–30 ml with prewarmed MEM (Fig. 1D). As this volume is below the paddle blades, place the flask on a gyratory shaker and allow it to rotate gently for the 2-hr adsorption period at 35° (Fig. 1E). Following adsorption infection, top up the culture with 95–100 ml complete MEM [with or without cycloheximide (0.5 μg/ml)] and return the flask to the magnetic stir plate in a 35° incubator for the infection to proceed (Fig. 1F). Monitor the infection by examining a few drops of the culture by light microscopy. The inclusions will appear as "blisters" in the McCoy cells (Fig. 2D). The curvature of the beads restricts the area of the monolayer examined, but with practice, patience, and continuous refocusing of the microscope, a good estimate of the number of cells infected as well as the stage of chlamydial development can be determined.

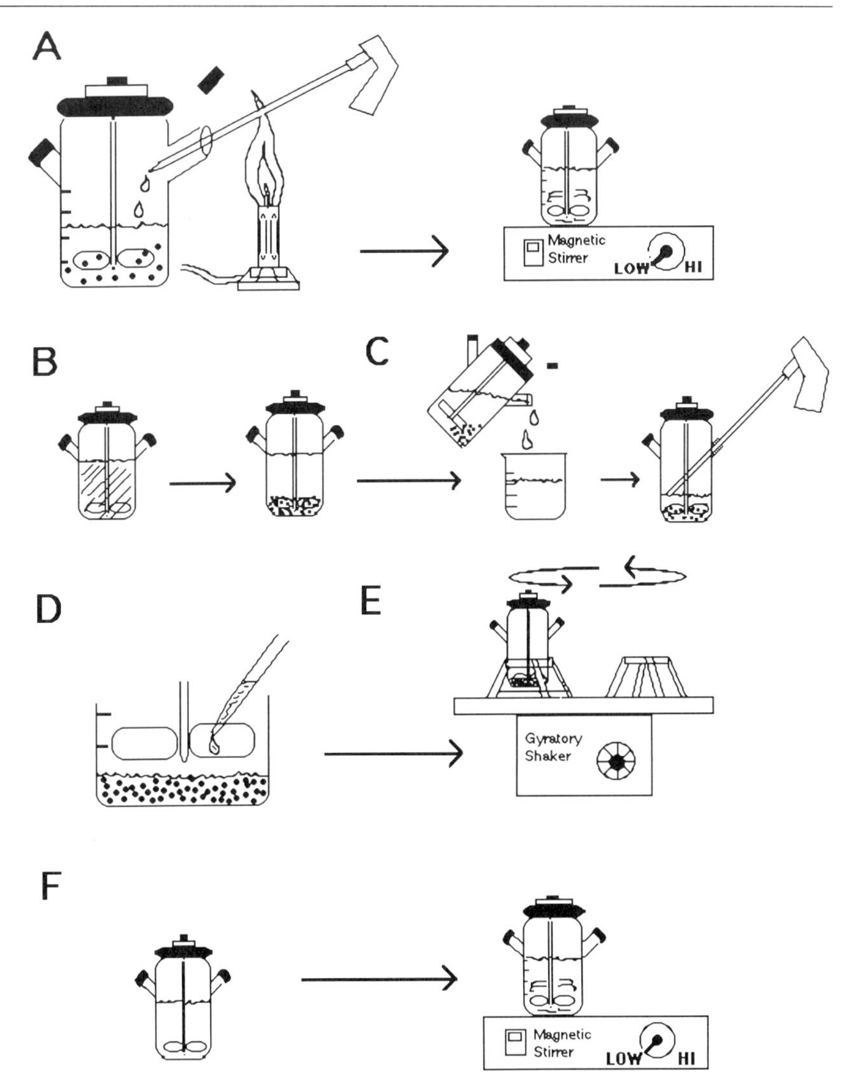

FIG. 1. Diagrammatic scheme for processing microcarrier bead cultures. See text for details.

Harvesting Elementary Bodies from Infected Bead Cultures

Allow the beads to settle out and pour off the culture supernatant into a cold sterile container placed on ice. Transfer the McCoy-coated beads to a 50-ml culture tube. Wash the spinner flask with a small volume of PBS devoid of Ca^{2+} and Mg^{2+} and combine the wash fluid with the culture supernatant. Draw off the excess liquid from the beads in the tube, add

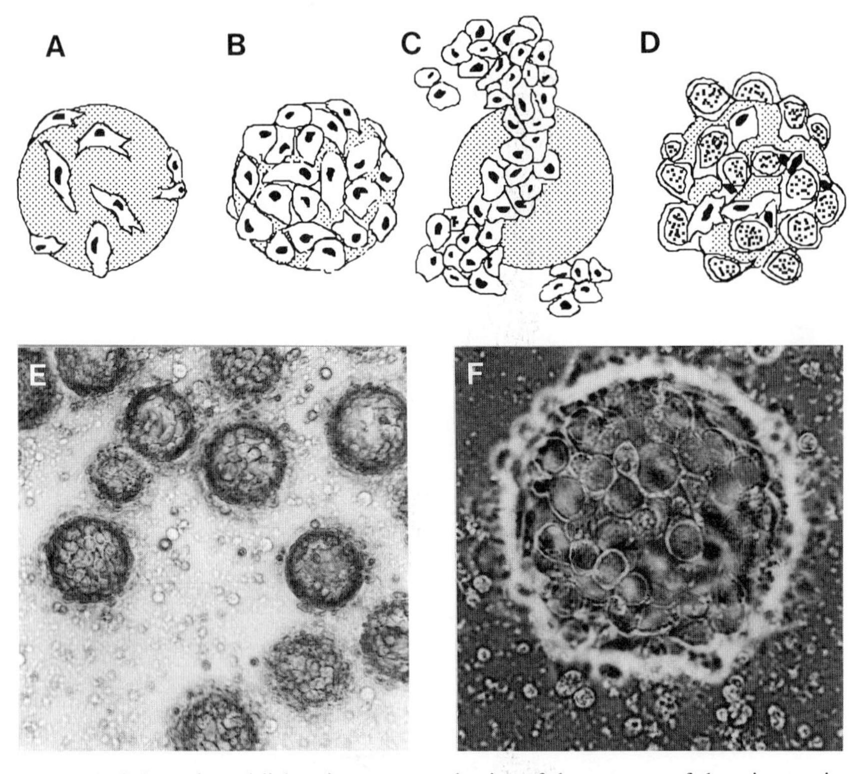

FIG. 2. Schematic and light microscope evaluation of the progress of the microcarrier bead cultures. See text for details.

a mixture of collagenase/dispase [Boehringer-Mannheim Biochemicals; working concentration: collagenase (0.1 unit/ml) and dispase (0.8 unit/ml), prepared in PBS devoid of Ca^{2+} and Mg^{2+} and filter sterilized], and adjust the total volume of the slurry to 25 ml in the PBS. Shake the tube gently for 10–15 min at 37° to allow the enzymes to contact the cells. The enzymes are then activated by the addition of $CaCl_2$ (0.9 m*M*) and $MgCl_2$ (0.5 m*M*). Continue gentle shaking of the tube for ~10 min. As the enzymes act on the collagen matrix, the McCoy cells are released from the beads. The slurry becomes noticeably turbid. Remove the infected cell suspension and combine it with the previous supernatant and wash. Wash the denuded beads several times with PBS and combine these washes with the cells. The infected cells are then washed and sonicated to release the chlamydiae, and the EBs are harvested as described earlier for harvest and purification of L929 suspension cells.

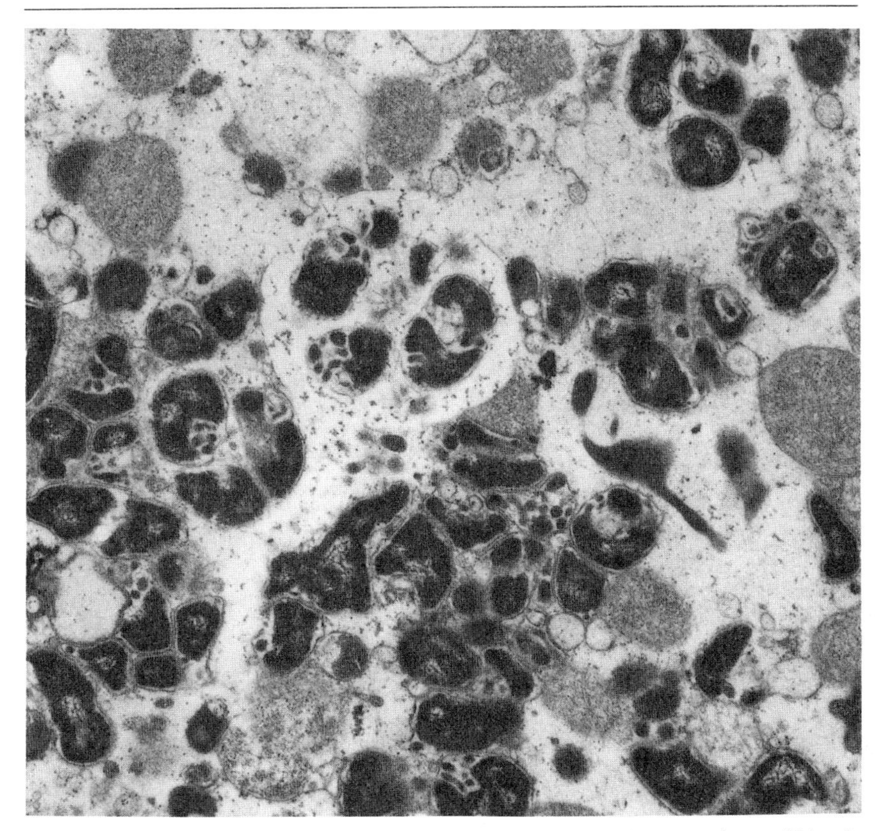

FIG. 3. Electron photomicrograph of isolated chlamydial inclusions at early to middevelopmental cycle during harvest of reticulate bodies.

Harvest and Purification of Reticulate Bodies

For harvest of metabolically active, osmotically fragile *C. trachomatis* RBs, infected McCoy cells are collected at 36 hr, detached from the beads as described, and washed twice in cold PBS, pH 7.0.[18-20] The final pellet of infected McCoy cells is resuspended in a sucrose (0.25 *M*)–EDTA (1 m*M*)–Tris (10 m*M*) buffer and placed on ice for 60 min. The cell suspension should be shaken gently every 10–15 min. Following centrifugation of the cell suspension at 500 *g* for 10 min, the infected cells are resuspended in cold HBSS containing $CaCl_2$ (0.14 g/liter) and $MgCl_2 \cdot 6H_2O$ (0.1 g/liter) and subjected to 15–20 strokes in a tight-fitting Dounce homog-

[18] A. Matsumoto, *J. Bacteriol.* **145,** 605 (1981).
[19] T. P. Hatch, E. Al-Hossainy, and J. A. Silverman, *J. Bacteriol.* **150,** 662 (1982).
[20] T. Hackstadt, W. J. Todd, and H. D. Caldwell, *J. Bacteriol.* **161,** 25 (1985).

enizer. The lysate is centrifuged for 5 min at 500 g. Supernatants containing released RBs are pooled; pellets containing whole cells are rehomogenized. A drop of the homogenate should be examined by light microscopy to monitor the effectiveness of cell breakage and to ensure release of intact RBs. The homogenization cycle may need to be repeated two or three times for optimal release of RBs from the infected cell population. The pooled RBs are exposed to DNase (1 mg/ml prepared in HBSS containing Ca^{2+} and Mg^{2+}) for 30 min at 4° and then layered onto a 30% Renografin cushion (prepared in PBS) for centrifugation for 1 hr at 95,000 *g* (SW27 rotor). This step effectively removes much of the contaminating eukaryotic cytoplasmic material from the RB pellet (Fig. 3). The bouyant density of RBs is 1.16–1.18, which is almost identical to that of mitochondria, and the two are difficult to separate by density centrifugation. Thus, there is always some mitochondrial contamination of RB preparations. The RBs are then subjected to discontinuous Renografin gradient centrifugation, as described for the purification of EBs, and collected from the diffuse band between the 40 and 44% layers. The diffuse band is mixed with an equal volume of Tris (10 m*M*)–sucrose (0.25 *M*) buffer, pH 7.4, and centrifuged for 40 min at 30,000 *g* (Sorval RC-5). The final RB pellet is resuspended in the desired solution and should be used immediately for metabolic and biochemical studies.

Acknowledgments

We thank Jane Raulston for Figs. 1 and 2A–D, Stephen Knight for the Figs. 2E and F, and Carolyn Davis for the RB preparation in Fig. 3.

[27] Culture of *Treponema pallidum*

By DAVID L. COX

A Brief History of Efforts to Cultivate *Treponema pallidum in Vitro*

Many attempts have been made to cultivate *Treponema pallidum in vitro* since it was discovered by Schaudinn and Hoffmann[1] in 1905 to be the etiological agent of human syphilis. As early as 1913, scientists recognized the need for the presence of host tissue to increase the probability of successful cultivation. In 1948, Perry[2] reported a tissue culture

[1] F. Schaudinn and E. Hoffmann, *Arb. Gesund.* **22,** 527 (1904–1905).
[2] W. L. M. Perry, *J. Pathol. Bacteriol.* **60,** 339 (1948).

method from which viable, virulent treponemes could be recovered for 5 to 7 days, but evidence for treponemal growth and replication was not documented. Wright[3] published a report of an extensive study using a variety of tissue cultures including cells from human (HeLa, amnion, and embryonic liver) and rabbit (testis, liver, and kidney) tissue. His studies made the first attempts to reduce exposure of the cultures to oxygen, a condition that later proved to be critical for *in vitro* cultivation. He was also the first to observe that *T. pallidum* firmly attached to the surface of the tissue culture cells.

In 1975, Fitzgerald *et al.*[4] reported the rapid attachment of treponemes to primary rabbit testicular cells. Nonpathogenic spirochetes, *T. denticola* and *T. phagedenis* biotype Reiter, failed to attach. Transmission electron micrographs documented their results, which were followed by another study using scanning electron microscopy.[5] In addition, Fitzgerald *et al.*[6] investigated the ability of *T. pallidum* to attach to 19 different mammalian cells. They demonstrated that attachment could be blocked with immune syphilitic rabbit serum. In an attempt to characterize optimal conditions for incubation, Fitzgerald *et al.*[7] investigated the effects of oxygen, reducing agents, and serum supplements on the viability of *T. pallidum in vitro*. Tissue culture medium containing millimolar concentrations of glutathione, cysteine, and dithiothreitol in an atmosphere of 3% oxygen maintained the number of virulent *T. pallidum* for 6 days.

Also in the mid-1970s, another laboratory headed by Dr. A. H. Fieldsteel was investigating conditions for the *in vitro* survival and growth of *T. pallidum*. Several discoveries came from these studies. First, they were one of the first to incorporate dithiothreitol (DTT), a potent reducing agent, into their medium.[8] Second, they developed a gradient culture system in which monolayers of Sf1Ep cells on glass coverslips were placed vertically in Leighton culture tubes. This established an oxygen gradient in which they could monitor the depth at which the treponemes best survived.[9] They were able to recover virulent treponemes from these cultures for up to 21 days. A small degree (3- to 5-fold) of replication was observed, but because of the difficulties in substantiating these small

[3] M. I. Wright, *Int. Congr. Ser.—Excerpta Med.* **55,** 884 (1963).

[4] T. J. Fitzgerald, J. N. Miller, and J. A. Sykes, *Infect. Immun.* **11,** 1133 (1975).

[5] T. J. Fitzgerald, P. Cleveland, R. C. Johnson, J. N. Miller, and J. A. Sykes, *J. Bacteriol.* **130,** 1333 (1977).

[6] T. J. Fitzgerald, R. C. Johnson, J. N. Miller, and J. A. Sykes, *Infect. Immun.* **18,** 467 (1977).

[7] T. J. Fitzgerald, R. C. Johnson, J. A. Sykes, and J. N. Miller, *Infect. Immun.* **15,** 444 (1977).

[8] A. H. Fieldsteel, Annual Progress Report to NIH: "Studies on in vitro Cultivation of *Treponema pallidum*." SRI International, Menlo Park, CA, 1975.

[9] A. H. Fieldsteel, F. A. Becker, and J. G. Stout, *Infect. Immun.* **18,** 173 (1977).

increases and the earlier report by Sandock *et al.*[10] of unsustained multiplication of *T. pallidum* cultures in the presence of oxygen, successful cultivation was not claimed. Third, they identified a cell line of cotton-tail epithelium, Sf1Ep (NBL-11), which appeared to support the maintenance of virulent *T. pallidum* longer than other cell lines.[11] Lastly, they discovered that only certain lots of fetal bovine serum (FBS) were capable of supporting the survival of *T. pallidum in vitro* and that some lots were very toxic.[12] These discoveries set the stage for the development of the first successful cultivation of *T. pallidum* after 75 years of unsuccessful attempts.

In 1979, Dr. Fieldsteel's laboratory began a study to substantiate that some multiplication was occurring in the gradient cultures by duplicating the conditions that existed at the optimal level in the gradient (17 mm below the surface). With a polarigraphic oxygen-sensing probe, it was determined that the oxygen concentration at this depth was between 1.5 and 2.0%. This environment was then duplicated in horizontal Sf1Ep monolayers. A 6- to 10-fold increase in the numbers of viable treponemes was observed with an inoculum of 10 million organisms. High concentrations of cysteine and glutathione were removed from the basal reduced medium (BRM), which also contained DTT, because these compounds are easily oxidized in the presence of oxygen. This consumed the reducing potential of the DTT in the medium. After it was discovered that the cultures became depleted of glucose after 7 days, the glucose concentration was raised from 100 to 250 mg/dl. This new medium was named BRMM (basal reduced medium modified). In tissue cultures with BRMM, substantial increases in treponemes were observed: 12-fold from an inoculum of 10 million treponemes, 27-fold from an inoculum of 2.5 million treponemes, and 49-fold from an inoculum of 1 million.[13] These increases were documented by concurrent increases in treponemal DNA. Because of past problems encountered by other investigators to reproduce cultivation claims in outside laboratories, these methods were corroborated by Dr. Steven Norris who was then at the University of California at San Diego.[14]

Additional studies[15] demonstrated that (1) the cultures would support treponemal growth under a wide range of oxygen concentrations (1.5–5.0%); (2) the optimum cultivation temperature was 34°; (3) the optimum serum concentration was 20% (v/v); and (4) an extract made from

[10] P. L. Sandock, H. M. Jenkin, H. M. Matthews, and M. S. Roberts, *Infect. Immun.* **19,** 421 (1978).

[11] A. H. Fieldsteel, J. G. Stout, and F. A. Becker, *Infect. Immun.* **24,** 337 (1979).

[12] A. H. Fieldsteel, J. G. Stout, and F. A. Becker, *In Vitro* **17,** 28 (1981).

[13] A. H. Fieldsteel, D. L. Cox, and R. A. Moeckli, *Infect. Immun.* **32,** 908 (1981).

[14] S. J. Norris, *Infect. Immun.* **36,** 437 (1982).

[15] A. H. Fieldsteel, D. L. Cox, and R. A. Moeckli, *Infect. Immun.* **35,** 449 (1982).

TABLE I
USE OF VARIOUS TISSUE CULTURE VESSELS FOR CULTIVATION

Vessel	Surface area (cm^2)	Sf1Ep	Treponemes	TpCM (ml)	Maximum yield
24-well plate	2.0	40,000	300,000	1.5	2.5×10^7
6-well plate	10.0	250,000	1,250,000	5.0	1.0×10^8
25-cm^2 flask	25.0	500,000	2,500,000	10.0	2.0×10^8
75-cm^2 flask	75.0	1,500,000	10,000,000	25.0	5.0×10^8
150-cm^2 flask	150.0	2,500,000	20,000,000	50.0	1.0×10^9
125-ml Magna-flex[a]	180.0	2,500,000	20,000,000	60.0	1.0×10^9

[a] Wheaton Instruments, Milleville, NJ.

testicular tissue stimulated the replication of *T. pallidum in vitro*. Although substantial increases were observed in primary cultures, subsequent passaged cultures yielded fewer and fewer treponemes with each passage. Even though several improvements have been made to increase the yield of treponemes in primary passage, this problem remains today.

Cultivation Methods

Cultivation of Tissue Culture Cells

All cell lines are cultivated and passed in Eagle's minimal essential medium (EMEM, GIBCO, Grand Island, NY) containing 10% FBS unless otherwise noted. It is critical that the medium be free of any antibiotics because they will prevent infection of the monolayers with *T. pallidum*. Sf1Ep cells grow slowly compared with other cell lines which may contribute to their ability to support the growth of *T. pallidum*. Approximately 4×10^6 cells in 25 ml of EMEM are seeded into a 150-cm^2 tissue culture flask. The medium is removed and replaced with fresh EMEM after 5 and 9 days of incubation. The cells are harvested and passed every 2 weeks.

Establishment of Tissue Culture Monolayers for Treponema pallidum Cultivation

Treponema pallidum can be cultivated in a variety of tissue culture vessels; 24-, 12-, and 6-well culture plates, as well as 25-, 75-, and 150-cm^2 tissue culture flasks, have been used successfully to cultivate *T. pallidum* (Table I). Cultivation in suspension cultures of Sf1Ep cells in Magna-Flex flasks has also been successful.[16] The amount of cells varies for each type

[16] B. S. Riley and D. L. Cox, *Appl. Environ. Microbiol.* **54,** 2862 (1988).

of vessel. Usually, the cultures are set up 2 days before infection and confluency is between 20 and 25%.

Preparation of Basal Reduced Medium Modified and Treponema pallidum Cultivation Medium

The formulations for BRMM and *T. pallidum* cultivation medium (TpCM) are given in Table II. The major difference is that TpCM contains more antioxidants than did its precursor, BRMM. BRMM with 10% FBS was used for the extraction of treponemes from rabbit testicular tissue, and TpCM, for cultivation with tissue culture cells. The rationale is that because the treonemes are in the extraction medium such a short time, it is not necessary to have all of the antioxidants present. The base solution for TpCM and BRMM consists of the first eight components (Earle's salt solution through MOPS buffer). Usually 500 ml of TpCM base is prepared the day before tissue cultures are to be infected. The pH of the base is adjusted to 7.4 before sterilization. Five hundred-milliliter filtration units

TABLE II
FORMULATION OF BRMM AND TpCM

Component	BRMM	TpCM	Units
Earle's salt solution, 10×	100.0	100.0	ml/liter
Amino acids, 50× [a]	20.0	10.0	ml/liter
NEAA, 100× [a]	10.0	10.0	ml/liter
Vitamins, 100× [a]	10.0	10.0	ml/liter
Glucose [b]	2500.0	2500.0	mg/liter
L-Glutamine, 200 m*M* [a]	10.0	10.0	ml/liter
$NaHCO_3^-$, 7.5% (w/v) [a]	27.0	27.0	ml/liter
MOPS buffer, 1 *M* (pH 7.4) [b]	25.0	25.0	ml/liter
Sodium pyruvate [b]	100.0	100.0	mg/liter
Dithiothreitol [b]	100.0	150.0	mg/liter
$CoCl_2$ [b]	0.0	5.0	μg/liter
Cocarboxylase [b]	0.0	2.0	μg/liter
Mannitol [b]	0.0	0.1	g/liter
Histidine [b]	0.0	0.05	g/liter
Catalase [b]	0.0	10,000.0	U/liter
Bovine superoxide dismutase [b]	0.0	25,000.0	U/liter
FBS [c]	200.0	100–200.0 [d]	ml/liter
Pyrogen-free water [e]	Fill to 1 liter		

[a] Flow Laboratories, Rockville, MD.
[b] Sigma Chemical Company, St. Louis, MO.
[c] HyClone Laboratories, Logan, UT.
[d] Depends on each lot; some lots support better growth at 10% (v/v).
[e] Abbott Laboratories, Abbott Park, IL.

equipped with membranes with 0.2-μm pores (Nalgene, Rochester, NY) are used for sterilization. For extraction, 45 ml of base and 5 ml of FBS are placed in a sterile 250- or 500-ml suction (side-arm) flask and sealed. The flask is alternately evacuated and gassed with a 5% CO_2 and 95% (v/v) N_2 mixture three times and stored. The next day a fresh solution (5 mg in 2 ml of base) of DTT is prepared, filter sterilized using a syringe filter with 0.2-μm membrane pores, and added to the flask. The flask is regassed as described above and stored for the extraction of treponemes.

It must be noted here that the lot of FBS used for extraction and cultivation must first be screened for its ability to support treponemal growth; many lots of FBS are toxic.[15] Screening is accomplished by obtaining a lot whose potential has already been established. Several samples can then be obtained from various serum companies and compared with the reference lot. In our experience, about one out of every four or five is suitable for cultivation. All lots used in our experiments supported growth that was 90% or greater than that of the reference lot.

Preparation of Inoculum of Treponema pallidum

New Zealand White male rabbits (10–12 pounds) are used for testicular passage and as a source of treponemes for inoculation into tissue cultures. For passage, 10^8 treponemes in 0.25 ml of TpCM are inoculated into rabbit testis. This inoculum originates from a frozen suspension of treponemes (4×10^8/ml) from a previous rabbit harvest. Ten days after infection, the rabbits are sacrificed and the testes aseptically removed. To extract the treponemes, the testis are trimmed of any fatty tissue and minced in a sterile petri dish. The minced tissue is placed into a sterile 125-ml Erlenmeyer flask and 12 ml of fresh BRMM (preparation described above) is added to it. The flask is gassed with a mixture of 93% N_2/5% CO_2/3% O_2 (v/v) for 30 sec and sealed with a sterile silicone stopper. The flask is placed on a shaking machine (Eberbach, Ann Arbor, MI) operating at 120 oscillations per minute for 30 min. The BRMM containing treponemes and testis extract is removed from the minced testicular tissue with a pipette and transferred into a 50-ml polypropylene conical centrifuge tube. The tube is briefly gassed with the gas mixture described above and sealed. The extract is then centrifuged at 500 *g* for 10 min at 4° to remove gross debris. After centrifugation, the extract is removed from the debris pellet and placed into a new sterile tube. One milliliter is removed from the treponemal suspension and 1 : 10 and 1 : 100 dilutions of the treponemal suspension are prepared and counted by dark-field microscopy. The remaining treponemal suspension is gassed and sealed for later use. This suspension is used as a source of inoculum and also for the preparation of testis extract.

Preparation of Testis Extract for Cultivation

Testis extract (TEx) is prepared by removing most of the treponemes from a portion of the treponemal suspension. This process involves centrifugation at 12,000 *g* for 10 min. The supernate is removed with a pipette and placed into a 50-ml polypropylene tube, gassed with the CO_2/N_2 mixture, and sealed. The TEx is heat inactivated at 56° for 30 min. Any precipitate is removed by centrifugation 12,000 g for 10 min.

Preparation of TpCM for Cultivation

TpCM is made from the base medium by adding the appropriate amounts of DTT, sodium pyruvate, $CoCl_2$, cocarboxylase, mannitol, histidine, catalase, bovine SODase, and FBS for the volume of medium prepared. This is usually prepared the day the tissue cultures are infected. The medium is alternately evacuated and gassed with a 5% CO_2 and 95% N_2 mixture three times. The appropriate number of treponemes is added to the TpCM before infection of the tissue cultures.

Infection of Tissue Culture Cells

To infect tissue cultures, the EMEM is removed with a pipette and replaced with the appropriate volume of TpCM containing the treponemes. If the tissue culture vessel is a flask, it is briefly gassed with 5% CO_2 and 95% N_2 and sealed. In contrast, tissue culture plates are placed into a vacuum chamber (Coy Laboratory Products, Inc., Ann Arbor, MI), evacuated (to -10 psi), and gassed with a 5% CO_2 and 95% N_2 mixture three times. This is especially critical for 24-well plates because it rapidly reduces the atmospheric oxygen in each well, especially the wells in the center of the plate. Otherwise, these wells would become microaerophilic much more slowly than the outside rows of wells because the moisture retention rings in the lid of the plate would reduce air flow. When all of the cultures have been prepared, the caps on tissue culture flasks are loosened and the tissue culture plates removed from the vacuum chamber. All cultures are placed in a Tri-Gas incubator (Forma Scientific, Marietta, OH) where a microaerophilic environment (3.5% O_2, 5% CO_2, and 91.5% N_2) is maintained throughout the cultivation period.

Quantitation of Treponemal Growth

Table III lists the amounts of trypsin–Versene (TV) solution and PBS used to harvest treponemes cultivated in vessels of varying size. Treponemal growth is routinely monitored by harvesting the cultures using a mixture of trypsin–Versene (0.05% trypsin–0.53 m*M* EDTA) suspended

TABLE III
HARVESTING TREPONEMES FROM VARIOUS TISSUE CULTURE VESSELS

Vessel	PBS rinse (ml)	TV solution (ml)
24-well plate	0.5	1
6-well plate	1	2
25-cm^2 flask	2	5
75-cm^2 flask	10	15
150-cm^2 flask	20	25
125-ml Magna-flex[a]	20	25

[a] Wheaton Instruments, Millville, NJ.

in PBS. To harvest the cultures, the TpCM is removed and placed into a 15-ml conical centrifuge tube. The cultures are then rinsed with 2 ml of PBS. The PBS is removed and placed in the conical centrifuge tube with the TpCM. Three milliliters of the TV solution is then added to the cultures and they are briefly gassed with 5% CO_2–95% N_2. The centrifuge tubes are also gassed. After the cultures are incubated at 34° for 5 min, the culture monolayers are repeatedly rinsed with the TV solution to dislodge all the tissue culture cells. This is then combined with the TpCM in the centrifuge tube. If the sample is stored for more than 5 min before counting, it is gassed with CO_2/N_2.

The total number of treponemes in a culture is determined by dark-field microscopy. Four microliters of the harvest suspension is placed on a Hawksley counting slide and counted (unfortunately these slides are no longer manufactured by Hawksley). An alternate method is to place 10 μl of harvest suspension onto a glass slide and carefully drop a 22 × 22-mm coverslip onto the liquid. The difficulty with this method is that the slide must float on the liquid droplet without protruding from the edges of the coverslip. The number of treponemes in 10 fields, selected randomly over the entire slide, is counted at a magnification of 800×. Both the number of treponemes and the number of motile treponemes are recorded. The number of treponemes per field is calculated and multiplied by a conversion factor (4.72×10^5) to obtain the number of treponemes per milliliter of harvest suspension. [The conversion factor will vary from microscope to microscope and must be determined using a calibration slide (Zeiss, Germany).] This number is multiplied by the volume of harvest suspension to determine the total number of treponemes from the culture. Cultures are routinely counted after 5, 7, 10, 12, 14, and 17 days of incubation to obtain a complete growth curve.

Growth of Treponema pallidum in Suspension Cultures Using Microcarrier Beads

Methods have been developed to cultivate *T. pallidum* on Sf1Ep cells attached to microcarrier beads.[16] The rationale for this development was to provide a better vehicle for serial passage experiments. Suspension cultures would be much easier to transfer from flask to flask because harvesting using TV would not be required. Thus, the cells and the treponemes would experience minimal disturbance during the suspension culture passage process as compared with more rigorous harvesting using TV.

Either polystyrene (Cytospheres, Lux, Newbury Park, CA) or collagen-coated dextran beads (Cytodex-3, Pharmacia Fine Chemicals, Piscataway, NJ) can be used to cultivate Sf1Ep cells for infection with *T. pallidum*. The polystyrene beads that come presterilized are easier to use, but cannot be viewed by dark-field microscopy because of their high refractivity. The collagen-coated dextran beads can be viewed by dark-field microscopy, but extra steps are required in their preparation. The polystyrene beads are suspended in EMEM with 10% FBS at a concentration of 120 mg/ml. The cell cultures with polystyrene beads contain 600 mg of beads, 30 ml of EMEM, and 4 ml of FBS.

In contrast, the dextran beads come as a powder and are hydrated in 10 m*M* PBS (pH 7.4) for 4 hr at room temperature. The beads are then washed three times with PBS by allowing the beads to settle, carefully removing most of the liquid above the beads, and adding more PBS. They are then sterilized by autoclaving. After the suspension cools, the beads are washed twice with EMEM and brought to a final concentration of 4 mg/ml. They may be stored at 4° until needed. Before the cells are added, 40 mg of the beads is incubated with 3 mg of rabbit plasma fibronectin in 30 ml of EMEM in a siliconized (Sigmacote, Sigma Chemical Company, St. Louis, MO) 125-ml MagnaFlex (Wheaton Instruments, Millville, NJ) stirring flask. Then, 5 ml of FBS is added to the flask, and the flask is stirred at 40 rpm for 1 hr on a variable-speed stirring table (MicroStir, Wheaton Instruments, Millville, NJ).

The desired number of Sf1Ep cells (see Table I) is added to the flask and the atmosphere in the flask is purged with 5% CO_2–95% N_2. The flask is then placed on the stirring table in a 34° incubator and stirred at 40 rpm for 1 min out of every 45 min. After 24 hr, 10 to 15 ml of fresh EMEM is added to the flask.

After the cells are incubated for 48 hr, half of the EMEM is removed and replaced with TpCM. This is repeated three times. The total volume of medium is adjusted to 60 ml by final addition of TpCM. Then 2 ml of heat-inactivated testis extract containing 2×10^7 treponemes is added to the flask and the atmosphere in the flask is purged with 5% CO_2–95% N_2.

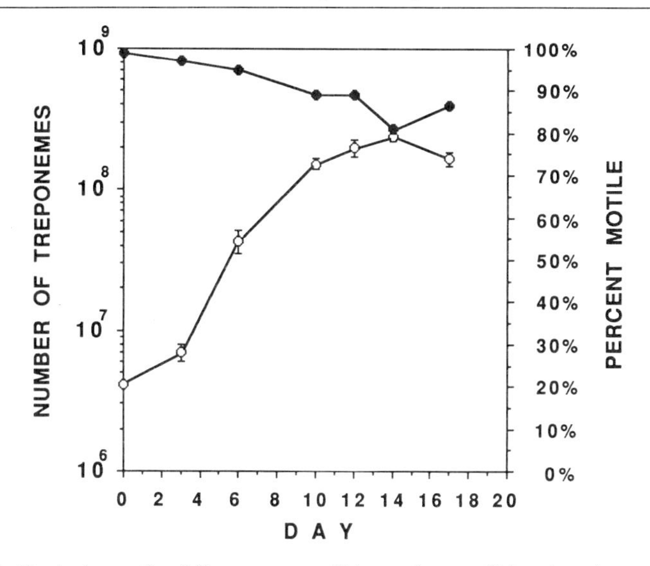

FIG. 1. Typical growth of *Treponema pallidum* subsp. *pallidum* in cultures of Sf1Ep cells containing TpCM as the growth medium. The open circles represent the number of treponemes in a T-25 tissue culture flask. The solid circles represent the percentage of total organisms that were motile.

The flasks are placed on a stirring table in a Tri-Gas incubator (Forma Scientific, Marietta, OH) at 34° in an atmosphere of 5% CO_2, 4% O_2, and 91% N_2 (v/v). The cultures are stirred at 35 rpm for 1 min out of every 45 min.

Typical Growth Curve of *Treponema pallidum in Vitro*

Figure 1 illustrates the typical growth curve for *T. pallidum* cultivated *in vitro*. When incubated at 34° in a microaerophilic atmosphere of 3–4% oxygen, the typical lag period for treponemal growth is about 2 days. Small increases (2- to 3-fold) can be detected as early as 3 days. The cultures remain in exponential growth from day 3 to day 12 of incubation; some cultures may remain in exponential growth until day 14 or 17. In rabbits, the generation time of *T. pallidum* was estimated to be between 30 and 33.[17] In tissue cultures, generation times between 35 and 40 hr are routinely observed. The stationary phase of growth in treponemal cultures is usually relatively short, typically lasting only 1 to 2 days. This is followed by a rapid decrease in motile, viable treponemes.

[17] T. B. Turner and D. H. Hollander, *W. H. O. Monogr. Ser.* **35** (1957).

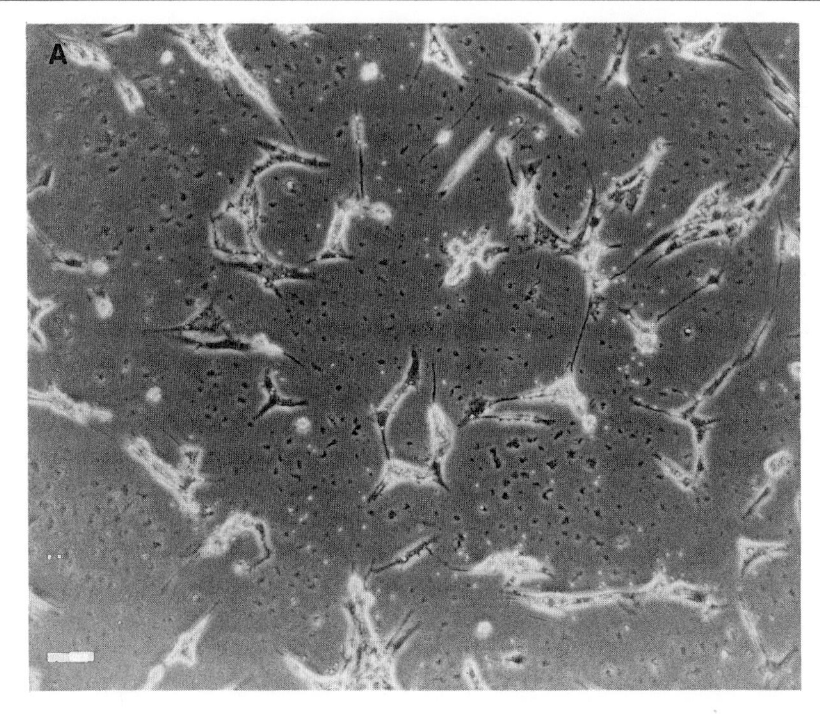

FIG. 2. Pathological effects of the growth of *Treponema pallidum* on Sf1Ep cells. (A) Phase-contrast photomicrograph of the cultures at low magnification. Bar: 10 μm. (B) Scanning electron micrograph of the same culture. Bar: 2 μm. (C) High-magnification scanning electron micrograph of the same culture. Bar: 1 μm. (D) High magnification scanning electron micrograph of a heavily infected cell. Bar: 1 μm.

Effects of *Treponema pallidum* on Host Cells *in Vitro*

Fitzgerald and co-workers have published several reports about the pathological effects of *T. pallidum* on various host tissues. In the first study,[18] rabbit testicular, HEp-2, human foreskin, rat cardiac, and rat skeletal muscle cells were incubated with viable treponemes (1 and 2 $\times$ 10^8/ml), heat-inactivated (56° for 10 min) treponemes, and high-speed supernate from infected testis extract. Viable *T. pallidum* caused rat testicular, human foreskin, and HEp-2 cells to become highly vacuolated and detached after 48 hr of incubation. Cultured cardiac cells rapidly lost their ability to beat within 14 hr of the addition of viable treponemes. The myotubules of rat skeletal muscle cells also became degraded and the cells retracted after 24 hr of incubation with *T. pallidum*. These pathologi-

[18] T. J. Fitzgerald, L. A. Repesh, and S. G. Oakes, *Br. J. Vener. Dis.* **58,** 1 (1982).

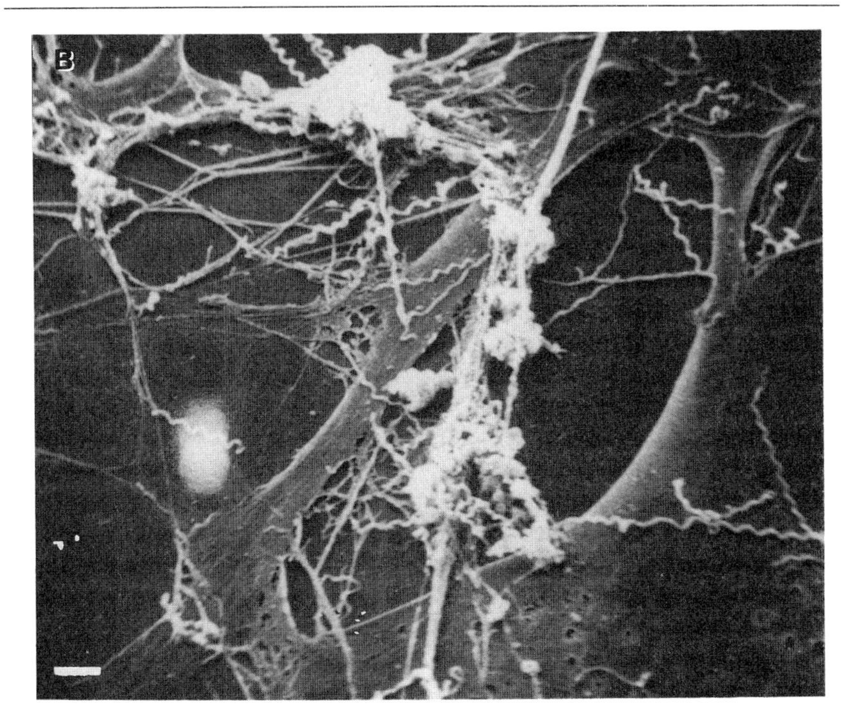

FIG. 2. (*continued*)

cal effects were not observed with the heat-killed treponemes or infected testis extract alone. In a similar study,[19] *T. pallidum* was incubated with cultured nerve cells from rat ganglia. The electrophysiological response became abnormal after 13 hr and completely failed after 18 hr of incubation. Lastly, *T. pallidum* was incubated with isolated rabbit capillary tissues.[20] After 48 hr of incubation, capillary destruction was evident. The endothelial cells became deformed and emaciated and lost their nuclear detail. Other species of nonpathogenic treponemes failed to attach to capillaries. In addition, the attachment of *T. pallidum* could be blocked with immune rabbit serum. Although the treponemes in these three studies were not replicating, these studies demonstrated an innate capability of *T. pallidum* to damage host tissue without the presence of an immune system.

[19] S. G. Oakes, L. A. Repesh, R. S. Pozos, and T. J. Fitzgerald, *Br. J. Vener. Dis.* **58,** 220 (1982).

[20] E. E. Quist, L. A. Repesh, R. Zeleznikar, and T. J. Fitzgerald, *Br. J. Vener. Dis.* **59,** 11 (1983).

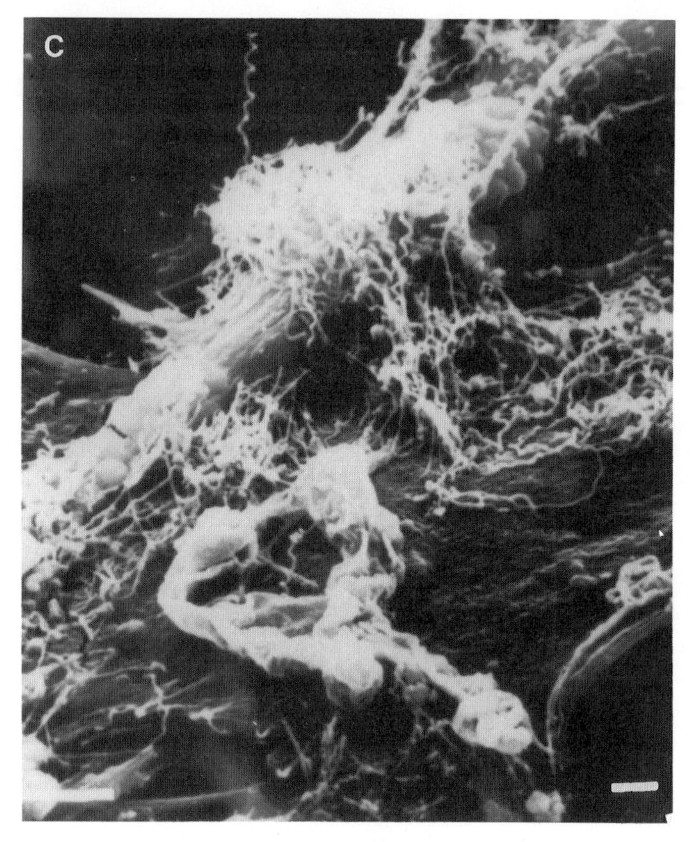

FIG. 2. (*continued*)

In the initial *in vitro* cultures of *T. pallidum,* no pathological effects were observed. The treponemal density at the peak of growth routinely fell between 40 and 50 treponemes per Sf1Ep cell. With the development of TpCM and other improvements in cultivation,[21] the treponemal density at the peak of growth was increased to around 150 to 200 treponemes per Sf1Ep cell. In these cultures, the morphology of the tissue monolayers deteriorated as the cultures grew and never became confluent (Fig. 2A). The cells in the monolayers are relatively sparse and numerous foci of deteriorated tissue can be seen. Scanning electron micrographs showed that many cells appear emaciated, highly vacuolated, and detached (Figs. 2B, C) like those described by Fitzgerald *et al.* They also showed that

[21] D. L. Cox, B. Riley, P. Chang, S. Sayahtaheri, S. Tassell, and J. Hevelone, *Appl. Environ. Microbiol.* **56,** 3063 (1990).

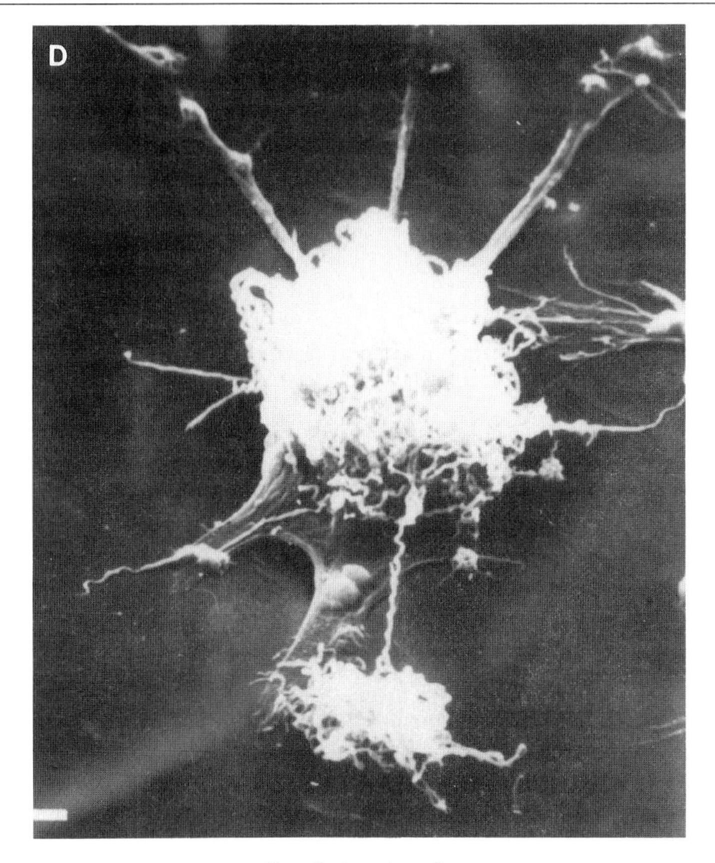

FIG. 2. (*continued*)

the focus of destroyed tissue seen Fig. 2A was a knotted mass containing numerous treponemes (Fig. 2D).

The cytopathic effect appeared sometime between 7 and 10 days of cultivation when the treponemal density would reach 100 organisms per cell or more. This demonstrated that *T. pallidum* does have some innate capability to cause tissue damage without the contribution of an intact immune system. Heretofore, many treponematologists have believed that most of the tissue damage associated with syphilis was due to the immune response of the host and not to any specific pathogenic mechanism possessed by *T. pallidum*. The major difference between these experiments and those previously done was that the cultivation model more closely resembles that of a true infection. The treponeme-to-cell ratio initially was about 8 : 1 and, over a period of 10 to 14 days, increased to 200 : 1.

The pathogenic effects, initially absent, did not appear until the later stages of treponemal growth. It also demonstrated that large numbers of viable treponemes were required to produce pathogenic effects on host tissue because little or no deterioration was observed at concentrations of less than 100 treponemes per cell. Furthermore, in separate experiments large numbers (200 per cell) of heat-inactivated (45° for 30 min) treponemes did not cause any pathological effects. This eliminates the possibility that these effects were caused by a toxic substance.

This *in vitro* model for treponemal pathogenesis can be used to initiate studies on a molecular level. For example, the effects of treponemal infection on various organelles of the host cell and their function might yield critical information on why *T. pallidum* appears to require mammalian cells for survival and replication. Several cell parameters, such as membrane potential, intracellular pH, intracellular Ca^{2+} concentration, and mitochondrial activity, could also be monitored over the time course of infection. It is hoped these studies would produce a clearer understanding of how *T. pallidum* causes cell destruction and greater knowledge of its metabolic requirements.

Summary of *in Vitro* Studies

In one study, Norris and Edmondson[22] investigated the changes that occur in culture conditions over a period of 12 days. Redox potential, pH, and the reduced sulfhydryl, glucose, and dissolved oxygen concentrations were monitored. At the peak of growth, the glucose was almost depleted and the pH had become acidic. In attempts to maintain treponemal growth, cultures were serially passaged after 3, 6, 9, and 12 days. In these cultures, subculturing after 6 days of primary growth resulted in decreased treponemal replication.

In a separate study, Norris and Edmondson[23] investigated the potential of different fractions of FBS to support treponemal growth. They determined that the growth-promoting activity provided by FBS was in the protein fraction. ultrafiltrate (consisting of components with a molecular weight of <10,000) from FBS did not support treponemal growth nor retention of motility.

In vitro cultivation has also been used to determine the minimum inhibitory concentration (MIC) and minimum bactericidal concentration (MBC) of some antibiotics used in the treatment of syphilis.[24] Standard

[22] S. J. Norris and D. G. Edmondson, *Infect. Immun.* **53,** 534 (1986).

[23] S. J. Norris and D. G. Edmondson, *Sex. Transm. Dis.* **13,** 207 (1986).

[24] S. J. Norris and D. G. Edmondson, *Antimicrob. Agents Chemother.* **32,** 68 (1988).

cultivation methods with BRMM were used, except that various concentrations of the antibiotics were added to the tissue cultures at the time of inoculation. Cultures were harvested after 7 days and the total number of treponemes and the number of motile treponemes were counted. The MICs of penicillin, tetracycline, erythromycin, and spectinomycin were determined to be 0.0005, 0.2, 0.005, and 0.5 μg/ml, respectively. MBC determinations were performed by intradermally injecting samples of each culture fluid into the skin on the shaved backs of rabbits. The injection sites were monitored for 45 days to see if lesions developed. All lesions were examined by dark-field microscopy for motile treponemes to verify an active treponemal infection. The MBCs of penicillin, tetracycline, erythromycin, and spectinomycin were 0.0025, 0.5, 0.005, and 0.5 μg/ml, respectively.

Acknowledgments

I dedicate this article to two scientists who made significant contributions toward the *in vitro* cultivation of *T. pallidum:* Dr. A. Howard Fieldsteel (1918–1982), the director of the laboratory that was first to achieve *in vitro* cultivation of *T. pallidum* in the spring of 1980, and Dr. Tom Fitzgerald (1944–1992), who published several articles beginning in the mid-1970s that characterized and detailed the interaction of *T. pallidum* with several mammalian cell lines. These studies proved critical in understanding the culture conditions necessary to provide an environment conducive for treponemal replication *in vitro*.

[28] Measurement of Invasion by Gentamicin Resistance

By ERIC A. ELSINGHORST

Introduction

The gentamicin survival assay is frequently used for the measurement of eukaryotic cell penetration by bacteria because of its simplicity and sensitivity relative to techniques such as Giemsa staining and direct observation. A search of the literature on bacterial invasion will attest to the growing use and popularity of this assay. The principle of the assay is based on the limited penetration of the aminoglycoside antibiotic gentamicin into eukaryotic cells.[1] Organisms that have penetrated the eukaryotic membrane are intracellular and are therefore protected from the bactericidal effects of gentamicin, whereas extracellular organisms are rapidly

[1] P. Vaudaux and F. A. Waldvogel, *Antimicrob. Agents Chemother.* **16,** 743 (1979).

killed by the antibiotic. After removal of the gentamicin, the eukaryotic cells are lysed to release the intracellular bacteria, which are then enumerated by plate count.[1,2] This methodology has been adapted to measure events that occur subsequent to invasion, such as intracellular survival and replication,[3] as well as intercellular spread.[4] The emphasis of this article is on the bacterial penetration of cultured epithelial cells, but gentamicin also has been used to study bacterial interactions with other eukaryotic cell types, such as macrophages.[5]

The following protocol describes a standard assay that was developed to measure the penetration of the human epithelial cell line Henle 407 by *Salmonella typhi;* however, it has been used successfully for measuring invasion of several different epithelial cell lines by several different enteric pathogens, including *Salmonella* species other than *S. typhi, Shigella* species, and various *Escherichia coli* (enterotoxigenic, enteropathogenic, and enteroinvasive *E. coli*). The protocol can serve only as a guideline as the conditions for optimum invasiveness by a particular pathogen must be empirically determined. There are numerous ways to modify this assay, from the treatment and growth conditions of the eukaryotic and prokaryotic cells to the parameters of the assay. Some of the most useful modifications are described following the standard assay.

Standard Invasion Assay

Day Prior to Assay

Preparation of Epithelial Cell Monolayers. Tissue culture techniques and methodology are not discussed in this article. For a thorough discussion of tissue culture techniques, the reader should consult other references.[6,7] The information given here is intended to present details that are relevant for invasion assays.

A single-cell suspension of the desired epithelial cells is prepared by trypsinization. It is important that a uniform suspension is prepared. Epithelial cell clumps result in an increased background (see Results, below). The suspension is diluted in tissue culture growth medium that is appropriate for the cell line being studied. One milliliter of the diluted cell suspension is added to each well of a 24-well tissue culture plate (1.77

[2] E. Kihlström, *Infect. Immun.* **17,** 290 (1977).
[3] J. A. Devenish and D. A. Schiemann, *Infect. Immun.* **32,** 48 (1981).
[4] E. V. Oaks, M. E. Wingfield, and S. B. Formal, *Infect. Immun.* **48,** 124 (1985).
[5] C. R. Lissner, R. N. Swanson, and A. D. O'Brien, *J. Immunol.* **131,** 3006 (1983).
[6] R. I. Freshney, "Culture of Animal Cells." Alan R. Liss, New York, 1987.
[7] W. B. Jakoby and I. H. Pastan, eds., this series, Vol. 58.

cm^2/well). The number of wells to prepare depends on the number of replicates desired; however, at least three wells should be used for each bacterial strain or specific condition to be studied. The density of the diluted cell suspension is dependent on the cell line being used for the assay. The goal of the dilution is to reach a point such that after overnight incubation, the epithelial cells will be 95 to 100% confluent. The number of cells constituting such a confluent monolayer is dependent on the cell line being used and can range from approximately 7×10^4 to 5×10^5. The number of cells in a monolayer for a specific epithelial cell line can be determined by trypsinization of confluent wells and direct counting by hemocytometer. The epithelial cells are incubated overnight at 37° in a 5% CO_2 humidified atmosphere.

Preparation of Bacterial Cultures. Typically, bacterial cells are grown overnight in LB medium (L-broth)[8] at 37°, shaking at 200 rpm. As discussed below, however, invasion may require specific growth conditions and media. Bacterial cultures should be prepared accordingly.

Day of Assay

Preparation of Epithelial Cell Monolayers. The monolayers can be washed prior to inoculation of the wells, although these washings may not be necessary (see Washes and Inoculation of Monolayers, below). If an antibiotic-containing tissue culture medium is used for growth of the monolayers, the wells must be washed before inoculation. Washings can be performed with antibiotic-free tissue culture medium, balanced salt solution (e.g., Earle's balanced salts), or phosphate-buffered saline (PBS), and are most easily accomplished by aspirating the supernatants from the wells, adding 0.5 to 1.0 ml of washing solution, briefly and gently shaking the plates on a rotating platform, and then repeating the washing procedure. After the final wash, 1.0 ml of balanced salt solution or tissue culture medium is added to each well.

Preparation of Bacterial Cultures. Overnight bacterial cultures are diluted 1:20 to 1:50 into 1 ml of fresh L-broth in 13 × 100-mm glass tubes and grown at 37°, 200 rpm, to midlogarithmic (midlog) phase [$\approx 10^8$ colony-forming units (cfu)/ml]. If bacterial growth conditions or media are other than those described above, the cultures should be passaged so that mid-log-phase organisms are used in the assay.

Assay. A 25-μl aliquot of a mid-log-phase culture is added directly to the tissue culture medium bathing the monolayers. This inoculum contains about 2 to 5×10^6 cfu resulting in a multiplicity of infection (m.o.i.) of

[8] J. H. Miller, "Experiments in Molecular Genetics," p. 433. Cold Spring Harbor Lab., Cold Spring Harbor, NY, 1972.

approximately 10 bacteria per epithelial cell. After inoculation of all wells, the 24-well plate is briefly and gently shaken on a rotating platform to distribute the inoculum throughout the tissue culture medium. If desired or necessary, the plate can be centrifuged to initiate contact between the bacteria and epithelial cells. Centrifugation is typically performed for 5 min at 600 *g* and room temperature in a swinging bucket rotor fitted with microtiter plate adaptors. The 24-well plate is then incubated at 37° for 2 hr in a 5% CO_2 humidified atmosphere. This incubation is referred to as the *invasion incubation*. At the time of inoculation, a quantitative plate count is performed on the mid-log-phase culture to determine the total number of input bacteria.

At the end of the invasion incubation the monolayers are washed three times with balanced salt solution or PBS. The purpose of these washes is to remove bacteria that have not adhered to, or invaded, the epithelial cell monolayer. The washes are performed as described above (Preparation of Epithelial Cell Monolayers). After the final wash, 1.5 ml of tissue culture medium or balanced salt solution containing 100 μg/ml gentamicin is added to the monolayer. The 24-well plates are then incubated at 37° for 2 hr in a 5% CO_2 humidified atmosphere. This incubation is referred to as the *gentamicin-kill incubation*. A larger volume of medium is added to the wells during the gentamicin-kill incubation than during the invasion incubation to ensure that bacteria that may have adhered to the walls of the wells are killed by the antibiotic.

After the gentamicin-kill incubation, the monolayers are washed twice with Earle's balanced salt solution or PBS to remove the gentamicin. After aspiration of the final wash, 1 ml of 0.1% Triton X-100 in deionized water is added to the wells to lyse the monolayers. To aid lysis, the plates are shaken on a rotating platform for 5 min. Liberated bacteria are quantitated by dilution and plating on an appropriate medium. The extent of dilution required to achieve countable numbers depends on the epithelial cells used in the assay and the invasiveness, or lack thereof, of the organism being studied, but can range from 10^0 to 10^{-5}.

Results

The results of an invasion assay are often presented as the percentage of the input number of bacteria that have survived the bactericidal action of gentamicin, i.e., (the total cfu recovered from a well/divided by the CFU in the inoculum) $\times 100$. Performing the experiment in triplicate (i.e., three replicate wells for each organism) allows calculation of a mean and a range for each assay. As discussed below, the assay is variable on a daily basis; therefore, invasion data are frequently reported as the results of a single experiment. These results should, however, be reflective and consistent with those obtained in replicate experiments.

Invasion assays can be highly variable on a day-to-day basis. This variability may be related to the overall physiological state of both the prokaryotic and eukaryotic cells. Therefore, it is important to include known invasive and "noninvasive" strains as positive and negative controls, respectively. These inclusions are particularly important if attempts are being made to establish invasiveness by a pathogen or by mutant strains. Typical noninvasive bacteria include laboratory strains of *E. coli*, such as HB101,[9] C600,[10] and MC4100,[11] that are used in molecular biological techniques. It must, however, be noted that noninvasive strains can survive the gentamicin-kill incubation, albeit at a low level, thereby establishing the background, or sensitivity limit, of the assay. It is uncertain if survival of noninvasive strains represents actual internalization or the sequestration of organisms in an external environment that is protected from gentamicin or its bactericidal effects (see Limitations of Assay, below). It is, however, important to include a negative control as an internal measure for the sensitivity of the assay on a particular day. Although varible on a daily basis, background recovery from negative controls can typically range from 0.001 to 0.05% depending on the cell line used in the assay.

A problem with calculating invasion as the percentage of input bacteria that have survived gentamicin treatment is the growth of the inoculum during the invasion incubation. Many organisms are capable of replicating in the tissue culture media or salt solutions that are used in the assay. Because of this growth, the final number of bacteria available to invade the monolayer by the end of the invasion incubation will have increased relative to that at the start of the assay. Additionally, the growth rates of various organisms in these media are not identical. To correct for bacterial growth during the invasion incubation, quantitative plate counts can be performed for each organism at the end of the invasion incubation. A method for accomplishing this count is to add Triton X-100 to a final concentration of 0.1% to an infected well. After allowing for lysis of the monolayer, a quantitative plate count will include all bacteria in the well (i.e., bacteria that are not associated with the monolayer, as well as those that have either adhered to or invaded the monolayer). Calculations of percentage invasion using the number of bacteria present in the well at the end of the invasion incubation allow for a closer comparison of the

[9] H. W. Boyer and D. Roulland-Dussoix, *J. Mol. Biol.* **41,** 459 (1969).
[10] R. K. Appleyard, *Genetics* **39,** 440 (1954).
[11] C. A. Kumamoto and J. Beckwith, *J. Bacteriol.* **154,** 253 (1983).

relative invasiveness of different pathogens for a particular epithelial cell line.

Another measure of invasiveness is the invasion index,[12,13] which is the number of invaded organisms taken as a percentage of the number of adhered organisms. To calculate the invasion index, adherence and invasion must be simultaneously determined. For measurement of adherence, a parallel set of wells are treated as described for the invasion assay; however, at the end of the invasion incubation, the wells are washed six times and then lysed with Triton X-100 as described for the invasion assay. As these wells have not been exposed to gentamicin, recovered bacteria represent those that have adhered to or invaded the monolayer. The invasion index is a measure of an organism's ability to adhere to a eukaryotic cell and how likely it is that that interaction will lead to internalization. Adherence is most likely a prerequisite for invasion of epithelial cells by any bacterial pathogen; however, that pathogen may bind to an epithelial cell membrane without subsequent internalization. Therefore, the index is most useful when comparing invasiveness between similar pathogens or isogenic strains or when measuring the capacity of a particular pathogen to invade different cell lines.

Intracellular Replication. The number of organisms recovered from an invasion assay may represent not only those organisms that have invaded the monolayer, but also those that have undergone replication after penetration of the epithelial cell. The extent of intracellular replication is dependent on both the bacterial strain and the cell line being studied in the assay (Fig. 1). For example, *Shigella flexneri* replicates rapidly inside many cell lines; however, intracellular replication of *Salmonella* species is dependent on the cell line examined, with rapid intracellular replication occurring within cells derived from specific tissues or organs, such as kidney. The extent to which intracellular replication contributes to the overall recovery from an invasion assay must be determined for each organism and epithelial cell line studied.

For measurement of bacterial intracellular multiplication, the standard invasion assay is modified by extending the length of the gentamicin-kill incubation over several hours. Separate 24-well plates are prepared for each time point with triplicate wells for each organism to be studied. The plates are infected simultaneously and treated as described for the standard assay. After the invasion incubation, all plates are washed and treated with gentamicin; however, the length of the gentamicin-kill incubation is varied over several time points. Time points can be taken at 1, 3, 6, 10,

[12] G. W. Jones, L. A. Richardson, and D. Uhlman, *J. Gen. Microbiol.* **127,** 351 (1981).

[13] E. A. Elsinghorst and D. J. Kopecko, *Infect. Immun.* **60,** 2409 (1992).

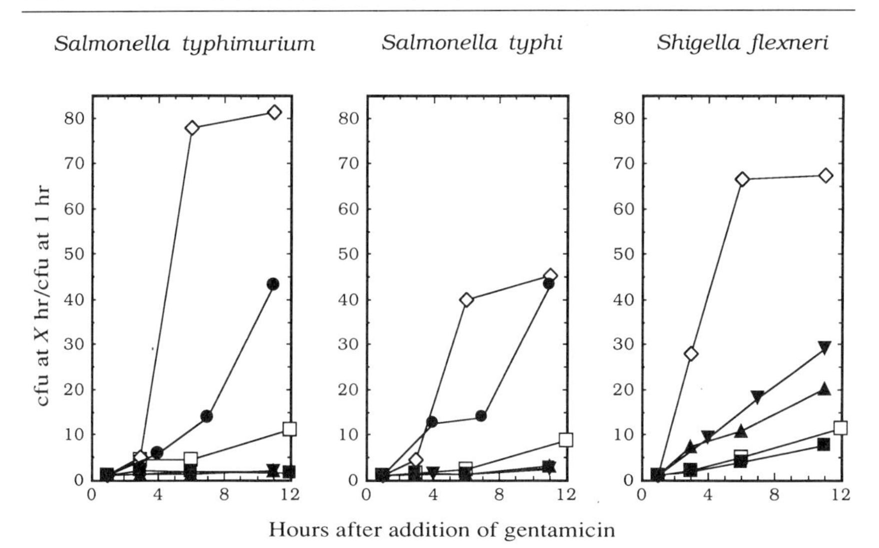

FIG. 1. Intracellular multiplication of enteric pathogens in several different epithelial cell lines. Multiplication assays were performed as described in the text, with the first time point taken after 1 hr exposure to gentamicin. The symbols represent the following cell lines (which are further described in Table I): (◇) A-498, (▲) HCT-116, (■) HCT-8, (▼) HeLa, (●) LLC-MK2, (□) T84.

16, and 24 hr of exposure to gentamicin, but the placement of time points will be determined by the rate of intracellular replication and the effects this replication has on the monolayer. At a time point, a 24-well plate is washed, the monolayers are lysed, and the bacteria are quantitated by plate count. The bacterial recoveries from the various time points are compared with the recovery at the initial time point. As gentamicin kills extracellular bacteria, an increase with time in the recovery of bacterial cells would reflect intracellular growth of the organism.

This methodology cannot distinguish between bacterial turnover within the eukaryotic cell (i.e., concurrent multiplication and death) and static maintenance of bacterial numbers (i.e., no or limited growth without bacterial death). Therefore, if the intracellular number of organisms does not increase with time, the possibility of intracellular replication still exists.

Irrespective of the capacity to replicate, intracellular bacteria can result in toxicity and lysis of the epithelial cells, particularly during the extended incubation times of a multiplication assay. To minimize these problems if they are encountered, the assay can be further modified by decreasing the multiplicity of infection, decreasing the length of the invasion incubation, and decreasing the length of the initial gentamicin time point. Additionally, as gentamicin may slowly penetrate the eukaryotic membrane, more reproducible replication curves may be obtained by decreasing the gentamicin concentration to 20 μg/ml.

Viability of Monolayer. Epithelial cell attachment and viability may be influenced during the course of an invasion assay by several factors such as extracellular and intracellular bacterial growth, medium acidification, and bacterial production of toxins. Death of epithelial cells or their release from the monolayer decrease bacterial recovery. The relative viability and integrity of the monolayers should be evaluated at the end of an invasion assay for each epithelial cell line and organism examined. If the monolayer is compromised, its stability can be increased by decreasing the m.o.i. and the lengths of the invasion and gentamicin-kill incubations. Epithelial cell viability and monolayer integrity can be measured by methods including trypan blue exclusion,[14] acridine orange staining,[15] crystal violet staining,[16] and direct observation under an inverted microscope.

Assay Modifications

There are numerous ways in which this assay can be modified depending on the specific organism that is being studied and the specific cell lines that are being used as hosts for bacterial invasion. These modifications can occur at virtually every step of the assay and in the preparation of both the prokaryote and eukaryote.

Growth of Epithelial Cell Line

The choice of cell line can be critical for establishing invasiveness by a particular pathogen. The physiological state of that cell line can also have a major influence on the outcome of an invasion assay. These variables are related to the presence and distribution of receptors that are necessary for bacterial invasion.

Intuitively, the cell lines used to establish or study bacterial invasion would come from the host tissue or organ that would be a likely target for this activity. Bacterial entry into epithelial cells could be overlooked if an inappropriate cell line was examined. For example, although *Shigella flexneri* and *Salmonella typhi* is capable of invading many epithelial cell lines, enterotoxigenic *E. coli* is specific for the cell lines it will penetrate

[14] M. K. Patterson, this series, Vol. 58, p. 151.

[15] M. R. Melamed, L. R. Adams, A. Zimring, J. G. Murnick, and K. Mayer, *Am. J. Clin. Pathol.* **57,** 95 (1972).

[16] M. S. Donnenberg, A. Donohue-Rolfe, and G. T. Keusch, *J. Infect. Dis.* **160,** 452 (1989).

TABLE I
INVASION OF EPITHELIAL CELLS BY ENTEROTOXIGENIC *Escherichia coli* H10407 AND *Salmonella typhi* Ty2[a]

Cell line	Source	% Invasion[b]		% Relative invasion[c]
		H10407	Ty2	
HuTu80	Human duodenum	0.03	5.14	0.58
HCT-8	Human ileocecum	0.41	1.57	26.11
HCT-116	Human colon	1.06	1.20	88.33
T84	Human colon	0.25	0.29	86.21
HeLa	Human cervix	<0.01	9.66	0.05
Chang	Human liver	0.05	1.84	2.72
A-498	Human kidney	0.18	0.67	18.40
LLC-MK_2	Monkey kidney	<0.01	1.90	0.26

[a] Reproduced with permission from Elsinghorst and Kopecko.[13]
[b] Standard assays were performed as described in the text; however, the length of the invasion incubation was increased to 3 hr for H10407.
[c] Invasion relative to *S. typhi*, representing 100%.

(Table I).[13] Until the specific eukaryotic receptor(s) recognized by the bacterium is identified, a range of cell lines may be required to characterize the invasive phenotype of an organism.

As described in the above protocol, the epithelial cells are grown as a 95 to 100% confluent monolayer. As such, these monolayers would be nonpolar. In a polarized monolayer, such as the intestinal epithelium, cell surface components are localized to specific plasma membrane domains, such as the apical, lateral, or basal cell surfaces.[17] The distribution of a receptor(s) necessary for bacterial invasion may be affected by this localization, thereby influencing the invasive phenotype of a pathogen. Many cell lines that are capable of polarization are available,[17] and the use of these cell lines as polarized monolayers may represent a more accurate model of the epithelial cell as seen by a bacterium. The use of polarized epithelial cell monolayers is discussed in depth in [30] of this volume.

Growth of Bacterial Inoculum

The expression of factors required for the invasive phenotype may be affected by the growth conditions used for the preparation of the bacterial inoculum. Conditions that have been shown to affect invasiveness include

[17] K. Simons and S. D. Fuller, *Annu. Rev. Cell Biol.* **1,** 243 (1985).

variables such as osmolarity,[18] growth phase,[19,20] oxygen availability,[19,20] and temperature.[21] Numerous other modifications could be considered. Varying conditions or growth media in ways that have been shown to affect the regulation of virulence factors (e.g., iron, calcium, pH)[22,23] represents a logical starting place in the search for conditions that may either allow for or increase invasiveness in a specific pathogen. Growth conditions also may affect the expression of additional adherence factors that are needed for optimal invasiveness. Invasion of bacteria grown as shaken or static liquid cultures, as well as cells grown on solid media, can be examined.

Washes and Inoculation of Monolayers

Washing the monolayers and bacterial cells before starting an assay may be necessary if components of media or metabolic products are inhibitory or toxic to either the eukaryote or prokaryote. These washings may not, however, be necessary for optimal invasion (Table II). Monolayer washing was described in the standard assay procedure. Bacterial cultures can be washed by centrifugation and resuspension to the original volume in fresh growth medium, PBS, or a tissue culture solution identical to that bathing the monolayers. The washed bacteria can then be used to inoculate the monolayers as described in the standard assay.

Alternately, the mid-log-phase bacterial culture can be harvested by centrifugation, then resuspended to a density of 2 to 5 $\times$ 10^6 cfu ml in tissue culture medium or balanced salt solution. If washing by centrifugation results in decreased invasiveness, the mid-log-phase culture can be diluted directly into fresh growth medium to the desired final density. After aspiration of the growth medium from the monolayers, 1 ml of diluted bacteria is added to each well, resulting in an m.o.i. of about 10. The monolayers are then treated as described for the standard assay.

Although the standard assay uses an m.o.i. of about 10 bacteria per epithelial cell, this ratio may need to be changed for optimal invasion by a specific organism or during certain experiments such as intracellular multiplication or invasion inhibition assays. It should be noted that increasing the m.o.i. often results in higher background levels (i.e., increased recovery of invasion negative controls).

[18] J. E. Galan and R. Curtiss, III, *Infect. Immun.* **58,** 1879 (1990).
[19] R. K. Ernst, D. M. Dombroski, and J. M. Merrick, *Infect. Immun.* **58,** 2014 (1990).
[20] C. A. Lee and S. Falkow, *Proc. Natal. Acad. Sci. U.S.A.* **87,** 4304 (1990).
[21] A. T. Maurelli, B. Blackmon, and R. Curtiss, III, *Infect. Immun.* **43,** 195 (1984).
[22] J. F. Miller, J. J. Mekalanos, and S. Falkow, *Science* **243,** 916 (1989).
[23] J. W. Foster, *J. Bacteriol.* **173,** 6896 (1991).

TABLE II
EFFECT OF VARIOUS WASHINGS PRIOR TO INOCULATION OF MONOLAYERS ON HeLa CELL INVASIVENESS OF *Salmonella typhi* Ty2

Bacterial washing[a]	HeLa cell washing[b]	% Invasion
Unwashed	−	12.04 ± 0.35
Unwashed	+	13.72 ± 2.80
I	−	13.37 ± 0.63
I	+	14.40 ± 2.60
II	−	11.79 ± 0.34
II	+	12.60 ± 0.60

[a] Bacteria were grown as described for the standard assay, then split into three samples that were treated as follows: unwashed, bacteria were used directly without additional treatment; I, cells were pelleted by centrifugation (5 min, 3000 *g*) and then resuspended in an equal volume of L-broth; II, cells were pelleted by centrifugation (5 min, 3000 *g*) and then resuspended in an equal volume of HeLa cell tissue culture medium. After the indicated treatment, standard assays were performed. Quantitative plate counts were performed on each sample and used in percentage invasion calculations to correct for loss of bacterial cells during washes.

[b] HeLa cell washings: One hour prior to inoculation, HeLa cells monolayers were either (−) untreated or (+) washed once with HeLa cell tissue culture medium.

Centrifugation

For either mode of inoculation, the plates can be centrifuged as described in the standard assay. The primary purpose of this step is to initiate and synchronize contact between the bacteria and the monolayer, with the assumption that any intimate contact or close association required for bacterial penetration can be provided by centrifugation. Centrifugation may not result in optimal invasion, and in some cases it may be detrimental (Table III); however, percentage invasion by some organisms, such as *Shigella flexneri,* can be increased by this step. Therefore, it must be determined for each organism of interest whether or not to include centrifugation as a routine step in a standard assay. In some experiments centrifugation may be desired to synchronize invasion, such as those to determine the optimal length of the invasion incubation. Additionally, centrifugation

TABLE III
CENTRIFUGATION OF INOCULATED MONOLAYERS VS PERCENTAGE RECOVERY FROM HeLa CELLS[a]

Organism	% Invasion	
	Centrifuged	Not centrifuged
Salmonella typhi Ty2	2.58 ± 0.28	4.45 ± 0.54
Salmonella typhi 101018	4.55 ± 0.78	6.64 ± 1.19

[a] Invasion assays were performed on 24-well plates in parallel. One plate was incubated at room temperature and the second plate was centrifuged as described in the standard assay, after which the plates were treated as described for the standard assay.

can be used in experiments studying the effects of various inhibitors on invasion to obviate the concern that the presence of these compounds may be altering an intimate contact or close association step that is required for invasion but not the invasion step itself.

Length of Invasion Incubation

Several elements contribute to the optimal length of the invasion incubation: Is there a need for induction of invasion factors? What specific interactions are required for internalization to occur (e.g., loose adherence, tight adherence)? What are the availability and abundance of receptors? Also, as penetration of the eukaryotic membrane could be considered an enzymatic reaction, the kinetic parameters and constants of that reaction also contribute to the optimal length of the invasion incubation. In consideration of these factors, the length of the invasion incubation can be increased or decreased from the 2-hr incubation described in the standard assay. The results of these changes vary with the organism (Table IV). In any length of invasion incubation, the effect of the bacteria on the viability of the monolayer as a result of extracellular and intracellular growth of the inoculum, medium acidification, toxicity from the release of cytotoxins, or other factors must be considered. By increasing the length of the invasion incubation any negative effects become more pronounced. The viability and attachment of the monolayer should be assessed at the end of the assay.

Gentamicin-Kill Incubation

The gentamicin-kill incubation can be modified by changing the antibiotic concentration and length of the incubation period. These variables

TABLE IV
LENGTH OF INVASION INCUBATION VS PERCENTAGE RECOVERY FROM HeLa CELLS[a]

Organism	Length of invasion incubation (min)	% Invasion
Salmonella typhi Ty2	15	1.85 ± 0.13
	30	2.71 ± 0.54
	60	9.37 ± 0.91
	120	10.24 ± 0.45
Salmonella typhi 101018	15	1.34 ± 0.32
	30	1.26 ± 0.09
	60	4.31 ± 0.11
	120	6.01 ± 0.41
Salmonella typhi I 14	15	1.06 ± 0.12
	30	1.41 ± 0.14
	60	3.51 ± 0.28
	120	3.60 ± 0.62
Salmonella typhimurium C5	15	0.61 ± 0.23
	30	1.43 ± 0.27
	60	1.63 ± 0.22
	120	1.97 ± 0.26
Shigella flexneri M90TW	15	0.32 ± 0.04
	30	0.51 ± 0.03
	60	1.90 ± 0.30
	120	3.11 ± 0.91

[a] The standard invasion assay was modified by altering the length of the invasion incubation. Preparation of bacterial cultures, monolayer inoculation and washings, and gentamicin-kill incubations were performed as described for the standard assay. The total number of bacteria per well was determined at the end of each time point as described under Results to correct for growth of the inoculum during the experiment. These numbers were used in calculating the percentage invasion for each organism and time point, respectively.

are dependent on the sensitivity of the test organism to gentamicin and the type of assay being performed. The concentration of gentamicin should at least equal the minimum bactericidal concentration (MBC) for the organism being studied. The MBC varies depending on the growth medium in which it is determined. It should therefore, be measured in the medium used during the invasion incubation.

Because gentamicin may slowly penetrate the eukaryotic membrane,[1] the antibiotic concentration and incubation times can be decreased if there is concern that gentamicin-mediated death of intracellular bacteria is occurring. The influence of gentamicin leakage on percentage invasion may, however, be negligible. The gentamicin MBC for *Salmonella typhi* Ty2 is 0.25 to 1.56 μg/ml depending on the growth medium used for the determination. The invasion of HeLa cells by Ty2 was measured in standard assays that were modified by performing the gentamicin-kill incubations with media containing either 5, 20, or 100 μg–ml gentamicin. The percentage invasion of Ty2 was 3.46 ± 0.38, 3.76 ± 0.23, and 3.55 ± 0.89, respectively. Decreasing the length of the gentamicin-kill incubation from 2 to 1 hr did not have an effect on the percentage invasion of HeLa cells by Ty2 in an otherwise standard assay. Gentamicin-mediated intracellular death of bacteria is most likely to be encountered during prolonged incubations, such as those found in intracellular multiplication and intercellular spread assays. For these types of experiments, it is useful to decrease the gentamicin concentration to 20 μg/ml. When altering the parameters of the gentamicin-kill incubation, bear in mind that the length of the incubation at the selected drug concentration must be sufficiently long to ensure death of extracellular bacteria.

Lysis of Monolayer

Intracellular bacteria surviving the gentamicin treatment are released by lysing the epithelial cells with a detergent; however, as these detergents also can lyse bacteria, it is important to evaluate bacterial viability under the conditions used to lyse the monolayers (i.e., type and concentration of detergent and length of incubation). Bacterial viability can be measured by diluting a mid-log-phase culture in lysis solution, incubating for the prescribed length of time, and then plating on appropriate growth media. Recovery of cells diluted in detergent should be compared with recovery of cells diluted in water. PBS can be substituted for water in lysis solutions if the dilution of bacteria in water decreases cell viability. Bacteria that have penetrated the epithelial cell may be physiologically stressed and more susceptible to detergents than a mid-log-phase culture. Therefore, the epithelial cell lysis conditions resulting in the maximal recovery of bacteria may need to be determined empirically.

The ease of eukaryotic cell lysis varies with cell line and degree of polarization. Some cell lines are difficult to lyse and may require physical assistance (such as repeated pipetting of the lysate) to achieve disruption of the monolayer. The extent of lysis can be examined by direct observation with an inverted microscope. The type and concentration of detergent, and length of incubation can be altered depending on the sensitivity of the cell line. Detergents that can be used for lysis include Triton X-100 and sodium deoxycholate at concentrations ranging from 1.0 to 0.01% in distilled water. The length of the lysis incubation can be increased from

5 min depending on the sensitivity of the eukaryote and prokaryote to the concentration of detergent used for lysis.

Limitations of Assay

Two major questions regarding the limitations of the gentamicin survival assay arise from the principles on which it is based: that there is limited penetration of gentamicin into eukaryotic cells, and that only organisms that are removed from the external environment (i.e., having penetrated the eukaryotic cell) are protected from the bactericidal effects of the antibiotic. These two questions are: Does gentamicin kill any intracellular organisms? and Does gentamicin kill all extracellular organisms?

Does gentamicin kill any intracellular organisms? Vaudaux and Waldvogel[1] studied the penetration of gentamicin into human polymorphonuclear leukocytes and showed that these cells did not accumulate the antibiotic while resting or during active phagocytosis; however, their gentamicin binding studies did not preclude slow penetration of the antibiotic through the eukaryotic membrane. If gentamicin is entering the epithelial cells during the course of an invasion assay, is it reaching concentrations sufficient to decrease bacterial viability? The answer to this question depends partly on the rate at which gentamicin penetrates the eukaryotic membrane and the sensitivity of the test organism to the antibiotic. As described earlier, varying the concentration of the antibiotic or length of the gentamicin-kill incubation has little or no effect on the percentage invasion of HeLa cells by *Salmonella typhi,* suggesting that in this specific example, gentamicin-mediated intracellular killing is not occurring; however, gentamicin might permeate sufficiently to decrease bacterial viability during prolonged gentamicin-kill incubations, such as those used during intracellular multiplication assays. The effect of gentamicin on intracellular bacterial populations needs to be examined on a case-by-case basis.

Does gentamicin kill all extracellular organisms? This question can be asked in the following way: Is there an extracellular environment that is sequestered from gentamicin or one that prevents gentamicin from exerting its bactericidal effects? Although it is not known if there are extracellular environments that can exclude gentamicin, there are conditions that are known to reduce its efficacy. Gentamicin is an aminoglycoside antibiotic, and initial uptake of these antibiotics by bacteria is strongly influenced by bacterial membrane potential and requires active protein synthesis.[24] Aminoglycosides are less effective at low pH[25] or under anaerobic condi-

[24] B. D. Davis, *Microbiol. Rev.* **51,** 341 (1987).
[25] M. Barber and P. M. Waterworth, *Br. Med. J.* **1,** 203 (1966).

tions (due to depressed bacterial membrane potential).[24,26] Extracellular microenvironments that reduce the efficacy of the drug and/or reduce bacterial uptake could theoretically diminish the effective concentration of the antibiotic to a level below that required to kill bacteria. These microenvironments could be formed at eukaryotic cell surfaces by folds or ruffles of the cell membrane or in intracellular spaces. Bacteria contained within such microenvironments could potentially survive gentamicin treatment and, following lysis of the eukaryotic cell, be counted as invasive. This extracellular protection may account for the recovery of "noninvasive" controls from invasion assays.

Recovery of noninvasive controls illustrates the need to confirm the invasive phenotype by other methods. This confirmation is particularly important when identifying and characterizing invasion by a pathogen for which the invasive phenotype has not been described previously, or if the percentage recovery of an organism is low relative to the control strains. Electron microscopy has frequently been used to verify bacterial penetration of eukaryotic cells. Measurement of invasion by alternate methods is discussed in depth in [29] of this volume.

[26] M. Kogut, J. Lightbown, and P. Isaacson, *J. Gen. Microbiol.* **39,** 155 (1965).

[29] Measurements of Invasion by Antibody Labeling and Electron Microscopy

By JOS P. M. VAN PUTTEN, JAN F. L. WEEL, and HEIKE U. C. GRASSMÉ

Introduction

Many bacterial pathogens have the capability to enter and reside within eukaryotic cells.[1-3] As described in [28], various methodologies have been developed to assess bacterial invasion into host cells *in vitro*, using infection models in combination with microscopy, metabolic labeling assays, or measurement of the intracellular survival of the bacteria. The purpose of this article is to outline and discuss practical strategies and technical details for successful immunomicroscopic measurement of bacterial inva-

[1] B. B. Finlay and S. Falkow, *Microbiol. Rev.* **53,** 210 (1989).
[2] M. J. Wick, J. L. Madara, B. N. Fields, and S. J. Normark, *Cell (Cambridge, Mass.)* **67,** 651 (1991).
[3] S. Falkow, *Cell (Cambridge, Mass.)* **65,** 1099 (1991).

sion in monolayers of cells. The advantages of microscopic analysis of infected eukaryotic cells are that it provides direct information about the number of adherent and ingested bacteria at the single cell level, and that it enables localization of the bacteria within specific cells or cellular compartments independent of their viability. This approach is essentially different from and complements intracellular survival assays, which involve the selection and counting of the viable intracellular bacteria isolated from a batch of host cells (see [28]). The basis of immunomicroscopic assessment of bacterial invasion is the differential staining of bacteria that are internalized from those that remain on the cell surface using immunoreagents. The methods to achieve this are described in this article and include double-fluorochrome immunolabeling and immunogold–silver staining procedures for use in combination with bright-field light microscopy, fluorescence microscopy, and confocal scanning microscopy. In the last section, electron microscopy methods are described for similar studies, as well as for immunochemical labeling of host cell antigens.

Principles and General Considerations in Immunostaining

The principle of the measurement of bacterial invasion by antibody labeling using light microscopy is that antibodies cannot pass through the plasma membrane of eukaryotic cells and, therefore, bind only to extracellular adherent bacteria and not to intracellular bacteria. This enables exclusive immunolabeling of the extracellular microorganisms. Counterstaining of the intracellular bacteria then provides the desired differential staining of adherent and ingested bacteria. This approach means that the essential requirements of immunomicroscopic measurement of bacterial invasion are (1) impermeability of the host cell plasma membrane to antibodies, (2) retainment of the antigenicity of the bacterial antigens during handling of the specimen, (3) a specific antibody that binds avidly to the antigen, (4) a means of labeling the specifically bound antibody, and (5) a suitable counterstain that diffuses freely into the cells and preferentially stains the intracellular bacteria.

Immunostaining Procedure

Existing immunomicroscopy methods to measure bacterial invasion all follow the above-mentioned principle and differ mainly in their strategy to stain the bacteria. The basic procedure starts with the washing of the monolayer of infected eukaryotic cells to remove nonadherent bacteria. The cells are then fixed with a chemical fixative which leaves the plasma membrane impermeable to antibody, and the extracellular bacteria are labeled using a bacteria-specific antibody. The most effective approach is to label the antigens indirectly, i.e., first incubate with the bacteria-specific antibody and then visualize this antibody with the help of a second fluorochrome- or gold-coupled ligand. This indirect labeling avoids potential damage to the first antibody by the conjugation procedure and enables the use of a limited number of conjugates with a variety of antibodies. Conjugates are commercially available but can also be prepared in the laboratory.[4,5] After immunostaining of the extracellular bacteria, the samples are counterstained to visualize the intracellular bacteria either with a conventional chemical dye (see immunogold–silver staining method) or by specific immunolabeling of the intracellular bacteria after treatment of the cells with detergent solution to allow antibodies access to the cell interior (double immunolabeling method, see next section). Dependent on the type of immunostaining applied, the specimens are viewed with conventional light microcopy or fluorescence microscopy. Irrespective of the staining method used, it is recommended the eukaryotic cells be cultured on coverslips to facilitate the handling of the specimen during the staining procedure and to enable viewing of the infected cells in the microscope at a high magnification (100× objective). A simple and inexpensive procedure for growing the cells on coverslips is to flame individual circular glass coverslips (e.g., 12 mm in diameter) taken from a batch kept in 96% ethanol on glass beads in a petri dish, using tweezers, and to insert them in, for example, a 24-well tissue culture tray. The growth on coverslips means that after infection of the cells and fixation of the samples, all subsequent incubation steps can be carried out with the coverslips floating cell-side-down onto 50 μl of incubation solution on Parafilm. This procedure reduces the volume of the immunoreagents required, ensures equal distribution of the solution over the coverslip, and prevents drying of the specimen. The Parafilm can be fixed on a film of water. The coverslips can easily be lifted from the tissue culture wells using a needle with a hooked tip and fine tweezers. When the coverslip is transferred to the next solution, excess buffer should be removed by touching the edge of the coverslip with filter paper. Drying of the samples should be avoided as this irreversibly damages the cells and increases nonspecific staining. The whole immunostaining procedure is carried out at room temperature (20°).

Fixation

Chemical fixation of the infected cells, which meets the above-mentioned criteria of retaining antigenicity and leaving the plasma membrane

[4] P. Brandtzaeg, *Scand J. Immunol.* **2,** 273 (1973).
[5] J. W. Goding, *J. Immunol. Methods* **13,** 215 (1976).

impermeable to antibodies while at the same time preserving cellular morphology, can be achieved in most instances with a fixing solution of phosphate-buffered saline (PBS: 143 mg $Na_2HPO_4 \cdot 2H_2O$/20 mg K_2HPO_4/800 mg NaCl/20 mg KCl in 100 ml of distilled water, pH 7.4) containing 1–2% (w/v) formaldehyde. Glutaraldehyde can be added to this solution, but should be used with caution, as certain antigens do not tolerate the strong crosslinking effect of this compound even at very low concentrations.[6] Both compounds stabilize the protein components of membranes by the crosslinking of free amino groups. The time required for proper fixation of monolayers of cells is at least 15 min (20°), but can for most antigens be extended to several days (4°) without impairment of antibody binding. The fixative is prepared by heating a weighed quantity of paraformaldehyde in water with continuous stirring at 55–60° until the solution clears, at which point paraformaldehyde is depolymerized. If the solution remains cloudy one or more drops of 1 *N* NaOH should be added. After cooling, the paraformaldehyde solution is mixed with an equal volume of double-concentrated PBS. $CaCl_2$ and $MgCl_2$ are added at a final concentration of 0.1 m*M* and the solution is adjusted to pH 7.2–7.4.[7] It is essential that all steps are carried out in a fume cupboard to avoid contact with the aldehydes. Buffered solutions appear to be stable for some weeks at 4°, but can also be kept at −20° which extends their stability.

Antiserum

The first requirement of the antiserum to be used in the immunomicroscopy assay is that it must recognize the whole population of bacteria. In other words, the antibodies should be directed against surfaced-exposed epitopes present on all of the extracellular and, when double immunolabeling is applied, all of the intracellular bacteria. As many bacterial membrane constituents are variably surface exposed or subject to phase or antigenic variation,[8] this can be achieved best with polyclonal antibodies that react with different antigens, but a monoclonal antibody directed against a conserved surface-exposed epitope will do as well. When double immunolabeling is used, it is essential that the same antibody but combined with a different fluorochrome be used to stain the extra- and intracellular bacteria. The antibodies are diluted in PBS and the solution is centrifuged just before use at 10,000 *g* at 20° for 2 min to remove aggregates. In general, nonspecific binding of antibodies is not a major problem in immunostaining the bacteria, because of the relatively high density of antigens

[6] J. F. L. Weel, C. T. P. Hopman, and J. P. M. van Putten, *J. Exp. Med.* **173,** 1395 (1991).
[7] A. M. Glauert, "Practical Methods in Electron Microscopy," Vol. 3, Elsevier, Amsterdam, 1991.
[8] B. D. Robertson and T. F. Meyer, *Trends Genet.* **8,** 422 (1992).

at the bacterial cell surface and the often prolonged exposure of the cultured cells to bovine serum (FBS) during their growth. Nonspecific immunostaining can be reduced further by including bovine serum albumin (BSA, 0.5%), FBS (5%), gelatin (0.1%), or Tween 20 (0.05%) in the immunoreagent solutions. When antiserum or ascitic fluid is being used as a source of antibody, it is often helpful to purify the immunoglobulins or, more simply, to delipidize the solutions. The latter can easily be done by vigorous mixing of equal volumes of serum (or ascites) and serum cleaner (Walter Gmbh, Kiel, Germany) for 15 min, followed by phase partition in an Eppendorf centrifuge (10,000 *g*, 3 min). The delipidized antiserum or ascites fluid partitions into the upper phase. The optimal concentration of antibody to be used is determined empirically by staining the specimen with a series of dilutions in comparison with control serum.

Measurement of Bacterial Invasion by Immunofluorescence Microscopy

A classic method to locate bacteria inside or outside infected eukaryotic cells uses a double-immunolabeling procedure in combination with fluorescence microscopy. The method involves washing and fixing of the infected cells as already described, followed by first an immunostaining of the extracellular bacteria. After permeabilization of the plasma membrane, a second immunostaining is performed to label specifically the intracellular microorganisms using a different fluorochrome. The methodology of staining the bacterial antigens essentially follows standard immunofluorescence microscopy techniques[9] with some exceptions. One major difference is that to label the adherent and ingested bacteria differentially the same antiserum is used in both immunoincubations but in combination with fluorochromes with distinct emission spectra. Use of the same antiserum is essential to prevent renewed staining of the extracellular organisms during the second immunoincubation. Commonly applied fluorochromes are fluorescein isothiocyanate (FITC), which results in yellow–green staining of the bacteria, in combination with tetramethylrhodamine isothiocyanate (TRITC) or sulforhodamine 101 acid chloride (Texas Red), which give the bacteria an intense red appearance. Important in staining the intracellular bacteria is that the reagents used to allow the antibodies access to the cell interior do not affect the primary staining result or unmask antigens on the already stained extracellular bacteria, which results after double staining of the adherent microorganisms in the second immunoincubation. Moreover, the antigenicity of the intracellular antigens should be retained. The reagents used for permeabilization of the plasma membrane are mild detergents such Triton X-100 (0.2–0.5% for 10–15

[9] K. Wang, J. R. Feramisco, and J. F. Ash, this series, Vol. 85, p. 514.

min) or saponin (0.05% for 15 min),[10] but cold acetone (3 min at −20°) can also be used. These compounds do not influence the primary antibody labeling, preserve cellular morphology, and only rarely influence the immunoreactivity of the bacterial antigens. At the end of the double-labeling procedure the cells are mounted on a microscopic slide in mounting medium, e.g., 90% glycerol/10% PBS, and viewed in a fluorescence microscope with the appropriate filters. To reduce photobleaching during microscopy, UV quenching inhibitors such as *N*-propylgallate,[11] *p*-phenylenediamine,[12] and Mowiol 4-88[13] can be used in the mountants.

Although the double-immunolabeling fluorescence microscopy method for measurement of bacterial invasion is easy to perform, several pitfalls should be kept in mind. It is critical that the antigen probed on the extracellular bacteria is fully saturated with antibody before staining the intracellular bacteria; i.e., an excess of antibody should be used. Incomplete saturation of the extracellular antigen with the primary antibody or of this antibody with the conjugate will result in double staining of extracellular bacteria during the second immunolabeling step. The same effect may occur as a result of the permeabilization procedure. Solubilization of plasma membrane lipids may lead to an unmasking of previously inaccessible bacterial antigens at the site of adherence with a concomitant double labeling of the extracellular bacteria with the two fluorochromes. Poor or negative staining of intracellular bacteria occurs when the intracellular antigens are not stably expressed, are masked by host cell components, or become degraded during the intracellular processing of the bacteria. There are additional disadvantages to the method: Photobleaching occurs; i.e., the fluorescence signals fade; when conventional fluorescence microscopy is being used only one fluorochrome can be viewed at a time; the unstained eukaryotic cells are barely visible in the microscope. Despite these limitations, the method is valuable and one of the best tools to study bacterial invasion.[14-16]

Procedure

All steps are carried out at 20°.

1. Wash the infected cells three times with PBS containing 1 m*M* $CaCl_2$ and 1 m*M* $MgCl_2$.

[10] I. Ohtsuki, R. M. Manzi, G. E. Palade, and J. D. Jamieson, *Biol. Cell.* **31,** 119 (1978).
[11] H. Giloh and J. W. Sedat, *Science* **217,** 1252 (1982).
[12] G. D. Johnson and G. M. de Nogueira Araujo, *J. Immunol. Methods* **43,** 349 (1981).
[13] G. V. Heimer and C. E. Taylor, *J. Clin. Pathol.* **27,** 254 (1974).
[14] P. Clerc and P. Sansonetti, *Infect. Immun.* **55,** 2681 (1987).
[15] S. G. Casey, W. M. Shafer, and J. K. Spitznagel, *Infect. Imun.* **52,** 384 (1986).
[16] S. M. Makino, J. P. M. van Putten, and T. F. Meyer, *EMBO J.* **10,** 1307 (1991).

2. Replace with PBS containing 1–2% paraformaldehyde for at least 15 min.
3. Rinse twice with PBS.
4. Quench the remaining free aldehyde groups with PBS containing 50 m*M* NH_4Cl for 5 min.
5. Rinse twice with PBS.
6. Incubate the coverslips cell-side-down on 50 μl of antibody solution on Parafilm for 30–60 min (time depends on the quality of the antibody).
7. Remove unbound antibody by three 5-min washes with PBS.
8. Incubate the coverslips cell-side-down on 50 μl of fluorochrome-conjugated ligand (e.g., protein A–FITC) for 30 min.
9. Remove unbound fluorochrome by three 5-min washes with PBS.
10. Incubate the coverslips cell-side-down on PBS containing 0.2% (v/v) Triton X-100 for 10–15 min to permeabilize the eukaryotic plasma membrane.
11. Replace with PBS for 5 min.
12. Incubate the coverslips with the same antibody as used in step 6 for 30–60 min.
13. Remove unbound antibody by three 5-min washes with PBS.
14. Incubate with the second fluorochrome-conjugated ligand (e.g., protein A–TRITC) for 30–45 min to label the intracellular bacteria.
15. Remove unbound conjugate by three 5-min washes with PBS.
16. Mount the coverslips in 90% glycerol/10% PBS (or another mounting medium) on a microscopic slide and seal with nail polish. View the specimen with a fluorescence microscope with the appropriate filters. The slides can be stored in a lightproof box at 4°.

Use of Immunogold–Silver Staining to Measure Bacterial Invasion by Conventional Light Microscopy

Most of the inconveniences of the fluorescence microscopy-based immunostaining procedure were overcome with the introduction of the immunogold–silver staining method of labeling antigens. This method differs from conventional immunocytochemical techniques in that the staining of the antigens is based on the principle of autometallography: the precipitation of silver on the surface of metallic particles.[17,18] According to this principle, ligands coupled to colloidal gold particles and bound to the antigen are enlarged by silver amplification of the colloidial gold to a size

[17] G. Danscher, *Histochemistry* **71,** 1 (1981).

[18] G. Danscher, J. O. R. Nörgaard, and E. Baatrup, *Histochemistry* **86,** 465 (1987).

visible in the light microscope. The silver enhancement occurs at a low pH and in the presence of a reducing agent (hydroquinone) by the reduction of silver ions to metallic silver. This immunogold–silver staining procedure, developed by Roberts[19] and Danscher,[17] is extremely sensitive, results in a stable signal with no problems of fading, and gives a high contrast that can be combined with conventional counterstains. The results can be viewed by conventional bright-field light microscopy.

Application of this procedure to locate bacteria inside or outside infected epithelial cells involves a single immunolabeling of the extracellular bacteria after washing and fixing of the infected cells, followed by counterstaining of the intracellular bacteria and the eukaryotic cells using a conventional chemical dye. Permeabilization of the plasma membrane and a second immunoincubation are avoided, and the extra- and intracellular bacteria can be viewed at the same time in the light microscope. An additional advantage is that fixation of the sample can be carried out after the immunostaining, which is beneficial when fixation-sensitive antigens have to be probed. After staining, the extracellular bacteria appear intensely black, clearly different from the rest of the specimen when examined by bright-field transmitted light illumination. Crystal violet (0.005% dissolved in water and incubated for at least 2 hr) is the recommended counterstain for most types of bacteria. This dye stains the intracellular bacteria deep violet, with the faint violet cytoplasm of the eukaryotic cells as background. The violet bacteria are in focus on a level different from that of the silver-stained extracellular microorganisms. The intensity of the crystal violet staining can be modulated by changing the concentration or the length of exposure to the dye. The specimens can be viewed immediately after staining by inverting the coverslip on a microscopic slide on a film of water. Background staining of the cytoplasm of the eukaryotic cells can be reduced by replacing the crystal violet solution with water to allow diffusion of the dye from the cytoplasm. The immunogold–silver staining method has been successfully used to evaluate bacterial invasion for a number of pathogens including *Neisseria* species, *Escherichia coli,* and *Helicobacter pylori.*[20,21]

Specific Reagents

Colloidal Gold Conjugates. Colloidal gold reagents are commercially available in various sizes, but they can also be prepared in the laboratory (see immunoelectron microscopy section). Prior to use in the immuno-

[19] W. J. Roberts, *Proc. K. Ned. Akad. Wet., Ser. C* **38,** 540 (1935).
[20] J. P. M. van Putten, C. T. P. Hopman, and J. F. L. Weel, *J. Med. Microbiol.* **33,** 35 (1990).
[21] J. P. M. van Putten, *J. Histotechnol.* **16,** 271 (1993).

gold–silver staining assay, the gold conjugates are diluted in PBS containing 1% gelatin to an optical density at 512 nm of 0.12 (for 5- to 10-nm gold particles), and microfuged (10,000 *g*, 3 min, 20°) to remove any gold aggregates. Incubations are carried out for 30 min at 20°. Prolonged incubations may increase background.

Silver Staining Reagents. All reagents to amplify the gold spheres to a light microscopically visible size can be prepared simply and cheaply from standard chemicals and clean water, but commercial silver enhancement kits are also available. The developing solution is prepared by mixing 70 ml of citrate buffer (citric acid monohydrate 2.55 g + trisodium citrate dihydrate 2.35 g/70 ml of water, pH 3.5) and 15 ml of reducing agent (hydroquinone 0.85 g/15 ml of water). Immediately before use 15 ml of silver ion solution (silver acetate 0.11 g/15 ml of water) is added. All solutions should be freshly prepared in carefully cleaned glassware using deionized water. A brown-colored developing solution indicates precipitation of silver, probably because of the unclean glassware or impure water. The silver enhancement reaction takes 2–10 min and can be viewed in the light microscope. The reaction is stopped by several rinses with water. Silver lactate (0.11 g/15 ml of water) can replace silver acetate but then the silver ion solution should be protected from light. The speed of the autocatalytic reaction can be reduced by the addition of gum arabic,[17] but in our hands this only lengthens the staining procedure without influencing results.

Procedure

Steps 1–6 are essentially the same as for the double-immunolabeling procedure described in the previous section. All incubations are carried out at 20°.

1. Wash the infected cells three times with PBS containing 1 m*M* $CaCl_2$ and 1 m*M* $MgCl_2$.
2. Replace with PBS containing 1–2% paraformaldehyde for at least 15 min.
3. Rinse twice with PBS.
4. Quench the remaining free aldehyde groups with PBS containing 50 m*M* NH_4Cl for 10 min.
5. Rinse twice with PBS.
6. Transfer the coverslips cell-side-down onto 50 μl of antibody solution on Parafilm and incubate for 30–45 min.
7. Remove unbound antibody by three 5-min washes with PBS.
8. Incubate the coverslips on 50 μl of gold-conjugate solution for 30 min.

9. Remove unbound gold spheres by two 5-min washes with PBS.
10. Rinse twice with deionized water to remove free ions.
11. Transfer the coverslips to a clean tissue culture dish, add 1 ml of physical developer, and incubate for 2–10 min. The reaction can be followed in the light microscope (silver acetate only). The developing solution should remain clear.
12. Stop the reaction with two rinses with deionized water.
13. Counterstain with 0.005% crystal violet (C.I. 42555) (or other suitable counterstain) for at least 2 hr.
14. Invert the coverslips on a water film on a glass slide and view the specimen using a conventional light microscope.
15. Store the specimen at 4° in staining solution or in water (stable for months), or mount in aqueous medium and seal with nail polish.

Confocal Scanning Microscopy and Bacterial Invasion

The recent introduction of confocal laser scanning microscopy[22] into the field of microbiology has opened up new ways to study the process of bacterial invasion. In this method a laser beam is focused at a single point in a fluorochrome-labeled specimen and the emitted light is focused through a pinhole in front of a photomultiplier, digitized, and stored in a computer for image analysis. The light emitted from locations either in front of or behind the focal point in the object is not focused in the pinhole and is therefore not detected; i.e., out-of-focus contributions from the object are not visible. In this way and by scanning through the sample, confocal microscopy can be used to visualize fluorochrome-labeled structures in a specific plane within the cells, i.e., to make optical sections of intact cells. Moreover, the digitization of the emitted signals enables image processing to enhance the confocal image. Functions that can be carried out by the image processor include changing of the intensity of the signals; the pointwise addition, subtraction, multiplication, or division of images; and, in principle, the analysis of three-dimensional structures. Another advantage of the technique compared with conventional fluorescence microscopy is dual-wavelength recording, which enables the user, for instance, to view adherent and intracellular bacteria at the same time or to relate directly intracellular bacteria and their surrounding host cell structures. The method has been successfully applied to localize bacteria inside eukaryotic cells, but is more suited to examine the mechanisms behind the entry and the intracellular processing of bacteria in eukaryotic

[22] D. M. Shotton, *J. Cell. Sci.* **94,** 175 (1989).

cells.[23,24] For measurement of bacterial invasion, the same immunostaining procedures as already described in the previous sections can be used.

Electron Microscopic Localization of Bacteria

Electron microscopy enables the localization of bacteria inside infected eukaryotic cells at the ultrastructural level. The procedure essentially involves fixation of the infected cells *in situ* using chemical fixatives such as glutaraldehyde and paraformaldehyde, followed by postfixation with osmium tetroxide, which largely crosslinks tissue lipids. The samples are then dehydrated in a graded series of ethanols and incubated with 100% propylene oxide (or xylene) which is miscible with the water-insoluble embedding media such as Epon and Araldite. After infiltration of the epoxy resins into the specimen and proper polymerization of the material, ultrathin (80-nm) sections are cut with a microtome, contrasted with heavy metal solutions such as uranyl acetate and/or lead citrate, and viewed in the electron microscope. The whole procedure usually takes several days and can be applied for ultrastructural localization of bacteria both in monolayers of cells and in infected organ culture models. For routine evaluation of bacterial invasion conventional, electron microscopy is not very attractive when compared with the established light microscopy methods. The method is laborious, quantitation of the number of bacteria is nearly impossible, and it is difficult to exclude that apparently vacuolated bacteria are truly located intracellularly unless serial sections are made or the surface membranes are lined with small-sized (1–3 nm) electron-dense material like colloidal thorium (1% suspension in 3% acetic acid, pH 2.5)[25] or colloidial gold particles[6] before sectioning of the specimen. These particles mark all compartments that are in an open connection to the extracellular space. Because of these limitations, conventional electron microscopy is mainly used in relation to bacterial invasion to confirm light microscopic observations or to localize bacteria in infected tissues that cannot be studied light microscopically. The technique has much more value in determining the exact location of the bacteria in relationship to the ultrastructure of the eukaryotic cells, although also for this application novel methods have become available (see [34]).

[23] B. B. Finlay, S. Ruschkowski, and S. Dedhar, *J. Cell. Sci.* **99,** 283 (1991).
[24] V. B. Young, S. Falkow, and G. K. Schoolnik, *J. Cell. Biol.* **116,** 197 (1992).
[25] M. E. Ward, J. N. Robertson, P. M. Englefield, and P. J. Watt, *in* "Microbiology—1975" (D. Schlesinger, ed.), p. 188. Am Soc. Microbiol., Washington DC, 1975.

Procedure

The method described here is typical of the available techniques and is meant to outline the basic procedure for electron microscopic examination of infected cells. For the general principles of tissue processing and for detailed discussions of electron microscopic techniques the reader is referred to standard texts.[7,26] In the procedure described, it is essential that all fixation and embedding steps be carried out in a fume cupboard with good ventilation, as many of the reagents used are very harmful.

1. Wash the infected cells cultured on plastic tissue culture dishes or on filters three times with PBS to remove nonadherent bacteria.
2. Fix the cells with 2% paraformaldehyde and 2% glutaraldehyde in PBS or cacodylate buffer (0.1 *M*, pH 7.4) for at least 15 min at 20°.
3. Wash several times with buffer over a period of at least 1 hr, postfix with 1% osmium tetroxide in the same buffer for 15–30 min at 20°, and either collect the cells in a glass vial by gently detaching them using a rubber policeman or directly proceed to step 4 with the cells still attached to the plastic culture dish.
4. Dehydrate the samples by passing them through a graded series of ethanols in water (50%, 70%, 95%, 100%, 100%, 5 min each) and then through 100% propylene oxide (2 × 10 min). Carefully remove each solution (at least 10 times the specimen volume) from the specimen with a Pasteur pipette before adding the next one. Cells not harvested in step 3 can now be collected in a more native state (i.e., without mechanical detachment) as they are released by propylene oxide which dissolves the plastic culture dishes.[27] Note that water-free (100%) ethanol is obtained by adding molecular sieve grains (type 4A, BDH Chemicals, Poole, UK) to 96% ethanol.
5. Prepare the embedding mixture,[7] e.g., Epon 812 (Fluka, Ronkonkoma, NY) by thoroughly mixing 5 ml Epon 812 with 8 ml of the hardener DDSA and 8 ml Epon 812 with 7 ml of the hardener MNA. Mix both solutions very thoroughly, and add prior to use 400 μl of the accelerator DMP-30 (the solution turns slightly red–brown).
6. Infiltrate the specimen with embedding medium by two 30-min incubations with a mixture of propylene oxide and the epoxy resin at ratios of 1 : 1 and 1 : 3, respectively, followed by a 1-hr incubation in pure epoxy resin. (The procedure may differ with different resins.)
7. Transfer the specimen to oven-dried (60°) gelatin capsules, fill the capsules with embedding medium, and place the sample at the bottom

[26] P. R. Lewis and D. P. Knight, *in* "Practical Methods in Electron Microscopy" (A. M. Glauert, ed.), Vol. 14. Elsevier, Amsterdam, 1992.

[27] P. Biberfield, *J. Ultrastruct. Res.* **25,** 158 (1968).

using fine tweezers. Polymerize by heating the capsules at 60° for at least 24 hr.

8. Dissolve the gelatin capsules in warm water, position the embedded specimen in the mounting block of the microtome, trim the block, and cut ultrathin sections with the microtome. Collect the sections from the trough on Formvar–carbon-coated grids using fine tweezers.

9. Contrast the sections with 5% uranyl acetate in methanol for 5–10 min, wash with 100% methanol, and let the grids dry or double-stain with lead citrate by floating them (section-side-down) on a drop of lead citrate solution (1–5 min), on 0.01 *N* NaOH (2 × 30 sec) and on distilled water (2 × 30 sec) in a closed petri dish.[26,28]

10. View the sections in the electron microscope at 50–80 kV. The grids can be stored at a dry place for years.

Immunoelectron Microscopy and Bacterial Invasion

A development generating increasing interest is the study of bacterial invasion using postembedding immunoelectron microscopy, which combines both electron microscopy and antibody labeling. In this technique, infected cells are fixed, embedded under nondenaturating conditions to preserve antigens, and cut into ultrathin sections, which are then incubated with antibodies and a gold-labeled conjugate.[29,30] The electron density of the gold spheres allows easy detection of the probed antigens in the electron microscope. This novel approach enables localization of the bacteria in infected host cells and, in addition, provides information about the antigenic characteristics of the bacteria, the host cell compartments in which they are contained, and the eukaryotic structures involved in the entry and intracellular processing of the bacteria, thus giving electron microscopy a more functionally oriented direction. The postembedding immunoelectron microscopic procedure with the highest immunolabeling efficiency is the cryosectioning method developed by Tokuyasu[29]; however, by use of low-denaturation resins such as Lowicryl K4M and HM20 or LR White, satisfactory results can be obtained as well, particularly when the antigens to be probed exist in large quantities within the cells.[30–33]

[28] E. S. Reynolds, *J. Cell. Biol.* **17,** 208 (1963).

[29] K. T. Tokuyasu, *J. Microsc. (Oxford)* **143,** 139 (1986).

[30] J. Roth, *J. Microsc. (Oxford)* **143,** 125 (1986).

[31] E. Carlemalm, R. M. Garavito, and W. Villiger, *J. Microsc. (Oxford)* **126,** 123 (1982).

[32] J. A. Hobot, *in* "Colloidal Gold: Principles, Methods and Applications" (A. M. Hayat, ed.), Vol. 2, p. 76. Academic Press, San Diego, 1989.

[33] G. R. Newman, *in* "Colloidal Gold: Principles, Methods and Applications" (A. M. Hayat, ed.), Vol. 2, p. 48. Academic Press, San Diego, 1989.

Because of the higher labeling efficiency, the low toxicity of the reagents, and the shorter procedure we favor cryosectioning. In the cryosectioning procedure, fixed specimens are enclosed in gelatin and infused with sucrose for cryoprotection and to endow the specimen with the plasticity that is required for sectioning. The samples are then frozen in liquid nitrogen and cryosectioned using an ultracryomicrotome. The sections are immunolabeled using antibodies and a colloidal gold conjugate, which because of their electron density can easily be traced in the electron microscope. A double (or triple)-labeling procedure can be applied using gold spheres of various sizes to mark different antigens at the same time.[7,34] The whole procedure takes 1–2 days (including overnight fixation) and has been successfully applied to investigate the entry and intracellular processing of various pathogenic bacteria.[6,35,36] The following section deals with those aspects of the cryosectioning technique that are directly relevant for immunoelectron microscopic examination of bacterial invasion.

Specific Reagents

Fixatives. The fixative to be used depends on the specific requirements of the tissue and the antigens to be probed. In most cases, the same fixation protocol can be used as already described for the light microscopy, except that at least 2% paraformaldehyde should be used to stabilize cell structure further. Cacodylate buffer (0.1 *M*, pH 7.2) can be used instead of PBS. Glutaraldehyde (0.01–0.1%) can be added if not detrimental to the antigens. All fixative solutions should be prepared in a fume cupboard. Fixed material can be stored for up to several months in formaldehyde solution at 4°, although with some epitopes gradual loss of antigenicity may occur. Occasionally, the chemical fixation procedure described damages the molecular features of the antigen, with concomitant poor immunolabeling of the bacteria. Then, a different antigen should be probed, or if this is not possible, the use of milder fixation procedures such as periodate–lysine–paraformaldehyde (PLP) fixation may improve results,[29,36,37] although this generally results in poorly preserved ultrastructure.

Embedding Reagents. Gelatin solution (5%) is freshly prepared by mild heating of 5 g of gelatin in 100 ml of PBS. For preparing 2.3 *M* sucrose, dissolve 78.7 g of sucrose in 20 ml of PBS with mild heating and

[34] G. Griffiths, K. Simons, G. Warren, and K. T. Tokuyasu, this series, Vol. 96, p. 466.

[35] J. F. L. Weel, C. T. P. Hopman, and J. P. M. van Putten, *Infect. Immun.* **57,** 3395 (1989).

[36] J. F. L. Weel, C. T. P. Hopman, and J. P. M. van Putten, *J. Exp. Med.* **173,** 1395 (1991).

[37] I. E. McLean and P. K. Nakane, *J. Histochem. Cytochem.* **22,** 1077 (1974).

make up to a final volume of 100 ml with PBS. The solution can be stored at 4°.

Colloidal Gold Reagents. Both commercially available and laboratory prepared gold-coupled reagents can be used.[38–41] A satisfactory method to prepare gold sols, the gold spheres that form the basis of the gold conjugate, is by the reduction of tetrachloroauric acid (Merck, Darmstadt, Germany) with tannic acid (Aleppo tannine, Mallinckrodt Inc. Paris, KY) and trisodium citrate, as developed by Slot and Geuze[40]:

1. Clean all glassware to be used by boiling with distilled water.
2. Prepare a gold chloride solution (1 ml 1% $HAuCl_4$ + 79 ml distilled water) and a reducing mixture [4 ml of 1% trisodium citrate dihydrate + 0.01–5 ml of 1% aqueous tannic acid (volume dependent on the particle size desired: 10 μl for 17-nm gold spheres, 5 ml for 3-nm gold spheres) + distilled water to a final volume of 20 ml; if more than 1 ml of 1% tannic acid is used, the pH of the reducing solution should be adjusted by adding an equal volume of 25 m*M* K_2CO_3 to that of tannic acid].
3. Bring both solutions to 60° and quickly add the reducing mixture to the gold chloride solution with vigorous stirring. The gold sol forms within 30 sec to 60 min, the time depending on the amount of tannic acid. Formation is finished when the combined solution has a wine-red color. A black or gray color indicates precipitation of the gold probably as a result of unclean glassware or impure water.
4. Boil the gold sols for 5 min while stirring (the red color should remain), cool on ice, and store at 4° or directly complex the gold spheres to protein.

The preparation of the gold conjugate basically involves the adsorption of protein to the gold sols (commercially obtained or laboratory prepared as described above) in several steps.[38,39] First, the pH of the gold sol is adjusted to above the isoelectric point of the protein to be coupled to achieve optimum adsorption.[42] Then, the minimal concentration of protein required to stabilize the sols is determined using the flocculation test. This test is based on the flocculation of the gold sol in the presence of electrolytes as long as the protein concentration is below the stabilization point. Flocculation causes a red-to-blue color change of the gold sol.[43] In the

[38] J. W. Slot and H. J. Geuze, *J. Cell. Biol.* **90,** 533 (1981).

[39] J. Roth, *in* "Techniques in Immunocytochemistry" (G. R. Bullock and P. Petrusz, eds.), Vol. 2, p. 217. Academic Press, London, 1983.

[40] J. W. Slot and H. J. Geuze, *Eur. J. Cell. Biol.* **38,** 87 (1985).

[41] D. A. Handley, *in* "Colloidal Gold: Principles, Methods and Applications" (A. M. Hayat, ed.), Vol. 1, p. 13. Academic Press, San Diego, 1989.

[42] W. Geoghehan and G. Ackerman, *J. Histochem. Cytochem.* **25,** 1187 (1977).

[43] M. Horisberger and J. Rosset, *J. Histochem. Cytochem.* **25,** 295 (1977).

final step the appropriate amounts of protein and gold sol are mixed and the solution is purified by gradient centrifugation. In practice, the adsorption procedure is as follows:

1. Adjust the pH of the gold sol by adding the appropriate volume of 0.2 *M* K_2CO_3 (or 0.2 *M* H_3PO_4) using pH indicator paper. Do not use a pH-meter, unless pH is measured from an aliquot of the gold sol that has been saturated with protein (e.g., 50 μg protein per 5 ml of gold sol) to prevent damage to the electrode.
2. Mix 50 μl of linear dilutions of protein solution (e.g., 50 μg/ml protein A in distilled water) with 250 μl of gold sol and, after 5 min, add 50 μl of 10% NaCl. Determine the minimum amount of protein required to prevent the red-to-blue color change and calculate the amount of protein needed to prepare the gold conjugate.
3. Add the minimum stabilizing amount of protein to the gold sol and stir rapidly and continuously for 2 min. Then, add 0.1% BSA (final concentration) to stabilize the complex maximally. Concentrate the gold conjugate by centrifugation (for 10-nm gold particles: 45,000 *g*, 45 min, 4°) and resuspend the loose pellet in a small volume of PBS.
4. Purify the gold conjugate over a continuous 10–30% glycerol gradient in PBS containing 0.1% BSA (for 10-nm gold particles: 45,000 *g*, 30 min, 20°) and store in aliquots at −70°.

Prior to use in immunoelectron microscopy the gold reagents are diluted in PBS containing 1% gelatin to an optical density at 520 nm of 0.06 (10-nm particles) and microfuged (10,000 *g*, 3 min) to remove any gold aggregates.

Methylcellulose–Uranyl Acetate Solution. This solution is critical for contrasting the sections and for maintaining ultrastructural morphology at the end of the immunolabeling. Methylcellulose 1.5% (Methocel MC, 15 mP · sec, Fluka, Ronkonkoma, NY) is prepared by adding 1.5 g of methylcellulose to 100 ml of water at 95°. After immediate cooling to 4°, the solution is kept at 4° for 3 days, with stirring during the first 4–8 hr, and then centrifuged (100,000 *g*, 60 min, 4°) to clear the solution, which can then be stored at 4°. Prior to use the methylcellulose is mixed with acid uranyl acetate (3% in water, pH 3) at a ratio of 9 : 1 (v/v) and kept on ice until use.

Procedure

The whole procedure is carried out at 20°, unless indicated otherwise.

Fixation and Embedding of the Infected Cells

1. Wash the infected cells (approximately 5×10^6 cells suffices) three times with PBS to remove non-cell-associated bacteria.

2. Replace with PBS containing 2–4% paraformaldehyde for at least 1 hr.

3. Collect the cells in 0.75 ml of fixing solution in an Eppendorf tube by gently detaching the monolayer using a rubber policeman.

4. Mix the cells gently with an equal volume of 5% gelatin (hand-warm), and pellet them by centrifugation (10,000 *g*, 10 sec, 4°). Remove excess gelatin, and centrifuge again (10,000 *g*, 10 sec, 4°) just before the gelatin solidifies. Keep the embedded material in fixative for at least 2 hr (4°).

5. Cut the Eppendorf tube just above the gelatin pellet with a razor blade and transfer the specimen-containing part to fixing solution for at least several hours to harden the specimen.

6. Transfer the embedded pellet to Parafilm using a blunt needle, cut it into small pieces (1 × 1 mm), and store these in fixing solution for at least 2–3 hr (overnight is better) at 4° until further processing. Prolonged storage, i.e., weeks, may result in loss of distinct epitopes.

Ultracryomicrotomy

1. Trim the embedded specimen and remove excess gelatin using a razor blade.

2. Transfer the specimen to 2.3 *M* sucrose after blotting excess fixing solution using filter paper, and incubate at 20° with occasional tilting until the specimen stops floating on the sucrose (approximately 30 min).

3. Place the sucrose-infused specimen with the pelleted cells on top of a polished specimen holder on a little drop of sucrose. Remove excess sucrose and quickly freeze the specimen by rapidly agitating the specimen holder in liquid nitrogen using tweezers. Place the specimen into a small metal container inside a Dewar containing liquid nitrogen. Avoid body contact with the nitrogen. Keep the specimen under nitrogen until cryosectioning. The specimen can be stored in liquid nitrogen for up to several weeks.

4. Transfer the small container with the frozen specimen under liquid nitrogen from the jar to the cryochamber (−120°) using tweezers, and position the holder carrying the specimen into the precooled (−100°) mounting block of the microtome.

5. Trim a flat surface on the frozen specimen and cut ultrathin sections (70–100 nm) with a precooled glass knife (−85° to −95°). Recover the sections by gently touching the sections on the knife with a wire loop (2 mm in diameter) filled with 2.3 *M* sucrose solution. Make sure that the sections are transferred to the sucrose before it freezes.

6. Transfer the thawed sections from the sucrose onto Formvar-coated nickel grids by touching the droplet onto the film. Then invert the grids

(section-side-down) and place onto 2% gelatin plates on ice until the start of the immunolabeling.

Immunolabeling of Cryosections

All steps are carried out at 20°.

1. Place the 2% gelatin-containing petri dishes with the grids on a warm plate (40°) until the gelatin melts and let the grids float on the molten gelatin for 10 min to facilitate diffusion of the sucrose into the gelatin.
2. Remove residual sucrose and block nonspecific binding of antibody by three 3-min incubations by floating the inverted (i.e., section-side-down) grids on a drop of PBS containing 0.15% glycine, 0.1% BSA, and 0.1% gelatin placed on Parafilm.
3. Incubate the grids on 5–25 μl of PBS containing diluted antibody for 30–60 min (dependent on the quality of the antibody).
4. Remove unbound antibody by four 5-min washes with PBS.
5. Incubate the grids on 5–25 μl of PBS containing 1% gelatin and the gold conjugate for 30 min.
6. Remove unbound gold spheres by three 5-min washes with PBS.
7. Remove all salts by three 1-min washes with distilled water to allow contrasting with uranyl acetate.
8. Contrast and cover the sections by incubating the grids on ice-cold drops of 0.3% uranyl acetate in 1.5% methylcellulose for 10 min, catch them with a 4-mm copper or platinum wire loop, carefully remove excess methylcellulose using filter paper, and let them air-dry with the loop in a vertical position. The dried film of the embedding mixture should give a gold–blue interference color. The grids are stored in a dry place until viewed in the electron microscope.

If double labeling of bacterial and/or host cell antigens is performed the first gold labeling (step 6) should be followed by an incubation with unconjugated ligand (e.g., protein A, 10 μg/ml for 30 min at 20°) to saturate binding sites before the second antibody and conjugate are applied (repeat steps 3–6). The second antibody is marked with conjugated gold spheres of larger size. Obviously, control experiments in which the second antibody is omitted from the protocol are essential to ensure that the primary antibody is completely saturated with the first conjugate.

Acknowledgments

We gratefully acknowledge Dr. Brian D. Robertson for critical reading of the manuscript. H. U. C. Grassmé received a grant from the Bundesminister für Forschung und Technologie (Grant 01-KI-8920).

[30] Polarized Epithelial Monolayers: Model Systems to Study Bacterial Interactions with Host Epithelial Cells

By M. GRACIELA PUCCIARELLI and B. BRETT FINLAY

Introduction

Many pathogenic microorganisms use different routes to gain access to various locales in the susceptible host. One such route of entry is adherence to and passage through an epithelial cell barrier. This property is commonly shared by several bacteria and represents a crucial step in dissemination of the pathogen to underlying tissues, blood, and the reticuloendothelial system.[1,2] Epithelial cells grown under normal tissue culture conditions are missing several features that epithelial cells in tissues of the host exhibit. They are not polarized and thus fail to form a well-defined brush border. They also do not form tight junctions and are incapable of forming impermeable barriers; however, a few epithelial cell lines have been identified that can be grown on porous membranes, thereby forming impermeable polarized monolayers.[3,4] Cells grown in this manner have defined apical (top) and basolateral (bottom) surfaces that are separated by tight junctions. These polarized epithelial monolayers are impermeable to ions, have a measurable transepithelial electrical resistance, and have several defined apical and basolateral surface markers. The most extensively characterized polarized epithelial cell line is the Madin–Darby canine kidney (MDCK) cell line, originally isolated from a dog kidney epithelium.[5] Three human adenocarcinoma cell lines, Caco-2, HT-29, and T84, can also form polarized monolayers when grown on permeable substrates.[6–8] HT-29 cells have the additional feature that when they are grown under certain conditions, they differentiate into different epithelial cell types.[7] Table I compares some of the salient features of these various cell lines that should be considered when choosing a model system. Fac-

[1] J. W. Moulder, *Microbiol. Rev.* **49,** 298 (1985).
[2] B. B. Finlay and S. Falkow, *Microbiol. Rev.* **53,** 210 (1989).
[3] K. Simons and S. D. Fuller, *Annu. Rev. Cell Biol.* **1,** 243 (1985).
[4] E. Rodriguez-Boulan and W. J. Nelson, *Science* **245,** 718 (1989).
[5] J. C. Richardson, V. Scalera, and N. L. Simmons, *Biochim. Biophys. Acta* **673,** 26 (1981).
[6] M. Pinto, S. Robine-Leon, M. Appay, M. Dedinger, N. Triadou, E. Dussaulx, B. Lacroix, P. Simon-Assmann, K. Fogh, J. Haffen, and A. Zweibaum, *Biol. Cell.* **47,** 323 (1983).
[7] C. Huet, Merino C., Sahuquillo, E. Coudrier, and D. Louvard, *J. Cell Biol.* **105,** 345 (1987).
[8] K. Dharmsathaphorn, J. A. McRoberts, K. G. Mandel, L. D. Tisdale, and H. Masui, *Am. J. Physiol.* **246,** 204 (1984).

TABLE I
PHENOTYPIC CHARACTERISTICS OF DIFFERENT POLARIZED EPITHELIAL CELL LINES

Cell line	Source	Support for polarization	Morphology of polarized cells	Presence of microvillar hydrolases
MDCK	Dog kidney epithelium	Permeable filters; impermeable substrates	Apical microvilli	Aminopeptidase
Caco-2	Human colon carcinoma	Permeable filters; impermeable substrates; accelerated with collagen	Well-developed apical microvilli and brush border	Two disaccharidases; two peptidases
T-84	Lung metastasis from human colon carcinoma	Permeable filters; impermeable substrates; accelerated with collagen	Lack of well-developed brush borders	No
HT-29	Human colon carcinoma	Permeable filters; uncoated impermeable glass	Brush borders only when glucose absent, capable of differentiation	Alkaline phosphatase, sucrose isomaltase, aminopeptidase N, dipeptidyl-peptidase IV

tors to be considered for a given bacterial pathogen include host specificity, tissue specificity, and the presence/absence of a brush border with microvilli.

Reports have described the use of polarized epithelial monolayers to study bacterial interactions with these epithelial barriers and have illustrated the utility of these models to study some aspects of bacterial pathogenesis. *Salmonella* species[9,10] and *Campylobacter jejuni*[11] can penetrate through epithelial monolayers grown on permeable filters. *Salmonella* preferentially interact with the apical surface and cause significant morphological alterations in the microvilli. In contrast, it has been reported that *Shigella flexneri* infects polarized epithelial cells only from the basolateral surface.[12] Enterotoxigenic *Escherichia coli* (ETEC) interact with HT-29 cells, binding to surface carbohydrates.[13]

Here we describe methods that are used to grow polarized epithelial cells in culture. In addition, various procedures are described that can be used to study various aspects of pathogen interactions with polarized monolayers. These methods are described for MDCK and Caco-2 cells, as our laboratory has used these two cell lines; however, the procedures are readily adaptable to any cell line that can be grown under conditions facilitating polarization. Additionally, modifications to these procedures will be needed for bacteria requiring special growth conditions, although these assays should, in principle, be applicable to most bacterial pathogens.

Growth and Bacterial Infection of Polarized Epithelial Monolayers

The procedure used to grow polarized monolayers of MDCK or Caco-2 cells is very similar, and minor modifications are mentioned below. Strain I MDCK cells (which have a transepithelial electrical resistance greater than 1000 Ω cm^2), strain II MDCK cells (which have a resistance of 100 Ω cm^2; ATCC CCL34), or Caco-2 cells (ATCC HTB 37) are used between passages 18 and 50. They are grown in Eagle's minimal essential medium (MEM) with Earle's salts, nonessential amino acids, and 10% fetal calf serum. Cells are passed (twice a week for MDCK, once a week for Caco-2) by diluting 1 : 5 to 1 : 10 after removal from the culture flask with an

[9] B. B. Finlay, B. Gumbiner, and S. Falkow, *J. Cell Biol.* **107,** 221 (1988).

[10] B. B. Finlay and S. Falkow, *J. Infect. Dis.* **162,** 1096 (1990).

[11] M. E. Konkel, D. J. Mead, S. F. Hayes, and W. Cieplak, Jr., *J. Infect. Dis.* **166,** 308 (1992).

[12] J. Mounier, T. Vasselon, R. Hellio, M. Lesourd, and P. J. Sansonetti, *Infect. Immun.* **60,** 237 (1992).

[13] J. R. Neeser, A. Chambaz, M. Golliard, A. H. Link-Amster, V. Fryder, and E. Kolodziejczyk, *Infect. Immun.* **57,** 3727 (1989).

ethylenediaminetetraacetic acid (EDTA)–trypsin solution (GIBCO, Gaithersburg, MD).

Polarized epithelial monolayers are cultured by adding trypsinized cells to the apical chamber of Transwell filter units (Costar, Cambridge, MA). These units contain a 0.33-cm^2 porous filter membrane that is available in several pore sizes. If bacterial penetration is to be studied, $3.0\text{-}\mu m$ pores are used (Costar No. 3415). Alternatively, $0.4\text{-}\mu m$ pore filters can be used if bacterial penetration is not being examined (Costar No. 3413). These filters have been treated to enhance epithelial cell adherence to the membrane. Trypsinized MDCK or Caco-2 cells ($150\ \mu l$, 1.5×10^5 cells) are added apically to each Transwell unit and the filter unit is placed in 1 ml of fresh medium bathing the basolateral surface. Filter units are grown under standard tissue culture conditions (37°, 5% CO_2). Both apical and basolateral media are replaced every 2–3 days by *gently* aspirating off the apical fluid and placing the unit in a new well containing fresh medium. Monolayers are fully polarized after 4–6 days (MCDK cells) or 10–14 days (Caco-2 cells). Several methods can be used to ensure polarization. Perhaps the simplest is to measure transepithelial electrical resistance, which is described below.

Prior to bacterial infection, if antibiotics were used in the medium, fluid is removed from both surfaces and rinsed with MEM medium containing 10% fetal calf serum without antibiotics. Fresh medium is then added to the Transwell units ($150\text{–}200\ \mu l$) and to the well (1 ml). Approximately $5\text{–}10\ \mu l$ of bacteria (5×10^7 bacteria) can be added to the apical or basolateral surfaces and the units incubated for the desired times. For basolateral infections, filters can be inverted and bacteria added directly to the basolateral surface prior to placing upright in the well.

Measurement of Transepithelial Electrical Resistance

One of the most important functions of epithelial cells *in vivo* is to form impermeable barriers, which are mediated by intercellular junctions known as tight junctions. Polarized epithelial monolayers also have the capacity to form such barriers. Transepithelial electrical resistance of polarized epithelial monolayers can be measured. Treatments that disrupt tight junctions, such as removing Ca^{2+} from the media or inhibiting actin filament function with cytochalasins, result in the loss of monolayer electrical resistance properties and cell polarity. The effect of pathogenic organisms on the monolayer integrity can also be determined by measuring electrical resistance after bacterial addition. For example, *Salmonella* species disrupt transepithelial resistance of MDCK and Caco-2 monolayers within 2–4 hr of bacterial infection.[9,10] In contrast, data obtained with *C. jejuni* demonstrate that although these bacteria penetrate through Caco-2 monolayers, no loss of electrical resistance is detected in infected monolayers.[11] In these assays apical or basolateral infection of nonadherent, noninvasive *E. coli* (such as HB101 or DH5α) should not produce any loss in electrical resistance of the monolayers.

Transepithelial electrical resistance measurements of polarized monolayers can be performed on epithelial cells grown in Transwell units. These measurements are performed with a Millicell-ERS resistance meter, which is designed for such measurements (Millipore, Bedford, MA). One electrode is inserted into the apical medium (being careful not to touch the monolayer) and the other is inserted through one of the three slots into the basolateral medium. Electrical resistance values are expressed as the area $\times$ resistance (Ω cm^2) and calculated by multiplying the measured resistance (in ohms) by the area of the filter (0.33 cm^2 for the Transwells). Background resistance values are also subtracted. The expected resistance of uninfected polarized Caco-2 monolayers is usually about 200–250 Ω cm^2, whereas MDCK monolayers (strain 1) have higher values, in the range of 1000 Ω cm^2.

Measurement of Bacterial Association with Polarized Monolayers

There are several possible outcomes when bacteria interact with polarized epithelial cells. Bacteria may adhere to the monolayer. Alternatively, they may enter into (invade) the epithelial cells. Once internalized, they may remain within the cell or penetrate through the monolayer to the opposite surface. Each of these outcomes can be determined experimentally, as can the polarity (apical versus basolateral) of infection of the monolayer.

The first monolayer association method described below relies on radioactive labeling of bacteria prior to addition to epithelial monolayers.[14] This method of labeling avoids the problems associated with bacterial growth, as the total counts remain constant. The percentage of radioactivity that remains in the filter after extensive washing represents the percentage of bacteria that are associated with the monolayer, either adherent or internalized. An alternative procedure uses detergent lysis of the eukaryotic cells after extensive wahsing to liberate the adherent and intracellular bacteria. In this method, viable organisms are measured. Both methods are complementary, but neither distinguishes between adherent and intracellular organisms. An additional method, which uses gentamicin to kill

[14] B. B. Finlay, J. Fry, E. P. Rock, and S. Falkow, *J. Cell Sci., Suppl.* **11,** 99 (1989).

adherent but not internalize organisms, can be used to quantitate the number of intracellular bacteria (see [28]).

Radioactive Monolayer Association Assay

Freshly grown bacteria are washed and resuspended in a methionine-free medium (Difco, Detroit, MI). After 30 min incubation (37°), bacteria are centrifuged and resuspended in the same assay medium containing 50 μCi/ml [^{35}S]methionine (New England Nuclear, Boston, MA) and incubated for an additional 30 min. Bacteria are washed three times in L-broth and 5 μl of bacteria is added to either surface of a polarized monolayer. After appropriate incubation times monolayers are washed several times in cold PBS, and the filter is cut out from the plastic support with a scalpel and placed in 5 ml of an aqueous counting scintillation (ACS II, Amersham, Arlington Heights, IL). The measured radioactivity is a measure of the number of bacteria associated (adherent and internalized) with the monolayer.

This assay has been used to measure *Salmonella cholerae suis* association with polarized MDCK monolayers.[9,14,15] With this organism, 15% of added radioactivity was bound to the monolayer 4 hr after infection. Different infection conditions can be used in this assay, including alterations in temperature and addition of bacterial metabolic inhibitors.[9]

A variation of this assay, which measures only adherence, uses glutaraldehyde-fixed monolayers. Polarized monolayers in Transwell filters are washed and fixed in 2% glutaraldehyde for 2 hr on ice. Filter units are washed (three times) before adding fresh medium and then radiolabeled bacteria.[15] Glutaraldehyde may or may not alter the appropriate host cell receptor, thereby affecting adherence levels of some bacterial pathogens.

Nonradioactive Association Assay

Another alternative procedure to determine the association of bacteria with polarized monolayers uses the detergent Triton X-100, which at relatively low concentrations lyses eukaryotic cells without affecting bacterial viability. In this method, infected monolayers are rinsed extensively with PBS and the infected filter is cut out and treated with 100 μl 1% Triton X-100 for 5–10 min at room temperature. PBS (400 μl) is added and mixed vigorously. All the viable bacteria associated with the monolayer are titered by plating appropriate dilutions on selective plates.

[15] B. B. Finlay, M. N. Starnbach, C. L. Francis, B. A. D. Stocker, S. Chatfield, G. Dougan, and S. Falkow, *Mol. Microbiol.* **2,** 757 (1988).

Intracellular versus Adherent Organisms

One method to quantitate adherent organisms using glutaraldehyde-fixed monolayers is described above. An alternate method uses gentamicin to kill adherent organisms. Infected monolayers are rinsed with PBS and fresh medium containing gentamicin (100 μg/ml) is added to both apical and basolateral fluids and incubated for at least 1 hr. This antibiotic kills extracellular bacteria without significantly affecting viability of intracellular organisms. When this assay is used in conjunction with the nonradioactive association assay, the difference between total associated bacteria and intracellular bacteria represents the proportion of adherent bacteria.

Quantitation of Bacterial Penetration through Polarized Epithelial Monolayers

As with the bacterial association assays, bacterial penetration through the monolayer (often called transcytosis) can be quantitated by two alternative methods. Both methods assume that the pore size of the filter is large enough to allow bacteria to pass through (i.e., 3 μm). The simplest method to quantitate penetration consists of removal of aliquots, at various times, of the medium opposite that to which bacteria were added.[9,10] These aliquots are diluted and titered by plating on selective plates and determination of viable counts. This method does not distinguish between bacteria that have recently penetrated and those that have been growing in the medium. An alternative procedure, which uses radiolabeled organisms, is based on quantitation of radioactivity present in both media (apical and basolateral), as well as in the monolayer.[14] This method gives a more detailed profile of the efficiency of penetration. This value can be compared with those values that represent bacteria that remain in the monolayer or that return to the same surface from which they entered.

Nonradioactive Bacterial Penetration Assay

After apical or basolateral infection of polarized monolayers (using 5×10^6 to 1×10^7 bacteria per filter unit), the medium of the opposite side is removed every hour and replaced with fresh medium prewarmed at 37°. These hourly changes limit the amount of bacterial growth in the opposite medium. Various dilutions of the harvested medium are plated on selective agar plates and the bacterial colonies counted. Nonadherent and noninvasive *E. coli* are used as controls of the penetration process. Motile *E. coli* such as DH5α should be used. These *E. coli* do not penetrate epithelial monolayers unless there is some physical damage to the epithelial cells or there are significant disruptions of the tight junctions. This bacterium can be added together with bacterial species that penetrate polarized epithelial monolayers to determine when the monolayer is significantly disrupted.[9-11] This method has been used to determine penetration rates of *Salmonella* species and *C. jejuni* and to screen for mutants in *S. cholerae suis* that are unable to penetrate the monolayer.[15]

Radioactive Bacterial Penetration Assay

Bacteria are radiolabeled with [^{35}S]methionine, as described above for the radioactive bacterial association assay, and added to several polarized epithelial monolayers. After an incubation period to establish invasion, the monolayers are washed with fresh medium and a few filters are removed for scintillation counting (time 0); the remaining filters are incubated for an additional period prior to harvesting. The radioactivity present in the filter at time 0 establishes a base level of counts representing bacteria associated with the monolayer (i.e., those adherent or internalized). At later times, radioactivity present in both apical and basolateral media as well as the filter are measured to monitor the percentage of bacteria that remain associated with the monolayer versus those that either penetrate the monolayer or return to the surface to which they were originally added. This assay has been described for apical infection of MDCK polarized cells by *S. cholerae suis*.[14]

Electron Microscopy of Infected Polarized Monolayers

One of the main features of polarized epithelial monolayers (when compared with nonpolarized cells) is that they produce a well-defined brush border containing microvilli on their apical surface. Electron microscopy is a direct visual confirmation of experimental results and yields morphological information about bacteria–epithelial cell interactions. These studies should be performed to complement other methods. We have briefly outlined methods used to prepare samples of polarized epithelial monolayers for electron microscopy, concentrating on special procedures needed for filter preparation, as the methods used differ little from standard electron microscopy procedures.

Transmission Electron Microscopy

Polarized epithelial monolayers grown in Transwell units are infected as described above. The filters are excised with a scalpel and washed several times with PBS and fixed in cold (4°) 2% glutaraldehyde in 0.1 *M* sodium phosphate buffer (pH 7.4) overnight. After further washing with phosphate buffer, filters are postfixed in cold 1% OsO_4 in 0.1 *M* phosphate buffer for 90 min and then stained with cold 0.25% uranyl acetate overnight. Samples are dehydrated in a graded series of ethanol and embedded in firm Spurr's plastic. Embedding in the plastic causes the filters to "roll up." Sample blocks are prepared such that sectioning cuts a cross section through these rolls, yielding a spiral cross section of the filter. Sections are stained with uranyl acetate and lead citrate using standard procedures before examination in the transmission electron microscope.

Scanning Electron Microscopy

Scanning electron microscopy (SEM) preparation follows standard methods. Infected monolayers are washed and excised with a scalpel. They are then fixed as described above and dehydrated in a series of alcohols. Samples are then dehydrated in a critical-point apparatus and mounted on stubs using double-sided tape, making certain that the apical surface is facing upward. Filters are then coated with gold and examined with a scanning electron microscope. Determining the apical and basolateral surface is often difficult with filters. As a cautionary note, apically seeded Caco-2 cells can grow on both surfaces of the filter.

Electron microscopic analysis of infected polarized cells has been done with *Salmonella* species, *C. jejuni,* and enterotoxigenic *E. coli* (ETEC).[9–11,13] These reports describe in more detail the methods used and also describe several morphological changes that occur on bacterial infection.

Discussion

We have described here several methods to study the interactions of bacterial pathogens with polarized epithelial monolayers. Polarized monolayers have several advantages over nonpolarized cell lines as a model system. These include, (1) the polarized distribution of epithelial surface proteins and microvilli, (2) measurement of epithelial permeability, (3) the ability to define which surface a pathogen interacts with, and (4) the ability to measure bacterial penetration. Thus these models more closely mimic the epithelial cell surfaces bacteria would interact with in a host and provide more suitable models. For example, the interactions of *Salmonella* species with polarized MDCK and Caco-2 parallel closely those observed in the intestine of live animals. These include effects on the brush border and cytoplasmic extrusions associated with the invasion of the cell by the bacteria. Despite these advantages, some pathogens may not interact with polarized epithelial monolayers. For example, we have been unable to detect *Salmonella typhi* and *Yersinia* species interactions with the apical surfaces of Caco-2 and MDCK cells, presumably because of differences in host receptor specificity and polarity.

In conclusion, polarized epithelial cells hold much promise in the study of bacteria–host epithelial cell interactions, and will yield new information that cannot be obtained from nonpolarized cells. We hope that the cells and techniques described here can be applied to other human pathogenic bacteria. This information will undoubtedly lead to a better knowledge of the pathogen–host interactions in the diseases produced by these microorganisms and lead to other applications of polarized epithelial monolayers in the study of microbial pathogenesis.

Acknowledgments

Work in B.B.F.'s laboratory is supported by operating grants from the British Columbia Health Care Research Foundation, the Medical Research Council of Canada, the Canadian Bacterial Diseases Center of Excellence, and a Howard Hughes International Research Scholars Award.

[31] Interjunctional Invasion of Endothelial Cell Monolayers

By DAVID A. HAAKE and MICHAEL A. LOVETT

Introduction

The pathogenesis of infections caused by spirochetes involves rapid and widespread dissemination throughout the host. The pathogenic spirochetes include *Treponema pallidum* subsp. *pallidum* (causative agent of syphilis), the pathogenic *Leptospira* (causative agents of leptospirosis), *Borrelia burgdorferi* (causative agent of Lyme borreliosis), and the relapsing fever *Borrelia*. The ability of these highly invasive bacteria to disseminate to essentially every organ of the body early in the course of infection has been noted in studies of human disease as well as in animal models. For example, viable *T. pallidum* can be recovered from the cerebrospinal fluid in 30% of patients with primary syphilis.[1]

The presence of *T. pallidum* in the bloodstream of 84% of patients with primary syphilis implies a hematogenous route of dissemination.[2]

[1] S. A. Lukehart, E. W. Höök, S. A. Baker-Zander, A. C. Collier, C. W. Critchlow, and H. H. Handsfield, *Ann. Intern. Med.* **109,** 855 (1988).

[2] J. H. Stokes, H. Beerman, and N. R. Ingraham, "Modern Clinical Syphilology." Saunders, Philadelphia, 1944.

Hematogenous spread requires that pathogenic spirochetes cross the endothelial barrier from the vascular lumen to the surrounding tissue; however, native endothelial cells have intercellular junctions that are impermeable even to small molecules. For spirochetes in the vascular lumen to invade adjacent tissue, they must have a specific mechanism for traversing the endothelial cell barrier.

We developed a method for studying the interaction of pathogenic spirochetes with endothelial cell monolayers *in vitro*.[3] To do this, we adapted a technique that has been used for studying macromolecular transport and leukocyte diapedesis. This techniques involves cultivation of endothelial cells on a polycarbonate filter. The filter is mounted on a chemotaxis chamber and suspended in tissue culture medium. This arrangement allows assessment of intercellular junction integrity by means of transendothelial electrical resistance measurement. The presence of upper and lower chambers makes it possible to quantitate transendothelial movement of bacteria. Using this technique it is possible to distinguish between bacterial attachment to endothelial cells and invasion of the endothelial cells monolayer.

Preparation of Endothelial Cell Monolayers

In this section we describe the method used to prepare endothelial cell monolayers on polycarbonate filter membranes for use in spirochete invasion experiments. Using human umbilical vein endothelial cells (HUVECs) has been a convenient approach. Human aortic endothelial cells may be obtained from aorta trimmed from donor heart transplant material but are less frequently available and more difficult to obtain. Rabbit aortic endothelial cells (RAECs) are a readily obtainable alternative; however, results obtained with RAECs have less certain relevance to human infection.

Isolation of Human Umbilical Vein Endothelial Cells

Human umbilical vein endothelial cells are isolated according to the method of Jaffe[4] with modifications. An umbilical cord is obtained at the time of delivery and promptly placed on ice. The outside of the cord is cleansed with 70% (v/v) ethanol and inspected, and sections free of clamp marks are chosen and sectioned with a sterile scalpel. The umbilical vein

[3] D. D. Thomas, M. Navab, D. A. Haake, A. M. Fogelman, J. N. Miller, and M. A. Lovett, *Proc. Natl. Acad. Sci. U.S.A.* **85,** 3608 (1988).

[4] E. A. Jaffe, R. L. Nachman, C. G. Becker, and C. R. Minick, *J. Clin. Invest.* **52,** 2745 (1973).

is first dilated and then cannulated at each end with a blunt needle with a flange at the tip. The flange allows the needle to be secured to the umbilical cord with a tie-on clamp. Three-way stopcocks are then connected to the needles. The umbilical vein is flushed with M199 containing antibiotics (100 U/ml penicillin, 100 mg/ml streptomycin, and 2.5 mg/ml Fungizone). Collagenase 0.03% in M199 containing antibiotics is then infused into the vein. Once the vein has been filled with collagenase solution, the stopcock on one end of the vein is closed, additional collagenase solution is added under pressure to dilate the vein, and then the second stopcock is closed. The vein is suspended in a U-shape in a 50-ml tube containing warm phosphate-buffered saline (PBS) in a 37° water bath for 15–30 min. The collagenase solution is removed and the umbilical vein is vigorously flushed 10 times with M199 containing antibiotics. The collagenase solution and the flush solution are centrifuged at 100 *g* for 5 min. The pellet is resuspended in growth medium: M199 containing 20% low-endotoxin, defined fetal bovine serum (e.g., Hyclone Laboratories, Inc., Logan, Utah), 2 m*M* L-glutamine (M. A. Bioproducts, Walkersville, MD), 1 m*M* pyruvate (M. A. Bioproducts, Walkersville, MD), and antibiotics as above). Resuspended cells are then plated in a tissue culture flask that has been pretreated with solution of PBS containing 0.1–0.5% gelatin (w/v) (Sigma, St. Louis, MO). The flask is incubated overnight at 37° in 5% CO_2. The next day the flask is rinsed several times with M199 to remove nonadherent contaminating cell types, and the endothelial cells are given fresh growth medium containing 1000 U/ml heparin (GIBCO BRL, Gaithersburg, MD) and 20 μg/ml endothelial cell growth supplement (Collaborative Research, Inc., Bedford, MA). Prior to confluence, HUVECs are releasing using 0.1% trypsin, pH 7.2, containing 2 m*M* ethylenediaminetetraacetic acid (EDTA) and are propagated at a one to three ratio (into flasks with three times the surface area). Smooth muscle cells may contaminate endothelial cell cultures. Therefore, cells should be identified as endothelial with either fluorescently labeled antibody to factor VIII-related antigen on DiI-acetylated LDL (Biomedical Technologies, Inc., Cambridge, MA).

Chemotaxis Chamber

We use polycarbonate filters (Nuclepore Corp., Pleasanton, CA) mounted onto ADAPS chemotaxis chambers (Model PC-2, ADAPS, Inc., Dedham, MA) for our spirochete invasion experiments. This allows measurement of transendothelial electrical resistance (TEER) using the device shown in Fig. 1. An alternative, which others have used, is the Transwell polycarbonate filter unit (Costar, Cambridge, MA), in which case TEER

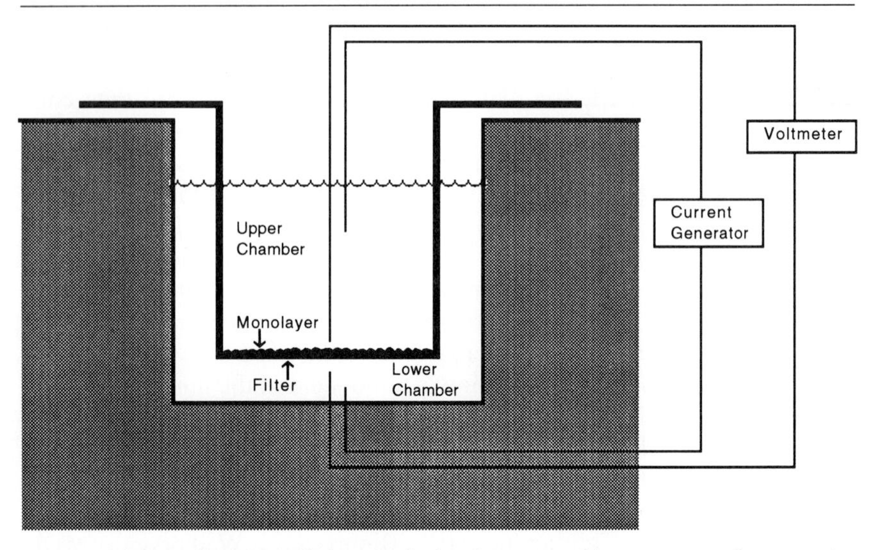

FIG. 1. Schematic of ADAPS chemotaxis chamber mounted in an apparatus for measuring transendothelial electrical resistance (see text for details). An endothelial monolayer is present on the filter at the bottom of the ADAPS chemotaxis chamber.

may be measured with a Millicell ERS Epithelial Voltohmeter (Millipore, Bedford, MA).[5] Polycarbonate filters are available in a vareity of pore sizes. We have avoided using filters with pore diameters greater than 3 μm because endothelial cells are able to "crawl" through larger pores. We have generally used filters with of 1-μm pore size because the number of pores per square centimeter is significantly greater than in filters with larger pores. The diameter of the largest spirochete, *B. burgdorferi,* is less than 0.3 μm.

Polycarbonate filters with a 1-μm pore diameter and 13-mm filter diameter (Nuclepore Corp.) are treated for 15 min with 70% (v/v) ethanol, rinsed extensively in double-distilled water, and boiled for 30 min in a solution of 0.1% (w/v) gelatin in PBS. The filters are gently rinsed in PBS to remove excess gelatin and then placed on top of wells in a 48-well tissue culture plate and dried at 60° for 30 min.[6] The filters are mounted on the bottom of ADAPS chambers using nontoxic silicon adhesive (Dow Corning, Midland, MI) and then cured for 1 hr at 60°. Filters are examined for ridges or other types of irregularities that could result in uneven seeding of endothelial cells. Twelve filter-chamber units are placed in a 24-well tissue

[5] B. S. Riley, N. Oppenheimer-Marks, E. J. Hansen, J. D. Radolf, and M. V. Norgard, *J. Infect. Dis.* **165,** 484 (1992).

[6] M. Territo, J. A. Berliner, and A. M. Fogelman, *J. Clin. Invest.* **74,** 2279 (1984).

culture plate in an alternating diagonal arrangement, gas-sterilized with ethylene oxide, and thoroughly aired (>24 hr) to remove residual toxic ethylene oxide.

Seeding Polycarbonate Filters with Human Umbilical Vein Endothelial Cells

Growth medium is placed in the empty wells of the 24-well tissue culture plate containing twelve filter-chamber units. The filters are transferred to the wells containing medium with a sterile forceps and allowed to incubate in 5% CO_2 for an hour. Each filter is seeded with 1.0×10^5 cells in 0.5 ml of growth medium (this is a seeding density of 1.5×10^5 HUVECs/cm^2; each filter has a surface area of 0.66 cm^2). After addition of the HUVEC suspension, the tissue culture plate is gently agitated side to side, back and forth. Swirling should be avoided because this could concentrate cells in the center of the filter. To minimize movement of medium through the filter, enough medium should be used so that the levels of medium inside and outside of the chamber are the same. Growth medium is added to the empty wells and the tissue culture plate is incubated at 37° for aproximately 2 hr. The plate should not be disturbed during this time to allow cells to attach. Medium is changed inside the chamber and the chambers are placed in adjacent wells with fresh medium. Changing medium inside the chamber should be done carefully so as not to dislodge cells. Roughly 10–20% of the medium inside the chamber should be left behind when changing medium to prevent touching the filter with the pipette.

Assessment of Intercellular Junction Integrity

Use of polycarbonate filters mounted on chemotaxis chambers makes it possible to assess the intercellular junction integrity of the endothelial cell monolayer. Although the rate at which the monolayer becomes confluent is determined largely by seeding density and cell growth, the extent of intercellular junction formation is dependent on the condition of the endothelial cells in the monolayer and their passage number HUVECs that have been passaged more than five to seven times after isolation should not be used. Overall monolayer permeability may also be affected by the presence of nonviable cells, contaminating cell types, and damage to the monolayer. Therefore, it is not safe to interpret the results of spirochete invasion experiments without some quantitative measure of endothelial cell monolayer integrity.

The standard method for assessing intercellular junction integrity of endothelial cell monolayers is measurement of the TEER. TEER is an excellent measure of the ability of small ions such as sodium to move across the filter–monolayer barrier.[7] We use the device shown in Fig. 1, which was constructed to accommodate the ADAPS PC-2 chemotaxis chamber. It is possible to maintain relative sterility during the process of measuring TEER. This is done by first cleansing the well with 95% ethanol. The device is then rinsed with sterile water, sterile PBS, and then M199 prewarmed to 37°.

Prior to initiating measurement of TEER, the device must be calibrated using a filter-chamber unit without a monolayer. The pores of the control filter should be hydrated in M199 for 30 min before use, because microbubbles trapped in the pores can create artificially high resistance. A constant current of 0.1 mA is generated, and the vertical position of the voltage electrode is adjusted to obtain a measurement of 1 mV. When filter-chamber units with monolayers are subsequently measured, any additional voltage difference will be due to the monolayer alone (i.e., TEER).

Transendothelial electrical resistance is expressed in terms of $\Omega \cdot cm^2$. For example, if a filter-chamber unit with a monolayer has a measured voltage of 2.8 mV, subtracting the voltage due to the filter and chamber without the monolayer (1.0 mV) yields the voltage attributable to the endothelial monolayer: 2.8 mV − 1.0 mV = 1.8 mV. The electrical resistance of the monolayer is determined by applying Ohm's law. Dividing the voltage difference attributable to the monolayer (1.8 mV) by the applied current (0.1 mA) yields the electrical resistance of the monolayer:

$$\frac{1.8\ \text{mV}}{0.1\ \text{mA}} = 18\ \Omega$$

The electrical resistance of the monolayer (18 Ω) is then multiplied by the surface area of the filter (0.66 cm^2) to generate a value that is independent of the size of the monolayer: $18\ \Omega \times 0.66\ cm^2 = 12\ \Omega\ cm^2$. It may seem counterintuitive to multiply rather than divide the electrical resistance by the surface area of the filter; however, a simple example can explain the reason for expressing TEER in terms of $\Omega\ cm^2$. If the same 12 $\Omega\ cm^2$ monolayer were grown on twice as large a filter (1.32 cm^2), the ability to resist the flow of electrical current would be half as much, and the measured electrical resistance would also be half: $9\ \Omega \times 1.32\ cm^2 = 12\ \Omega\ cm^2$. Expressing TEER in terms of $\Omega\ cm^2$ makes its value independent of filter surface area. This allows comparison of TEER obtained by different investigators growing endothelial monolayers on filters of different sizes. For spirochete invasion experiments we accept HUVEC monolayers with a TEER of 10 $\Omega\ cm^2$ or greater.

[7] M. Navab, G. P. Hough, J. A. Berliner, J. A. Frank, A. M. Fogelman, M. E. Haberland, and P. A. Edwards, *J. Clin. Invest.* **78,** 389 (1986).

It is also possible to assess intercellular junction integrity of endothelial monolayers using radiolabeled molecules such as bovine serum albumin (BSA). The rate of protein translocation from the upper chamber to the lower chamber in an intact endothelial cell monolayer should in general be less than $2\%/cm^2/hr$.[8] The advantage of this technique is that less manipulation of the monolayer is involved and the potential for loss of monolayer sterility is avoided. The disadvantage of using radiolabeled molecules is that extensive washing would be necessary if a subsequent measurement of protein translocation were required or if the monolayer were to be used to study the invasion of radiolabeled spirochetes.

A rapid screening method of evaluating tight junction integrity of a monolayer is to add a higher level of medium in the upper chamber and observe the rate of transfer to the lower chamber and the time required for the levels of medium inside and outside the chamber to reach equilibrium. In aortic endothelial monolayers with a TEER greater than 10 Ω cm^2, the transfer rate is less than 100 $\mu l/hr/cm^2$. It is possible to screen the monolayers rapidly in this manner, eliminating the leaky ones, followed by more quantitative determination of permeability.

Amnion

Human amniotic membrane is an alternative to synthetic polycarbonate filters. The amniotic membrane must first be separated from the chorion and surrounding placental tissue, followed by extensive washing in HEPES-buffered saline (HBS) (137 m*M* NaCl, 4 m*M* KCl, 11 m*M* glucose, 10 m*M* HEPES, pH 7.4) containing high concentrations of antibiotics (penicillin 500 U/ml and streptomycin 200 μg/ml). The amniotic epithelium is removed by treating the amnion with sterile 0.25 *M* NH_4OH for 2 hr on a rocking platform at room temperature followed by careful scraping with a rubber policeman.[9] Amnion may then be mounted on sterile ADAPS chambers using an O ring. A groove for the O ring on the outer surface of the ADAPS chamber helps to fix the O ring in place and prevent detachment of the amnion from the ADAPS chamber.

Amnion provides a natural surface for growth of endothelial cell monolayers, and reportedly allows formation of intercellular junctions between endothelial cells.[8] It is also possible to measure TEER of monolayers grown on amnion using the device shown in Fig. 1. The interaction of *B. burgdorferi* and endothelial monolayers grown on amnion has been

[8] S. M. Albelda, P. M. Sampson, F. R. Haselton, J. M. McNiff, S. N. Mueller, S. K. Williams, A. P. Fishman, and E. M. Levine, *J. Appl. Physiol.* **64,** 308 (1988).

[9] M. B. Furie, E. B. Cramer, B. L. Naprstek, and S. C. Silverstein, *J. Cell Biol.* **98,** 1033 (1984).

studied.[10] The major disadvantage of this technique is that after crossing the endothelial cell monolayer, spirochetes must negotiate the relatively dense collagen network of the amniotic membrane, which is several micrometers thick. Although *T. pallidum* is able to penetrate amnion and appear in the lower chamber, this process takes several hours.[11] For this reason, quantitation of spirochetes in the lower chamber would underestimate the degree of invasion of endothelial cell monolayers grown on amnion. Electron microscopy of thin sections of monolayers at the conclusion of an invasion experiment is useful in determining the route of invasion; however, electron microscopy cannot be used to follow the time course of invasion in a single monolayer.

Invasion of Endothelial Cell Monolayers by Spirochetes

Cultivation of Spirochetes

Cultivation of spirochetes is difficult relative to other bacteria. Spirochetes require enriched media and their doubling times are slow. Furthermore, it should be remembered that biohazard safety precautions are important when working with virulent spirochetes. *Borrelia burgdorferi* is cultivated in Barbour–Stoenner–Kelly medium at 34°C.[12] *Leptospira* species are cultivated at 30° in a polysorbate albumin medium which can be purchased commercially (PLM-5, Intergen) or made from primary components (EMJH medium).[13] The virulence of *B. burgdorferi* and *Leptospira* species becomes attenuated in about 10 serial passages after isolation from an animal host. Nonpathogenic treponemes, such as *Treponema phagedenis* biotype Reiter, are grown at 34° in Spirolate broth (GIBCO BRL) supplemented with 10% heat-inactivated normal rabbit serum.

Unlike other spirochetes, *T. pallidum* cannot be cultivated *in vitro* and, instead, must be maintained and passaged by intratesticular inoculation of adult male New Zelanad White rabbits that are seronegative with the Venereal Disease Research Laboratory test.[14] Infected rabbits are housed individually, maintained at 18–20°, and given antibiotic-free food and water. Ten or eleven days after inoculation the rabbits are euthanized. The testes are removed using sterile technique, minced with a sterile pair of scissors, and placed in heat-inactivated (56° for 30 min) normal rabbit serum. Extraction is performed by shaking on a rotary shaker for 15–30

[10] A. Szczepanski, M. B. Furie, J. L. Benach, B. P. Lane, and H. B. Fleit, *J. Clin. Invest.* **85,** 1637 (1990).

[11] T. J. Fitzgerald and L. A. Repesh, *Infect. Immun.* **55,** 1023 (1987).

[12] A. G. Barbour, *Yale J. Biol. Med.* **57,** 521 (1984).

[13] R. C. Johnson and V. G. Harris, *J. Bacteriol.* **94,** 27 (1967).

[14] J. N. Miller, S. J. Whang, and F. P. Fazzan, *Br. J. Vener. Dis.* **39,** 195 (1963).

min at room temperature. The fluid is removed and centrifuged at 400 *g* for 10 min to remove tissue debris. Treponemes are enumerated using dark-field microscopy.

Experimental Conditions

Endothelial cell monolayer invasion can be studied using various pathogenic microorganisms. In this section we emphasize the experimental conditions relevant to spirochetes. Obviously conditions must be compatible with spirochetal viability. Antibiotics, especially penicillin, should be removed prior to addition of spirochetes to the endothelial monolayer. The onset of spirochete immobilization by penicillin is concentration dependent. Below 0.1 U/ml penicillin, *T. pallidum* immobilization does not begin until 4 hr has elapsed. We have also noted that certain lots of fetal bovine serum are toxic to spirochetes. *T. pallidum* is microaerophilic and should be incubated in an atmosphere of 5% O_2, 5% CO_2, 90% N_2. The TEER of endothelial monolayers is maintained under these conditions. This is not surprising, given that an atmosphere of 5% O_2 and 5% CO_2 results in partial pressures of oxygen and carbon dioxide similar to those found in the venous circulation.

Consideration should also be given to the temperature at which endothelial monolayers and spirochetes are to be coincubated. Spirochetes may be inactivated at temperatures at or above 37°. The optimal temperature for incubation of *T. pallidum* and *B. burgdorferi* is 34°. *Leptospira* species are generally cultivated at 30°. Whatever incubation temperature is chosen, it is essential that the percentage of motile spirochetes be determined both before addition to the monolayer and at the conclusion of the experiment.

Transendothelial electrical resistance should be measured at the beginning and end of each experiment to ensure that a drop in TEER has not occurred. The spirochete outer membrane does not contain lipopolysaccharide. *Leptospira* species have a lipopolysaccharide-like substance; however, this material is devoid of endotoxin activity. Because *T. pallidum* is harvested from testicular extract, washing is required to remove inflammatory mediators.[5] *T. pallidum* and other spirochetes may be centrifuged at 8000–9000 *g* for 10–15 min at 4°. Higher centrifugal forces result in improved recovery of spirochetes at the expense of significant losses in motility (i.e., viability).

Invasion Kinetics

Invasion kinetics may be determined by quantitating the number of organisms in the lower chamber by dark-field microscopy.[15] A dark-field

[15] J. N. Miller, "Spirochetes in Body Fluids and Tissues: Manual of Investigative Methods." Thomas, Springfield, IL, 1971.

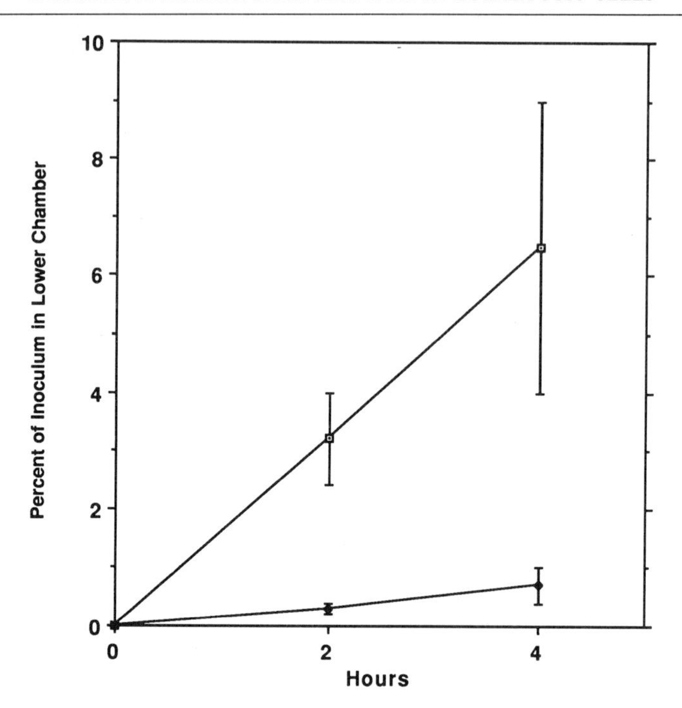

FIG. 2. Representative time course of endothelial monolayer invasion by (□) *Treponema pallidum* and (◆) *Treponema phagedenis* biotype Reiter. A concentration of 2×10^8 spirochetes per milliliter was applied to the endothelial monolayer at time zero. The invasion rate is linear during the first 4 hr.

microscope must be calibrated before it can be used to determine bacterial concentrations. Depending on the optics of the dark-field microscope, a 10-μl sample under a 1-cm^2 coverslip containing a spirochete concentration of 10^6/ml results in roughly one organism per high-power field. For this reason, it is statistically difficult to enumerate spirochetes accurately at concentrations below 10^5/ml. Radiolabeling of spirochetes may be required when using low concentrations of bacteria. *Leptospira* and *Borrelia* may be intrinsically labeled using either [^{3}H]thymidine or [^{35}S]methionine. Labeled nucleotides are not useful in the case of *T. pallidum* because the organism does not replicate *in vitro*. The standard procedure used for intrinsically labeling *T. pallidum* with [^{35}S]methionine was described by Stamm and Bassford.[16]

The rate at which spirochetes appear in the lower chamber is linear for the first 4 hr (Fig. 2). Organisms are detectable in the lower chamber

[16] L. V. Stamm and P. J. Bassford, *Infect. Immun.* **47,** 799 (1985).

within 1 hr. Invasion of the monolayer itself occurs more rapidly; spirochetes can be demonstrated electron microscopically below the monolayer (i.e., between the endothelial cells and the filter) within 15 min.

The ability to invade an endothelial monolayer is correlated with spirochete pathogencity. For example, the actively motile, nonpathogenic *T. phagedenis* biotype Reiter invades endothelial monolayers at a rate of roughly 10% that of *T. pallidum* (see Fig. 2).[2] An important control when comparing the invasion rates of different spirochetes is to measure invasion of filters without monolayers. Viable *T. phagedenis* biotype Reiter and *T. pallidum* invade filters without monolayers at nearly the same rate. Similar findings have been obtained when the avirulent culture-attenuated strain B31 of *B. burgdorferi* was compared with virulent strain 297 of *B. burgdorferi*.[17] Likewise, the nonpathogen *Leptospira biflexa* penetrates endothelial monolayers less than 10% as rapidly as virulent *Leptospira interrogans*.[18] These differences are remarkable considering that the morphology and motility of pathogenic and nonpathogenic spirochetes are similar. Nonmotile pathogens, usually obtained by heat inactivation for 10 min at 65°, are also noninvasive. Our conclusion is that motility is necessary but not sufficient for invasion of endothelial cell monolayers.

Statistics

Experiments involving spirochete invasion of endothelial monolayers are subject to multiple sources of intrinsic variability. Enumeration by dark-field microscopy is most accurate between 10^6 and 10^7 spirochetes/ml. More concentrated specimens should be diluted down to this range. We suggest counting 10–20 fields per sample and obtaining several samples at each time point. Different batches of endothelial monolayers show variation in TEER. Spirochete preparations vary in the degree of motility (i.e., viability). For these reasons, it is important that each arm of an invasion study involve several identical experiments performed in parallel. Results should be expressed as the mean ± the standard deviation of the mean. *P* values should be calculated where applicable.

Mechanism of Invasion

A number of important questions can be addressed in studies of spirochete invasion of endothelial monolayers. What role does the endothelial

[17] S. D. Thompson, C. D. Barkham, E. M. Walker, J. N. Miller, and M. A. Lovett, in preparation.

[18] D. A. Haake and M. A. Lovett, in preparation.

cell play in the process of endothelial monolayer invasion by spirochetes? One way to address this question is to look for changes in endothelial cells after exposure to spirochetes. Riley *et al.*[5] have found that *T. pallidum* activates HUVECs, with increased expression of intercellular adhesion molecule 1 (ICAM-1) and procoagulant activity on the endothelial cell surface. It is unclear whether endothelial cell activation promotes invasion by *T. pallidum*. We coincubated *T. pallidum* and the noninvasive spirochete *T. phagedenis* biotype Reiter with endothelial monolayers to address the question of whether endothelial changes induced by *T. pallidum* might allow *T. phagedenis* biotype Reiter to invade the monolayer more readily. We found that *T. phagedenis* biotype Reiter invaded endothelial monolayers no more rapidly in the presence of *T. pallidum* than in its absense, suggesting that whatever endothelial monolayer changes were induced by *T. pallidum,* they were not sufficient to allow invasion by *T. phagedenis* biotype Reiter. Electron microscopy of endothelial monolayers exposed to both types of treponemes confirmed that only *T. pallidum* was able to make its way below the monolayer and be found between the endothelial cells and the filter (Fig. 3).

Another way that the mechanism of endothelial monolayer invasion may be addressed is by comparing invasion in the presence or absence of EDTA. EDTA reversibly lowers the TEER of endothelial monolayers by opening intercellular junctions. Pathogens and nonpathogens alike are able to penetrate endothelial monolayers in the presence of EDTA. This type of experiment is also a useful control because it confirms both the barrier function of the intercellular junctions in the absence of EDTA and the motility of the nonpathogenic organism.

An approach to understanding the mechanism of invasion is to address the relationship of attachment to invasion. Spirochetes attach readily to endothelial cells, as they do to most eukaryotic cells in tissue culture. The extracellular matrix proteins fibronectin, collagen, and laminin all promote spirochete attachment. Although *T. pallidum* fibronectin-binding proteins have been identified, their role in pathogenesis is unclear.[19–21] The attachment ligands that are important in the process of endothelial invasion by spirochetes have not been elucidated. Like invasion, the ability of spirochetes to attach to endothelial cells correlates with pathogenicity. Nonpathogens such as *T. phagedenis* biotype Reiter are relatively nonadherent, whereas pathogenic spirochetes such as *T. pallidum* readily attach to endothelial cells.

[19] J. B. Baseman and E. C. Hayes, *J. Exp. Med.* **151,** 573 (1980).
[20] K. M. Peterson, J. B. Baseman, and J. F. Alderete, *J. Exp. Med.* **157,** 1958 (1983).
[21] D. D. Thomas, J. B. Baseman, and J. F. Alderete, *J. Exp. Med.* **161,** 514 (1985).

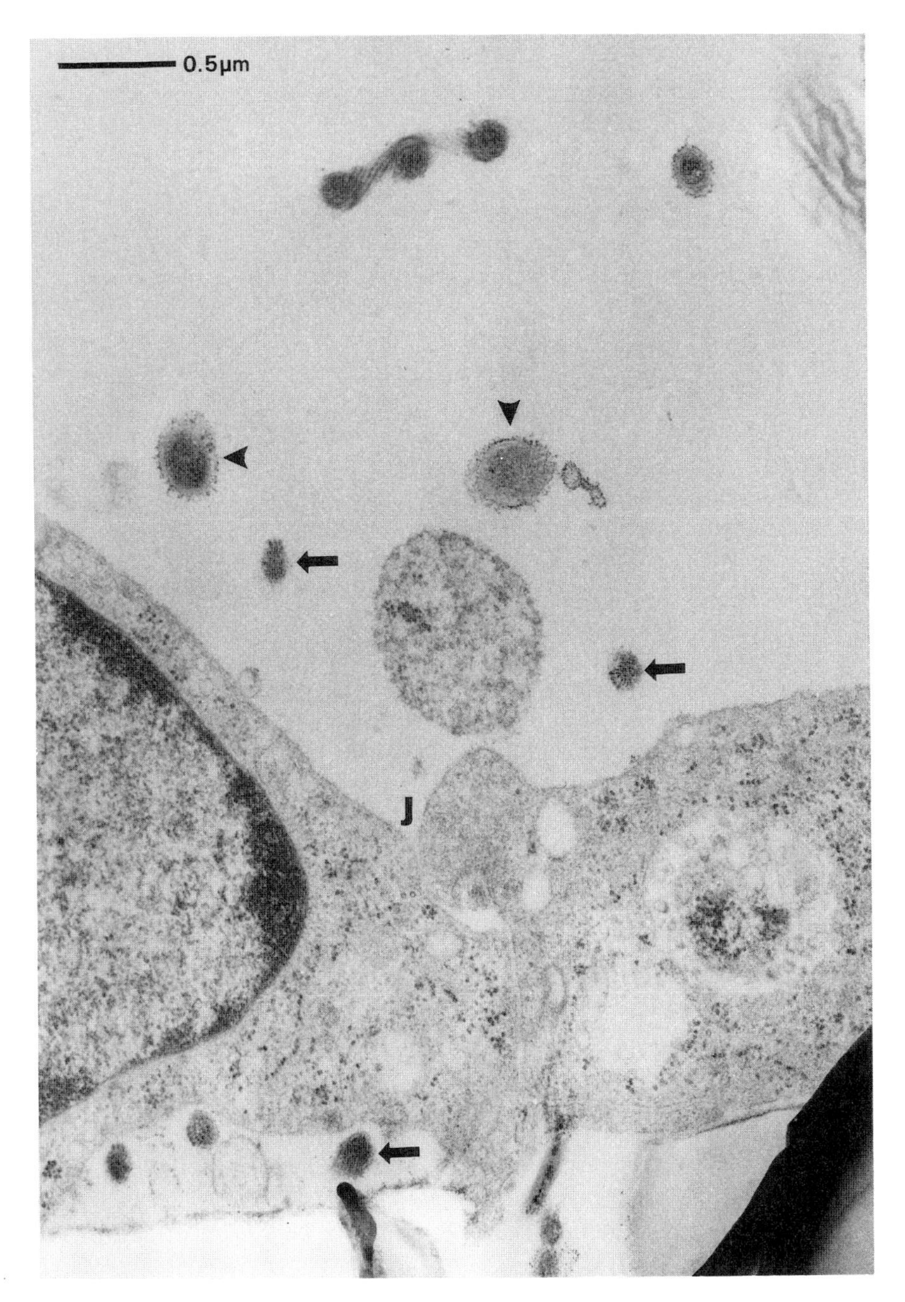

FIG. 3. Electron micrograph of *Treponema pallidum* (arrows) and *Treponema phagedenis* biotype Reiter (arrowheads), after simultaneous exposure to an endothelial monolayer. The two organisms can be distinguished readily on the basis of morphological differences. Both organisms are seen above the endothelial cells, whereas only *T. pallidum* is seen below. An intercellular junction is present (J).

Attachment and monolayer invasion studies may be performed either separately or simultaneously. Dark-field microscopy may be used to quantitate spirochetal attachment to endothelial cells grown on glass coverslips. The experiment is carried out with coverslips in wells of a 24-well plate. Attachment of endothelial cells to the coverslip is facilitated by pretreating the coverslip for 1 hr with a solution of PBS containing 1.0% gelatin. When the attachment of spirochetes is to be enumerated, the coverslip is inverted on a microscope slide and observed directly with the dark-field microscope. Attachment experiments can also be conveniently performed using Nunc chamber slides. Because polycarbonate filters are opaque, radiolabeling of spirochetes is necessary to study attachment and invasion of a single endothelial cell monolayers simultaneously. When using radiolabeled organisms it is important to remember that spirochetes may become associated with endothelial cells either by attaching to the cell surface or by entering into the endothelial cell. It is possible to differentiate intracellular and extracellular organisms by releasing the endothelial cells from the filter with trypsin, centrifuging at 100 *g* for 5 min at 4°, and counting the supernatant and cell pellet separately.

Electron microscopy of endothelial cell monolayers exposed to pathogenic spirochetes provides the most compelling evidence for the route of invasion. The method for preparing samples for electron microscopy is as follows: At the completion of an experiment involving spirochete invasion of an endothelial cell monolayer, monolayers are rinsed three times with medium without serum at 37°, are fixed in 2% glutaraldehyde in PBS at 4°, are postfixed with 1% osmium tetroxide in PBS, are dehydrated in ethanol, and are embedded. Sections are stained with uranyl acetate and lead acetate.

In studies of *T. pallidum* invasion, we have found an abundance of organisms in the intercellular junctions between endothelial cells (Fig. 4). Intercellular *T. pallidum* were observed in close proximity to tight junctions, indicating that the intercellular junction integrity of the endothelial monolayer had been maintained during the experiment. Nonpathogenic spirochetes, such as *T. phagedenis* biotype Reiter, are not able to invade endothelial monolayers and are not found in intercellular junctions. There is clearly a correlation between virulence, invasion of endothelial monolayers, and the ability to penetrate intercellular junctions. By contrast, the significance of spirochetes within endothelial cells is uncertain. Endothelial cells are capable of phagocytosis and are able to engulf noninvasive spirochetes such as *T. phagedenis* biotype Reiter (Fig. 5). Intracellular treponemes are found when either *T. pallidum* or the nonpathogen *T. phagedenis* biotype Reiter is exposed to endothelial monolayers. Virulent *B. burgdorferi* readily invades endothelial monolayers, and evidence has

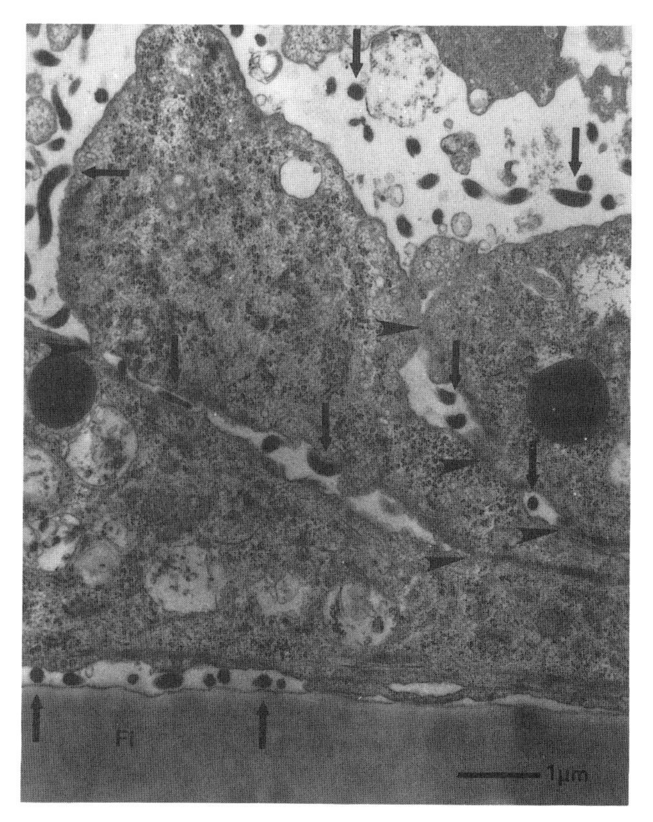

FIG. 4. Electron micrograph of *Treponema pallidum* (arrows) above, between, and below endothelial cells grown on a polycarbonate filter (Fi). Intercellular junctions are indicated by arrowheads.

been presented for either an intercellular[10,17] or intracellular[22] route of endothelial cell monolayer invasion. Electron microscopic studies have clearly demonstrated this organism in intercellular junctions.[10,17] On the other hand, the fate of spirochetes within endothelial cells is not understood.

Conclusion

Endothelial monolayer invasion studies provide a means to address questions related to spirochetal dissemination *in vitro*. Endothelial mono-

[22] L. E. Comstock and D. D. Thomas, *Infect. Immun.* **57,** 1626 (1989).

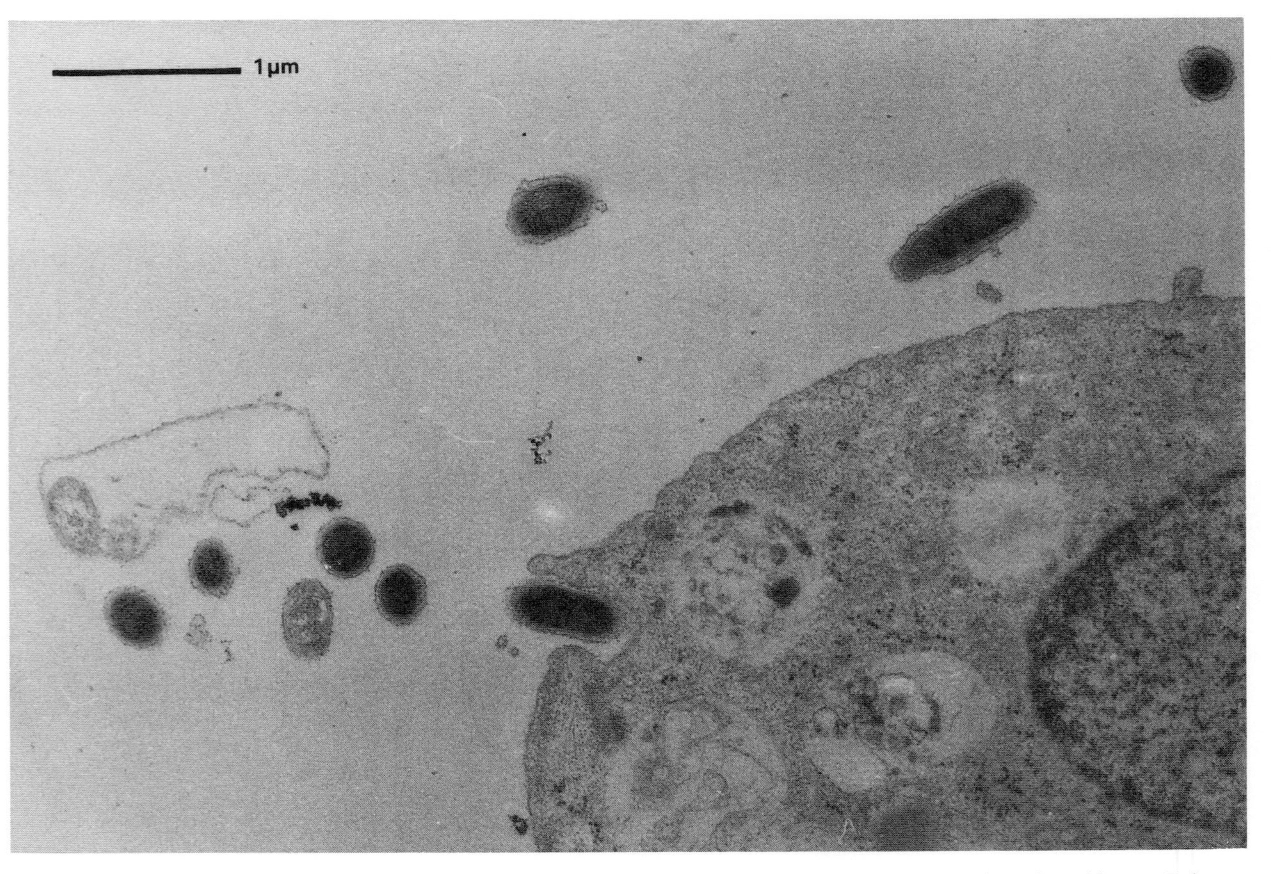

FIG. 5. Electron micrograph of an endothelial cell in the process of phagocytosis of *Treponema phagedenis* biotype Reiter, a nonpathogenic spirochete.

layers are prepared by seeding polycarbonate filters or amnion with human umbilical vein endothelial cells. The integrity of intercellular junctions can be assessed by measuring transendothelial electrical resistance. Electron microscopy of endothelial monolayers exposed to pathogenic spirochetes is used to determine the route of invasion. When interpreting electron micrographs it is important to note that endothelial cells are able to phagocytose pathogenic as well as nonpathogenic spirochetes. By using nonpathogenic spirochetes as controls a correlation can be found between spirochete virulence, ability to invade endothelial monolayers, and the presence of intercellular organisms on electron microscopy. The molecular mechanism by which intercellular invasion occurs remains a subject of investigation.

Acknowledgments

We thank Dr. Mahamad Navab for invaluable help in the preparation of this manuscript. This work was supported by Public Health Service Grants AI-21352 and AI-29733 to M.A.L. and a Veterans Administration Research Advisory Group Award to D.A.H. D.A.H. was a Burroughs Wellcome Fund Postdoctoral Research Fellow of the Infectious Diseases Society of America at the time some of this work was performed.

[32] Intracellular Growth of Bacteria

By SIAN JONES and DANIEL A. PORTNOY

Introduction

The following protocol has been developed to quantitate the intracellular growth of bacteria with particular reference to *Listeria monocytogenes* and has been applied successfully to the J774 mouse macrophage-like cell line as well as an epithelial cell line and primary cultures of murine macrophages.[1] A monolayer of eukaryotic cells grown on round glass coverslips is infected in the manner outlined in detail below. Only intracellular bacterial cell growth is measured because growth of extracellular bacteria is inhibited by the addition of gentamicin subsequent to bacterial internalization.

[1] D. A. Portnoy, P. S. Jacks, and D. J. Hinrichs, *J. Exp. Med.* **167,** 1459 (1988).

Preparation of Coverslips

Round glass coverslips (1 × 12 mm, Propper Manufacturing Co. Inc., Long Island City, NY) are kept in 70% (v/v) ethanol until ready for use. Manipulation of the coverslips is easily accomplished using jeweler's forceps. Prior to use, the coverslips are dipped in 95% (v/v) ethanol and quickly flamed using a Bunsen burner. It is helpful to blot off excess alcohol by gently touching the edge of the coverslip to a tissue, as flaming the coverslip in the presence of excess alcohol frequently leads to shattering of the coverslip. Immediately after flaming, the coverslips are placed into a sterile 60 × 20-mm petri dish (Lab-Tek; American Scientific Products, McGaw Park, IL). Fifteen coverslips can be fit into each dish. It is important to ensure that none of the coverslips overlap as this will prevent the formation of a uniform monolayer of cells on each coverslip.

Preparation of Eukaryotic Cells

Tissue culture cells are grown on coverslips in standard medium. Our laboratory routinely uses Dulbecco's Modified Eagle's Medium (DME/High, JRH Biosciences, Lenaxa, KS) supplemented with 10% (v/v) fetal calf serum and 292 μg/ml glutamine without antibiotics. Epithelial cells are passaged at 1/5 from a confluent monolayer onto the coverslips 2 days before use. J774 cells can be grown either in T-flasks or in suspension in spinner flasks, and are harvested the night prior to infection as described below. Both cell lines are maintained in the presence of 10 μg/ml streptomycin and 100 U/ml penicillin until ready for use.

The density of the J774 cells is determined, and the requisite volume to yield 2×10^6 cells/dish is removed. The cells are then subject to centrifugation at 1000 rpm at 4° for 10 min in a table-top centrifuge (Sorvall RT 6000B, DuPont Company, Wilmington, DE). The supernatant is removed and the appropriate amount of fresh medium without antibiotics is added to yield 2×10^6 cells in 6 ml. It is important to ensure that the cells are well suspended prior to addition to the coverslips in the petri dish. Failure to do so may lead to clumping. Addition of the cells to the dishes tends to disturb the coverslips and they should be gently tapped into place to ensure that none of the coverslips overlap. The cells are incubated overnight at 37° with 5% CO_2. Cells should be checked by light microscopy the following morning prior to infection to assess the adequacy of the monolayer.

Preparation of Bacteria

The prototypic bacterium used in these experiments is *Listeria monocytogenes* strain 10403S which belongs to serotype 1; however, the method is appropriate for all serotypes. An overnight culture of the bacterial strain to be used should be made by picking a colony from a fresh plate and inoculating a 2-ml culture of brain–heart infusion broth (BHI, Difco Laboratories, Detroit MI) in a 15-ml conical tube. Bacteria may be grown to logarithmic or stationary phase; however, in our laboratory we routinely grow the overnight culture lying flat at 30° to stationary phase (2×10^9 bacteria/ml). The next morning 1 ml of the culture is transferred to a 1.5-ml Eppendorf tube and subjected to centrifugation for 1 min at room temperature at 14,000 *g* in a microfuge. The supernatant is removed and the bacteria are suspended in 1 ml of phosphate-buffered saline (PBS) pH 7.4, by vortexing vigorously. The bacteria are then spun again and the pellet is suspended in 1 ml of PBS. The bacteria are washed in this manner to remove any secreted proteins which may have toxic effects on the host cells.

Procedure for Infection

The number of bacteria used to infect the monolayer varies depending on the experiment. We describe the number necessary to achieve 10^4 internalized bacteria per coverslip. J774 cells are infected with 1×10^5 bacteria per dish for 30 min, whereas Henle cells are infected with 2×10^7 bacteria for 1 hr. To perform the infection, the appropriate volume of washed bacteria is transferred to a sterile 15-ml conical tube. Five milliliters of medium is removed with a pipette from the petri dish containing the coverslips and transferred into the conical tube containing the bacteria. The contents of the tube are mixed quickly by vortexing and then distributed evenly over the coverslips which are then returned to the incubator, after making sure that none of the coverslips have been dislodged.

At 30 and 60 min after infection for J774 and Henle cells, respectively, the tissue culture medium is aspirated and the coverslips are washed three times with 5 ml of PBS prewarmed to 37°. After the final wash, 5 ml of the appropriate fresh medium without antibiotics at 37° is added to each dish. After an additional 30 min gentamicin sulfate is added to a final concentration of either 5 or 50 μg/ml and mixed by gently swirling. The minimum inhibitory concentration (MIC) of gentamicin for *L. monocytogenes* is 0.25–1 μg/ml,[2] however, although the growth of extracellular, but not intracellular, bacteria is inhibited by both 5 and 50 μg/ml of gentamicin, we have found that at lower concentrations the gentamicin is more slowly bactericidal and does not fully prevent spread of the bacteria by the extracellular route. Therefore, to absolutely preclude

[2] A. Kucers and N. McK. Bennett, "The Use of Antibiotics: A Comprehensive Review with Clinical Emphasis." Lippincott, Philadelphia, 1987.

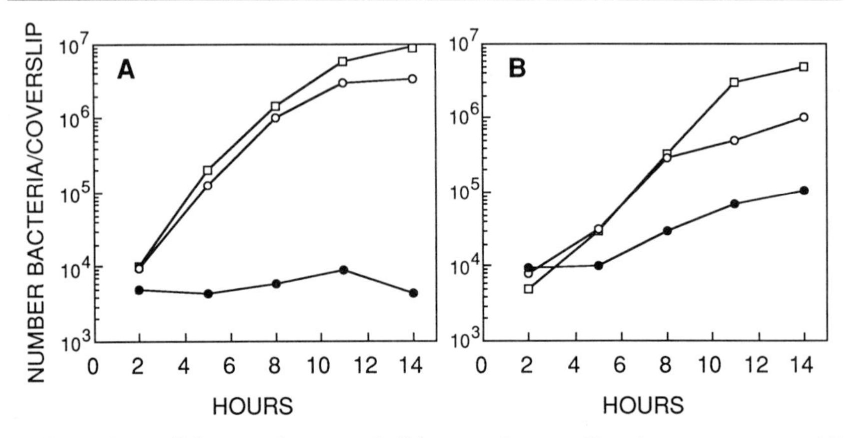

FIG. 1. Intracellular growth curves of wild-type and mutant *Listeria monocytogenes*. (A) J774 mouse macrophage-like cell line. (B) Henle 407 human epithelial cell line. Wild-type strain (□); DP-L1044 (●); DP-L1054 (○). Data represent averages for three coverslips. (Reproduced with permission from Sun *et al.*[6])

the possibility of extracellular spread, we recommend the use of gentamicin at 50 μg/ml. The exact reason for this differential killing by gentamicin is not known, and it should not be assumed to be applicable to other bacterial species. Factors that have been postulated to be of importance include (1) differential partitioning of gentamicin between extracellular and intracellular compartments[3]; (2) pH dependence of gentamicin activity,[4,5] i.e., gentamicin is relatively inactive at acidic pH; and (3) cytoplasmic location of the bacteria.

At regular intervals after infection, the number of bacteria per coverslip is assessed in the following manner. At each time point, coverslips are removed in triplicate from the petri dish in sterile fashion and transferred to 15-ml conical tubes containing 5 ml of sterile distilled H_2O. The monolayers are lysed by vortexing the tubes vigorously for 15 sec, thus releasing the bacteria. Appropriate aliquots are then plated onto Luria–Bertani agar plates (10 g tryptone, 5 g yeast extract, 10 g NaCl, pH 7.2/liter) and incubated overnight at 37°. The average colony count is determined for each time point and used to plot a growth curve. The result for a single coverslip should not differ by more than 25% from the mean. Typical growth curves for wild-type and mutant strains of *L. monocytogenes* in

[3] W. L. Hand and N. L. King-Thompson, *Antimicrob. Agents Chemother.* **29,** 135 (1986).

[4] R. S. Edson and C. L. Terrell, *Mayo Clin. Proc.* **66,** 1158 (1991).

[5] J. Blaser and R. Luthy, *J. Antimicrob. Chemother.* **22,** 15 (1988).

J774 mouse macrophage-like cells and Henle 407 epithelial cells are shown in Fig. 1.[6]

This method may be adapted to look at initial events, i.e., killing by the host cell, by using primary cultures of bone marrow macrophages which, unlike J774 cells, are bactericidal for *L. monocytogenes*. To do so, an additional time point should be performed at 30 min, i.e., just after the coverslips have been washed and prior to the addition of gentamicin.

It is often useful to visualize the bacteria directly by staining a fourth coverslip at the appropriate time point using Diff-Quick stain (Baxter Scientific Products, Miami, FL). Once the stained coverslips have air-dried, they can be mounted on a microscope slide using a drop of Permount (Fisher Scientific, Fair Lawn, NJ). The coverslip should be placed with the cell side down and then gently tapped to ensure uniform dispersion of the Permount. After 24 hr the coverslips set permanently and may be visualized under oil immersion.

[6] A. N. Sun, A. Camilli, and D. A. Portnoy, *Infect. Immun.* **58,** 3770 (1990).

[33] Inhibitors of Cytoskeletal Function and Signal Transduction to Study Bacterial Invasion

By ILAN ROSENSHINE, SHARON RUSCHKOWSKI, and B. BRETT FINLAY

Introduction

Many pathogenic bacteria are capable of entering (invading) host eukaryotic cells, and using this process as a pathogenic mechanism. Bacterial uptake usually involves exploitation of host cell functions to facilitate invasion. Thus, to study bacterial invasion, the role of host components and functions should be examined. Perhaps the easiest way to define the eukaryotic components involved in bacterial uptake is to use drugs that disrupt specific host functions. Inhibitor studies should, however, be interpreted with caution, as the drug may have other effects or targets. Instead, they should be used as an initial probe to identify the components involved, followed by further studies using other methodologies. This article details some of the available drugs that have been used to study bacterial invasion (Table I). These include inhibitors of cytoskeletal structures, such as microfilaments and microtubules, and drugs that block various host cell signal transduction pathways, including inhibitors of host protein kinases and calcium chelators. Attempts have been made to detail some of the

TABLE I
INHIBITORS THAT HAVE BEEN USED TO STUDY BACTERIAL INVASION

Inhibitor	Target	Stock	Source	Working concentration	Reversibility	Remarks	Ref.
Cytochalasin D	Actin filaments	1 mg/ml in DMSO	Sigma (St. Louis, MO)	1 μg/ml		More potent and specific than cytochalasin B	3, 5, 6
Cytochalasin B	Actin filaments	1 mg/ml in DMSO	Sigma	1–10 μg/ml		Has other effects	5, 7, 8
Nocodazole	Microtubules	10 mg/ml in DMSO	Sigma	10 μg/ml		Thought to be microtubule specific	3
Colchicine	Microtubules	1 mg/ml	Sigma	1–10 μg/ml		Has other effects	5, 6, 10
Vincristine	Microtubules	1 mg/ml	Sigma	1–10 μg/ml		Has other effects	5, 6
Vinblastine	Microtubules	1 mg/ml	Sigma	1–10 μg/ml		Has other effects	5, 6
Staurosporine	PKC, TPKs, PKA	1 m*M* in DMSO	Sigma	1 μM	Slow	Inhibits many kinases, potent	13
Genistein	TPKs	100 m*M* in DMSO	ICN (Costa Mesa, CA)	50–250 μM	Rapid	Reversibility is useful	13
Tyrphostin	TPKs	50–100 m*M* in DMSO	BRL (Gaithersburg, MD)	500 μM	Slow	Lengthy preincubation	13
EDTA	Ca^{2+} chelator	10 m*M*	Sigma	200 μM	Yes	Chelates only extracellular Ca^{2+}	18
EGTA	Ca^{2+} chelator	10 m*M*	Sigma	200 μM	Yes	Chelates only extracellular Ca^{2+}	18
BAPTA	Ca^{2+} chelator	10 m*M*	Calbiochem (La Jolla, CA)	200 μM	Yes	Chelates only extracellular Ca^{2+}	18
BAPTA/AM	$[Ca^{2+}]_i$ chelator	10 m*M* in DMSO	Calbiochem	200 μM	Yes	Chelates only intracellular Ca^{2+}	18

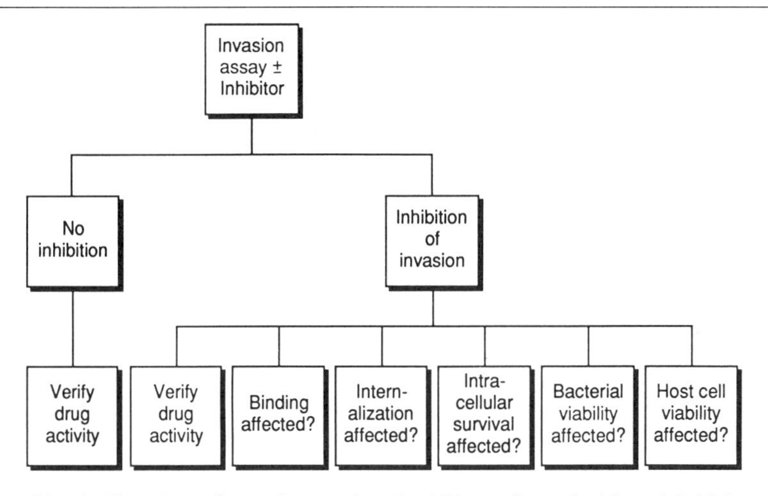

FIG. 1. Flowchart of experiments that should be performed with each inhibitor.

controls that should be performed, as these are important for correct interpretation of the results. Although there is a wide body of literature describing these various host components and functions, references that describe these functions in the context of invasive organisms have been preferentially cited.

In general, the principles followed for each inhibitor are similar and are outlined in Fig. 1. Initially, invasion is measured with and without the inhibitor, often using an invasion assay as described in [28].[1] If the inhibitor causes a decrease in the number of bacteria recovered after an invasion assay, several other experiments are performed to rule out additional effects that the drug may have that affect the results of the invasion assay (Fig. 1). Alternatively, if no effect is observed, the investigator must determine that the drug actually affected its predicted target. In addition, other drugs that affect the same target by a different mechanism (if available) should be tested to confirm the results of a given inhibitor.

Inhibitors of Cytoskeletal Function

Microfilaments are polymerized complexes of actin that participate in many active cell processes in eukaryotic cells, including cell movement and phagocytosis. Cytochalasins are a class of drugs that inhibit actin polymerization, thereby blocking microfilament function[2]; however, these

[1] E. A. Elsinghorst, this volume [28].
[2] S. B. Carter, *Nature (London)* **213,** 261 (1967).

drugs also have several other significant effects on cells.[2] Uptake of most, if not all, bacteria into eukaryotic cells (including nonphagocytic cells such as those of epithelial origin) can be blocked by the use of cytochalasins.

Cytochalasin D is considered to be the most specific and potent of the cytochalasins. To determine if this drug affects bacterial uptake, cultured cells are preincubated in 1 μg/ml cytochalasin D (Sigma) for 30 min prior to infecting with bacteria and measuring invasion levels as described by Elsinghorst.[1] Cytochalasin D is kept in the medium throughout the invasion assay. Stocks of cytochalasin D are made in dimethyl sulfoxide (DMSO, 1 mg/ml) and stored at −20°. If invasion is inhibited, various doses of drug should be tested to establish a dose dependence for invasion.

Several controls should also be performed. The effect of cytochalasin D on bacterial viability/growth and host viability should be established. The effect of cytochalasin D on bacterial adherence should be determined, as it should not affect adherence but should affect internalization. Visualization by phase or fluorescent microscopy is often the simplest way to determine if adherence is affected. Additionally, to establish that microfilaments have been disrupted, treated and untreated cells can be stained with a fluorescent phalloidin.[3,4] The uptake of many invasive organisms, including *Escherichia coli, Salmonella, Shigella,* and *Yersinia* species, can be blocked by cytochalasins.[4–9] One of these organisms may be used as a positive control for determining the effect of cytochalasin D on invasion.

Microtubules are filaments consisting of polymerized tubulin. Microtubules form an extensive network in the cytoplasm and participate in intracellular movement of organelles via several microtubule motors. Invasion of a few bacteria, including enteropathogenic *E. coli* and *Neisseria gonorrhoeae,* can be blocked by using microtubule inhibitors.[6,10]

Colchicine, vincristine, and vinblastine are drugs that can be used to depolymerize microtubules and determine the role of microtubules in bacterial invasion[3,5,6]; however, these drugs also have other effects on eukaryotic cells. A more specific and potent drug, nocodazole, has been described that depolymerizes microtubules and blocks their function.[11]

[3] B. B. Finlay, S. Ruschkowski, and S. Dedhar, *J. Cell Sci.* **99,** 283 (1991).
[4] P. Clerc and P. J. Sansonetti, *Infect. Immun.* **55,** 2681 (1987).
[5] B. B. Finlay and S. Falkow, *Biochimie* **70,** 1089 (1988).
[6] M. S. Donnenberg, R. A. Donohue, and G. T. Keusch, *FEMS Microbiol. Lett.* **57,** 83 (1990).
[7] G. Bukholm, *Acta Pathol. Microbiol. Immunol. Scand., Sect. B* **92B,** 145 (1984).
[8] E. Kihlström and L. Nilsson, *Acta Pathol. Microbiol. Scand., Sect. B* **85B,** 322 (1977).
[9] B. B. Finlay, I. Rosenshine, S. M. Donnenberg, and J. B. Kaper, *Infect. Immun.* **60,** 2541 (1992).
[10] W. P. Richardson and J. C. Sadoff, *Infect. Immun.* **56,** 2512 (1988).
[11] K. Parczyk, W. Haase, and K. C. Kondor, *J. Biol. Chem.* **264,** 16837 (1989).

Nocodazole (Sigma) stocks are made in DMSO, whereas colchicine, vincristine, and vinblastine can be dissolved in water or phosphate-buffered saline. To use colchicine, vincristine, or vinblastine, cultured cells are preincubated in 1 to 10 μg/ml of drug for 30 min to 1 hr at 37° prior to bacterial addition. With nocodozole, cultured cells are preincubated in 10 μg/ml drug for 1 hr on ice, then warmed to 37° for 30 min prior to bacterial addition. This incubation at 4° for nocodazole is essential for complete microtubule disruption. All the drugs are kept in medium throughout the duration of the invasion assay.

As with cytochalasin D, if an effect is seen, a dose dependence on invasion should be established. The drug's effect on bacterial and host cell viability should also be measured. Additionally, complete disruption of the microtubules by the drug can be confirmed by immunofluorescent staining and visualization of microtubules.[3] Finally, the ability of the drug to block enteropathogenic *E. coli* or *N. gonorrhoeae* could be examined.

Host Protein Kinase Inhibitors

It is becoming apparent that invasive bacteria have developed mechanisms with which they can exploit existing host signal transduction pathways to trigger cytoskeletal rearrangement and bacterial uptake. Many of these pathways use protein kinases to transmit signals, including kinases that phosphorylate serine and threonine residues, such as protein kinase C (PKC), and those that phosphorylate tyrosine residues, such as tyrosine protein kinases (TPKs). Agents are becoming available that inhibit these various enzymes, and thus they can be used to probe whether bacterial invasion involves a host kinase.

There are numerous PKC inhibitors which have varying potencies and permeability for inhibiting PKC activity in whole cells, including calphostin C, H7, H9, and staurosporine.[12] Staurosporine is a potent inhibitor of PKC that blocks PKC activity in intact cells; however, it also effectively blocks other serine/threonine protein kinases such as cAMP-dependent protein kinases (protein kinase A) and some classes of TPKs.[12] The broad spectrum of inhibition and potency of this drug make it a suitable choice with which to begin kinase inhibitor experiments.

A 1 m*M* stock solution of staurosporine (Boehringer-Mannheim, Mannheim, Germany, or Sigma) is made in DMSO. As with the cytoskeletal inhibitors, cells are preincubated in various concentrations of staurosporine, ranging from 0 to 1 μM, for 30 min. Invasion assays are then

[12] T. Tamaoki and H. Nakano, *Bio/Technology* **8,** 732 (1990).

done on treated and untreated cells, keeping the drug present for the duration of the experiment, and the level of inhibition of invasion is determined.[13]

There are several TPK inhibitors that have become commercially available. These include genistein[14] and various members of the tyrphostin family.[15] Genistein inhibits TPK activity by blocking the binding of ATP to the enzyme, whereas tyrphostins compete with the tyrosine residue binding of the kinase. Thus, both drugs are quite TPK specific, yet work by different mechanisms and have limited spectra within the TPK family.

Stocks of genistein (100 m*M*, ICN or UBI, Lake Placid, NY) and tyrphostin (50 m*M*, BRL) are made in DMSO. The effect of genistein is rapid, so preincubation periods are short (10 min); however, tyrphostins penetrate cell membranes slowly, and incubation periods of 12 hr may be required. Once cells have been treated with the drug, invasion assays are performed as usual, and invasion levels are compared with those of untreated cells.

It is extremely important to perform several controls before results of kinase inhibitors can be successfully interpreted. Initially, efforts should be made to confirm that the drug is having the desired effect. For the PKC inhibitors, this can include measuring PKC phosphorylation of histone H1 with or without the inhibitor present, a measure of PKC activity.[13] Alternatively, the effect of the inhibitor on whole cells can be tested by measuring phosphorylation of normal PKC substrates such as p80 (MARCS). This can be determined by prelabeling cultured cells with ^{32}P and then activating PKC with phorbol esters.[13] With the TPK inhibitors, Western blots of serum-starved cells that are treated with serum-supplemented medium can be probed with a monoclonal antibody to phosphotyrosine residues to determine if the TPK inhibitors are effectively blocking TPK induction.

As with any inhibitor study, kinase inhibitors may have other effects. Thus, viability of bacterial and host cells should be determined with the inhibitors. This can be done by measuring bacterial survival and growth curves and staining cultured cells with trypan blue. Alternatively, the inhibitors may block some bacterial function. Bacteria that have been treated with the drug and then washed prior to addition to host cells may give an indication of this activity. The inhibitor may also trigger some antibacterial activity once the organisms are internalized. To rule out this possibility, previously untreated cells containing internalized bacteria are

[13] I. Rosenshine, V. Duronio, and B. B. Finlay, *Infect. Immun.* **60,** 2211 (1992).

[14] T. Akiyama, J. Ishida, S. Nakagawa, H. Ogawara, S. Watanabe, N. Itoh, M. Shibuya, and Y. Fukami, *J. Biol. Chem.* **262,** 5592 (1987).

[15] P. Yaish, A. Gazit, C. Gilon, and A. Levitzki, *Science* **242,** 933 (1988).

then treated with the inhibitor, and the number of surviving intracellular bacteria is determined.[13] Adherence levels of the bacteria should also be examined to determine if uptake, but not adherence, is affected.

The inhibition of protein kinases by staurosporine is only slowly reversible. Thus, cells can be treated with staurosporine, the drug removed, and then bacteria added.[13] If inhibition of invasion is still observed, it indicates that the effect of the drug is on the host cell. Alternatively, effects of drugs such as genistein are rapidly reversible, and bound bacteria should be rapidly internalized following removal of the drug.[13] Perhaps the most convincing evidence that a drug is specifically affecting uptake, rather than paralyzing a cell, comes from comparison of two invasive organisms. For example, uptake of *E. coli* containing invasin from *Yersinia* is blocked by staurosporine, tyrphostin AG34, and genistein, yet uptake of *S. typhimurium* is not affected by any of these treatments.[13] These two organisms could also be used as controls in kinase inhibitor studies.

Calcium Chelators

Calcium fluxes are used in several signal transduction pathways by eukaryotic cells.[16] For example, phagocytosis may involve a localized Ca^{2+} flux beneath particles about to be internalized. Alternatively, intracellular Ca^{2+} ($[Ca^{2+}]_i$) levels may be increased by release of intracellular Ca^{2+} stores. Release of second messenger Ca^{2+} is often triggered by phospholipase C activity via inositol phosphate signaling. Calcium chelators provide convenient tools to probe the role of calcium in signaling bacterial invasion.

Several compounds are available that chelate calcium. Most of these compounds are not permeable to eukaryotic cells and, thus, chelate only extracellular Ca^{2+}. Three such compounds are BAPTA 1,2-bis(2-aminophenoxy)ethane-*N,N,N′,N′*-tetraacetic acid (BAPTA, Sigma) ethylene glycol bis(β-aminoethyl ether)-*N,N,N′,N′*-tetraacetic acid (EGTA), and ethylenediaminetetraacetic acid (EDTA). These chelators need to be used with Ca^{2+}-free tissue culture medium when probing the role of extracellular Ca^{2+} fluxes in bacterial invasion. (Normal tissue culture medium contains 100–200 mg/liter Ca^{2+}.) Ten millimolar stocks of these drugs can be made in Ca^{2+}-free phosphate-buffered saline (PBS).

Because removal of Ca^{2+} causes detachment of cultured cells, special precautions must be taken to ensure that the host cells remain adherent to the solid surface during the invasion assay. To set up an invasion assay, adherent eukaryotic cells (such as epithelial cells) are resuspended in

[16] H. Rasmussen and P. Q. Barrett, *Physiol. Rev.* **64,** 938 (1984).

tissue culture medium (containing Ca^{2+}) with 10% fetal calf serum (FCS), plated onto 24-well microtiter plates (1×10^5 cells/well) that have been previously coated with poly(L-lysine), and grown overnight. Poly(L-lysine) (0.1% w/v, Sigma) is diluted 1/50 in PBS, and 1 ml is added to each well, incubated overnight at 4°, washed twice with PBS, and incubated 2 hr at 37° in serum-free tissue culture medium containing 2 mg/ml heat-inactivated (70°, 1 hr) bovine serum albumin.

Prior to use, eukaryotic cells are washed with Ca^{2+}-free PBS and then preincubated with the appropriate chelator for 60 min in a serum-free Ca^{2+}-free tissue culture medium such as Spinner's medium (GIBCO, Grand Island, NY). Several concentrations of chelator should be tried, but a final concentration of 200 μM chelator is usually sufficient when Ca^{2+}-free medium is used. Bacteria should be washed and resuspended in Ca^{2+}-free medium prior to addition to eukaryotic cells. Invasion levels are then quantitated and compared with invasion levels with Ca^{2+} present.

1,2-bis(2-aminophenoxy)ethane-*N,N,N′,N′*-tetraacetic acid acetoxymethyl ester (BAPTA/AM, CalBiochem) is an acetoxy methyl ester of BAPTA that is readily permeable to eukaryotic cells. Once internalized, it is hydrolyzed by cytosolic esterases, releasing the active chelator BAPTA which then chelates $[Ca^{2+}]_i$, thereby blocking this second messenger.[17] A 10 m*M* stock solution of BAPTA/AM is made in DMSO. Invasion assays using BAPTA/AM are done similarly to that described for the extracellular Ca^{2+} chelators. Invasion levels with BAPTA/AM should be compared with those with BAPTA or Ca^{2+} present. It should be noted that when comparing the effects of these two drugs, the actual concentration of BAPTA inside cells varies, depending on permeability and esterase activity for each cell system.

Again, several controls should be done to determine if these chelators have any effect on bacterial adherence, bacterial viability, host cell viability, and intracellular bacterial viability. Additionally, reversibility of the chelator should be determined by washing it away and then adding Ca^{2+}-containing medium. Special attention should be directed toward ensuring that the poly(L-lysine) treatment prevents all eukaryotic cell detachment after chelator addition. This can be determined by trypan blue staining and cell counting. As an additional control, *S. typhimurium* uptake is blocked by BAPTA/AM but not by BAPTA treatment, and removal of BAPTA/AM reverses the inhibition.[18]

[17] V. Torre, H. R. Matthews, and T. D. Lamb, *Proc. Natl. Acad. Sci. U.S.A.* **83,** 7109 (1986).
[18] S. Ruschkowski, I. Rosenshine, and B. B. Finlay, *FEMS Microbiol. Lett.* **95,** 121 (1992).

Future Directions of Research

As mentioned above, inhibitor studies should be used as initial indicators of host processes that may be involved in bacterial uptake. Results from these studies can be used to indicate future research directions. For example, if cytochalasin D inhibits uptake (indicating that microfilaments may be involved), attempts can be made to visualize rearrangements in polymerized actin by staining cells with fluorescent phalloidins. If rearrangements are detected, antibodies to their cytoskeletal proteins can be used to obtain a more complete list of the cytoskeletal components involved. Similarly, rearrangement of microtubules may be triggered by the invading organism, as detected by microtubule staining. Microtubule rearrangement associated with invasion does not necessarily mean that microtubules are required for invasion. For example, *S. typhimurium* causes significant rearrangement of microtubules, but does not require intact microtubules for invasion.[3]

If host protein kinase inhibitors block invasion, several lines of research should be considered. If PKC appears to be involved, PCK activity can be downregulated by the use of phorbol esters. Alternatively, labeling cells with ^{32}P and infecting with an invasive organism may identify proteins that are preferentially phosphorylated during invasion. If TPK activity is required, tyrosine phosphorylated proteins can be identified by using monoclonal antibodies that recognized phosphotyrosines. Characterization of the host proteins that are phosphorylated may provide additional clues about the signal transduction pathway being used by the invading organism.

Should extracellular Ca^{2+} chelators inhibit invasion, and addition of Ca^{2+} removes this inhibition, experiments can proceed in several directions. For example, uptake of ^{45}Ca can be measured in the presence or absence of invading bacteria. Alternatively, several drugs that block various Ca^{2+} channels could be tested. If BAPTA/AM has an effect, the role of host phospholipase C should be considered. This includes measurement of inositol phosphate production.[18] If intracellular or extracellular chelators have an effect on invasion, Ca^{2+} fluxes in infected host cells should be measured using fluorescent Ca^{2+} probes such as Fura-2/AM[19,20] (see [35]).

In conclusion, the use of inhibitors of host cytoskeleton and signal transduction and judicious interpretations can provide an opening for the investigator to pursue in more detail the host components involved in

[19] P. L. Clerc, B. Berthon, M. Claret, and P. J. Sansonetti, *Infect. Immun.* **57,** 2919 (1989).
[20] T. J. Baldwin, W. Ward, A. Aitken, S. Knutton, and P. H. Williams, *Infect. Immun.* **59,** 1599 (1991).

bacterial invasion. As studies in these areas progress, much will be learned about the intimate (but not necessarily mutually beneficial) relationships invasive organisms have with host cells.

Acknowledgments

Work in B.B.F.'s laboratory is supported by operating grants from the British Columbia Health Care Research Foundation, the Medical Research Council of Canada, the Canadian Bacterial Diseases Center of Excellence, and a Howard Hughes International Research Scholars Award. I.R. is a recipient of a long-term fellowship from the European Molecular Biology Organization (EMBO) and a Canadian Association of Gastroenterology Industry Fellowship.

[34] Methods to Visualize Actin Polymerization Associated with Bacterial Invasion

By LEWIS G. TILNEY and MARY S. TILNEY

Introduction

Because actin is the single most abundant protein in animal cells, it is not surprising that invading microorganisms must interact with it, at least indirectly. How close the association is depends on whether the microorganism is attached to the cell surface, resides in the cytoplasm enclosed in a phagosomal membrane, or enters the cytoplasm proper after breaking out of the phagosomal vacuole. What has not been appreciated until recently is the extent to which some invading microorganisms have exploited the host cell actin. In short, they have made use of the actin for their own insidious purposes, not only to attach to cells, but to move around in the cytoplasm and to spread to new host cells.[1,2]

The exploitation of actin by invading pathogenic bacteria is not only fascinating to the microbiologist, but it has also attracted the attention of cell biologists interested in how actin filaments function in eukaryotic cells. For the latter, the pathogen can be thought of as a particle that, when introduced into the eukaryotic cytoplasm, elicits a prescribed, yet simplified behavior, a behavior that can readily be dissected by altering, via molecular biological techniques, the nature of the particle. Thus the pathogen can be used to explore how a eukaryotic cell uses its complex

[1] L. G. Tilney and D. A. Portnoy, *J. Cell Biol.* **109,** 1597 (1989).
[2] V. B. Young, S. Falkow, and G. K. Schoolnik, *J. Cell Biol.* **116,** 197 (1992).

cytoskeleton, a skeleton that is composed of actin as well as at least 50 separate actin-binding proteins. It is analogous to what has been learned about cytoplasmic transport, endocytosis, and secretion using viruses (e.g., Semliki Forest virus and vesicular stomatitis viruses) and/or viral proteins.

Clearly much can be and has been learned about the interaction of intracellular pathogens and the host cell actin at the light microscope level by using fluorescently labeled molecules that bind to actin filaments in permeabilized cells, e.g., phalloidin and phallicidin,[3] or the injection of labeled actin or labeled actin-binding proteins into living cells.[4–6] The latter techniques are particularly valuable as one can study living cells and assay by video microscopy exactly what the bacteria are doing. By control of the conditions and careful analysis of video sequences, what the bacteria are doing and how they are doing may be discerned. Particularly striking is the work of Thériot *et al.*[4] in which caged actin was microinjected into cells. By exposure of the actin tails of *Listeria* with a bar of UV light, not only can the disassembly of actin be followed, but also the site, relative to the bacterium where the actin is assembling can be determined. Furthermore, combining these methods with the addition and/or injection of materials that inhibit these actin assemblies, e.g., cytochalasin, and/or the use of mutant pathogens that fail to express certain gene products (see Section V of this volume) provides a means of determining what is controlling actin assembly during movement and spreading of the pathogens in the host cytoplasm.

As light microscopic techniques are covered by others in this volume (see [36]), we concentrate on methods to examine the actin filaments by electron microscopic techniques.

Philosophy behind Fixation for Electron Microscopy

Adequate preservation must include the following: (1) The cell must be immediately arrested—it has to be frozen in time. (2) All the components involved in the process of interest must somehow be cross-bridged together in the position they occupy and in the state they were at the moment of fixation. (3) The components of a cell that are not involved in the process under scrutiny must be eliminated from the final preparation.

[3] R. Ménard and P. J. Sansonetti, this volume [36].

[4] J. A. Thériot, T. J. Mitchison, L. G. Tilney, and D. A. Portnoy, *Nature (London)* **357,** 257 (1992).

[5] J. M. Sanger, F. S. Southwick, and J. W. Sanger, *J. Cell Biol.* **111,** 390a (1990).

[6] G. A. Dabiri, J. M. Sanger, D. A. Portnoy, and F. A. Southwick, *Proc. Natl. Acad. Sci. U.S.A.* **87,** 6068 (1990).

The last point is frequently overlooked. The concentration of protein in the cytoplasm is extraordinary; for example, the concentration of hemoglobin in erythrocytes is 340 mg/ml, and that of unpolymerized actin and actin-binding proteins in sperm is 370 mg/ml.[7] If all the protein molecules plus sugars, amino acids, nucleotides, etc., are cross-bridged in place, the cytoplasm becomes so dense that the process of interest, e.g., the relationship of the actin filaments to the invading bacterium, is obscured in a thin section. Thus, what is desired for a fixative is that it rapidly enter the cell, freeze whatever is to be preserved, and, at the same time, make holes in the limiting membrane and/or extracellular coat so that soluble components can diffuse away before being preserved by fixatives. The wealth of information gained in thin sections of detergent-extracted cells has made us acutely aware of the importance of clearing from view components not involved in a particular process. These preparations have revealed the "cytoskeleton" of cells. Unfortunately, in detergent-extracted cells membranes are not preserved so this technique is useful only for non-membrane-related events. Furthermore, because detergent penetration takes time, the structure of interest has to remain stable until fixation (cross-bridging) occurs.

Fixation of Actin Filaments

Superimposed on the difficulties with fixation just mentioned is the fact that actin filaments are notoriously sensitive to osmium tetroxide,[8] yet without osmium there is insufficient contrast in thin sections to recognize individual filaments. What the osmium does to actin filaments is truly horrendous, and unless extreme care is taken the actin filament bundles, as well as individual filaments, look like mush in thin sections. Maupin-Szamier and Pollard[8] demonstrated that if a solution containing actin filaments is "fixed" with osmium tetroxide, within seconds the filaments break into small pieces, as noted when the solution is examined by negative staining or viscometry. With longer periods of osmication the actin is cleaved into a series of peptides, as seen by sodium dodecyl sulfate gel electrophoresis. Accordingly these investigators tried various conditions to stabilize filaments, including prefixation with glutaraldehyde, different buffers, different pH conditions, and reduced temperature for osmication. They found that the least damage was induced by very brief osmication at 0° in the presence of a phosphate buffer at pH values around 6.0.

[7] L. G. Tilney and S. Inoué, *J. Cell Biol.* **93,** 820 (1982).

[8] P. Maupin-Szamier and T. D. Pollard, *J. Cell Biol.* **77,** 837 (1978).

Glutaraldehyde did not seem to inhibit damage by osmium. Our approach[9] was empirical: We tried different fixatives under various conditions on the same cell type, a cell type that contained a bundle of parallel actin filaments, hoping to find the optimal method with which to visualize actin filaments in thin sections of that cell. The best procedure was to fix cells by immersion in a solution containing 1% OsO_4, 1% glutaraldehyde, 50 m*M* phosphate buffer at pH 6.2 for no more than 45 min at 0°. It is not surprising that empirically derived conditions[9,10] and those derived from experiments involving viscometry and gel electrophoresis agree, leading to the conclusion that the least harmful conditions for preservation of actin filaments are low concentrations of OsO_4 in a phosphate buffer at pH 6.0–6.2 for the minimum time at 0°. The only discrepancy is that simultaneous fixation with glutaraldehyde and osmium tends to preserve the integrity of actin filaments in a cell, although this procedure does not reduce filament breakage *in vitro*.

It is also true that certain actin-binding proteins, namely, tropomyosin and myosin [or some of its proteolytic fragments, namely, subfragment 1 (S1) or heavy meromyosin (HMM)], stabilize actin filaments to osmication. Certain molecules when present in the fixative also tend to stabilize the actin filaments. These include tannic acid[11] and ruthenium red.[12]

A good fixative for an invading pathogenic bacterium, associated with actin filaments, is one that acts rapidly, preserves actin filaments, and at the same time releases soluble protein molecules so that the cytoskeleton is readily visible. Accordingly we use a fixative that contains both osmium and glutaraldehyde in phosphate buffer at pH 6.2 applied at 0° for 45 min. A low-osmolarity fixative, e.g., phosphate buffer, at only 50 m*M* is used so that soluble components rapidly diffuse out through the pores in the membrane created by osmium because water is pouring into the cell and diluting the soluble proteins not held in place. After fixation, which is as brief as possible, the tissue is washed with distilled water to remove any substance not strongly cross linked, as well as to remove phosphate ions, and stained *en bloc* in uranyl acetate overnight to maximize contrast in the specimen. The samples are then dehydrated in acetone as it is more polar (harsher) than ethanol and acts to eliminate soluble components. The thin sections are examined on uncoated grids using tiny objective apertures in the

[9] L. G. Tilney, *J. Cell Biol.* **69,** 51 (1976).

[10] M. S. Mooseker and L. G. Tilney, *J. Cell Biol.* **67,** 725 (1975).

[11] P. Maupin and T. D. Pollard, *J. Ultrastruct. Res. Mol. Struct. Res.* **94,** 92 (1986).

[12] L. G. Tilney, D. J. DeRosier, and M. S. Tilney, *J. Cell Biol.* **118,** 71 (1992).

electron microscope, e.g., 7 μm. The last two procedures greatly increase contrast and thus the visibility of the actin filaments.

Fixation of Intact Infected Cells

Solution A: 4 ml 2% OsO_4 in water

Solution B: 1 ml of 8% stock of glutaraldehyde, 2 ml 0.2 *M* phosphate buffer, pH 6.2, 1 ml H_2O

Both solutions are stored in an ice bucket and mixed just before use because the osmium and glutaraldehyde are thought to react with each other. Fixation is for 30–45 min in an ice bath. The 8% stock solution of glutaraldehyde obtained from Electron Microscope Sciences (Fort Washington, PA) contains glutaraldehyde as monomers and as intermediate and higher polymers together in the same solution. Pure monomeric glutaraldehyde is not a good fixative because crosslinking is not maximized.

Fixation of Detergent-Extracted Infected Cells

To study the cytoskeleton of detergent-extracted cells, extract infected cells with a solution of 1% Triton X-100, 3 m*M* $MgCl_2$, and 50 m*M* phosphate buffer at pH 6.8 for 10 min at 4°. Immediately following extraction, fix as described above.

Procedure for Ruthenium Red Fixation

Better fixation of all the actin filaments in detergent-extracted cells can be obtained if ruthenium red is used in the fixative. The preparation must be detergent extracted prior to fixation with ruthenium because this compound does not penetrate intact cells either before or during fixation with glutaraldehyde.

After detergent extraction the infected cells are washed briefly in cold 0.05 *M* cacodylate at pH 7.4. The extracted and washed ghosts are then fixed in 0.25% glutaraldehyde in 0.075 *M* cacodylate containing 1% ruthenium red for 1 hr, rinsed in 0.05 *M* cacodylate, and postfixed in 1% OsO_4 containing ruthenium red in cacodylate buffer for 30 min. The specimen is then rinsed and stained *en bloc* overnight with uranyl acetate and processed as before.[12]

Decoration of Actin Filaments with Subfragment 1 of Myosin

Because we have not obtained good results with commercially obtained subfragment 1 of myosin, we generally prepare our own. In the procedure[13]

[13] S. S. Margossin and S. Lowey, *J. Mol. Biol.* **74,** 301 (1973).

rabbit muscle is extracted with high salt (0.6 *M* NaCl), and the myosin is washed with low salt, resolubilized, and then precipitated a second time with ammonium sulfate, followed by papain digestion. We have followed procedure of Margossin and Lowey[13] without changes. Subfragment 1 can be frozen in liquid nitrogen in small aliquots that keep indefinitely. The samples we prepared more than 10 years ago in a very concentrated form, about 70 mg/ml, are still being used. The S1 solution was pipetted into tiny Eppendorf tubes with caps in 100-μl aliquots. No cryoprotectant was used. The S1 solution contained 50 m*M* phosphate buffer. The tubes were immersed and stored in liquid nitrogen until they were thawed before use. The thawed S1 can be kept on ice for 1–2 weeks, but it should not be refrozen.

To decorate actin filaments infected cells are detergent extracted (1% Triton X-100, 3 m*M* $MgCl_2$, 50 m*M* phosphate buffer at pH 6.8 for 10 min at 4°). The detergent solution is decanted and 5 mg/ml S1 in 0.1 *M* phosphate buffer is added at pH 6.8 for 30 min. The first 10 min of decoration takes place on ice; the last 20 min, at room temperature on an oscillating table. The S1 solution is decanted and the specimen washed in 0.1 *M* phosphate buffer for 20 min to remove unbound S1.

Fixation is carried out at room temperature in 1% glutaraldehyde with 2% tannic acid and 0.05 *M* phosphate buffer at pH 6.8 for 30 min. The preparation is then washed in 0.1 *M* phosphate buffer and postfixed in 1% OsO_4 in 0.1 *M* phosphate buffer at pH 6.2 for 30 min at 4°. After osmication the preparation is washed three times in water and stained *en bloc* before dehydration and embedding.

Addition of the tannic acid to the glutaraldehyde is the key to good decoration.[14–17] The tannic acid is fixed to the surface of the actin and S1 molecules which, because it has many groups that react with osmium, gives a kind of "negative-stained" appearance to the decorated filaments that makes the visualization of the arrowheads clear in thin sections.

To prepare the primary fixative (tannic acid and glutaraldehyde), 2% tannic acid is added to phosphate buffer at pH 6.8 and then heated with swirling until the tannic acid goes into solution (pale yellow color). The solution is allowed to cool to room temperature. The tannic acid solution is made up shortly before use and glutaraldehyde is added to it immediately before use. Fixation is carried out at room temperature because if the tannic acid/glutaraldehyde mixture is cooled to 4°, it will precipitate. No detergent should be present as it also precipitates the tannic acid.

[14] D. A. Begg, R. Rodewald, and L. I. Rebhun, *J. Cell Biol.* **79,** 846 (1978).

[15] L. G. Tilney, D. J. DeRosier, and M. J. Mulroy, *J. Cell Biol.* **86,** 244 (1980).

[16] L. G. Tilney and L. A. Jaffe, *J. Cell Biol.* **87,** 771 (1980).

[17] L. G. Tilney, D. J. DeRosier, A. Weber, and M. S. Tilney, *J. Cell Biol.* **118,** 83 (1992).

[35] Measurement of Free Intracellular Calcium Levels in Epithelial Cells as Consequence of Bacterial Invasion

By JOHN L. PACE and JORGE E. GALÁN

Introduction

It is increasingly apparent that stimulation of host cell signaling by microorganisms plays an important role in pathogenesis.[1-5] Therefore it is becoming more important to study host cell signaling events that the pathogen may alter or use to gain access to the host, to avoid host defences, or to produce clinically evident disease. Free intracellular calcium ($[Ca^{2+}]_i$) serves as a second messenger for a variety of eukaryotic cell events.[6,7] $[Ca^{2+}]_i$ has been demonstrated to link cell surface receptor stimulation with intracellular effectors and to modulate cytoskeletal structure, membrane fluidity, enzyme activity, transmembrane ion fluxes, and other cell functions.[6-9] Therefore, $[Ca^{2+}]_i$ fluxes are likely to be involved in the pathogenesis of several microorganisms. Enteropathogenic strains of *Escherichia coli* (EPEC), for example, cause an increase in $[Ca^{2+}]_i$ in HEp-2 cells which is accompanied by profound cytoskeletal rearrangements in the infected cells.[10] These bacterially induced changes are thought to be essential for the production of the attaching and effacing lesion of epithelial cells which is observed both *in vitro* and *in vivo*.[10]

Invasion of intestinal epithelial cells by *Salmonella* species is an important step in the pathogenesis of these microorganisms.[11,12] On contact with

[1] J. E. Galán, J. Pace, and M. Hayman, *Nature (London)* **357,** 588 (1992).

[2] T. J. Baldwin, S. F. Brooks, S. Knutton, H. A. Manjarrez Hernandez, A. Aitken, and P. H. Williams, *Infect. Immun.* **58,** 761 (1990).

[3] S. L. Weinstein, M. R. Gold, and A. L. DeFranco, *Proc. Natl. Acad. Sci. U.S.A.* **88,** 4148 (1991).

[4] I. Rosenshine, V. Duronio, and B. B. Finlay, *Infect. Immun.* **60,** 2211 (1992).

[5] J. B. Bliska, K. Guan, J. E. Dixon, and S. Falkow, *Proc. Natl. Acad. Sci. U.S.A.* **88,** 1187 (1991).

[6] R. Jacob, *Biochim. Biophys. Acta* **1052,** 427 (1990).

[7] R. W. Tsien and R. Y. Tsien, *Annu. Rev. Cell Biol.* **6,** 715 (1990).

[8] P. W. Marks and F. R. Maxfield, *Cell Calcium* **11,** 181 (1990).

[9] J. Meldolesi, E. Clementi, C. Fasolato, D. Zacchetti, and T. Pozzan, *Trends Pharmacol. Sci.* **12,** 289 (1991).

[10] T. J. Baldwin, W. A. Aitken, S. Knutton, and P. H. Williams, *Infect. Immun.* **59,** 1599 (1991).

[11] A. Takeuchi, *Am. J. Pathol.* **50,** 109 (1967).

[12] R. A. Gianella, O. Washington, P. Gemski, and S. B. Formal, *J. Infect. Dis.* **128,** 69 (1973).

the epithelial cell surface, *Salmonella* causes a transient disruption of the microvilli.[11,13,14] This is accompanied by profound rearrangements of the cell cytoskeleton, with accumulation of actin and other proteins at the site of bacterial entry.[14,15] Changes in the cell cytoskeleton may be effected by $[Ca^{2+}]_i$ through actin-associated proteins such as vinculin and gelsolin.[16] When $[Ca^{2+}]_i$ is at basal levels in the cell, vinculin acts as an actin-bundling protein and stabilizes microfilaments; however, when the cell is stimulated and $[Ca^{2+}]_i$ increases, this protein actually severs actin filaments.[16] The resulting cytoskeletal rearrangements may provide a membrane surface for endocytosis or free actin monomers for the construction of new cytoskeletal structures. As changes in $[Ca^{2+}]_i$ have been shown to have a profound impact on the cytoskeletal structure, we investigated whether *Salmonella* was able to alter the levels of $[Ca^{2+}]_i$ in infected cultured epithelial cells. We developed a system that allowed us to perform $[Ca^{2+}]_i$ measurements using a cuvette-based spectrofluorometric instrument. The system is based on the use of the fluorescent probe Fura-2/AM and Henle 407 cells grown on microcarrier beads. Utilization of the microcarrier allowed us to perform measurements on populations of *Salmonella typhimurium*-infected epithelial cells cultured under conditions of physiological calcium. The fluorescent probe Fura-2/AM exhibits increased fluorescence and a shift in excitation maxima on calcium binding. In addition, its membrane permeability allows the easy loading of epithelial cells for calcium measurements. What follows is a detailed description of the method.

Cell Culture

Henle 407 cells are cultured in 25-cm^2 tissue culture flasks using Dulbecco's modified Eagle's medium containing 10% bovine calf serum (DMEM), 100 μg/ml penicillin G, and 50 μg/ml streptomycin. When the cell cultures are 80–90% confluent, they are removed from the flask by trypsinization with 5 ml of 0.05% (w/v) trypsin–0.53 m*M* ethylenediaminetetraacetic acid (EDTA) in phosphate-buffered saline (PBS). Trypsin is then neutralized by adding 5 ml of the growth medium described above. After centrifugation (500 g, 5 min, 25°), the cells are resuspended in 5 ml of growth medium, and 2 ml of the cell suspension is added to 8 ml of growth medium in a flask containing 0.052 g sterile Ventreglas microcarrier

[13] B. B. Finlay and S. Falkow, *J. Infect. Dis.* **162,** 1096 (1990).

[14] C. Ginocchio, J. Pace, and J. E. Galán, *Proc. Natl. Acad. Sci. U.S.A.* **89,** 5976 (1992).

[15] B. B. Finlay, S. Ruschkowski, and D. Dedhar, *J. Cell Sci.* **99,** 283 (1991).

[16] M. S. Mooseker, *Annu. Rev. Cell Biol.* **1,** 209 (1985).

beads (1.03 g/cm^3 specific gravity, 90–150 μm, Ventrex Lab., Portland, ME). The surface area of this quantity of microcarrier is approximately 48 cm^2 and is equivalent to a 24-well tissue culture plate. The growth medium is replaced with fresh DMEM without antibiotics 24 hr after inoculating the microcarrier beads with Henle 407 cells. These samples are routinely used for $[Ca^{2+}]_i$ assays 48 hr after seeding.

Glass-coated microcarrier beads are superior to plastic beads, because they absorb less light at the wavelengths used for the spectrofluorometric measurements, and they allow better cell growth. We have found no differences in the ability of *S. typhimurium* to invade cultured epithelial cells grown on microcarrier beads or on tissue culture plates.[17]

Loading of Cells with Fura-2/AM for Assay of Free Intracellular Calcium

Cells cultured on microcarrier beads as described above are gently rinsed with growth medium and then transferred into a centrifuge tube. After centrifugation (400 *g*, 5 min, 25°), cells are washed once in assay buffer (20 m*M* HEPES, 137 m*M* NaCl, 5 m*M* KCl, 0.4 m*M* $MgSO_4$, 0.5 m*M* $MgCl_2$, 0.4 m*M* NaH_2PO_4, 1.0 m*M* $CaCl_2$, 1.0 m*M* glucose, and 1.0 m*M* glutamine, pH 7.4), and suspended in 10 ml of the same buffer.[18] Fura-2/AM (stock 50 μM, in anhydrous dimethyl sulfoxide) is then added to a final concentration of 1 μM. After 1 hr incubation at 37°, unabsorbed dye is removed by washing the cells in assay buffer three times. Dye-loaded cells are resuspended in 10 ml of assay buffer and cooled on ice.

Growth of Bacteria and Infection of Cell Microcarrier

Overnight cultures of bacteria grown in L-broth at 37° on a rotary wheel (20 rpm) are diluted in the same medium to an absorbance of 0.17 at 600 nm and incubated for 30 min.[19]

One-half milliliter of assay buffer is added to 0.5 ml of the dye-loaded cell suspension (approximately 1×10^5 cells). Samples are cooled on ice, and then bacteria are added at a multiplicity of infection (m.o.i) of 50 and incubated for 4 hr at 4°. Adsorption of bacteria to epithelial cells at 4° improves the kinetics of the experiment. Internalization of bacteria does not occur at 4°, although organisms can still adhere to the epithelial cells

[17] J. Pace, M. J. Hayman, and J. E. Galán, *Cell* **72,** 505–514 (1993).

[18] B. W. Hitchin, P. R. M. Dobson, B. L. Brown, J. Hardcastle, P. T. Hardcastle, and C. J. Taylor, *Gut* **32,** 893 (1991).

[19] J. E. Galán and R. Curtiss, III, *Proc. Natl. Acad. Sci. U.S.A.* **86,** 6383 (1989).

without causing $[Ca^{2+}]_i$ fluxes. After incubation, nonadhered bacteria are removed by washing three times with cold assay buffer by centrifugation (400 g, 5 min, 4°).

Spectrofluorometric Measurements

Fluorescence is measured in a spectrofluorometer capable of an excitation wavelength of 340 nm and an emission wavelength of 510 nm (for Fura-2). The instrument should be equipped with a temperature-regulated cuvette holder set at 37°. Ultraviolet transparent cuvettes with four optical sides are required, and a semimicro (1.5 ml) size is most useful. The cell microcarrier sample is very buoyant and only gentle stirring is needed to maintain the cells in suspension.

The effect on the levels of $[Ca^{2+}]_i$ in Henle 407 cells after infection with *S. typhimurium* SR11 and the noninvasive mutants SB111 (*invA*) and SB109 (*invE*) was evaluated.[14,20] Following adherence of bacteria to Henle 407 cells on ice as described above, periodic fluorescence measurements (F) were recorded. The maximum fluorescence (F_{max}) of the sample was determined by the addition of ionomycin to 10 μM which allowed saturating levels of calcium to enter the epithelial cells. Subsequently, the minimum fluorescence (F_{min}) was measured by adding $MnCl_2$ to 1 mM which quenched the dye's fluorescence.[21] The $[Ca^{2+}]_i$ of the Henle 407 cells was calculated using the equation[21]

$$[Ca^{2+}]_i = K_d[(F - F_{min})/(F_{max} - F)]$$

The dissociation constant, K_d, of Fura-2/AM is assumed to be 224 nM.[21] As shown in Fig. 1, *S. typhimurium* SR11 caused a marked increase in $[Ca^{2+}]_i$ in Henle 407 cells. $[Ca^{2+}]_i$ increased significantly by 15 min after initial measurements and reached levels in excess of 1 μM. Infection with the noninvasive *invA* (SB111) and *invE* (SB109) mutant strains of *S. typhimurium* did not cause any significant changes in $[Ca^{2+}]_i$ of Henle 407 cells. As $[Ca^{2+}]_i$ measurements were performed in a population of cells, the rate of $[Ca^{2+}]_i$ increase does not reflect the kinetics of the Ca^{2+} fluxes in individual cells, but rather, the kinetics of bacterial invasion.

Use of Chemical Agents to Complement Studies of Free Intracellular Calcium

Determining the source of Ca^{2+} may provide information about the mechanism by which $[Ca^{2+}]_i$ is increased. $[Ca^{2+}]_i$ levels can be increased

[20] J. E. Galán, C. Ginocchio, and P. Costeas, *J. Bacteriol.* **174,** 4338 (1992).

[21] G. Grynkiewicz, M. Poenie, and R. Y. Tsien, *J. Biol. Chem.* **260,** 3440 (1985).

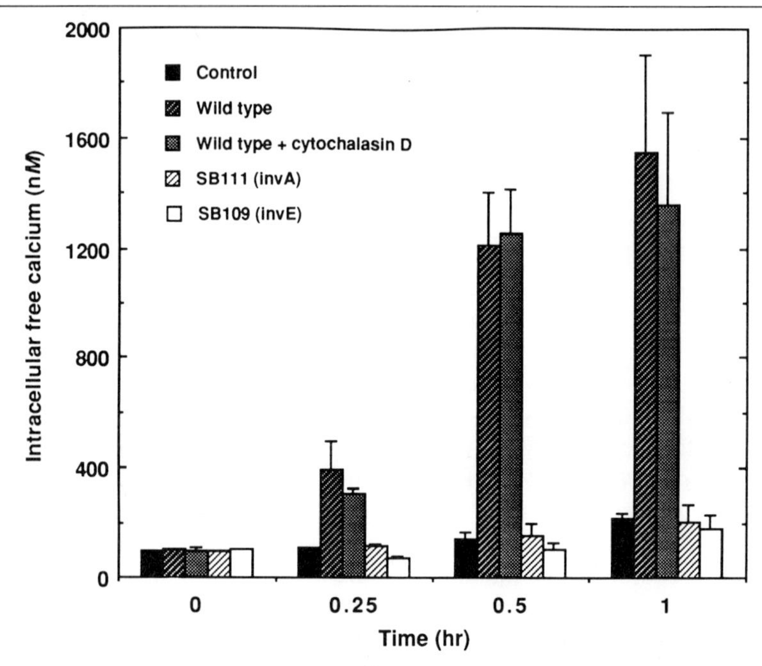

FIG. 1. Intracellular calcium levels ($[Ca^{2+}]_i$) in Henle 407 cells during infection by wild-type *Salmonella typhimurium* SR11 and the noninvasive mutants SB111 (*invA*) and SB109 (*invE*).

by opening of plasma membrane Ca^{2+} channels or by release of calcium from intracellular stores.[7-9] Several pharmacologically active agents can be used to help distinguish between these two possibilities.

Dantrolene, for example, is able to block release of calcium from intracellular caffeine-sensitive stores. Treatment of cells with this compound before bacterial infection can be used to determine if the $[Ca^{2+}]_i$ increase is due to release from these stores. In fact, dantrolene has been shown to prevent the increase in $[Ca^{2+}]_i$ of EPEC-infected cultured epithelial cells.[10]

To determine if calcium influx is required for $[Ca^{2+}]_i$ increases during bacterial infection, $[Ca^{2+}]_i$ levels of epithelial cells can be measured using an assay buffer without calcium. To do this cells are loaded with Fura-2/AM and infected with bacteria as described above, but during subsequent steps Ca^{2+} is removed from the assay buffer. Calcium is subsequently added (to 1 m*M*) back to the buffer for the determination of F_{max}.

Several agents have been shown to block efficiently specific types of voltage-regulated Ca^{2+} channels, preventing calcium entry, and could be used during assay of $[Ca^{2+}]_i$ to determine whether a specific class of ion channels is involved. For example, 1,4-dihydroxypyridine compounds such as nitrendipine or nifedipine, act on L-type Ca^{2+} channels but not on T- or N-type channels.[22] The phenylalkylamine verapamil or the benzothiazepine diltiazem may also block some of these channels; however, receptor-regulated calcium channels are often unaffected by these chemicals.[22] Lanthanum, another calcium antagonist, blocks entry of calcium into eukaryotic cells.[23] $[Ca^{2+}]_i$ levels of cells in the presence of lanthanum cannot, however, be determined. Lanthanides are fluorescent and interfere with fluorescent measurements of Fura-2/AM. In contrast, cadmium chloride, which also blocks entry of calcium, can be used during fluorescent assays of $[Ca^{2+}]_i$ by inclusion in the assay buffer.

Salmonella typhimurium-induced $[Ca^{2+}]_i$ increases of Henle 407 cells were unaffected by dantrolene.[17] Effects of nifedipine and verapamil on $[Ca^{2+}]_i$ of Henle 407 cells during *S. typhimurium* infection have not been determined, but these agents were unable to prevent invasion of the epithelial cells by the bacterium; however, lanthanum and cadium inhibited invasion of Henle 407 cells by *S. typhimurium*.[17]

In addition to determining the source of calcium in the $[Ca^{2+}]_i$ fluxes during bacterial infection, it may be useful to determine whether bacteria can trigger these changes from the outside or if $[Ca^{2+}]_i$ fluxes are a consequence of bacterial invasion. Epithelial cells can be treated with cytochalasin D (1 μg/ml) during Fura-2/AM loading and subsequent steps in the measurement of $[Ca^{2+}]_i$. Cytochalasin D disrupts actin microfilament structure and prevents entry of some bacterial pathogens. If $[Ca^{2+}]_i$ increases occur in cells treated with cytochalasin D, then the infecting bacteria must trigger the $[Ca^{2+}]_i$ rise from the epithelial cell surface. *S. typhimurium* increases $[Ca^{2+}]_i$ of Henle 407 cells treated with cytochalasin D, suggesting that a pathogen–host cell surface interaction must trigger the calcium increase.[17]

Alternative Methods for Measurement of Free Intracellular Calcium

Other calcium-sensitive dyes and instruments can be used for measurement of $[Ca^{2+}]_i$. Quin-2 (2-{[2-bis(carboxymethyl)amino-5-methylphenoxy]methyl}-6-methoxy-8-bis(carboxymethyl)aminoquinoline) and Fluo-3 (9-{4-bis-4(carboxymethyl)amino-3-[2-(bis(carboxymethyl)amino-5-methylphenoxy)ethoxy]phenyl}-2,7-dichloro-6-hydroxy-3*H*-xanthin-3-one) both have been used for the fluorescent measurement of $[Ca^{2+}]_i$.[10,24]

[22] M. M. Hosey and M. Lazdunski, *J. Membr. Biol.* **104,** 81 (1988).

[23] G. B. Weiss, *Annu. Rev. Pharmacol.* **14,** 343 (1974).

[24] D. A. Williams, *Cell Calcium* **11,** 589 (1990).

Like Fura-2/AM, the esterified forms quin-2/AM and fluo-3/AM are membrane permeable. Quin-2 is less fluorescent and more photolabile and has reduced affinity for calcium than Fura-2, and its use has largely been superceded by the latter dye.[25] Fluo-3 has the advantage that both its excitation and emission maxima are in the visible range, and it can be used for microscopic observations.[24] Other workers have suggested that Fluo-3/AM could be used to screen for bacterial mutants that cannot cause $[Ca^{2+}]_i$ increases in epithelial cells.[10] The esterified form of Fluo-3 is not fluorescent, and the free form is fluorescent only when calcium is bound. These characteristics prevent the ratiometric measurement of $[Ca^{2+}]_i$ with Fluo-3, but reduce many of the artifacts associated with dye leakage or overloading.

Cells are most often cultured as adherent monolayers for Ca^{2+} studies; however, some investigators have measured $[Ca^{2+}]_i$ of cells grown in suspension.[26] This can be done by reducing the level of calcium in the growth medium to 100 μM, which inhibits adhesion of the cells to the substratum and to other cells. Studies with cells cultured in suspension may not be appropriate in some cases, because of the subphysiological levels of calcium used. The characteristics of cells cultured in low-calcium medium may be different, and responses requiring influx of extracellular calcium would be diminished. To determine if the epithelial cells are exhibiting normal behavior under low-calcium conditions, a control stimulus should be used. Histamine has been used as a positive control during measurements of $[Ca^{2+}]_i$ in cells cultured using low-calcium medium.[26] This control may, however, be inappropriate for the measurement of some responses. Histamine causes release of calcium from intracellular stores which are closely regulated, but cannot serve as a control for signaling processes that require influx of extracellular calcium.

Other instruments are available for the measurement of $[Ca^{2+}]_i$. Using a spectrofluorometer with a large sample port, levels of $[Ca^{2+}]_i$ of epithelial cells cultured on a glass coverslip can be determined by fluorescent methods. Cells on the coverslip are loaded with Fura-2/AM and then placed in a 3-ml cuvette for measurement. The disadvantage of this method is that only $[Ca^{2+}]_i$ of cells in the excitation light path is measured.

Dual-wavelength spectrofluorometers have the ability to monitor fluorescence simultaneously with two excitation wavelengths. By use of this capability, ratiometric measurement of $[Ca^{2+}]_i$ can be conducted.[21] Fura-2 can be used for ratiometric assay of $[Ca^{2+}]_i$ using the equation[21]

$$[Ca^{2+}]_i = K_d[(R - R_{min})/(R_{max} - R)] \times (S_{f2}/S_{b2})$$

[25] P. H. Cobbold and T. J. Rink, *Biochem. J.* **248,** 313 (1987).

[26] P. L. Clerc, B. Berthon, M. Claret, and P. J. Sansonetti, *Infect. Immun.* **57,** 2919 (1989).

where R is the ratio of fluorescence at 510 nm using the 340 and 380 nm excitation wavelengths, with R_{max} and R_{min} at saturating and zero $[Ca^{2+}]_i$ levels, respectively.[21] S_{f2}/S_{b2} is the ratio of free to calcium-bound dye with excitation at 380 nm.[21] Fluorescent images of individual cells may be collected rapidly and digitized for computer analysis. Processed images can be generated that provide information about the temporal and localized increase in $[Ca^{2+}]_i$ in response to cell stimulation.[27] Ratiometric imaging has the advantage of correcting for artifacts. Analysis of single-cell $[Ca^{2+}]_i$ provides a greater amount of information than can be obtained with measurements of cell populations, but the disadvantage of single-cell measurements is that often only 15–20% of the cells produce an observable response.

Confocal microscopy offers the advantage of providing information on the localized nature of $[Ca^{2+}]_i$ increases in response to some stimuli.[24] By use of Fluo-3, $[Ca^{2+}]_i$ changes have been observed in some cells by confocal microscopy. Processing of data collected by confocal microscopy can provide information about the three-dimensional localization of $[Ca^{2+}]_i$ in stimulated cells.

Conclusion

It is now evident that microorganisms are able to send and receive signals from the host. The understanding of the details of this molecular cross-talk is in its infancy, but it is clear that it will be a subject of intensive research in the field of bacterial pathogenesis in the coming years. The study of host cell signaling mechanisms that may be altered or used by microorganisms to cause disease is likely to lead to a better understanding of microbial pathogenicity. Ca^{2+} is of central importance in a great variety of eukaryotic cell signaling processes. It is therefore becoming increasingly important to measure changes in the levels and/or distribution of $[Ca^{2+}]_i$ as a consequence of microbe–host cell interactions. The methods described in this paper should be helpful in carrying out these studies. If pathogens are able to modulate the levels of $[Ca^{2+}]_i$ in the host cell, protocols can be designed to isolate mutants that failed to cause these responses. This, in turn, may help to better understand the language of the host–parasite cross-talk.

Acknowledgments

We thank Dr. E. London, Dr. N. Ulbrandt, and Dr. N. Marrion (State University of New York, Stony Brook, NY) for assistance in the use of the spectrofluorometer and

[27] A. J. O'Sullivan, T. R. Cheek, B. B. Moreton, M. J. Berridge, and R. D. Burgoyne, *EMBO J.* **8**, 401 (1989).

helpful discussions concerning the use of Fura-2/AM, M. Zierler for careful review of this manuscript, and Ventrex Laboratories for providing the microcarrier beads. This work was supported by Public Health Service Grant AI-30492 from the National Institutes of Health and a grant from the Sinsheimer Foundation (to J.E.G.). J.E.G. is a Pew Scholar in the Biomedical Science and a Searle–Chicago Community Trust Scholar.

Section V

Identification of Genes Involved in Invasion

[36] *Shigella flexneri:* Isolation of Noninvasive Mutants of Gram-Negative Pathogens

By ROBERT MÉNARD and PHILIPPE J. SANSONETTI

Introduction

The capacity to invade epithelial cells is a key determinant of virulence for a number of pathogenic bacteria. This property to enter, to grow within, and eventually to kill epithelial cells that are not professional phagocytes allows these pathogens to colonize host mucosal surfaces and cause a range of epithelial destruction. Most invasive pathogens then gain access to submucosal tissues, and some may proceed to systemic dissemination. In the study of virulence processes, genetics and molecular biology remain the primary approaches, relying on techniques that have progressed considerably since the first entry-associated DNA sequences were identified in *Shigella flexneri*[1] and in *Yersinia pseudotuberculosis*.[2] The development of these techniques has allowed the unequivocal characterization of some of the bacterial products that mediate access of the major invasive bacteria to the intracellular compartment. Here, we first consider the various strategies that may be undertaken to study the genetic basis of an invasive process, and we then focus on the methods that have been applied in the identification and characterization of the genes necessary for the invasive process in *S. flexneri* infection.

Genetic Approaches for Isolation of Noninvasive Mutants

Identification of Invasion-Associated Genes

To identify invasion-associated genes of invasive organisms, two principal approaches may be followed: direct cloning and transposon mutagenesis.

In employment of the cloning strategy, it is assumed that proper expression, export, and organization of the invasion products are obtained at the surface of the noninvasive recipient strain, thereby conferring the invasive phenotype. Isolating invasive clones in a cell invasion assay can be facilitated by an enrichment procedure that positively selects for

[1] A. T. Maurelli, B. Baudry, H. d'Hauteville, T. L. Hale, and P. J. Sansonetti, *Infect. Immun.* **49,** 164 (1985).

[2] R. R. Isberg and S. Falkow, *Nature (London)* **317,** 262 (1985).

invasive recombinants, by killing extracellular bacteria with an antibiotic that does not penetrate significantly into cells.[3] Invasion gene(s) carried by the cloned sequence are ultimately identified by transposon mutagenesis and isolation of mutations that abolish invasion; however, because the invasion phenotype could be determined by multiple gene products, the technique may be unsuccessful when using classic vectors and may require cosmid vectors that permit the cloning of large fragments of DNA.[1,4,5] In addition, the invasion-associated gene(s) may not be expressed in the bacterial host that is used as a recipient of the cloned sequence, and hosts more closely related to the pathogen, or a nonvirulent derivative of the organism under investigation, may be needed.

By use of transposon mutagenesis, transposon insertions are obtained in many different sites in the genome of the virulent pathogen, and noninvasive clones are identified by applying the appropriate assay. Screening for mutants that have lost a virulence property leads to efficient identification of genes involved in the corresponding phenotype, but this approach requires testing of each of thousands of mutants, in contrast to the positive selection that is often available using recombinant DNA techniques. Whenever its use is possible, transposon Tn*5* remains the "gold standard," owing to its relatively low insertional specificity and to the stability of the induced mutations[6]; however, other transposable elements have been constructed to simplify the screening procedure. For example, transposon Tn*phoA*, a derivative of transposon Tn*5*, affords a powerful screening criterion for mutants in virulence-associated genes.[7] Tn*phoA* insertions generate hybrid proteins composed of alkaline phosphatase fused to amino-terminal sequences of proteins into whose genes the transposon is inserted. Such hybrids display alkaline phosphatase activity if the target protein contributes a sequence that promotes export from the cytoplasm. Because most proteins involved in invasion are expected to be membrane, periplasmic, or secreted proteins, their respective genes can theoretically be identified by first screening for active *phoA* fusions on appropriate media. Alternatively, transcriptional gene fusions that join promoters to the *Escherichia coli* lactose operon or to a promoterless antibiotic resistance gene provide a simple method to isolate a gene fusion while recovering an insertional mutation. If an environmental cue, such as temperature,

[3] V. L. Miller and S. Falkow, *Infect. Immun.* **56,** 1242 (1988).

[4] E. A. Elsinghorst, L. S. Baron, and D. J. Kopecko, *Proc. Natl. Acad. Sci. U.S.A.* **86,** 5173 (1989).

[5] J. E. Galán and R. Curtiss, III, *Proc. Natl. Acad. Sci. U.S.A.* **86,** 6383 (1989).

[6] D. E. Berg, *in* "Mobile DNA" (D. E. Berg and M. M. Howe, eds.), p. 185. Am. Soc. Microbiol. Washington, DC, 1989.

[7] C. Manoil and J. Beckwith, *Proc. Natl. Acad. Sci. U.S.A.* **82,** 8129 (1985).

regulates expression of the invasive phenotype, mutants may first be screened for temperature regulation of their fusion and subsequently be tested for their ability to invade cells, thus reducing the number of clones to be assayed *in vitro*. The invasion tests can also be streamlined by using multititer plaques. Finally, once invasion mutants have been identified, the interrupted sequences are cloned, using the antibiotic resistance marker of the transposon as a tag, and are in turn used as a probe to clone a functional copy of the gene from the wild-type strain.

Characterization of Invasion-Associated Genes

Once an invasion-associated gene has been identified, characterization of the role of the gene in invasion relies on comparison of the virulence phenotype of the wild-type strain with that of the isogenic derivative bearing the mutated gene. Analysis of the phenotype of random insertion mutants is usually complicated, however, by the polar effect that the inserted element may exert on the transcription of downstream genes.[8] Site-directed mutagenesis of a virulent strain allows a more reliable analysis of the mutant phenotype. This method requires cloning of the gene to construct a mutation, usually by deletion of part of the gene and insertion of a selectable fragment. The *in vitro* constructed mutated gene is then used to replace its wild-type allele in the virulent parental strain by homologous recombination via a *recA*-dependent event. This strategy thus enables the creation of a nested mutation and the insertion of a nonpolar fragment into the gene. Ultimately, to ensure that the virulence defect of the mutant is indeed due to inactivation of the gene, complementation of the mutant by the intact gene must restore the wild-type phenotype of the strain. This general scheme fulfills the "molecular Koch's postulates" proposed by Falkow.[9]

Several gene replacement strategies have been used to introduce mutations into virulence-associated genes of gram-negative pathogens. The mutated gene can be introduced into the parental strain by means of either a replicating or a nonreplicating DNA molecule. In the former case, the vector that delivers the mutation must be lost after the recombination event has occurred. Three methods are described briefly.

In the strategy developed by Ruvkun and Ausubel,[10] a P-group conjugative plasmid (p1) that confers resistance to antibiotic A1 and carries the target gene, inactivated by insertion of a cassette that mediates resistance to antibiotic A2, is conjugated into the recipient strain, in which it repli-

[8] D. E. Berg, A. Weiss, and L. Crossland, *J. Bacteriol.* **142,** 439 (1980).

[9] S. Falkow, *Rev. Infect. Dis.* **10,** Suppl. 2, S274 (1988).

[10] G. B. Ruvkun and F. M. Ausubel, *Nature (London)* **289,** 85 (1981).

cates stably. After conjugation of a second plasmid (p2) that belongs to the same incompatibility group as p1 and confers resistance to antibiotic A3, cells in which the plasmidborne mutation has replaced the genomic region are selected on plates containing antibiotics A2 and A3. Spontaneous curing of plasmid p2 is then obtained by subculturing the recombinant strain in the absence of A3. This marker exchange procedure was used to construct cholera toxin-deficient mutants by recombination of an *in vitro*-constructed deletion of the *ctx* operon into each of the two resident copies of the *ctx* operon carried by a wild-type *Vibrio cholerae* strain.[11]

Marker exchange has also been obtained in *E. coli* strains carrying *recBC sbcBC* or *recD* mutations, which do not degrade incoming linear DNA while remaining recombination proficient.[12,13] After transformation of these strains with a linear fragment that consists of DNA homologous to the target gene straddling a selectable marker, without a plasmid origin of replication, cells that have acquired the antibiotic resistance marker should have undergone the expected double-crossover event. For example, an *invA* insertional mutation carried by a linear fragment was transformed into an *E. coli recD* strain containing the cloned *S. typhimurium* wild-type locus to generate a recombined locus, which was subsequently introduced into the chromosome of a wild-type *S. typhimurium* strain by phage transduction.[14]

Most mutations in virulence-associated genes of gram-negative pathogens have been constructed by use of *pir*-dependent replicating plasmids, i.e., suicide plasmids. Vector pJM703.1[15] and its derivative pGP704 contain the P-type-specific recognition site for *trans* active RP4 transfer functions. They also contain the origin of replication from plasmid R6K,[16] which requires the *pir* gene-encoded π protein for its function. Such plasmids can be maintained in and mobilized from *E. coli* SM10λ*pir*,[17] which supplies the π protein and provides factors necessary to mobilize the suicide plasmid derivatives to any gram-negative bacterium. These tools allow the generation of two kinds of mutations. The first corresponds to the integration of a suicide plasmid construct that contains an internal fragment of the gene into the wild-type locus of the genome. The second corresponds to the replacement of the intact copy of the genome by the

[11] J. J. Mekalanos, D. J. Swartz, G. D. Pearson, N. Harford, F. Groyne, and M. de Wilde, *Nature (London)* **306,** 551 (1983).

[12] M. Jasin and P. Schimmel, *J. Bacteriol.* **159,** 783 (1984).

[13] C. B. Russell, D. S. Thaler, and F. W. Dahlquist, *J. Bacteriol.* **171,** 2609 (1989).

[14] J. E. Galan, C. Ginocchio, and P. Costeas, *J. Bacteriol.* **174,** 4338 (1992).

[15] V. L. Miller and J. J. Mekalanos, *J. Bacteriol.* **170,** 2575 (1988).

[16] R. Kolter, M. Inuzuka, and D. R. Helinski, *Cell (Cambridge, Mass.)* **15,** 1199 (1978).

[17] R. Simon, U. Priefer, and A. Pühler, *Bio/Technology* **1,** 784 (1983).

disrupted copy carried by the suicide plasmid. Both techniques of allelic exchange and plasmid integration have been widely used in the genetic analysis of the *S. flexneri* invasive process, and are described below.

Phenotypic Analysis of Invasive Process of *Shigella flexneri*

As several other examples of invasive bacterial pathogens are covered in this volume, we focus on the invasion of epithelial cells by *S. flexneri*. Shigellosis, or bacillary dysentery, is an invasive disease of the human colon. Invasion assays, such as that of HeLa, Henle, and HEp-2 cells[18] or, more recently, human colonic cancerous cell lines, such as Caco-2 cells,[19] have been used to analyze the molecular and cellular basis of cell invasion. The entry process and the capacity to grow, move intracellularly, and spread from cell to cell are the major processes that have been studied.

Entry involves a process of directed phagocytosis. In addition to being evaluated by the gentamicin assay (see [28]), which provides essentially quantitative data (Fig. 1), it can also be characterized by staining infected cells with Giemsa stain to locate precisely intracellular bacteria (Fig. 2). It can also be assessed by labeling cells at early stages of infection with NBD–phallacidin, a dye specific for polymerized actin, to show the bulk of F-actin that participates in the ongoing phagocytic process (see Fig. 2).[20]

Suitable cell and tissue invasion assays are essential for the assessment of the extent of alterations in the entry phenotype caused by a given mutation. Mammalian cell cultures are now most often used according to the general protocol described in Fig. 1. Once entry has taken place, extracellular bacteria are killed by an antibiotic that poorly penetrates into eukaryotic cells. Cells are then lysed, and the colony-forming units (cfu) of the lysate correspond to the number of viable intracellular bacteria. Results should be expressed as the percentage of the initial inoculum recovered. This technique can be used to detect either recombinant clones expressing an invasive property or mutants that have lost this property; however, although the former situation allows direct positive selection for invasive clones, the latter does not, and each individual mutant needs to be assayed. Modifications can be introduced into this basic technique, but the general principles remain the same.

The ability of *Shigella* to spread from cell to cell is best assessed by studying their capacity to form plaques on confluent monolayers of HeLa

[18] T. L. Hale, R. E. Morris, and P. F. Bonventre, *Infect. Immun.* **24,** 887 (1979).

[19] J. Mounier, T. Vasselon, R. Hellio, M. Lesourd, and P. J. Sansonetti, *Infect. Immun.* **60,** 237 (1992).

[20] P. Clerc and P. J. Sansonetti, *Infect. Immun.* **55,** 2681 (1987).

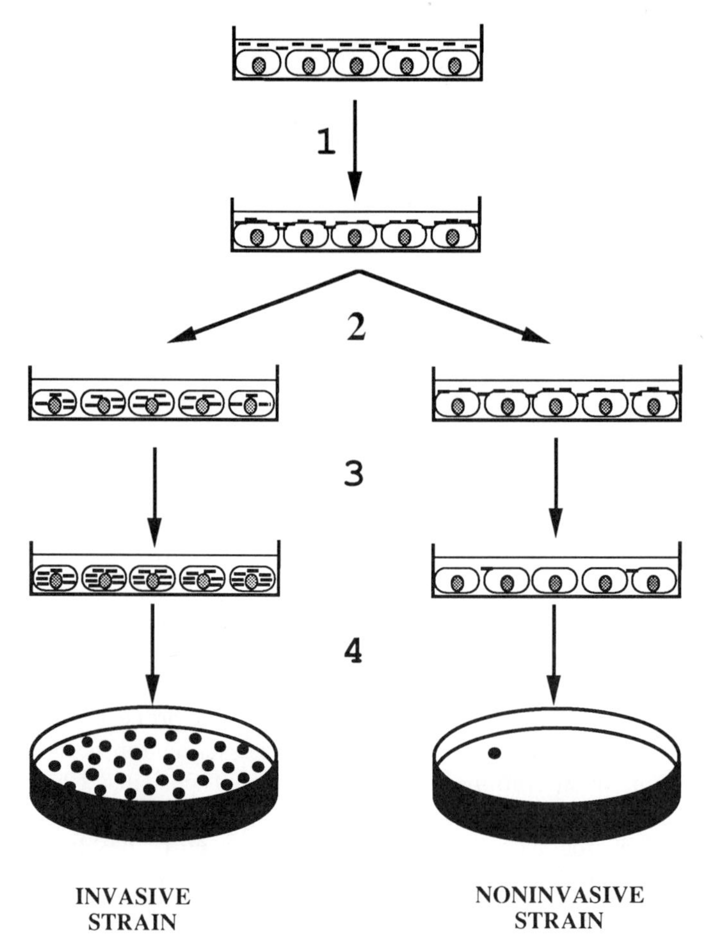

FIG. 1. HeLa cell invasion assay (gentamicin assay). *Materials:* Minimal essential medium (MEM, Eagle's, 1×) with Earle's salts containing glutamine; fetal calf serum (FCS); trypsin solution (1×), 0.25% (w/v); Earle's balanced salt solution (EBSS) without $NaHCO_3$; sodium bicarbonate solution (7.5%) (w/v); Giemsa staining solution; plastic dishes for tissue culture (usually 35 mm in diameter, but may be adapted to specific situations); sodium deoxycholate 0.5% in PBS (w/v). *Preparation of HeLa cells for invasion.* Inoculate 35-mm plastic dishes with 2 ml of a cell suspension at 2.5×10^5 cells/ml. Grow overnight at 37°, 6% CO_2, in MEM with 10% FCS (v/v). *Preparation of bacterial cultures for infection of cells.* Incubate bacteria overnight in trypticase soy broth (TSB) at 37° with aeration. Subculture at a dilution of 1/100 in TSB and grow for about 2 hr at 37° until an OD_{600} of 0.3 is reached. Centrifuge the bacteria at 5000 *g* for 10 min at room temperature and resuspend the pellet in MEM without FCS. *Invasion of the cells.* Step 1: 2 ml of the bacterial suspension in MEM is added to the cells (the multiplicity of infection is about 100 under such conditions). Centrifugation is carried out for 10 min at 1500 *g*, at room temperature. Step 2: Incubate at 37° for 45–60 min without CO_2. Step 3: Wash the dishes three times with EBSS, add 2 ml of MEM containing gentamicin at a concentration of 50 μg/ml, and incubate for 1 hr or longer. Step 4: Remove medium, wash three times in EBSS, lyse cells with 0.5% sodium deoxycholate, plate serial dilutions on trypticase soy agar, and incubate plates at 37° overnight. Count colony-forming units and present results as the percentage of the initial inoculum.

cells.[21] In this test, extracellular gentamicin prevents infection of uninfected cells by bacteria released into the extracellular medium. Bacteria present in infected cells grow rapidly and spread from cell to cell, thereby causing the formation of a cytopathic focus that is easily visualized in the confluent monolayer. This assay is described in Fig. 3.

The relevance of *in vitro* studies, however, must be assessed by testing the mutants that have been characterized in cell culture assays in more definitive *in vivo* assays, such as the guinea pig keratoconjunctivitis assay (Sereny test), which is the topic of a specific article in Volume 235.[22] The rabbit ligated ileal loop assay[23] or the intragastric inoculation of macaque monkeys, which subsequently develop a dysentery similar to shigellosis,[24] is sometimes used.

Genetic Analysis of Invasive Process of *Shigella flexneri*

The presence of a 220-kb plasmid is required for expression of the invasive phenotype of *S. flexneri*, as well as other *Shigella* species, and is sufficient to confer invasive capacities on *E. coli* K12.[25] The methods that have been used to identify and characterize the invasion-associated genes present on this plasmid have followed the general strategy described above.

Cloning of Invasion-Associated Genes

Because of the complexity of the *Shigella* invasion loci, invasive recombinants could not be obtained by using classic vectors; therefore, a cosmid cloning method was carried out. Partial digests of pWR100, the virulence plasmid of the serotype 5 isolate M90T, were shotgun-cloned into a multicopy cosmid vector.[1] Recombinant cosmids were transduced into a λ-sensitive, plasmidless *S. flexneri* 2a recipient, and transductants able to invade HeLa cells *in vitro* were selected. Analysis of each cloned insert that conferred the entry phenotype showed that they each contained a common core of ca. 35 kb, which thus defined the minimum sequence necessary for entry into epithelial cells. This DNA fragment encodes the four virulence-associated Ipa polypeptides (IpaA, IpaB, IpaC, and IpaD),

[21] E. V. Oaks, M. E. Wingfield, and S. B. Formal, *Infect. Immun.* **48,** 124 (1985).

[22] D. Kopecko, this series, Vol. 235 [3].

[23] S. B. Formal, D. Kundel, H. Schneider, N. Kunev, and H. Sprinz, *Br. J. Exp. Pathol.* **42,** 504 (1961).

[24] S. B. Formal, T. H. Kent, H. C. May, A. Palmer, S. Falkow, and E. H. LaBrec, *J. Bacteriol.* **92,** 17 (1966).

[25] P. J. Sansonetti, T. L. Hale, G. I. Dammin, C. Kapper, H. H. Collins, Jr., and S. B. Formal, *Infect. Immun.* **39,** 1392 (1983).

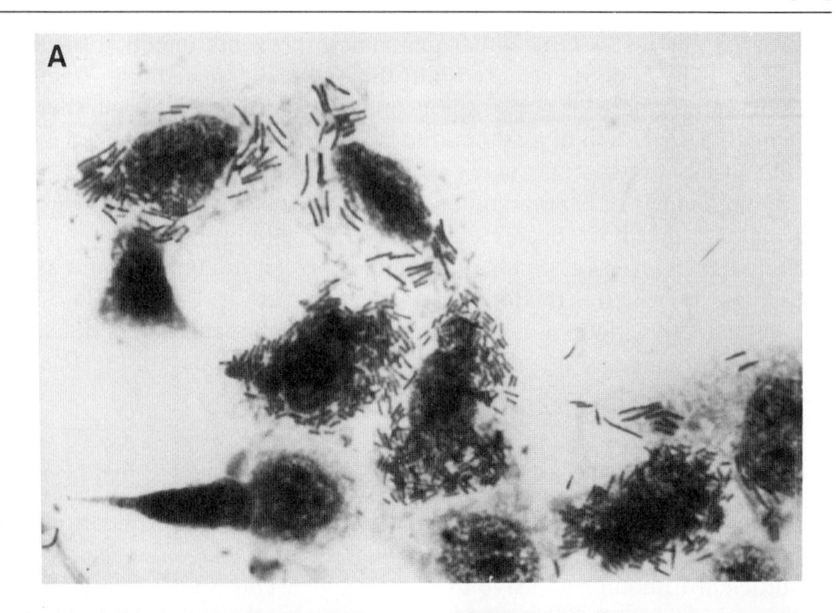

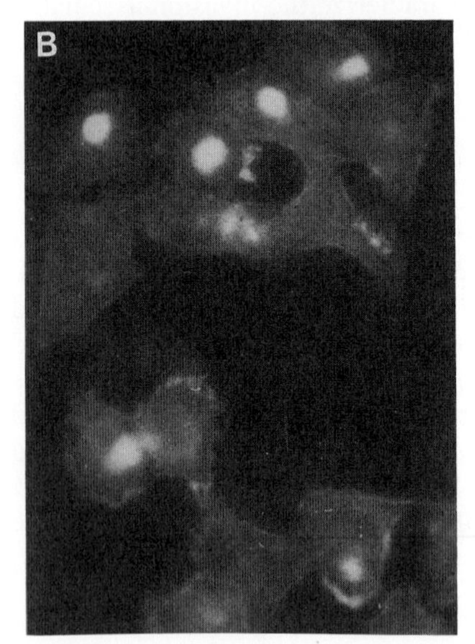

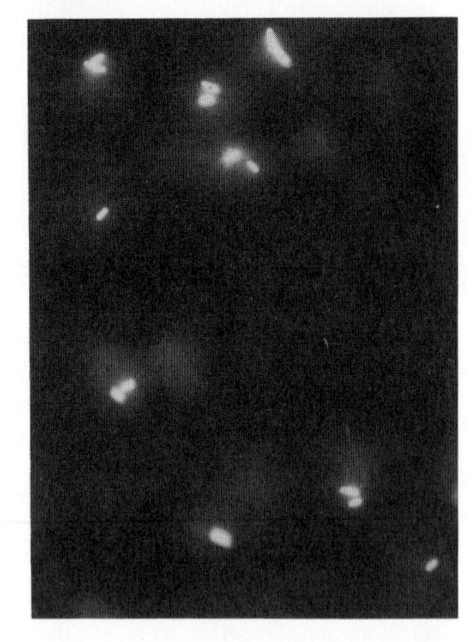

FIG. 2. Staining of infected cells. Invasion may be evaluated by either visualizing intracellular bacteria with standard Giemsa staining or by staining foci of polymerized actin that are observed at entry sites in association with the directed phagocytosis of the entering bacteria. *Giemsa staining:* After three washes in EBSS, cells that have been infected as described in Fig. 1 are fixed in 2 ml ethanol for 30 sec, washed twice in distilled water, and then stained for 10 min with Giemsa stain diluted at 1/20 in distilled water. Preparations are dried and covered with a drop of immersion oil, a glass coverslip, and a second drop of oil,

identified as the major antigens in the humoral immune response to shigellosis. The full spectrum of virulence properties associated with the wild-type strain was not, however, expressed by the recombinant strains in that they failed to produce a positive Sereny test. Further experiments demonstrated that they were impaired in their ability both to multiply within eukaryotic cells and to spread intracellularly and from cell to cell.

Random Insertion Mutations

Saturation Tn*5* mutagenesis was achieved on pMYSH6000, the large plasmid of *S. flexneri* 2a. The integration site of each of 304 independent Tn*5* insertions was located on a *Sal*I restriction map of the plasmid (Fig. 4),[26] and each mutant was tested for its capacity to invade LLC-MK2 cells (Inv) and to provoke keratoconjunctivitis in mice (Ser). One locus, lying within a 4.5-kb region of *Sal*I fragment G and named *virG*, was required for the Ser^+ phenotype, but not for the Inv^+ phenotype. The remaining avirulent mutants were affected in both phenotypes (Ser^-, Inv^-); their transposon insertion sites were located in five *Sal*I fragments of the plasmid, named B, P, H, D, and F. Transposon insertions in *Sal*I fragment F disrupted a 1-kb gene named *virF*, which turned out to be a transcriptional activator of the virulence-associated genes. Between *virG* and *virF*, five regions, clustered in a 31-kb segment that extends through the contiguous fragments B, P, H, and D, were also found to be necessary for the entry phenotype.[27] The restriction map of this segment was similar

[26] C. Sasakawa, S. Makino, K. Kamata, and M. Yoshikawa, *Infect. Immun.* **54,** 32 (1986).
[27] C. Sasakawa, K. Kamata, T. Sakai, S. Makino, M. Yamada, N. Okada, and M. Yoshikawa, *J. Bacteriol.* **170,** 2480 (1988).

prior to microscopic observation. (A) Invasive *Shigella flexneri*. Bacteria can be seen as large rods distributed throughout the cell cytoplasm. *Staining of foci of polymerized actin at S. flexneri entry sites with 7-nitrobenz-2-oxa-1,3-diazole (NBD)–phallacidin.* Cells grown on glass coverslips and infected as described, but for a short period (1 hr or less), are washed three times in EBSS and are fixed in PBS containing 3% paraformaldehyde, 0.1 μM $CaCl_2$, 0.1 μM $MgCl_2$ for 20 min. After three washes in PBS, the remaining paraformaldehyde is removed by treatment with 50 mM NH_4Cl in PBS for 10 min. Permeabilization of the cell membrane is then achieved by treatment with 0.1% Triton X-100 in PBS for 4 min. Cells are then treated with a 10 U/ml solution of 7-nitrobenz-2-oxa-1,3-diazole–phallacidin (NBD–phallacidin) in PBS, 0.2% gelatin, to stain specifically filamentous actin (F-actin) with a yellow–green fluorescence. They are then washed in PBS and mounted in glycerol and Moviol 4-88. NBD–phallacidin uses the same fluorescence filters as fluorescein derivatives (excitation at 490 nm and emission at 525 nm). (B) Invasive *S. flexneri*. Bacteria are labeled with anti-LPS rhodamine antiserum (right); F-actin, labeled by NBD–phallacidin, accumulates at sites of bacterial entry (left).

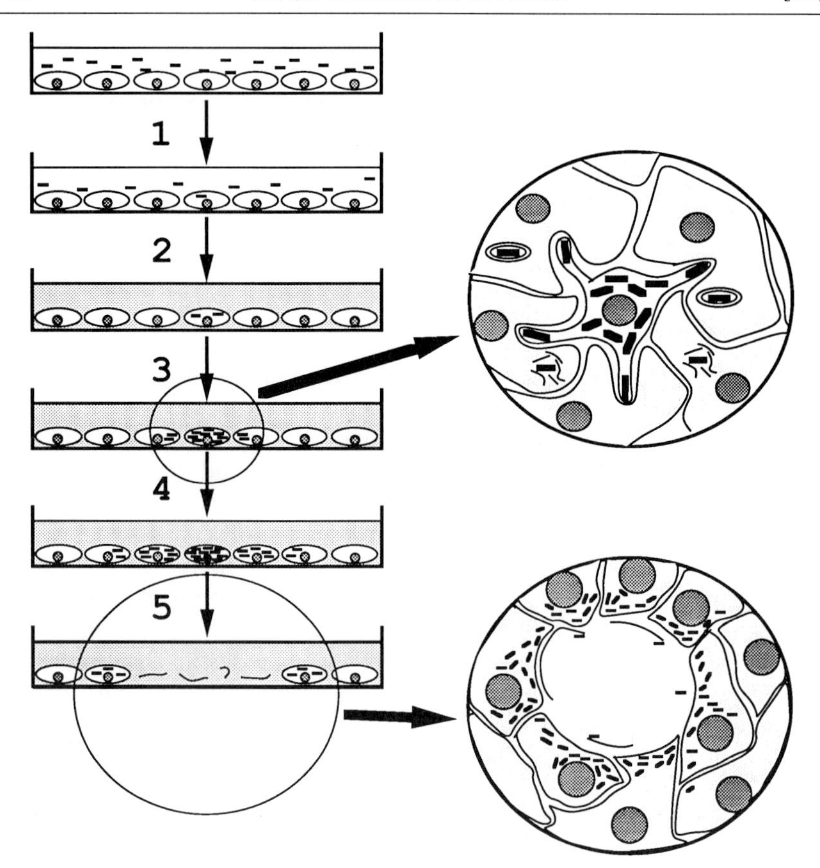

FIG. 3. Plaque assay. *Preparation of HeLa cells*. Cells are grown to confluency on 35-mm plastic dishes by adding 2 ml of a cell suspension at a concentration of 4×10^5 cells/ml in antibiotic-free MEM with 10% fetal bovine serum (FBS). *Preparation of bacterial suspensions*. Bacteria are grown as described in the legend to Fig. 1. Dilutions of 10^{-2} and 10^{-3} are usually added to the monolayer. *Plaque assay procedure*. Step 1: Confluent monolayers are washed three times with EBSS. After aspiration of the wash medium, 1 ml of the bacterial suspension is added to the cells and incubation is carried out for 120 min at 37° without CO_2. Step 2: After three washes in EBSS, 2 ml of agarose overlay is added. This overlay consists of MEM, 20% (v/v) FBS, gentamicin (50 μg/ml), and 0.5% (w/v) agarose. Steps 2–5: Plates are incubated for 2 to 4 days at 37° in 6% CO_2 and are examined daily for the appearance of plaques.

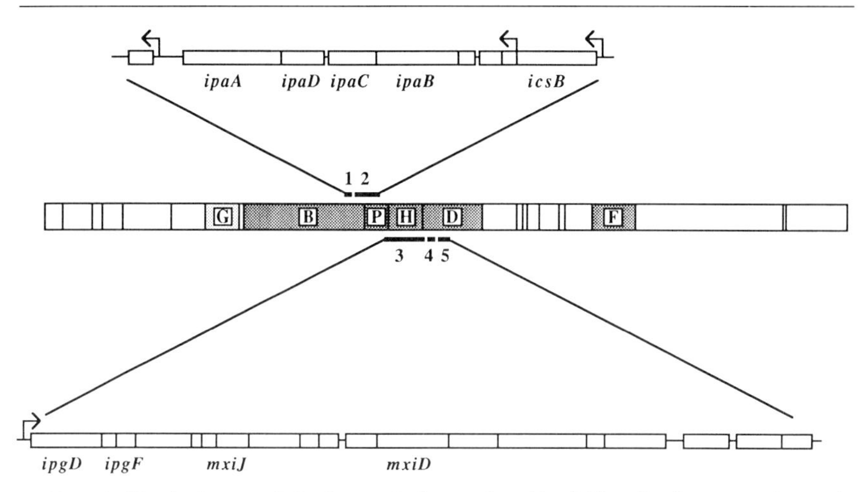

FIG. 4. Physical map of the large virulence plasmid of *Shigella flexneri* and genetic organization of the entry-associated regions. The 23 *Sal*I-generated fragments of the large plasmid pMYSH6000 of *S. flexneri* serotype 2a strain YSH6000 are shown. By Tn*5* mutagenesis of this plasmid, 304 transposon insertions were obtained, and each was assigned to one of the 23 *Sal*I fragments by restriction analysis of the recombinant plasmids. The phenotype of each pMYSH6000::Tn*5* mutant was assayed in the Sereny test (Ser) and in an invasion assay of epithelial cells (Inv). Each mutant that was found to be avirulent (Ser^-) contained an inserted transposon in *Sal*I fragment G, B, P, H, D, or F (labeled fragments). Of these, Tn*5* insertions in fragment G induced a Ser^- but Inv^+ phenotype, whereas the remaining avirulent mutants generated by insertions in fragment B, P, H, D, or F displayed a Ser^- and Inv^- phenotype. The precise physical location of independent Tn*5* insertions in fragments B, P, H, and D that abolished invasion was determined, thereby defining five entry-associated regions within a 31-kb DNA fragment of the plasmid, named regions 1 through 5 (indicated by bold horizontal bars). Open reading frames identified by sequencing of these regions are symbolized by open boxes. Identified promoters, symbolized by arrows, and names of some of the characterized genes are indicated.

to that of the 35-kb insert of the cosmid clone that conferred the invasive phenotype. Analysis of Tn*5*-induced mutants in region 2 indicated that the *ipaB, ipaC, ipaD,* and *ipaA* genes are clustered within an operon and that the IpaB, IpaC, and IpaD proteins are essential for the entry process.[28]

Another approach in the identification of virulence-associated genes used gene fusion technology and took advantage of the temperature-regulated virulence phenotype of *Shigella*. Wild-type strains grown at 37° are virulent and invade epithelial cells, whereas the same strains grown at 30° are noninvasive. Virulence-associated genes transcribed from tem-

[28] C. Sasakawa, B. Adler, T. Tobe, N. Okada, S. Nagai, K. Komatsu, and M. Yoshikawa, *Mol. Microbiol.* **3,** 1191 (1989).

perature-regulated promoters were identified by random insertions of the transposable bacteriophage λp*lac*Mu53 into a λ-sensitive virulent strain of *S. flexneri* 2a.[29] Ten thousand lactose-fermenting (Lac$^+$) transductants selected at 37° were screened for temperature-dependent production of β-galactosidase. Seven of these transductants appeared to be temperature regulated for lactose utilization, in that they expressed a Lac$^-$ phenotype at 30°. To ensure that this phenotype was due to a single insertion of the transposable element, each of the seven temperature-regulated fusions was transduced into the wild-type strain. Each of the resulting thermoregulated mutants was then tested for its ability to invade HeLa cells; only four *inv*::*lacZ* fusion mutants appeared noninvasive. One *inv*::*lacZ* fusion disrupted the *ipaB* gene (region 2), whereas the remaining three inactivated genes mapped to regions 3 and 5. The latter genes were later shown to be involved in proper surface expression of the Ipa proteins and were thus called *mxi* (membrane expression of invasion plasmid antigens).[30]

Site-Directed Mutations

Random mutations, generated either by transposon insertions or by *lac* operon fusions, have allowed the localization of five regions, clustered in a 31-kb fragment, that are necessary for the entry phenotype. Transcriptional analysis of these regions has shown that they contain genes clustered in operons.[27,30] As pointed out above, because of the frequent polarity of a transposon mutation, the phenotype of the mutant cannot be directly attributed to the dysfunction of the mutated gene. Strains generated by insertion of appropriate DNA fragments are more suitable for evaluating the contribution each gene makes to virulence. In addition to yielding much information about the encoded factors, determination of the nucleotide sequence of the genes facilitates the construction of site-directed mutants. Therefore, the nucleotide sequence of the five regions of the entry-associated fragment has been determined (see Fig. 4) and has allowed the construction of mutants using the following techniques.

Mutagenesis Procedure. PRINCIPLE OF METHOD. Both techniques of allelic exchange and plasmid integration, described in Fig. 5, make use of plasmid pGP704, which can be mobilized from *E. coli* SM10λ*pir* into *S. flexneri,* in which it is unable to replicate. The technique of allelic exchange, described below, is based on the following assumptions: (1) Recombinational events in the recipient strain can be identified by selecting for the mutation-mediated antibiotic resistance; (2) double recombina-

[29] A. E. Hromockyj and A. T. Maurelli, *Infect. Immun.* **57,** 2963 (1989).

[30] G. P. Andrews, A. E. Hromockyj, C. Coker, and A. T. Maurelli, *Infect. Immun.* **59,** 1997 (1991).

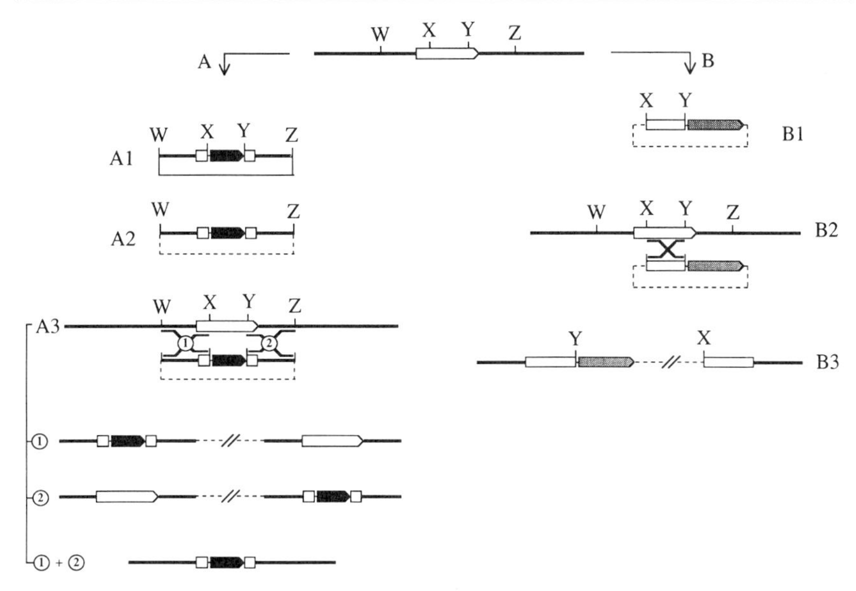

FIG. 5. Construction of mutants by allelic exchange (A) and suicide plasmid integration (B) procedures. The gene of interest, represented by an open arrow, and flanking regions on the genome, indicated by bold lines, are shown at the top. Two strategies to obtain mutants of the gene via *in vivo* homologous recombination are outlined. A selectable DNA fragment and a promoterless *lacZ* gene are represented by a black arrow and a stippled arrow, respectively. Vector DNA (not shown to scale) is indicated by thin (common vector) or dashed (*pir*-dependent replicating vector) lines. W, X, Y, and Z symbolize restriction sites. (A) *Exchange of a wild-type gene with an in vitro constructed mutation of the gene.* (A1) After cloning the gene and its flanking regions (fragment W–Z) into a replicative plasmid in *Escherichia coli*, an internal fragment of the gene (X–Y) is replaced by a cassette that confers resistance to kanamycin (Km). (A2) The DNA fragment encompassing the disrupted gene (W–Z) is cloned into the suicide vector pGP704, which confers resistance to ampicillin (Ap). (A3) The resulting plasmid is subsequently transferred by conjugal mating from *E. coli* to *Shigella flexneri*. Most transconjugants result from integration of the suicide plasmid through single recombinational events between regions located either upstream (event 1) or downstream (event 2) from the cassette on the suicide plasmid and homologous regions on the large plasmid of *S. flexneri*. Such recombinant strains are resistant to Ap. Clones in which a double recombinational event (events 1 + 2) has exchanged the intact gene for the disrupted copy have lost the suicide plasmid and are sensitive to Ap. (B) *Mutation by plasmid integration.* (B1) Plasmid pLAC1 is a derivative of pGP704 that contains the *E. coli lacZ* reporter gene. An internal fragment of the gene of interest (X–Y) is inserted upstream from the *lacZ* gene in pLAC1. (B2) The pLAC1 derivative is transferred by conjugal mating from *E. coli* to *S. flexneri*, and transconjugants are selected on Sm- and Ap-containing plates. (B3) Ap-resistant clones arise through homologous recombination between the identical sequences carried by the *Shigella* large plasmid and the pLAC1 derivative. The integration event duplicates the cloned internal fragment (X–Y), generating two truncated copies of the gene, and places *lacZ* under the control of the promoter of the mutated gene.

tional events, which exchange the wild-type locus with the mutated copy and lead to the loss of the plasmid vector, can be identified by the sensitivity of such mutants to ampicillin, the resistance marker carried by the suicide plasmid.

STRAINS AND PLASMIDS. *Escherichia coli strain DH5αλpir* {*endA1, hsdR17* ($r_k^- m_k^+$), *supE44, thi1, recA1, gyrA* (Nalr), *relA1*, Δ(*lacZYA–argF*)*U169*, F′[Φ80*dlac*Δ(*lacZ*)*M15*] (λ*pir*)}[31] is a highly transformable strain used as a host for constructs in the suicide plasmid vector. *E. coli* strain SM10λ*pir* [*thir, thr, leu, tonA, lacY, supE, recA*::RP4-2Tc::Mu (Km^r) (λ*pir*)][17] is used to transfer these constructs to *S. flexneri* M90T-Sm. *S. flexneri* M90T-Sm is a spontaneous, streptomycin (Sm)-resistant derivative of the wild-type serotype 5 strain M90T; this strain has no alteration in its ability either to invade and form plaques on HeLa cells or to provoke keratoconjunctivitis in guinea pigs.[32] Plasmid pGP704 (ori*R6K*, *mob*, Ap^r) and its derivative pLAC1,[32] which contains the *lacZ* reporter gene from Tn*917-lac*, are suicide plasmids.

MEDIA AND ANTIBIOTICS. *Escherichia coli* strains are grown on Luria–Bertani [Bactotryptone 1% (w/v), yeast extracts 0.5% (w/v), NaCl 1%, pH7] agar or broth at 37° and *S. flexneri* on trypticase soy (TCS, Diagnostics Pasteur, Marnes la Coquette, France) agar or broth at 37°. Antibiotics are used at the following concentrations: ampicillin (Ap), 100 μg/ml; kanamycin (Km), 50 μg/ml; streptomycin (Sm), 100 μg/ml.

SELECTABLE DNA FRAGMENTS. Two types of cassettes were used to create mutations in the target genes (Fig. 6). A 2-kb Ω-derived interposon, carried by plasmid pHP45Ω-Km, contains a kanamycin resistance gene (*aphA2*) transcribed from its own promoter and is flanked both by transcription and translation termination signals and by synthetic polylinkers.[33] The flanking ρ-independent transcriptional terminator sequences of bacteriophage T4D gene *32* account for the strong polarity of interposon-induced mutations.[34] To create nonpolar mutations, a cassette that respects natural downstream transcription was constructed.[35] This 850-bp cassette contains a kanamycin resistance gene (*aphA3*) and does not carry any promoter or transcription terminator. In addition, translation stop codons in all three reading frames are positioned upstream from the resistance gene, and a consensus ribosome binding site (GGAGG), as well as a start codon, is located downstream from the gene. In-frame cloning of the start codon

[31] D. M. Woodcock, P. J. Crowther, J. Doherty, S. Jefferson, M. Noyer-Weidner, S. S. Smith, M. Z. Michael, and M. W. Graham, *Nucleic Acids Res.* **17,** 3469 (1989).

[32] A. Allaoui, J. Mounier, M. C. Prévost, P. J. Sansonetti, and C. Parsot, *Mol. Microbiol.* **6,** 1605 (1992).

[33] R. Fellay, J. Frey, and H. Krisch, *Gene* **52,** 147 (1987).

[34] P. Prentki and H. Krisch, *Gene* **29,** 303 (1984).

[35] R. Ménard, P. J. Sansonetti, and C. Parsot, unpublished results.

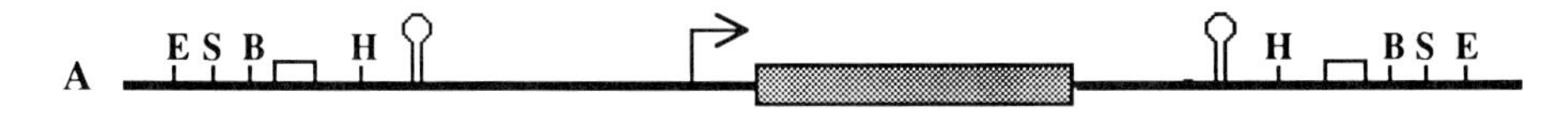

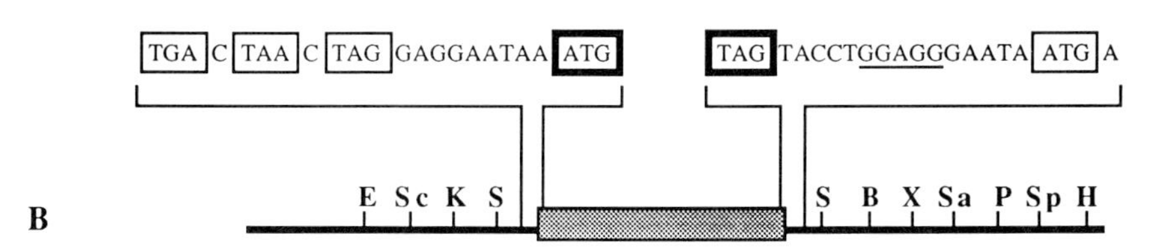

FIG. 6. Selectable DNA fragments used in site-directed mutagenesis of *Shigella flexneri* virulence-associated genes. (A) *Derivative of the Ω interposon.* A derivative of the Ω interposon, shown as a 2-kb *Eco*RI fragment, carries a kanamycin resistance gene (indicated by a stippled box) which is flanked, in inverted orientation, by transcription (symbolized by loops) and translation (symbolized by open boxes) termination signals and by synthetic polylinkers. The resistance gene (*aphA2*) is transcribed from its own promoter (represented by an arrow), present on the original 1.5-kb *aphA2*-containing fragment of transposon Tn*5*. This promoter allows selection of transconjugants that contain an interposon-induced mutation. (B) *Nonpolar cassette.* To create nonpolar mutations in genes clustered in an operon, a cassette that allows natural downstream transcription was used. This 850-bp cassette is represented here inserted in the *Sma*I site of plasmid pUC18. It contains a kanamycin resistance gene (*aphA3*), indicated by a stippled box, and its start and stop codons are shown in bold boxes. This cassette does not carry a promoter or a transcription terminator. In addition, the resistance gene is preceded by translation stop codons in all three reading frames (indicated by thin boxes) and is immediately followed by a consensus ribosome binding site, GGAGG (underlined), and a start codon (indicated in a thin box). Restriction enzyme sites: B, *Bam*HI; E, *Eco*RI; H, *Hin*dIII; K, *Kpn*I; P, *Pst*I; S, *Sma*I; Sa, *Sal*I; Sc, *Sac*I; Sp, *Sph*I; X, *Xba*I.

of the cassette with the stop codon of the mutated gene should ensure translation of the 3′ extremity of the mutated gene and should respect translational coupling, if any, of the truncated and downstream genes.

In vitro CONSTRUCTION OF MUTATION. A mutation of the gene of interest is constructed by replacing an internal fragment with a selectable DNA fragment. Experiments are carried out with usual vectors and *E. coli* hosts. The fragment carrying the mutated gene is subsequently cloned into plasmid pGP704. The pGP704 derivative carrying the mutation is ultimately transformed into $CaCl_2$-treated SM10λ*pir* cells, and transformants are selected on Ap-containing plates.

In vivo RECOMBINATION PROCEDURE. Several colonies of the donor strain (*E. coli SM10λpir* carrying the suicide plasmid derivative) and the recipient strain (*S. flexneri* M90T-Sm) (ratio 1 : 1) are mixed in 200 μl of TCS broth on a TCS plate and incubated at 37° for 8 hr. The mating mixture is then collected from the plate and resuspended in 300 μl of TCS broth, and 100 μl of serial dilutions is then plated on TCS agar containing Sm and Km and incubated at 37° for 16 hr. Transconjugants are then tested for their sensitivity to Ap by replica plating 500 individual colonies onto Ap- and Km-containing plates. Ap-sensitive, Km-resistant clones are further tested for correct allelic exchange by colony hybridization, using the deleted fragment as a probe. The predictable change in the restriction map of the region caused by homologous recombination is ultimately confirmed by Southern blot hybridization. The frequency of proper double-crossover events is highly dependent on the insert and varies, in part, with the amount of homologous DNA that flanks the selectable fragment. In our hands, higher recombination efficiencies were obtained with regions of homology greater than 1 kb in length on each side of the cassette. On the average, one mutant having undergone the expected event was found among 20 Ap-sensitive clones, out of 500 Km-resistant transconjugants. The nature of the events that occurred in the remaining Ap-sensitive strains has not been examined.

Construction of icsB Mutant. The Ω interposon was used to investigate the role of *icsB*, a gene located between regions 1 and 2. An *icsB* mutant was obtained by exchange of the interposon with an internal fragment of *icsB*.[32] When tested in cell invasion assays, the *icsB* mutant remained invasive, but formed small plaques on HeLa cell monolayers and was unable to provoke keratoconjunctivitis in guinea pigs. Electron microscopic examination of infected HeLa cells showed that the *icsB* mutant was able to lyse the phagocytic vacuole and to form protrusions at the surface of infected cells, but remained trapped within protrusions surrounded by two membranes. Complementation analysis with a plasmid bearing *icsB* alone showed that the mutant phenotype was indeed due to the *icsB* mutation. Moreover, the polar effect of the interposon-induced mutation led to the demonstration that *icsB* does not belong to the *ipa* operon, as Ipa proteins were normally produced by the mutant.

Construction of ipaB Mutant. To address the role of *ipaB* in the entry process, a cassette allowing reinitiation of transcription was used. A constitutive promoter was thus cloned downstream from the distal transcription termination signal of the Ω cassette, and the resulting fragment was inserted into the *ipaB* gene and was subsequently exchanged with the wild-type gene on the large plasmid.[36] This *ipaB* mutant was not invasive in HeLa cells and, when internalized in macrophages, could not lyse the phagosomal membrane. Although only low levels of IpaC were produced by the *ipaB* mutant, *in trans* expression of *ipaB* was sufficient to restore

[36] N. High, J. Mounier, M. C. Prévost, and P. J. Sansonetti, *EMBO J.* **11,** 1991 (1992).

invasion and intracellular multiplication, indicating that IpaB is involved in both invasion and lysis of the phagosomal membrane.

Construction of Mutants in ipa and mxi Operons. The partial polar effect observed in the *ipaB* mutant, despite the inserted constitutive promoter, pointed to the need for a cassette able to preserve the transcription of downstream genes from the wild-type promoter. For this purpose, the nonpolar cassette that does not contain any promoter or transcription terminator was used. The *ipaC* and *ipaD* genes were each inactivated by this cassette, and analysis of the mutant phenotypes confirmed that each is involved in the entry process.[35] The cassette was also inserted into genes in region 3, such as *ipgD, ipgF, mxiJ,* and *mxiD*. The *mxiJ* and *mxiD* mutants were unable to invade HeLa cells and to provoke keratoconjunctivitis, and secretion of several polypeptides, including IpaB and IpaC, was abolished in these mutants.[37,38] The *ipgD* and *ipgF* nonpolar mutants displayed the same invasive phenotype as the wild-type strain, in contrast to the noninvasive phenotype of mutants carrying transposon insertions in this region of the virulence plasmid of *S. flexneri* 2a.[39] This difference between the phenotypes of the two types of mutants is likely to reflect the polar and nonpolar nature of the mutations.

Acknowledgments

We thank Marcia B. Goldberg and Claude Parsot for their advice and careful reading of this document.

[37] A. Allaoui, P. J. Sansonetti, and C. Parsot, *Mol. Microbiol.* **7,** 59 (1993).
[38] A. Allaoui, P. J. Sansonetti, and C. Parsot, *J. Bacteriol.* **174,** 7661 (1992).
[39] A. Allaoui, R. Ménard, P. J. Sansonetti, and C. Parsot, *Infect. Immun.* **61,** 1707 (1993).

[37] Isolation of *Salmonella* Mutants Defective for Intracellular Survival

By FRANCES BOWE and FRED HEFFRON

Traditionally, the study of bacterial pathogenicity has encompassed the disciplines of bacteriology, immunology, and histopathology. In recent years, however, these studies have broadened to include investigations, at the molecular level, of the interaction of bacteria with host cells during infection.[1–3] The use of attenuated bacterial mutants in various *in vivo* and *in vitro* disease model systems has led to enormous advances in the understanding of the mechanisms underlying bacterial survival *in vivo*, especially within cells.[4–6] This is now a burgeoning area of investigation which requires familiarity with molecular biological techniques, cell culture, and microscopy.

For a number of years, a major focus of research in this laboratory has been elucidation of some of the mechanisms employed by *Salmonella typhimurium* to facilitate its survival and growth within murine macrophages. We have predominantly used attenuated mutants in our work, although we have also made some observations regarding the intracellular behavior of wild-type *S. typhimurium*. With insights acquired by past and present laboratory members, we have written this article with the hope that it may be of some help to investigators who are interested in bacteria–host cell interactions and particularly those whose experience of working with tissue culture models of bacterial infections may be limited. We describe a possible strategy one might use to isolate mutants of *S. typhimurium* that are unable to survive inside macrophages and subsequently to investigate the precise effect of the mutation on the kinetics of intracellular survival. Although the particular examples we discuss relate to our work with *Salmonella,* many of the general principles and indeed the overall strategy could be applied to the investigation of any bacterial pathogen.

Generation of Bacterial Mutant Bank

Selection of Appropriate Mutagen

The first step is to choose an appropriate random mutagenesis system. A number of options are available, some of which are briefly described below.

[1] N. Buchmeier and F. Heffron, *Infect. Immun.* **59,** 2232 (1991).
[2] L. G. Tilney and D. A. Portnoy, *J. Cell. Biol.* **109,** 1597 (1989).
[3] P. J. Sansonetti, A. Ryter, P. Clerc, A. T. Maurelli, and J. Mounrier, *Infect. Immun.* **51,** 461 (1968).
[4] J. L. Gaillard, P. Berche, C. Frehel, E. Gouin, and P. Cossart, *Cell* (*Cambridge, Mass.*) **65,** 1127 (1991).
[5] R. R. Isberg, *Infect. Immun.* **57,** 1998 (1989).
[6] B. B. Finlay, M. N. Starnbach, C. L. Francis, B. A. D. Stocker, S. Chatfield, G. Dougan, and S. Falkow, *Mol. Microbiol.* **2,** 757 (1988).

Historically, bacterial mutagenesis experiments were performed using chemical mutagens.[7,8] The major advantages of this approach are that (1) the mutations occur randomly in the chromosome and (2) they are quite subtle, i.e., mutants arise as a result of changes at specific bases rather than by the gross changes associated with insertion of large elements into the genome. Chemical mutagenesis might be useful if one were interested in identifying mutations that result in an enhanced level of intracellular survival compared with the wild-type, as mutants can be selected directly. The disadvantages of using chemical mutagens are that (1) there is no way to guard against the occurrence of multiple mutations in the same bacterium and (2) there is no direct method for identifying individual mutations. In such cases, identification of the relevant gene requires a time-consuming and sometimes difficult screen. Therefore, most genes involved in intracellular survival can more readily be studied using transposon insertions.

Transposons are very useful for a number of reasons; when the appropriate conditions are used, each mutant bears a single, independently derived insertion. Furthermore, transposons also carry a wide variety of selectable antibiotic markers, greatly simplifying subsequent manipulations. The choice of a particular antibiotic determinant should be considered in the light of likely subsequent constructions. Other factors worth considering include the insertion specificity of the transposon and the cellular location of the mutated gene product.

The bacteriophage-derived transposon Mu is thought to insert most randomly in the chromosome, i.e., without apparent "hot" or "cold" spots. Thus, employment of one of the Mu derivatives increases the probability that every gene will be mutagenized at least once. We have found the mutagenesis system described by Hughes and Roth, based on the use of a transposition-defective Mu derivative, MudJ,[9] extremely useful. A further advantage is that there is an excellent transposon delivery system based on P22 transduction, which reduces the possibility of isolating siblings. Essentially, the system is based on a strain of *S. typhimurium*, TT10288, that contains two Mud transposons, MudJ (inserted in *hisD*) and MudI (inserted in *hisA*). These transposons are separated by approximately 4 kb of chromosome containing the *hisB, hisC,* and *hisH* genes. MudJ carries kanamycin resistance and is defective for transposition; i.e., it does not encode transposase and therefore cannot mediate its own

[7] E. A. Edelberg, M. Mandel, and G. C. C. Chen, *Biochem. Biophys. Res. Commun.* **18,** 788 (1965).

[8] B. N. Ames and H. J. Whitfield, *Cold Spring Harbor Symp. Quant. Biol.* **31,** 221 (1966).

[9] K. T. Hughes and J. Roth, *Genetics* **119,** 9 (1988).

transposition. MudI has an ampicillin resistance gene and encodes transposase. A key feature of the system is that P22 cannot simultaneously package both transposons in their entirety because of size constraints. A P22 lysate, grown on TT10288, can be used to transduce a wild-type strain with selection for kanamycin-resistant, ampicillin-sensitive transductants. MudJ is then stably maintained in the chromosome as it cannot mediate further transposition events.

Two minor problems have arisen in our hands using this procedure; at a high multiplicity of infection (m.o.i.), some of the mutants generated may contain two transposon insertions. This can be detected by Southern blotting using Mu as a probe, and each mutation can be isolated by retransducing the parental strain with a lysate made from the mutant. We have also observed that some of the mutants generated have a rough lipopolysaccharide (LPS) which is unrelated to the transposon insertion. The importance of retransduction of interesting mutations into the smooth, parental strain is therefore doubly important.

There are, however, cases where Mu may not be the best option. The identification of genes that encode envelope or secreted proteins, for example, is greatly simplified by the use of another system, Tn*phoA* mutagenesis.[10] Many of the genes that influence intracellular survival encode such proteins.[11,12] Tn*phoA* contains the *Escherichia coli* alkaline phosphatase gene with a deletion of the upstream 5′ region that includes translational start signals and signal sequence. The transposon is arranged so that in-frame fusion with a secreted protein or the periplasmic loop of a cytoplasmic membrane protein results in alkaline phosphatase activity when the cognate gene is expressed. The transposon is delivered using pRT733, a pJM703.1-based "suicide" plasmid, which can be mobilized into the strain under study by RP4, but cannot replicate outside a specific *E. coli* host which supplies replication functions *in trans*.[13] The host strain, SM10 (λ *pir*), contains RP4 in the chromosome and a prophage encoding the *pir* gene which allows replication of the plasmid.[14] Antibiotic-resistant transconjugants (kan^r) arise by transposition of Tn*phoA* to the chromosome of the new host, and events that result in fusions to membrane proteins can be identified by their blue color on indicator plates containing X-phos (5-bromo-4-chloro-3-indoyl phosphate *p*-toluidine salt). Dougan

[10] C. Manoil and J. Beckwith, *Proc. Natl. Acad. Sci. U.S.A.* **82,** 8129 (1985).

[11] W. S. Pulkkinen and S. I. Miller, *J. Bacteriol.* **173,** 86 (1991).

[12] B. J. Stone, C. M. Garcia, J. L. Badger, T. Hassett, R. I. F. Smith, and V. L. Miller, *J. Bacteriol.* **174,** 3945 (1992).

[13] R. K. Taylor, C. Manoil, and J. J. Mekalanos, *J. Bacteriol.* **171,** 1870 (1989).

[14] R. Simon, U. Preifer, and A. Puhler, *Bio Technology* **1,** 784 (1983).

and collaborators[15] have found a high frequency of mutations that affect *Salmonella* virulence (10%) among random, unselected, Tn*phoA* fusions expressing alkaline phosphatase. A lower frequency of insertions affecting virulence is observed using MudJ, presumably because these insertions can be located in any gene.[16]

We have found two problems arise using Tn*phoA* to mutagenize *Salmonella*: The first is that *Salmonella* already contains a phosphatase (the product of *phoN* is an acid phosphatase) and therefore appears blue on indicator plates. The level of PhoN activity, however, varies considerably from one strain to another (V. Miller, personal communication, 1991). There are two possible ways to overcome the problem, either by using a strain that is low in endogenous alkaline phosphatase activity or, as we have done, by constructing a $phoN^-$ mutant strain to use as the parent in the mutagenesis.[17] The second problem with the system is that Tn*phoA* and other derivatives of Tn*5* have a relatively high insertion specificity. There are, however, numerous examples of Tn*phoA* insertions, not just within a specific membrane protein but within each transmembrane domain of the protein.[18,19] This suggests that the transposon may insert more randomly than was previously believed.

Several other transposition systems can be used for mutagenesis of *Salmonella,* including excellent Tn*10* derivatives specifying resistance to many different antibiotics, including chloramphenicol, tetracycline, ampicillin, and kanamycin.[20] We have previously constructed mutations with Tn*10*, using an F′::Tn*10* that is temperature-sensitive for replication, to deliver the transposon.[21]

Potential Difficulties with Mutagenesis

Three difficulties may be encountered in carrying out transposon mutagenesis. The first is the isolation of siblings. The advantage of the MudJ system is that this rarely occurs; a bacteriophage P22 lysate is made from a strain containing Mu which is then introduced into a recipient strain by infection with this P22 stock. Because the transductants do not grow until after they have been plated on solid medium, there is essentially no chance

[15] I. Miller, D. Maskell, C. Hormaeche, K. Johnson, D. Pickard, and G. Dougan, *Infect. Immun.* **57,** 2758 (1989).

[16] F. Bowe, J. Lipps, R. Tsolis, and F. Heffron, unpublished results, 1994.

[17] P. I. Fields, E. A. Groisman, and F. Heffron, *Science* **243,** 1059 (1989).

[18] H. Y. Song and W. A. Cramer, *J. Bacteriol.* **173,** 2935 (1991).

[19] J. B. van Beilen, D. Penninga, and B. Witholt, *J. Biol. Chem.* **267,** 9194 (1992).

[20] N. Kleckner, J. Bender, and S. Gottesman, this series, Vol. 204, p. 139.

[21] P. I. Fields, R. V. Swanson, C. G. Haidaris, and F. Heffron, *Proc. Natl. Acad. Sci. U.S.A.* **83,** 5189 (1986).

of isolating siblings. With other delivery systems, principally plasmids, there is a much greater chance of isolating identical mutations.

A second potential problem is instability of the transposon insertion. Transposons that encode their own tranposases, including essentially all those that occur naturally, may give rise to multiple insertions which can lead to rearrangements in the mutant. We have previously used the unmodified Tn*10* for mutagenesis experiments and have experienced such problems. The advantage of the newer derivatives is that the mutations are stable as the transposon does not contain a transposase and cannot therefore rearrange.

The third pitfall is the occurrence of multiple insertions in a single bacterium. We have experienced this problem when carrying out the Hughes mutagenesis procedure, although this was most likely due to use of an inappropriate multiplicity of infection, as mentioned previously.

Storage of Mutant Banks

We do not store entire mutant banks, as we have found it easier to generate new ones with each mutant hunt; however, if storage at −70° is preferred, glycerol at concentrations of 15–50% and dimethyl sulfoxide (DMSO) at 15% are appropriate. Storage of avirulent mutants in agar stabs should, however, be avoided. We have found that our *Salmonella* strains spontaneously lose virulence if they are stored for prolonged periods in this manner.

Evaluation of a Mutant Bank

Once the mutagenesis is completed, it is important to test whether a representative and random bank has been generated. If chemical mutagens are used, then reversion of the appropriate auxotrophic marker, perhaps from one of the Ames tester strains, should be used as a positive control.[22] If mutagenesis was carried out using transposons, the frequency and distribution of auxotrophic mutations should be checked and compared with those found by other workers. For example, Hughes and Roth[9] report a 5% frequency of nonhistidine auxotroph isolation.

Identification of Mutants Unable to Survive within Macrophages

Prescreening

Before the screening commences, it is useful first to eliminate those mutants that have been generated as an artifact of the mutagenesis sys-

[22] J. McCann, E. Choi, E. Yamasaki, and B. N. Ames, *Proc. Natl. Acad. Sci. U.S.A.* **72,** 5153 (1975); J. McCann and B. N. Ames, *ibid.* **73,** 950 (1976).

tem, together with any that are likely ultimately to prove either uninteresting or irrelevant. For example, when using the Hughes system, one can expect a high frequency of homologous recombination into the *his* operon. Hughes and Roth[9] report that such recombinants are all phenotypically His$^-$. The MudJ cassette in the original strain is inserted in *hisD*; thus, some of these chromosomal sequences are packaged by P22 while the transducing lysate is being prepared. During the mutagenesis, therefore, a bias toward recombination of the Mud elements into the *hisD* gene of *S. typhimurium* is observed. In our experience, the incidence of this event can be as high as 50% and, thus, prescreening or plating on minimal medium to eliminate these *his* auxotrophs is a necessary prerequisite to screening for any other phenotypes. This can be accomplished by replica plating on minimal medium with and without histidine.

The color-based selection of Tn*phoA* mutants allows preselection of potentially interesting mutants; i.e., a blue color indicates insertion of the transposon into a gene encoding an envelope or secreted protein so that white colonies can immediately be discarded. It should, however, be kept in mind that not all genes are expressed on standard laboratory medium and growth under conditions that mimic the intracellular milieu may be necessary for expression of some.[23]

Designing Suitable Screening Procedure

In choosing a cell type to use for screening a mutant bank there are a number of factors to consider. Cell lines are more convenient to use during the initial screening. The choice of cell line should take into account both the cell type and animal species from which it was derived. In our studies, for example, we use a mouse model of *Salmonella* infection, as the kinetics of infection for *Salmonella* in a mouse model have been well characterized.[24] It therefore makes most biological sense to use murine cells where possible in our experiments. In the absence of a suitable mouse cell line, cells derived from another natural host for the pathogen should be employed, particularly cells derived from tissues that are usually infected by *Salmonella*. Some cell lines that may be appropriate include J774 (mouse, monocyte–macrophage-like),[25] U937 (human, monocyte-like),[26] and Caco-2 (human, colon).[27]

[23] N. Buchmeier and F. Heffron, *Science* **248,** 730 (1991).
[24] P. B. Carter and F. M. Collins, *J. Exp. Med.* **139,** 1189 (1974).
[25] P. Ralph, J. Prichard, and M. Cohn, *J. Immunol.* **114,** 898 (1975).
[26] C. Sundstrom and K. Nilsson, *Int. J. Cancer* **17,** 565 (1976).
[27] J. Fogh, *J. Natl. Cancer Inst.* (*U.S.*) **58,** 209; **59,** 221 (1977).

Another important consideration is how the assay will be scored; i.e., how far reduced should its survival level be, relative to the wild type, before a mutant is considered to have an impaired ability to survive intracellularly? We have found the best strategy is to set this level quite low (i.e., a twofold difference in survival compared with the wild type) during the first round of screening and then to raise it somewhat later. This avoids the possibility of missing a potentially interesting mutant.

The procedure for screening in J774 macrophages is outlined below, together with some information on the general care of the cells. We obtain tissue culture media and supplements from GIBCO (Gaithersburg, MD) and sera from Hyclone (Logan, UT).

Maintenance of J774 Macrophages. The growth medium we use for J774 cells is made with Dulbecco's modified Eagle's medium (DMEM, containing 4500 mg/ml glucose), supplemented with 10% (v/v) heat-inactivated equine serum, 1% (v/v) nonessential amino acids (Gibco stock solution), 1% (v/v) glutamine (Gibco stock solution). It is filter-sterilized and stored at 4°.

We routinely maintain cells in 75-cm^3 tissue culture flasks (Costar, Cambridge, MA, or Falcon, Lincoln Park, NJ) or in plastic petri dishes at 37° with 5% CO_2. Cells do not adhere as well to the latter and are therefore easier to remove by simply scraping the monolayer. The tissue culture medium should be changed at least twice per week, and when necessary, the density of growth can be adjusted by removing some cells. We usually move cells to a new flask or dish every 2 to 3 weeks. Frozen cell stocks are maintained at −150° or in liquid nitrogen, using 40% DMEM, supplemented with 50% heat-inactivated fetal calf serum (FCS) and 10% DMSO.

Assessment of Macrophage Function. A precaution that should be taken when using J774 cells is to check periodically that they are capable of generating an oxidative burst, i.e., production of reactive oxygen intermediates following exposure to pathogens. This is a property of the cells that can be lost after passage *in vitro*. If cells that are unable to generate a respiratory burst are used for screening, therefore, interesting mutants may be missed. The following is a convenient assay for oxygen radical production and is essentially that described by Damiani *et al.*[28] (B. Bloom, personal communication).

1. Seed 24-well plates with 5×10^6 (in a 0.5-ml volume) cells per well in tissue culture medium and allow to adhere overnight at 37°.

[28] G. Damiani, C. Kiyotaki, W. Soeller, M. Sasada, J. Peisach, and B. R. Bloom, *J. Exp. Med.* **152,** 808 (1980).

2. Wash cells with warm (37°) phosphate-buffered saline, pH 7.4 (PBS).
3. Add 1 ml of nitroblue tetrazolium (NBT, Sigma, St. Louis, MO) and 1 or 0.1 μg/μl phorbol myristate acetate (PMA) solution.
4. Cover the plate with aluminum foil and incubate at 37° for 1 hr.
5. Wash plates with warm PBS and fix for 1 hr in 3.7% cold acetaldehyde.
6. Count as positive those cells that contain a dark precipitate. Note that the control wells without PMA should yield approximately 5% positive cells.

NBT SOLUTION. 25% fetal calf serum, 75% Krebs–Ringer phosphate with glucose (KRPG) (Sigma), 20 mg/40 ml NBT. Dissolve with vigorous mixing at 37° for 30 min and filter-sterilize. (This solution should be freshly made up each time.)

PMA SOLUTION. Stock: 10 mg/ml in DMSO, store at −70° in 20-μl aliquots. For the assay, dilute to a final volume of 200 μl (1 μg/μl) or 2 ml (0.1 μg/μl) with PBS.

Screening in Professional Phagocytes. The screening procedure, described by Fields *et al.*[21] is best carried out in 96-well plates (Fig. 1). J774 cells are grown to confluence in petri dishes (150 mm), each of which normally yields approximately 10^7 cells.

1. Seed a 96-well plate with 2×10^5 cells per well in antibiotic-free tissue culture medium and incubate at 37° for 2 hr to allow cells to adhere.
2. Replace the tissue culture supernatant with fresh medium (without antibiotics) after washing each well with Hanks' balanced salt solution.
3. Centrifuge the bacterial mutants onto the cells (10^5 organisms in 50-μl of DMEM, centrifuge at 250 *g* for 10 min).
4. After incubation at 37° for 1 hr, add gentamicin at 200 μg/ml to each well in a volume of 200 μl of DMEM.
5. After 2 further hr of incubation, replace the supernatant with DMEM containing 10 μg/ml gentamicin.
6. Following overnight incubation, remove the supernatants and lyse the macrophages with 0.5% sodium deoxycholate in water or normal saline.
7. Plate an aliquot of the lysis mixture on LB agar plates containing the appropriate antibiotic. Mutants with a survival rate at least twofold less than that of the wild-type parent can be used for the next round of testing.

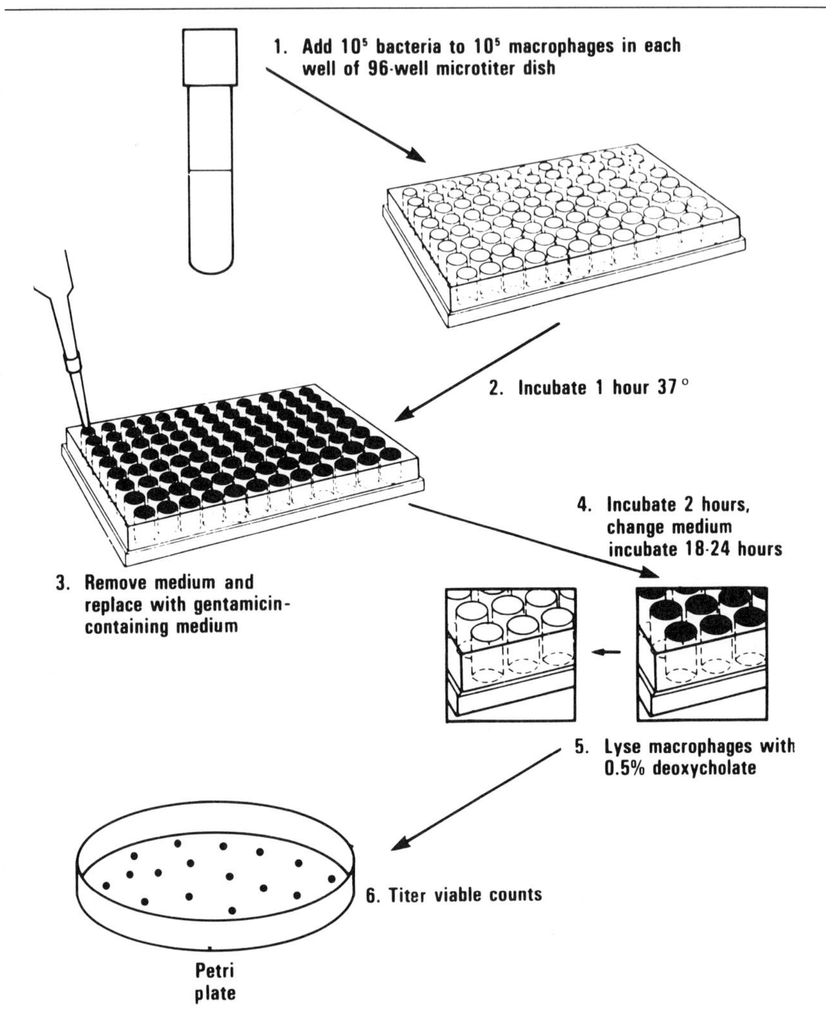

FIG. 1. Identification of *Salmonella typhimurium* mutants that are unable to survive within macrophages.

Verification of Mutant Phenotype

When mutants that are apparently unable to survive inside cells have been identified, further precautions should be taken before they are characterized in detail. It is important to eliminate those mutants whose apparent inability to survive intracellularly is actually an artifact of the screening procedure. A number of tests may be carried out, depending on the particular screening system used. Some of these are described below.

TABLE I
SENSITIVITY OF *Salmonella typhimurium* MudJ MUTANTS TO SODIUM DEOXYCHOLATE *IN VITRO*[a]

Strain	Survival ratio[b]	Sensitivity
CL 230	22.41	S
CL 278	2.52	I
CL 520	0.85	R
CL 547	1.19	R
14028s	0.90	R

[a] For each mutant, 10^7 bacteria/ml were incubated in parallel in either PBS or 0.5% sodium deoxycholate in PBS, at 37° for 2 hr. The viability in each tube was then determined by plating. R, Resistant; I, intermediate; S, sensitive.
[b] Results for each mutant are expressed as survival in PBS/survival in sodium deoxycholate.

1. If a mutant were sensitive to the lysing agent used it would be recovered in low numbers, falsely suggesting an inability to survive inside cells. To eliminate this possibility, all mutants that are apparently unable to survive inside cells should be rechecked using at least one other lysis method, e.g., water. Mutants may also be incubated with the lysing agent and then plated to check for sensitivity. Table I shows the results of an experiment to test the sensitivity of a group of MudJ mutants to sodium deoxycholate. All of these mutants initially appeared sensitive to the bactericidal effects of J774 macrophages in assays where deoxycholate was used to lyse cells. In the case of CL 230, however, it is clear that the apparent inability of the mutant to survive within macrophages was in fact a result of its sensitivity to the lysis agent used in the assays.
2. A reduced growth rate is another possible reason for the apparent inability of a mutant to survive intracellularly. This possibility can be tested very simply by comparing the *in vitro* growth rates of the mutant and parental strain in Luria–Bertani (LB) broth.
3. An important precaution is to check efficient uptake of bacteria into cells at the start of the assay, as a poor survival rate at 24 hr may also be attributable to the low numbers of bacteria taken up by cells initially. This can be done by taking a sample at the zero time point.
4. It has been shown that cells in culture can sequester the antibiotics used in their routine maintenance.[29] As *Salmonella* is somewhat resistant to streptomycin, there is a possibility that a given mutant may simply be

[29] P. M. Tulkens, *Eur. J. Clin. Microbiol. Infect. Dis.* **10,** 100 (1991).

more susceptible to intracellular antibiotics and not to the intracellular environment as such. To eliminate this possibility a macrophage survival assay can be performed on all mutants, using cells that have been maintained in the absence of antibiotics for several days. In addition, the minimum inhibitory concentration (MIC) of antibiotics to which mutants are exposed should be determined for each and compared with that of the parent.

Confirmation of Mutant Phenotype

There are two final tests that must be carried out on what should, at this stage, be a much reduced group of putative intracellular survival mutants. The first is to retransduce all mutations into the wild-type background. This confirms that the mutant phenotype results from the transposon insertion. The second is to perform Southern analysis using the transposon as a probe to verify that each mutant is a result of a single transposon insertion and also to eliminate duplicates from the bank.

Detailed Characterization of Intracellular Survival Mutants

Once a bank of bona fide intracellular survival mutants has been established, detailed analysis of each can begin. The function of a mutated gene can be investigated using two complementary approaches, genetic and phenotypic.

Genetic Mapping

In the absence of DNA sequence data, it is still possible to obtain genetic clues to the identity of a gene by defining its location on the bacterial chromosome. Defining a map location that is identical, or at least very close, to that of a known gene can indicate the direction further experiments should take. Traditional transductional mapping experiments can be carried out, although for *S. typhimurium* mutants two simpler approaches are available. One of these is a system developed by Benson and Goldman[30] in conjunction with the current physical map. The Benson system is based on a series of *S. typhimurium* mapping strains, each of which bears a MudP22 hybrid inserted at a defined location in the chromosome. The phage is defective for tail formation and is "locked" in position on the chromosome; however, on induction with mitomycin C, the phage begins to package adjacent DNA. It is capable of packaging as much as a 3-min interval (approximately 100 kb), although the efficiency

[30] N. R. Benson and B. S. Goldman, *J. Bacteriol.* **174,** 1673 (1992).

of packaging decreases with distance from the insertion. Thus, 70% of packaged DNA is from the first minute, the second constitutes 25% of the total, and the remainder is from the third minute. DNA can be extracted from the phage and transferred to a nylon filter. By use of a small fragment of the mutated gene of interest, the ordered DNA can be probed by Southern hybridization and the map location of the gene determined with reasonable accuracy. A further advantage of this system is that the DNA isolated from the mapping strains can later be used to clone the gene identified.

The second procedure for mapping *Salmonella* mutants was developed by Liu and Sanderson[31] and by Wong and McClelland[32] and uses pulsed-field gel electrophoresis (PFGE). The approach these groups took exploits the relatively infrequent occurrence of *Xba*I and *Bln*I sites in the *Salmonella* genome. Thus, clear reference digestion patterns have been obtained for *S. typhimurium* with each of these enzymes following PFGE. The correspondence between the digestion fragments and their location on the physical map of *S. typhimurium* has also been determined. If the transposon used to generate the mutants also contains *Xba*I and *Bln*I sites (Tn*10* contains both) or if the sites can be introduced into the transposon, the restriction digest of mutant chromosomal DNA would be different from that of the parent; i.e., one would expect disappearance of the restriction fragment containing the transposon with the concomitant appearance of two new fragments. Comparison of parental and mutant digests therefore identifies the position of the transposon and, thus, the location of the mutated gene.

Phenotypic Analysis

Determination of the *in vitro* phenotypes of a mutant in a variety of functional assays is useful in ascribing a role in pathogenicity to the gene in question. A number of convenient tests can be carried out before more laborious assays of the interaction of mutants with cells are attempted. These assays also eliminate mutants that may not be so interesting, such as LPS mutants and auxotrophs.

Preliminary Assays of Mutant Phenotypes

1. The first and most convenient phenotype to test is auxotrophy, as mutations in aromatic amino acid and purine biosynthesis pathways, for example, have been shown to reduce the virulence of *Salmonella* species and have also been associated with a reduced ability to grow intracellu-

[31] S. T. Liu and K. E. Sanderson, *J. Bacteriol.* **174,** 1662 (1992).

[32] K. K. Wong and M. McClelland, *J. Bacteriol.* **174,** 1656 (1992).

larly.[33] Streaking mutants out on minimal agar plates (made with Noble agar) demonstrates whether they require supplements for growth. If a mutant appears auxotrophic the nature of the auxotrophy can be determined by replica-plating mutants on defined medium in the presence or absence of particular supplements. The details of many of the appropriate tests are described elsewhere.[34]

2. The role of the virulence plasmid of *S. typhimurium* during infection is not fully understood although it is required for full expression of virulence.[35] Loss of the plasmid should be checked by performing a plasmid extraction according to the method of Kado and Liu[36] and comparing the plasmid profiles of the mutant with those of the parental strain.

3. An incomplete LPS usually results in avirulence of *Salmonella* species.[37] The level of O side-chain expression is a very important determining factor in this association.[38,39] Rough mutants may also have a reduced ability to survive inside macrophages.[40] LPS profiles can be checked by silver-staining polyacrylamide gels of boiled, proteinase K-treated whole cell lysates.[41,42] A quick test can also be carried out by checking for plaque formation by bacteriophage P22.[35] This phage is smooth strain specific, but it is still possible to observe some plaque formation if the mutant strain is only slightly rough. Very rough strains are quite sensitive to detergents and these can be detected by plating mutants on solid medium incorporating detergents such as bile salts and checking for particularly poor growth.

Interaction of Mutants with Eukaryotic Cells. The interaction of mutants with eukaryotic cells can be examined on three levels: (1) the kinetics of the interaction (i.e., the numbers of bacteria that survive together with their location in the cell), (2) the identification of the particular cell components involved in the inhibition survival of a given mutant, and (3) the regulatory signals that normally control expression of the inactivated gene.

[33] D. O'Callaghan, D. Maskell, F. Y. Liew, C. S. F. Easmon, and G. Dougan, *Infect. Immun.* **56,** 419 (1988).

[34] R. W. Davis, D. Botstein, and J. R. Roth, "Advanced Bacterial Genetics." Cold Spring Harbor Lab., Cold Spring Harbor, NY, 1980.

[35] G. W. Jones, D. K. Robert, D. M. Svinarich, and H. J. Whitfield, *Infect. Immun.* **38,** 476 (1982).

[36] C. I. Kado and S. T. Liu, *J. Bacteriol.* **145,** 1365 (1981).

[37] M. Nakano and K. Saito, *Nature (London)* **222,** 1085 (1969).

[38] R. J. Roantree, *Annu. Rev. Microbiol.* **21,** 443 (1967).

[39] M. B. Lyman, J. P. Steward, and R. J. Roantree, *Infect. Immun.* **13,** 1539 (1976).

[40] N. Buchmeier and F. Heffron, *Infect. Immun.* **57,** 1 (1989).

[41] C. M. Tsai and C. E. Frasch, *Anal. Biochem.* **119,** 115 (1982).

[42] P. J. Hitchcock and T. M. Brown, *J. Bacteriol.* **154,** 269 (1983).

KINETICS OF INTERACTION. Although the initial screen identifies mutants that are defective for intracellular survival after overnight incubation, the precise kinetics of this interaction should be examined in detail. This can be achieved by isolating bacteria from cells over time and determining the level of survival by plating. The assay used in this laboratory to determine the kinetics of intracellular survival is very similar to the original screening procedure described earlier.

Microscopy studies can also be performed to complement these kinetic experiments. Many characteristic features of *Salmonella* survival within macrophages have been identified in this way; e.g., Buchmeier and Heffron[40] used electron microscopy to compare the survival of wild-type *S. typhimurium* in different populations of macrophages and determined that the bacteria are always found in membrane-bound phagosomes, even while dividing (Fig. 2). Thus, electron or confocal microscopy may be very useful in providing further clues to the block in intracellular survival of mutants.

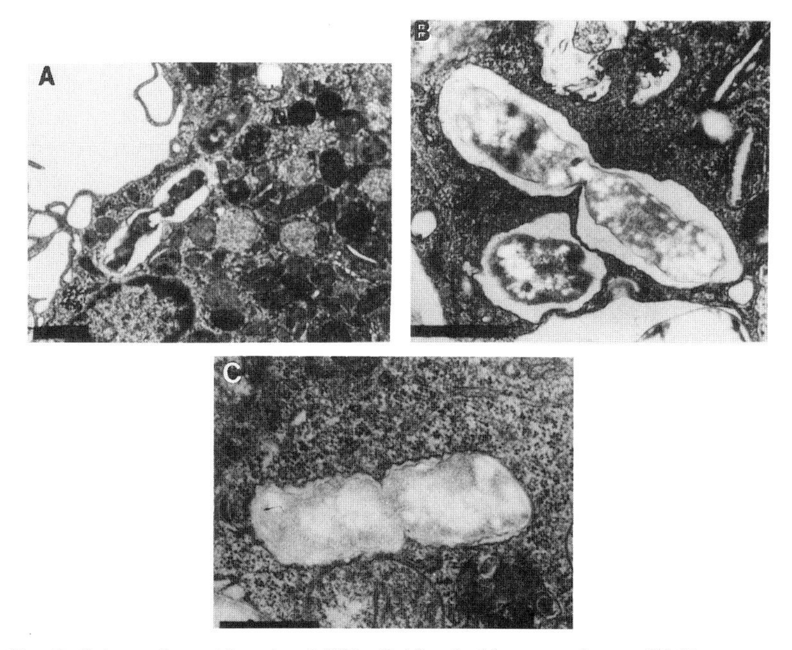

FIG. 2. *Salmonella typhimurium* 14028s dividing inside macrophages. (A) Bone marrow-derived macrophages (BALB/c), 14 hr postinfection. (B) Bone marrow-derived macrophages (SWR/J), 20 hr postinfection. (C) J774 macrophages, 4 hr postinfection. Bars: 1 μm. (Reproduced with permission from Buchmeier and Heffron.[40])

The phagocytes used can be from either a cell line or a primary culture. It is often useful to perform the assay with both for comparison, as it has been shown that *Salmonella* is more susceptible to the bactericidal effects of certain primary cell cultures than to cell lines; i.e., peritoneal macrophages are more efficiently bactericidal than those derived from the spleen and both are more effective at killing *Salmonella* than are J774 cells.[40] Primary cells can be isolated according to procedures outlined in Buchmeier and Heffron[40] and in [8].

1. Cell monolayers are prepared in 24-well plates by aliquoting a suspension of 5×10^5 cells per well in a 0.5-ml volume of antibiotic-free medium and allowing the cells to adhere and spread for 1–2 hr at 37° and 5% CO_2. (Immediately after the cells have been added to the plate, agitate very gently for a few seconds to ensure an even distribution in each well.) The formation of a confluent monolayer with complete spreading of the individual cells on the plastic is essential to the successful performance of the assay, as cells that have not adequately spread demonstrate a less efficient ability to phagocytize bacteria.
2. The macrophages are then infected with a washed suspension of opsonized bacteria at a multiplicity of infection of approximately 10 bacteria per cell. It is convenient to take 100 μl of an overnight broth culture, resuspend in 5 ml of PBS, and wash the cells by centrifugation. To opsonize the bacteria, the pellet should be resuspended in 20 μl of normal mouse serum and incubated at 37° for 15 min. The bacteria are then diluted to a volume of 10 ml with fresh tissue culture medium.
3. The tissue culture supernatant in each well is replaced with fresh antibiotic-free medium before infection of the monolayer with an aliquot of 250 μl of the bacterial suspension. Organisms are then centrifuged onto the cells at 250 *g* for 10 min and incubated for 20 min at 37°. Between 10 and 20% of the bacterial inoculum is phagocytized depending on the precise multiplicity of infection used.
4. Monolayers are then washed three times with PBS to remove non-cell-associated bacteria, and fresh tissue culture medium containing 12 μg/ml of gentamicin is added to kill all extracellular organisms. This is the zero time point in the assay.
5. We collect samples at 0, 1, 4, and 20 hr. Sample collection is carried out by first aspirating the supernatant off the monolayer and then lysing the cells with 0.5 ml of either a detergent (e.g., 1% deoxycholic acid, 1% Triton X-100) or water. The well is then rinsed with 0.5 ml of PBS. The lysate and rinse from each well are pooled and stored on ice before plating.

These kinetic studies may highlight the mechanism underlying the clearance of the bacteria from the cells; e.g., if a particular mutant were rapidly cleared with no apparent growth or plateau phase, it might suggest that it was unusually susceptible to an early host defense mechanism such as oxidative killing.

IDENTIFICATION OF PARTICULAR CELL COMPONENTS. Particular components of the macrophage can be isolated and the sensitivity of the mutant to each can be determined. For example, it is possible to prepare extracts of cytoplasmic granules and to test the sensitivity of mutants to these.[43] It may not always be necessary to isolate subcellular fractions. For example, sensitivity to toxic oxygen intermediates produced during the respiratory burst could be tested by comparing the growth kinetics of mutants in normal macrophages with those in macrophages that cannot generate an oxidative burst, e.g., D9.[44]

REGULATORY SIGNALS. It may also be useful to compare one- and two-dimensional outer membrane protein gels of the parent and mutant strains to determine whether the mutation results in the loss of one important protein or whether the insertion is in a regulator, leading to the loss of several proteins involved in survival.[23,45] Protein profiles should be checked for mutants grown *in vitro* and within macrophages.[23]

Once a given mutation has been identified, it is still necessary to verify that the phenotype of the mutant is a direct result of the mutation. This can be achieved by using a molecular version of Koch's postulates, i.e., to complement the mutation by introducing a copy of the cloned gene into the mutant and determining that the resulting phenotype is wild type.

Final Remarks

Once a potentially interesting macrophage survival gene has been identified, *in vitro* assays can provide strong and useful predictive indications of its function *in vivo*. It is, however, dangerous to draw firm conclusions from *in vitro* data, without strong supportive evidence from *in vivo* models; e.g., from gut pathogen data, it is presumably easier to invade a cell monolayer in a stationary bath of tissue culture fluid than to overcome the considerable obstacles encountered *in vivo*, such as a churning intestinal millieu containing stomach acid, mucus, IgA and resident flora (and all this before an approach to a lining epithelial cell can even be attempted). Similarly, isolation *in vitro* of one component of the immune system, such as the macrophage, does not take account of interactions between macrophages and other immune defenses that occur *in vivo*.

[43] C. A. Parkos, R. A. Allen, C. G. Cochrane, and A. J. Jesaitis, *J. Clin. Invest.* **80,** 732 (1987).
[44] M. Goldberg, L. S. Belkowski, and B. R. Bloom, *J. Clin. Invest.* **85,** 563 (1990).
[45] K. Hantke, *Mol. Gen. Genet.* **182,** 288 (1981).

In an earlier section we dealt with the importance of carefully choosing cell lines. Although many general processes will be the same regardless of the cell type used, the origin of the line or primary culture is significant and must be taken into account when interpreting data. A study by Eisenhauer and Lehrer[46] describing differences in defensin profiles between murine and human neutrophils highlights this point. In this case for example, any conclusions drawn regarding bacterial susceptibility to the bactericidal effects of neutrophils would have to take account of these differences.

In conclusion, *in vitro* systems allow the identification of many genes involved in bacterial pathogenicity and can often suggest their role in infection. They may also help to elucidate the relative importance of different host cell types in resisting the progression of disease and thus highlight the stage of infection at which a mutant may be blocked by the host. Where possible, all such conclusions can be greatly strengthened by subsequent *in vivo* testing.

Acknowledgments

The authors express their appreciation to the members of the Heffron laboratory and to Dr. Nancy Buchmeier, Dr. Peadar O Gaora, and Dr. Vivian Hwa for their helpful comments and critical reading of the manuscript.

[46] P. B. Eisenhauer and R. I. Lehrer, *Infect. Immun.* **60,** 3446 (1992).

[38] Small Plaque Mutants

By SIAN JONES and DANIEL A. PORTNOY

Introduction

Since Dulbecco's[1] discovery of viral plaques on monolayers of tissue culture cells in 1952, this procedure has undergone a variety of modifications and has been used extensively as a tool to study viral pathogens. This technique has been employed for the study of intracellular bacterial pathogens such as *Shigella*.[2] In this article we describe the modifications to the plaque assay we have developed for use with the facultative intracellular bacterial pathogen *Listeria monocytogenes* and how this assay has

[1] R. Dulbecco, *Proc. Natl. Acad. Sci. U.S.A.* **38,** 747 (1952).
[2] E. V. Oaks, M. E. Wingfield, and S. B. Formal, *Infect. Immun.* **48,** 124 (1985).

been used to isolate mutants defective in intracellular growth and cell-to-cell spread.[3,6]

The technique can be used to screen populations of mutants or can be adapted to screen potential mutants individually. The former technique has the advantage of screening large numbers of plaques, whereas the latter has the advantage of being able to detect mutants that are unable to plaque or that have decreased plaque-forming efficiency. These two techniques are described separately.

Method 1: Screening Populations of Bacteria

Preparation of Tissue Culture Cells

A variety of adherent cells can be used, but for large, stable, homogenous plaques we use mouse L2 fibroblasts.[4] Our cell line was obtained from Susan Weiss (University of Pennsylvania School of Medicine, Philadelphia). Cells are grown in Dulbecco's Modified Eagles medium (DME/High, JRH Biosciences, Lenexa, KS) supplemented with 10% (v/v) fetal calf serum, 292 μg/ml glutamine, 10 μg/ml streptomycin, and 100 U/ml penicillin. Cells are passaged at 1/10 in T-flasks until ready for use. Approximately 4 days before infection, a confluent monolayer of cells is removed by trypsinization and resuspended in 10 ml of the same medium without antibiotics. One milliliter of this suspension of cells is added to 11 ml of medium without antibiotics in a 100 $\times$ 20-mm petri dish (Falcon 3003, Becton Dickinson Labware, Lincoln Park, NJ). Cells are evenly distributed by gently swirling the dish. The cells are typically incubated at 37° and 5% CO_2 until confluent, although plaquing can also be performed using subconfluent monolayers.

Bacterial Strains and Growth Conditions

Listeria monocytogenes strain 10403S is the parental strain used in these experiments. It is a member of serogroup 1 and is resistant to streptomycin in concentrations up to 1 mg/ml. Transposon libraries of bacteria containing random insertions of a derivative of Tn*917*, Tn*917-lac*, were generated as previously described.[5] Permanent stocks of bacterial strains are stored in 50% Luria–Bertani medium/glycerol at −70°.

Overnight cultures for infection are prepared by inoculating bacteria from a frozen glycerol stock into 2 ml of brain–heart infusion broth (BHI,

[3] A. N. Sun, A. Camilli, and D. A. Portnoy, *Infect. Immun.* **58,** 3770 (1990).

[4] K. H. Rothfels, A. A. Axelrad, L. Siminovitch, F. A. McCulloch, and R. C. Parker, *Proc. Can. Cancer Res. Conf.* **3,** 189 (1959).

[5] A. Camilli, D. A. Portnoy, and P. Youngman, *J. Bacteriol.* **172,** 3738 (1990).

[6] M. Kuhn, M. C. Prévost, J. Mounier, and P. J. Sansonetti, *Infect. Immun.* **58,** 3477 (1990).

Difco Laboratories, Detroit, MI) in a 15-ml conical tube. The cultures are grown overnight at 30° lying flat to stationary phase (2×10^9 cells/ml). The next day, 1 ml of the culture is transferred to a 1.5-ml Eppendorf tube and subjected to centrifugation at 14,000 *g* in a microfuge for 1 min at room temperature. The supernatant is removed, and the pellet resuspended in 1 ml of phosphate-buffered saline, pH 7.4 (PBS), by vortexing vigorously. The bacteria are then pelleted again and suspended in 1 ml of fresh PBS.

Procedure for Infection

The following infection results in approximately 300–500 plaques per 100-mm dish. Thirty microliters of a 1 : 10 dilution, in PBS, of the washed bacterial suspension is used per petri dish. The bacteria are directly inoculated into the petri dish, uniformly distributed by gently swirling the dish, and incubated at 37° for 1 hr. It should be noted that L2 cells are relatively refractory to infection; hence, a large number of *L. monocytogenes* are added. The infection varies, however, with different cell types and bacterial strains. One hour after infection the medium is removed and the monolayer is washed three times with 5 ml of PBS that has been prewarmed to 37°. After the last wash, all of the PBS is removed, and 10 ml of DME in 0.7% (w/v) agarose containing 10 μg/ml gentamicin sulfate is added to each dish again with gentle swirling to ensure uniform distribution. The DME is made as a 2× stock containing 10% (v/v) fetal calf serum (FCS) and prewarmed to 37°. The agarose is kept as a 1.4% stock in distilled H_2O and is melted and kept in a 56° water bath until ready for use. Immediately prior to use, the 2× medium and agarose are combined in equal volumes along with gentamicin to a final concentration of 10 μg/ml and mixed. The agarose mixture should be allowed to cool until it is just warm to the touch, because if it is too hot it will kill the monolayer, and if it is too cool it will begin to solidify. Once the agarose has set, the petri dishes are again returned to the incubator where they are incubated right side up for 2 days to 1 week; however, we routinely incubate for 3 days.

Staining

After 3 days, the monolayers are stained with a 10-ml overlay of DME/agarose containing neutral red (GIBCO, Grand Island, NY). The agarose/DME mixture is prepared as previously described with the exception that gentamicin is omitted and neutral red is added to a final concentration of 200 mg/ml. The neutral red should be added last, because addition of the neutral red while the agarose is still hot tends to cause precipitation of the neutral red with resultant obscuring of the plaques.

Once the agarose overlay has set, the dishes are again returned to the incubator. Plaques begin to appear as clear zones on a background of red approximately 4–8 hr later. Since each plaque represents a focus of infection, mutants that are defective in intracellular growth or cell-to-cell spread form plaques of diminished size which may range from only modest reductions to barely visible pinpoint plaques. Plaques of interest may be picked using a sterile toothpick by inserting the toothpick through the agarose overlays into the center of the plaque. The toothpick is then used to inoculate a 2-ml culture of BHI which is grown as previously described.

Plaque Purification

To ensure that the plaques that have been picked are pure, repeated cycles of infection are performed in a manner analogous to that outlined above with the following modifications.

L2 cells are split from a confluent monolayer at a dilution of 1/5 in medium without antibiotics. Two milliliters of the cell suspension is then added per well to a 6-well cluster dish (Costar, Cambridge, MA). The monolayers should be confluent on the day of infection which takes approximately 2–4 days. Each individual well can be infected with 6 μl of a 1 : 10 dilution of washed bacteria. The rest of the procedure is as previously described except that 2 ml of PBS is used for each of the washes, 3 ml of DME/agarose/gentamicin is used for the first overlay, and 1 ml is used per well of the neutral red mixture. This procedure of plaque picking is continued until the plaques are purified to homogeneity. Figure 1A depicts

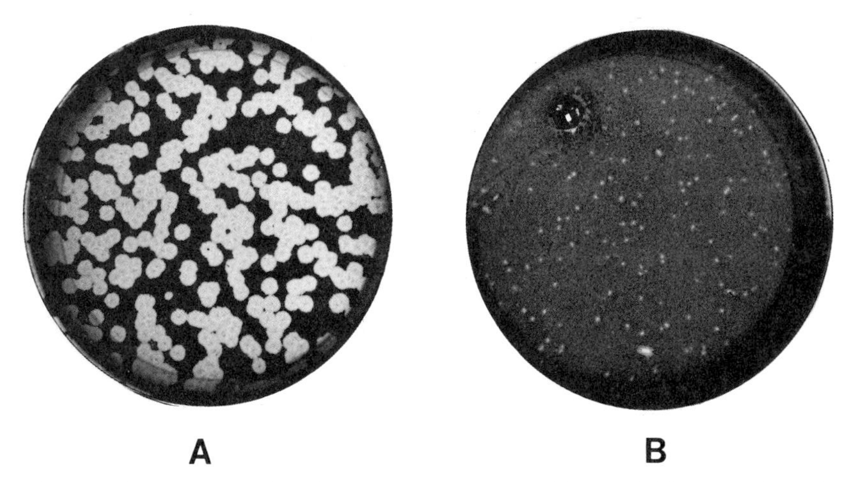

FIG. 1. Plaque formation by *Listeria monocytogenes* in L2 cells. (A) Wild-type bacteria. (B) Small plaque mutant. (Reproduced with permission from Sun *et al.*[3])

wild-type *L. monocytogenes* plaques, whereas Fig. 1B reveals infection by a small plaque mutant.

Method 2: Screening of Individual Colonies

As previously stated, the above protocol has been further adapted to enable individual colonies to be assayed for plaquing ability. This modification offers the advantage of being able to detect mutants that fail to plaque or those that have decreased plaquing efficiency. It should be noted that the modifications in this second protocol also obviate the need for washing the bacteria prior to infection.

Preparation of Cells

L2 cells are grown as previously described. Four days prior to infection they are passaged to a 24-well cluster dish (Costar, Cambridge, MA) by cutting cells at a dilution of 1/10 in medium without antibiotics. One milliliter of the cell suspension is then added to each of the wells. Cells are incubated as before at 37° and 5% CO_2 until confluent.

Bacterial Growth Conditions

Individual colonies are inoculated into 2 ml of 1/10 strength BHI in 13 × 100-mm disposable glass culture tubes (Fisher Scientific, Pittsburgh, PA). Cultures are grown upright, without shaking, at 30° overnight. *L. monocytogenes* remains viable at 4° for several months, and the tubes should be stored at 4° after the infection has been performed until the results of the plaque assay are known.

Infection and Staining

L2 cells should be confluent on the day of infection. Two microliters of the overnight bacterial culture is inoculated directly into the correspondingly numbered well. Uniform distribution of the bacteria is ensured by gently swirling the dish prior to incubation. One hour after infection, the tissue culture medium is removed and 0.5 ml of DME in 0.7% agarose containing 10 μg/ml of gentamicin sulfate (prepared as previously described) is added per well. It should be noted that the washing of the monolayers with PBS is omitted in this method. The dishes should be returned to the incubator after the agarose has set and incubated upright for the next 3 days. Three days after infection, the plaques are stained with 0.5 ml/well neutral red in DME/agarose prepared as previously outlined.

Plaques begin to appear over the next several hours but are optimally visualized at 12–14 hr.

Interpretation of Results

Approximately 20–30 plaques/well are obtained routinely using wild-type *L. monocytogenes*. If bacteria are defective in intracellular growth or cell-to-cell spread then the plaques in that well will be uniformly diminished in size, as each well is infected from a single clone of bacteria. Differences in size vary from barely detectable pinpoint plaques to those that are just slightly smaller than wild type. Although there is some variation in number of plaques per well, a 2- to 10-fold reduction should be considered significant. Culture tubes corresponding to wells with decreased plaque size or number should be saved. The others may be discarded at this point.

It is helpful to confirm the more subtle plaque size variations using 6-well cluster dishes because the differences are often more apparent using larger wells. L2 cells should be cut at 1/5 and 2 ml of the cell suspension added per well, as previously described. Each well is infected with 4 μl of an overnight culture (grown in 1/10 BHI) inoculated with the appropriate refrigerated specimen. Alternatively, the culture may be used directly from the refrigerator if stored for 1 week or less without significant loss in bacterial count; however, if the culture is several weeks old, then a slightly larger inoculum size may be needed to yield a similar number of plaques, i.e., 6 μl. Furthermore, if the culture tube is taken from the refrigerator, it is important to mix the tube well prior to removing the aliquot for infection because the bacteria tend to sediment with time. Confirmation that a mutant has decreased plaquing efficiency is best performed using 100 $\times$ 20-mm petri dishes and the protocol outlined under Method 1.

[39] Isolation of Hyperinvasive Mutants of *Salmonella*

By CATHERINE A. LEE and STANLEY FALKOW

Rationale

Genetic Techniques for Identification of Invasion Factors

Numerous techniques are used to characterize the properties of bacterial pathogens. These properties distinguish them from nonpathogenic bacteria and permit them to cause infection and disease which are described in this volume and in Volume 235 of this series. For example, the ability of certain pathogens to enter mammalian cells contributes to their pathogenicity. Several *in vitro* assays have been developed to study this process (see [26]–[35]). To elucidate the mechanism of invasion, as well as to establish the role of specific bacterial determinants by "molecular Koch's postulates," investigators have sought to identify mutations and genes that specifically affect invasion. Other articles in this section ([36]–[38], [40]–[42]) describe different experimental methods to identify invasion genes. For example, noninvasive mutants have been sought. Noninvasive mutations, by definition, alter genes that affect invasion; however, noninvasive mutations are commonly isolated by tedious screening techniques and may affect invasion nonspecifically. Also, invasion genes have been identified by their ability to confer the invasive phenotype to noninvasive *Escherichia coli* bacterial strains. Such cloned invasion genes are readily characterized by molecular techniques; however, it may not be possible to clone concomitantly all of the genes required for invasion from an invasive pathogen if they are not contiguous or not expressed in *E. coli*.

This article describes an alternative approach to identify genes involved in invasion. These methods allow positive selection of mutations that affect invasion and, in addition, are applicable to microorganisms that require many genes for invasion. Basically, the goal is to isolate bacterial mutants, called *hyperinvasive mutants*, that have increased ability to enter epithelial cells. This procedure could be used for the study of a variety of invasive pathogens, but the specific methods described here focus on the isolation of hyperinvasive *Salmonella* mutants.

Salmonella Invasion into Epithelial Cells

Salmonella cross the mucosal barrier during infection.[1,2] They also enter both polarized and nonpolarized cultured epithelial cells.[3,4] Traditional genetic approaches have identified many noninvasive *Salmonella* mutants.[5–7] Although certain noninvasive mutations appear to be specific for invasion, others appear to affect invasion nonspecifically by altering

[1] A. Takeuchi, *Am. J. Pathol.* **50,** 109 (1967).

[2] K. J. Worton, D. C. Candy, T. S. Wallis, G. J. Clarke, M. P. Osborne, S. J. Haddon, and J. Stephen, *J. Med. Microbiol.* **29,** 283 (1989).

[3] B. B. Finlay, B. Gumbiner, and S. Falkow, *J. Cell Biol.* **107,** 221 (1988).

[4] B. B. Finlay and S. Falkow, *J. Infect. Dis.* **162,** 1096 (1990).

[5] B. B. Finlay, M. N. Starnbach, C. L. Francis, B. A. Stocker, S. Chatfield, G. Dougan, and S. Falkow, *Mol. Microbiol.* **2,** 757 (1988).

[6] J. E. Galán and R. Curtiss, III, *Proc. Natl. Acad. Sci. U.S.A.* **86,** 6383 (1989).

[7] V. L. Miller and S. Falkow, *Infect. Immun.* **56,** 1242 (1988).

general properties of the bacterial cell surface or bacterial metabolism.[5] Four invasion genes from *Salmonella typhi* have been cloned by their ability to allow *E. coli* recombinants to enter cultured epithelial cells. Interestingly, the homologous genes from *Salmonella typhimurium* introduced into *E. coli* do not confer invasiveness.[8] It is not known if this difference is due to lack of expression of the *S. typhimurium* genes in *E. coli* or the failure of genes to encode homologous functions.

Salmonella invasiveness is regulated by osmolarity and the availability of oxygen during bacterial growth. High osmolarity and oxygen limitation increase *Salmonella* entry into cultured epithelial cells.[9-11] Bacterial growth state also affects invasion. *Salmonella* grown to stationary phase are severely compromised in their ability to enter epithelial cells.[9,10] The expression of many pathogenic properties of bacteria is regulated by environmental conditions.[12,13] In this way, bacteria appear to limit the expression of pathogenic factors to the appropriate time and location during infection. One site within the host at which *Salmonella* might grow under conditions of high osmolarity and low oxygen availability is the lumen of the small bowel. Thus, it is possible that the *Salmonella* invasion factors that are regulated by osmolarity, oxygen, and growth state are those that allow *Salmonella* to cross the epithelial barrier.

Identification of Invasion Factors by Isolation of Hyperinvasive Mutants

The modulation of *Salmonella* invasiveness by environmental conditions could be due to the transcriptional regulation of essential invasion factors. Thus, one approach to identify invasion factors is to isolate mutants with altered expression of invasion factors. For example, a mutant might constitutively express a normally regulated invasion factor. Such a mutant might be able to enter epithelial cells even when grown under repressing conditions.

As described in [28], the aminoglycoside antibiotic gentamicin does not readily cross mammalian membranes. Extracellular bacteria are killed by addition of gentamicin, whereas intracellular bacteria are spared from its action and remain viable. This technique allows the positive selection

[8] E. A. Elsinghorst, L. S. Baron, and D. J. Kopecko, *Proc. Natl. Acad. Sci. U.S.A.* **86,** 5173 (1989).

[9] R. K. Ernst, D. M. Dombroski, and J. M. Merrick, *Infect. Immun.* **58,** 2014 (1990).

[10] C. A. Lee and S. Falkow, *Proc. Natl. Acad. Sci. U.S.A.* **87,** 4304 (1990).

[11] D. A. Schiemann and S. R. Shope, *Infect. Immun.* **59,** 437 (1991).

[12] J. J. Mekalanos, *J. Bacteriol.* **174,** 1 (1992).

[13] J. F. Miller, J. J. Mekalanos, and S. Falkow, *Science* **243,** 916 (1989).

for invasive bacterial strains from populations of noninvasive bacteria. Similarly, hyperinvasive mutants can be positively selected from pools of mutagenized bacteria after growth under repressing growth conditions. After incubation with mammalian cells, the hyperinvasive bacteria are able to enter the cells to become gentamicin resistant, whereas the normal, repressed bacteria remain extracellular and are killed by gentamicin.

The basic steps for isolation of hyperinvasive mutants are (1) to prepare a pool of bacterial mutants, (2) to grow the pool of mutants under repressing growth conditions, (3) to incubate the pool of mutants with mammalian cells to allow the hyperinvasive mutants to enter the cells, (4) to treat the infected mammalian cells with gentamicin to kill the noninvasive, extracellular bacteria preferentially, and (5) to recover the viable intracellular hyperinvasive bacteria that have survived the gentamicin treatment. There are several aspects of the isolation procedure that should be optimized: the mutagenesis technique, the repressing growth conditions, the number of enrichment cycles for selection of hyperinvasive mutants, and the identification of independent mutations.

Experimental Considerations

Mutagenesis

The isolation of hyperinvasive mutations depends on the mechanisms modulating microbial invasion. If simple disruption of a repressor function results in constitutive expression of an invasion factor(s), hyperinvasive mutants should be easy to isolate. Alternatively, if expression of invasion factors is regulated by a transcriptional activator, hyperinvasive mutations might be quite rare and might only result from alteration of the activator function or enhancement of the invasion factor promoter element. Thus, the isolation of hyperinvasive mutations also depends on the mutagenesis method.

The most random method for mutagenesis is use of chemical mutagens, for example, ethylmethane sulfonate and nitrosoguanidine. The isolation of a nitrosoguanidine-generated hyperinvasive mutant of *S. typhi* is described below. Unfortunately, even given the genetic and molecular tools available for study of *Salmonella,* chemically generated mutations are especially difficult to characterize as the phenotype can be detected only by an invasion assay. Mutations that affect invasion more commonly have been generated by transposon mutagenesis. For example, Tn*10,* Tn*5*, and Tn*phoA* have been used to generate insertion mutations that lie within and disrupt genes.[5,14] Tn*5*-derived transposons often are more useful than

[14] C. A. Lee, B. D. Jones, and S. Falkow, *Proc. Natl. Acad. Sci. U.S.A.* **89,** 1847 (1992).

Tn*10*-derived transposons for generation of random mutants because of the low sequence specificity of Tn*5* insertion.[15,16]

Traditionally, transposon mutagenesis had been limited to gene disruption mutations. Recently, transposons have been engineered to contain promoter elements at their ends. For example, Tn*5*B50, Tn*5*B60, and Tn*tac1* encode the *neo* or *tac* promoters.[17,18] In this way, although insertion of the engineered transposons within a gene would disrupt the gene, insertion of the engineered transposons upstream of a gene and in the proper orientation would result in expression of the gene from the exogenous transposon-encoded *neo* or *tac* promoter. The isolation of such transposon-generated *S. typhimurium* hyperinvasive mutants is described below. Interestingly, one class of transposon mutations was obtained only by mutagenesis using an engineered transposon containing an exogenous promoter.[14]

Growth Conditions

After pools of bacterial mutants have been generated, the isolation of hyperinvasive mutants may depend on the growth conditions selected for the assay. For example, it is possible that one set of *Salmonella* invasion genes is regulated by oxygen, another by osmolarity, and yet another by growth state. Thus, different hyperinvasive mutations might be selected after growth of the pool of mutants under aerobic conditions as compared with growth in stationary phase. For example, we have selected hyperinvasive *S. typhimurium* mutants on the basis of their ability to enter HEp-2 cells after aerobic growth.[14] Interestingly, the characteristics of these mutants are distinct from those of hyperinvasive *S. typhi* mutants which were isolated on the basis of their ability to enter cells after growth to stationary phase (described below).

Enrichment for Hyperinvasive Mutants

In principle, hyperinvasive mutants can be isolated from any invasive pathogen. As long as conditions can be defined in which invasion of bacteria into mammalian cells is not saturating, a mutation might be found that increases the efficiency of the basal level of invasion (Table I). Of course, in the case of regulated invasion factors, the efficiency of invasion is simply manipulated by growth of the bacteria. The degree to which

[15] D. E. Berg, M. Schmandt, and J. B. Lowe, *Genetics* **105,** 813 (1983).
[16] N. Kleckner, D. A. Steele, K. Reichardt, and D. Botstein, *Genetics* **92,** 1023 (1979).
[17] W.-Y. Chow and D. E. Berg, *Proc. Natl. Acad. Sci. U.S.A.* **85,** 6468 (1988).
[18] R. Simon, J. Quandt, and W. Klipp, *Gene* **80,** 161 (1989).

TABLE I
CYCLIC ENRICHMENT FOR HYPERINVASIVE MUTANTS[a]

	Sequential cycles of selection for mutants			
	Composition of inoculum of 10^7 bacteria		Composition of survivors after invasion assay	
Cycle	Hyperinvasive	Repressed	Hyperinvasive	Repressed
1	10^3	10^7	20	10^4
2	2×10^4	10^7	4×10^2	10^4
3	4×10^5	10^7	8×10^3	10^4
4	4.5×10^6	5.5×10^6	9×10^4	5.5×10^3
5	9.4×10^6	6×10^5	1.9×10^5	600
6	10^7	3×10^3	2×10^5	3

[a] Hypothetical parameters: Invasiveness of repressed bacteria, 0.1%; invasiveness of hyperinvasive mutant, 2.0%; number of hyperinvasive mutants per random mutants, $1/10^4$; number of independent mutants in starting pool, 10^4; selection for gentamicin survivors from an initial inoculum, 10^7.

invasiveness is affected by growth conditions defines the experimental protocol for selection of hyperinvasive mutants. For example, if incubation of stationary-phase *Salmonella* with mammalian cells never yields any gentamicin-resistant intracellular bacteria, it is possible that isolation of gentamicin-resistant bacteria after a single selection cycle with mammalian cells will yield the desired hyperinvasive mutants. If, however, the difference between the invasiveness of repressed bacteria and induced bacteria is only 20-fold with significant numbers of gentamicin-resistant survivors even after repression, several cycles of enrichment for hyperinvasive mutants may be required before a significant proportion of the gentamicin-resistant survivors represent hyperinvasive mutants.

An example of the effect of sequential cycles of enrichment for isolation of hyperinvasive mutants from a large pool of random mutants is shown in Table I. In this example, five or more cycles of enrichment are required to yield a population of gentamicin-resistant survivors consisting of, predominantly, hyperinvasive mutants. The successful enrichment for hyperinvasive mutants is evident after plating for colony-forming units (cfu) immediately after the fifth or sixth sequential invasion assay, as 20-fold more gentamicin-resistant survivors are recovered from this assay as compared with an assay with normal, repressed *Salmonella* (2×10^5 vs 10^4). Obviously, the number of enrichment cycles required for isolation of hyperinvasive mutants varies depending on the five parameters described in Table I.

Isolation of Independent Hyperinvasive Mutants

A major consideration in any mutant analysis is the need to isolate independent mutants. Only by isolation of different mutations can there be a comprehensive analysis of the factors involved in a particular phenotype. Thus, many independent pools of random mutants should be generated for analysis. Furthermore, only a single hyperinvasive mutant should be selected for subsequent analysis from each experimental pool. It is possible that the final population of hyperinvasive mutants would be siblings derived from a single hyperinvasive mutant in the original pool of random mutants (see Table I).

Isolation of Transposon-Generated *Salmonella typhimurium* Hyperinvasive Mutants

Transposon Mutagenesis of Salmonella typhimurium

Salmonella typhimurium strain EE251, a spontaneous streptomycin-resistant derivative of SL4012, can be mutagenized with Tn*5* or with Tn*5*B50 using a delivery system based on plasmid pRTP1.[14] The mutagenesis system is based on the fact that strain EE251 contains a streptomycin-resistant allele of *rpsL*, which is phenotypically recessive to the streptomycin-sensitive *rpsL* allele on the pRTP1 derivatives. In this way, transposition of Tn*5* or Tn*5*B50 onto the genome of strain EE251 and subsequent loss of the plasmid can be selected for by growth of such EE251::Tn mutants in the presence of streptomycin and kanamycin (for Tn*5*) or streptomycin and tetracycline (for Tn*5*B50). Details of the mutagenesis procedure are described in detail in the steps below and are schematically diagrammed in Fig. 1.

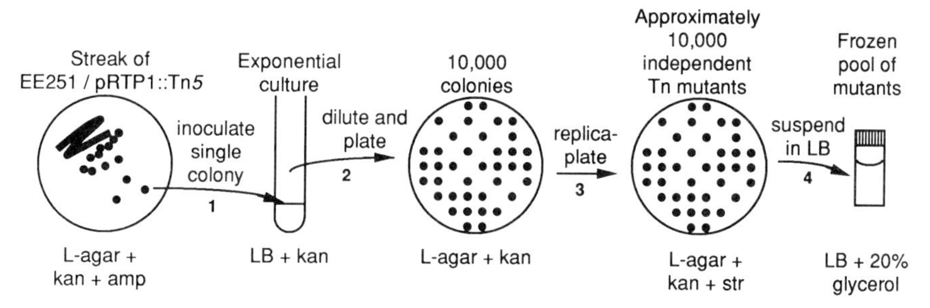

FIG. 1. Steps to generate one pool of EE251::Tn*5* mutants. Numbered steps are described in the text.

1. To generate 10 pools of independent transposon mutants, prepare 10 exponentially growing cultures derived from 10 different colonies of EE251/pRTP1::Tn.

2. Dilute each culture and plate 5000 to 10,000 cfu onto L agar containing the antibiotic appropriate for the transposon used, kanamycin (40 μg/ml) or tetracycline (10 μg/ml).

3. Incubate the 10 plates overnight at 30° to allow for growth of the colonies. Replica-plate the colonies from the 10 plates onto L agar plates containing streptomycin (100 μg/ml) and kanamycin/tetracycline to select for the EE251::Tn mutants within each colony.

4. Incubate the 10 replica plates overnight at 37° and then suspend the streptomycin-resistant, kanamycin/tetracycline-resistant colonies from each plate.

5. Store each pool frozen in 20% glycerol at −20° or −70°.

Using this procedure, almost every streptomycin-resistant, kanamycin/tetracycline-resistant colony on the replica plate will represent an independent transposon mutant. The surprisingly efficient selection of transposon mutants that have lost the dominant streptomycin-sensitive pRTP1 derivative is likely due to the random segregation properties of the ColE1 plasmid. A small percentage of the cells in each pool may remain ampicillin resistant and likely are streptomycin-resistant as a result of inactivation of the *rpsL* gene on the pRTP1 derivative.

Sequential Enrichment for Salmonella typhimurium Mutants That Enter Mammalian Cells Even When Grown Aerobically

The selection procedure for isolation of *S. typhimurium* mutants that can enter HEp-2 cells after aerobic growth is described below and is schematically diagrammed in Fig. 2. Duplication of the enrichment sequence for each pool increases the likelihood of obtaining a hyperinvasive mutant, as there is a chance that it might not be recovered from the initial enrichment cycles. Thus, two different aerobic cultures were prepared from each independent pool of *S. typhimurium* transposon mutants.

1. Inoculate a small amount of the thawed stock into two different 1-ml aliquots of Luria–Bertani (LB) to a final density of 10^8 cfu/ml. Agitate the culture overnight.

2. (a) Inoculate 1 μl of the overnight culture into 1 ml of LB broth. (b) Grow each culture with agitation to a final density of 10^8 cfu/ml, the aerobic culture condition that is repressing for *Salmonella* invasiveness.

3. Inoculate 0.1 ml of each aerobic culture (10^7 cfu) into the medium overlying HEp-2 monolayers.

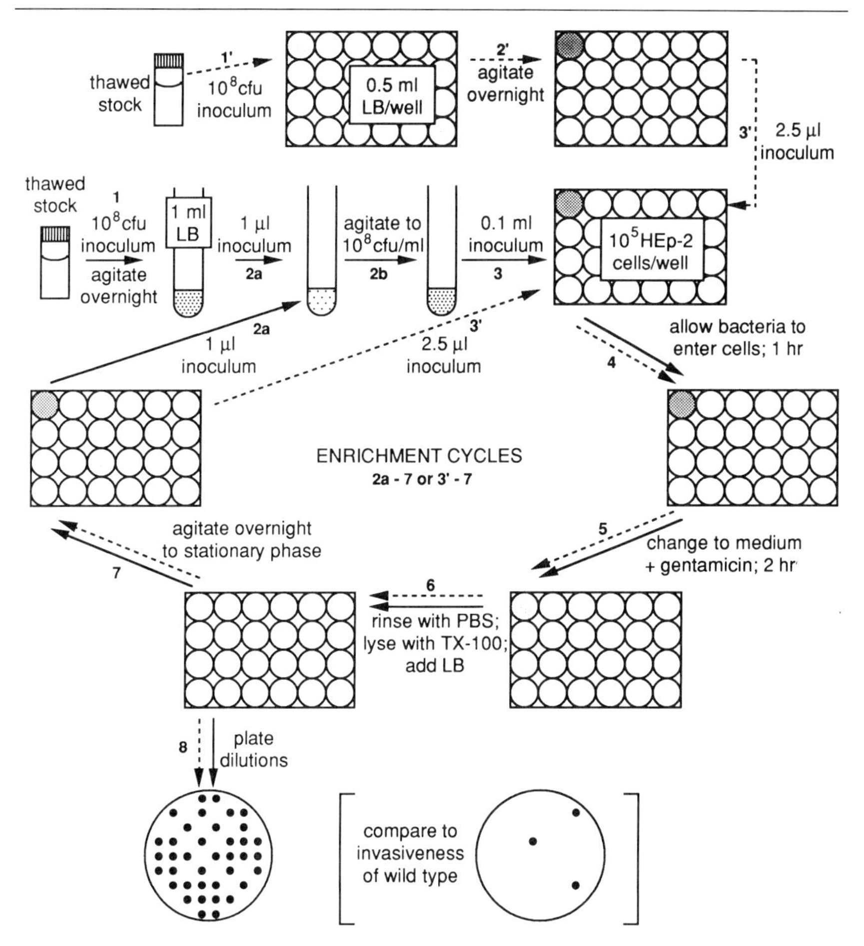

FIG. 2. Enrichment sequence for isolation of hyperinvasive *Salmonella* mutants. Numbered steps are described in the text. Dashed arrows allow enrichment for *Salmonella* mutants that can enter cells even when grown to stationary phase, steps $1' + 2' + [3' - 7]_n + [3' - 6] + 8$. Solid arrows allow enrichment for *Salmonella* mutants that can enter cells even when grown aerobically, steps $1 + [2a - 7]_n + [2a - 6] + 8$.

4. Allow the bacteria to enter the HEp-2 cells during a 1-hr incubation at 37° in 5% CO_2.

5. After the 1-hr, change the tissue culture medium to RPMI 1640 containing 5% fetal bovine serum (FBS) and 100 μg/ml gentamicin, so that the extracellular bacteria are preferentially killed during an additional 2-hr incubation.

6. Release any intracellular bacteria by rinsing the monolayers twice with phosphate-buffered saline (PBS) and incubating the monolayer with 50 μl 1% Triton X-100 for 10 min at room temperature.

7. Recover the viable bacteria by adding 1 ml of LB broth to each well and agitating the entire dish overnight to generate a saturated bacterial culture.

8. Repeat the selection procedure (steps 2–7) the next day by diluting 1 μl of the saturated culture into 1 ml of LB broth and again grow with agitation to obtain an aerobic culture.

9. After four or more sequential enrichment cycles, release any intracellular bacteria from the HEp-2 cells. This time, directly plate dilutions of the sample from step 6 onto LB agar. Enumeration of the colony-forming units released from each monolayer indicates, as discussed above, which wells contain mutant strains that are more invasive than the wild-type *S. typhimurium*.

10. Purify single colonies from such enriched samples for analysis.

Analysis of Salmonella typhimurium Hyperinvasive Mutants

Hyperinvasive mutations defined by transposon insertions are desired. Initially, test each potential hyperinvasive mutant in invasion assays to see if it is more invasive than the wild type. If so, transduce each transposon mutation into a wild-type strain. Then test resultant antibiotic-resistant transductants for their ability to enter HEp-2 cells. Mutants that appear to be hyperinvasive by the gentamicin assay should also be analyzed by other means (i.e., by microscopic visualization of bacterial internalization).

Results

Using these selection conditions, three classes of transposon mutants of *S. typhimurium* were identified that were better able to enter HEp-2 cells (Table II).[14] One class of mutations disrupted the *che* genes that are required for bacterial chemotaxis. Another class of mutations affected expression of *rho,* an essential gene. The third class of mutations were obtained only after mutagenesis with the engineered transposon Tn5B50, which contains the *neo* promoter at one end. This last class of mutations seems to be due to expression of an invasion gene from the exogenous promoter. The locus was named *hil* for hyperinvasion locus.

Further study of *che, mot,* and *fla* mutants suggests that the rotation and physical orientation of flagella around *Salmonella* affect the efficiency of the invasion process.[19] In this way, *che* mutants that are nonchemotactic

[19] B. D. Jones, C. A. Lee, and S. Falkow, *Infect. Immun.* **60,** 2475–2480 (1992).

TABLE II
EFFECT OF REPRESENTATIVE TRANSPOSON MUTATIONS ON *Salmonella typhimurium* SL1344 INVASIVENESS

SL1344 derivative	Relative invasiveness[a]: growth conditions	
	Aerobic	Low oxygen
Wild type	1 ± 0.2	43 ± 2
::Tn*10*-181 (*che*)	13 ± 4	54 ± 10
::Tn*10*-177 (*rho*)	16 ± 5	62 ± 8
::Tn*5*B50-378 (*hil*)	74 ± 7	373 ± 60
::Tn*5*B50-380 (*hil*)	18 ± 8	96 ± 9

[a] Mutations were introduced into the SL1344 strain background by P22-mediated transduction. Values represent the means ± standard error of multiple assays and were normalized such that the invasiveness of aerobically grown SL1344 equals 1. The actual percentage of the aerobic SL1344 inoculum that entered HEp-2 cells in 1 hr was 0.11 ± 0.018%.

because of their inability to tumble are better able to enter mammalian cells than wild-type bacteria that exhibit both swimming and tumbling behaviors. The increased ability of *rho* mutants to enter HEp-2 cells is presumably due to a change in invasion factor expression. Mutations in *rho* have pleiotropic effects on transcription termination and the state of DNA supercoiling. Further study is required to identify the specific change in gene expression that results in increased invasiveness of *rho* mutants.

Phenotypic and molecular analysis of *hil*::Tn*5*B50 mutants suggests that there is a gene(s) downstream of Tn*5*B50-378 that acts positively on expression of *S. typhimurium* invasiveness (Fig. 3). If this is the case, deletion or disruption of the *hil* locus might result in loss of invasiveness. Analysis of the effect of a *hil* deletion mutation on bacterial entry into HEp-2 cells shows that this region is essential for expression of *S. typhimurium* invasiveness. Even when grown under low oxygen conditions, the *hil* deletion mutant is 1000-fold less able to enter HEp-2 cells than the comparably grown parental strain (see Fig. 3).[14]

The corresponding change in *hil* expression and bacterial invasiveness suggests that the *hil* locus is specific for invasion. Furthermore, the deletion of *hil*, which reduces the invasiveness of bacteria grown under low-oxygen conditions, suggests that the *hil* mutants are affected in the oxygen-regulated pathway. Thus, as hoped, the selection for hyperinvasive

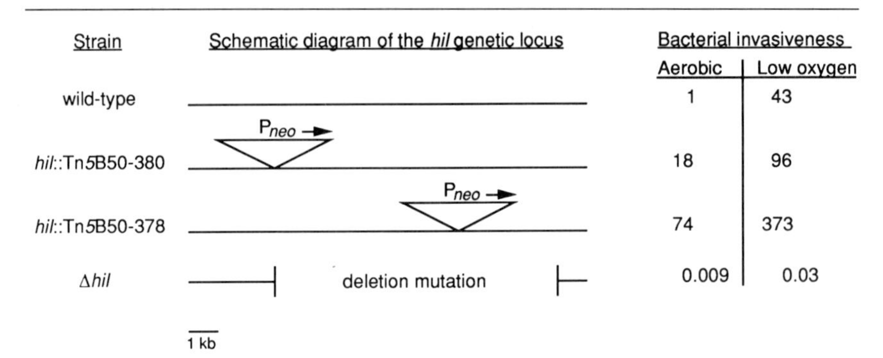

FIG. 3. Effect of *hil* mutations on *Salmonella typhimurium* invasiveness. The ability of bacteria to enter HEp-2 cells is normalized such that the invasiveness of aerobically grown wild type equals 1.

S. typhimurium mutants has identified *hil,* a gene(s) that is essential for oxygen-regulated invasion.

Isolation of Nitrosoguanidine-Generated *Salmonella typhi* Hyperinvasive Mutants

Salmonella typhi strain 404Ty was obtained from B. A. D. Stocker (Stanford University School of Medicine). 404Ty is a Δ*aroA148* derivative of an Indonesian strain, 3083/80, originally characterized by L. LeMinor (Institut Pasteur, Paris), which has two flagellar antigen phases, d and z66. 404Ty is grown in LB broth supplemented with 2,3-dihydroxybenzoic acid (DHB, 10 μg/ml), which is required for optimal growth of the Δ*aroA* mutant strain.

N-Methyl-N'-nitro-N-nitrosoguanidine Mutagenesis of Salmonella typhi

Treatment with nitrosoguanidine is described by Miller.[20] Details of a mutagenesis procedure that has been used to produce 48 independent pools of 10^6 potential mutants each are described below.[21] Mutagenesis should be optimized by determining the rate of killing, as well as the rate of mutation induced by chemical treatment. Mutagenized samples can be divided into independent pools immediately after mutagenesis, prior to outgrowth.

[20] J. H. Miller, "Experiments in Molecular Genetics." Cold Spring Harbor Lab., Cold Spring Harbor, NY, 1972.

[21] C. A. Lee and S. Falkow, unpublished observations (1989).

1. Grow a 5-ml culture of 404Ty in LB/DHB to exponential phase.
2. Wash the bacterial cells twice in 0.1 *M* sodium citrate and resuspend the cells in 2 ml 0.1 *M* sodium citrate.
3. Mix two 1-ml portions of the bacterial suspension with 0 and 40 μl of a nitrosoguanidine stock solution (2 mg/ml in 0.1 *M* sodium citrate), and incubate at 37° for 30 min. Nitrosoguanidine is an extremely hazardous chemical and should be handled with precaution as recommended by the manufacturer.
4. Wash the treated bacterial cells with PBS and resuspend in 1 ml LB/DHB.
5. To determine the effect of nitrosoguanidine treatment on bacterial viability, plate dilutions of the treated and untreated aliquots on L-agar/DHB. Killing due to mutagen treatment should be approximately 50%.
6. To divide the mutagenized sample into pools of independent mutants, inoculate 48 aliquots of the mutagenized sample into 48 tubes containing 1 ml LB/DHB and grow overnight. Prepare cultures of the untreated culture as a control.
7. To determine the increase in mutation due to the nitrosoguanidine treatment, plate 100 μl of several control and mutagenized overnight cultures on L-agar/DHB containing rifamicin (100 μg/ml).
8. Store each culture frozen in 20% glycerol.

Sequential Enrichment for Salmonella typhi Mutants That Enter Mammalian Cells Even When Grown to Stationary Phase

The selection procedure for isolation of *S. typhi* mutants that are able to enter MDCK cells even when grown to stationary phase is described below and is diagrammed in Fig. 2.

1′. Inoculate a small amount of the thawed stock into 0.5-ml aliquots of LB/DHB in a multiwell plate to a final density of 10^8 cfu/ml.

2′. Agitate the plate overnight to obtain stationary-phase cultures.

3′. Inoculate 2.5 μl of each overnight culture (10^7 cfu) into the medium overlying MDCK monolayers.

4. Allow the bacteria to enter the MDCK cells during a 1-hr incubation at 37° in 5% CO_2.

5. Change the tissue culture medium to RPMI 1640 containing 5% fetal bovine serum and 100 μg/ml gentamicin. The extracellular bacteria are preferentially killed during an additional 2-hr incubation.

6. Release any intracellular bacteria by rinsing the monolayers twice with PBS and incubating the monolayer with 50 μl 1% Triton X-100 for 10 min at room temperature.

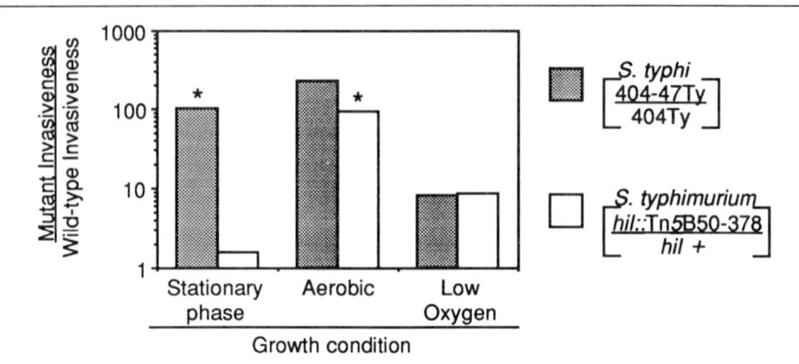

FIG. 4. Effect of growth conditions on the hyperinvasive phenotype of *Salmonella* mutants. Each bar represents the increase in invasiveness of the mutant strain relative to the comparably grown wild-type parent. The bars marked with an asterisk indicate the growth conditions under which the hyperinvasive mutant was selected.

7. Recover the viable bacteria by adding 0.5 ml of LB/DHB to each well and agitating the entire dish overnight to generate a saturated bacterial culture.

8. Repeat the selection procedure (steps 3′–7) the next day by inoculating 2.5 μl of the saturated bacterial culture into the medium overlying MDCK cells.

9. After four or more sequential enrichment cycles, release any intracellular bacteria from the MDCK cells. This time, directly plate dilutions of the sample from step 5 onto L-agar/DHB. Enumeration of the colony-forming units released from each monolayer indicates, as discussed above, which wells contain mutant strains that are more invasive than the wild-type *S. typhi*.

10. Purify single colonies from such enriched samples for analysis.

Analysis of Salmonella typhi Hyperinvasive Mutants

Test each potential hyperinvasive mutant in invasion assays to determine if it is more invasive than the wild type. Mutants that appear to be hyperinvasive by the gentamicin assay should also be analyzed by other means (i.e., by microscopic visualization of bacterial internalization). Genetic analysis of mutations in *S. typhi* is more difficult than that in *S. typhimurium;* however, generalized transducing phage are available for study of *S. typhi*.[22–24] Initially, a large pool of random transposon insertions might be prepared in the hyperinvasive strain. In this way, a transpo-

[22] L. S. Baron, S. B. Formal, and W. Spilman, *Proc. Soc. Exp. Biol. Med.* **83,** 292 (1953).

[23] L. S. Baron, S. B. Formal, and W. Spilman, *J. Bacteriol.* **69,** 177 (1955).

[24] K. Mise, M. Kawai, Y. Yoshida, and A. Nakamura, *J. Gen. Microbiol.* **126,** 321 (1981).

son linked to the hyperinvasive mutation by transduction could be isolated. Such linked transposons would allow all of the hyperinvasive mutations to be mapped and introduced into nonmutagenized strain backgrounds.

Results

Several hyperinvasive mutants of *S. typhi* 404Ty were isolated using the described technique.[21] The characteristics of one of these mutants, 404-47Ty, is reported below. None of the *S. typhi* mutants has been characterized genetically.

Salmonella typhi mutant 404-47Ty was selected for the ability to enter MDCK cells even when grown to stationary phase. Quantitation of the invasiveness of stationary-phase 404-47Ty shows that the mutant is 100-fold more invasive than wild-type 404Ty. The mutant is also hyperinvasive when grown aerobically, although this phenotype was not demanded by the selection (Fig. 4). Interestingly, the *S. typhi* hyperinvasive mutant differs from the *S. typhimurium* hyperinvasive *hil* mutant, which remains unable to enter epithelial cells when grown to stationary phase (see Fig. 4).[21] It is not known if the *S. typhi* mutation affects a factor that is specific for invasion. For example, the *S. typhi* mutant may express greater levels of an essential adherence factor that facilitates subsequent invasion. Alternatively, the mutations may result in a nonspecific change in bacterial surface properties that alter the efficiency of the invasion process. Further study of the *S. typhi* hyperinvasive mutant may distinguish these possibilities.

Acknowledgments

This work was supported by U.S. Public Health Service Grant AI-26195 (to S.F.) and a Stanford University Digestive Disease Center Award to C.A.L. (DK-38707). Bradley Jones identified and characterized the effect of *che* mutations on invasiveness. Michael Starnbach contributed to the characterization of the 404-47Ty hyperinvasive mutant.

[40] Molecular Cloning of Invasion Genes from *Yersinia* and *Salmonella*

By VIRGINIA L. MILLER and BARBARA J. STONE

Introduction

Many bacterial pathogens spend at least a portion of their time within cells of the host. Advances in tissue culture technology and cell biology have facilitated a closer examination of how bacteria enter and survive within the eukaryotic cell; however, further elucidation of these processes at a molecular level is greatly aided by the cloning and characterization of bacterial genes and their products which are necessary for adherence, entry, and survival within eukaryotic cells.

Two basic approaches, both of which use a tissue culture invasion assay to identify genes necessary for adherence and invasion, have proved fruitful in this regard. The first approach is to identify genes that can confer an invasive phenotype to a noninvasive, nonadherent organism such as *Escherichia coli* HB101. This approach has proved useful for organisms such as *Yersinia pseudotuberculosis*[1] and *Yersinia enterocolitica*[2] where a single gene (either *inv* or *ail*) can confer an invasive phenotype to *E. coli,* and should also work for pathogens where the genes necessary for an invasive phenotype are closely linked. The second approach clones genes by complementation of an invasion-defective mutant. This approach has proved particularly useful for pathogens such as *Salmonella*[3,4] that have several loci around the chromosome that contribute to the invasion process.

Both methods make use of a tissue culture invasion assay described below. First, the bacteria are added to a subconfluent monolayer of tissue culture cells (a monolayer that is 60–80% confluent works best for our system) grown in a microtiter dish (24-well). It is important to use similar multiplicities of infection for all samples that are to be compared. An optional step prior to incubation is to centrifuge the bacteria gently onto the monolayer (162 *g*, 10 min, room temperature), thereby increasing contact between the bacteria and the eukaryotic cell plasma membrane.

[1] R. R. Isberg and S. Falkow, *Nature (London)* **317,** 262 (1985).
[2] V. L. Miller and S. Falkow, *Infect. Immun.* **56,** 1242 (1988).
[3] B. J. Stone, C. M. Garcia, J. L. Badger, T. Hassett, R. I. F. Smith, and V. L. Miller, *J. Bacteriol.* **174,** 3945 (1992).
[4] J. E. Galan and R. Curtiss, III, *Proc. Natl. Acad. Sci. U.S.A.* **86,** 6383 (1989).

This step is not necessary but generally results in more reproducible results and a shorter lag between addition of bacteria and entry. Following addition of the bacteria and centrifugation, the microtiter tray is incubated at 37°, 5% CO_2 (the incubation time is variable depending on the system under study, but we routinely use 90 min to 3 hr). After incubation, the tissue culture medium is removed by aspiration and the monolayer is washed with phosphate-buffered saline (PBS, 0.128 *M* NaCl/2.7 m*M* KCl/ 1.5 m*M* KH_2PO_4/8 m*M* $Na_2HPO_4 \cdot 7H_2O$) several times (three to five) to remove bacteria that are not adherent or internalized. Fresh tissue culture medium containing the antibiotic gentamicin is then added to kill extracellular adherent bacteria (see [28]). The intracellular bacteria are protected because gentamicin does not penetrate the eukaryotic cell. Other aminoglycosides such as kanamycin have also been used, and in principle any approach that would selectively kill extracellular bacteria could be used.

After incubation at 37°, 5% CO_2 (again the time is variable depending on the individual system, but we routinely use 90 min), the monolayers are washed with PBS to remove the antibiotic. Addition of 200 μl of 1% Triton X-100 for 5 min lyses the eukaryotic cells and releases intracellular bacteria. The Triton X-100 is then diluted by addition of 800 μl of bacteriologic medium [i.e., Luria–Bertani (LB) broth], and the bacteria are diluted further and plated on the appropriate bacteriologic medium to determine viable counts. Some bacteria are sensitive even to short exposures to Triton X-100, in which case some other approach such as treatment with 0.5 m*M* ethylenediaminetetraacetic acid (EDTA) in PBS and agitation[5] should be used to disrupt the eukaryotic cells and release the bacteria. The *E. coli, Yersinia,* and *Salmonella* we have used do not, however, appear to be greatly affected by this treatment with Triton X-100. Results from these assays are usually expressed either as the number of intracellular bacteria per well or as percentage invasion, which is defined as 100 × (number of intracellular bacteria/number of bacteria added).

Method 1: Identification of Genes That Can Confer an Invasive Phenotype to Nonadherent, Noninvasive *Escherichia coli*

Successful use of this approach was first reported by Isberg and Falkow[1] for *Y. pseudotuberculosis,* and has subsequently been used for *Y. enterocolitica*[2] and *Salmonella typhi.*[6]

[5] J. H. Shaw and S. Falkow, *Infect. Immun.* **56,** 1625 (1988).

[6] E. A. Elsinghorst, L. S. Baron, and D. J. Kopecko, *Proc. Natl. Acad. Sci. U.S.A.* **86,** 5173 (1989).

Step 1

Construct a library of DNA fragments from the organism of interest (i.e., *Y. enterocolitica*). Transform the library into a noninvasive, nonadherent *E. coli* strain such as HB101. In general, a library containing fairly large fragments of genomic DNA is recommended in the event that the invasive phenotype is encoded by a large operon or by several linked genes. Either overexpression or lack of expression of invasion-promoting proteins is a potential problem related to choice of vector for construction of the library. Overexpression of genes encoding proteins that promote invasion due to use of a vector with a high copy number may adversely affect cell growth, so we recommend using a single or low copy vector for construction of the library. Alternatively, the invasion gene(s) may not be expressed in the heterologous *E. coli* strain, so an expression vector with an inducible promoter could also be used to construct the library. If two or more unlinked genes are necessary to confer invasion, then Method 2 may be a more appropriate approach.

Step 2

To enrich for invasive clones, pool the transformants from step 1. This pool can be either used directly or diluted and cultured overnight prior to the enrichment step. If the library was constructed in a cosmid vector with large inserts such that the complete genome is represented by relatively few clones, then one might consider using smaller pools of 20 to 100 clones. Approximately 10^7 (multiplicity of infection of ~100 : 1) pooled bacteria should be added to the tissue culture monolayer, and the invasion assay performed as above. The first reports using this approach did not kill extracellular bacteria with gentamicin during the enrichment.[1,2] Instead, the monolayers were washed 10–15 times with PBS and lysed with Triton X-100 immediately. This method recovers clones conferring both strongly adherent and invasive phenotypes.

Bacteria recovered from the enrichment step should then be tested individually (or first pooled and enriched again if necessary) in a quantitative invasion assay. If using pools of small numbers of clones for the enrichment step, it may be immediately obvious which pool contains the invasive clone, as significantly more bacteria will be recovered from the assay for that particular pool of clones. When cloning genes that confer invasion to *E. coli* HB101 from either *Y. pseudotuberculosis*[1] or *Y. entero-*

[7] J.-L. Gaillard, P. Berche, C. Frehel, E. Gouin, and P. Cossart, *Cell* (*Cambridge, Mass.*) **65,** 1127 (1991).

colitica,[2] approximately 50% of bacteria surviving one round of enrichment carried genes encoding determinants of host cell invasion.

Method 2: Identification of Genes Required for the Invasion Phenotype by Complementation of an Invasion-Defective Mutant

Successful use of this approach has been reported for *Listeria monocytogenes*,[7] *Salmonella typhimurium*,[4] and *Salmonella enteritidis*.[8]

Step 1

Identify an invasion-defective mutant(s) of the organism of interest (i.e., *S. enteritidis*) (see [36]). A noninvasive strain of the organism can be used, or mutagenesis can be employed to generate invasion-defective mutants. The use of a noninvasive strain may limit future studies if the organism has more than one locus involved in invasion as that particular strain may have a defect in only one locus. Moreover, it may be difficult to restore the invasion phenotype if this strain has defects in more than one distinct, unlinked invasion locus. Although chemical mutagenesis could be used to generate invasion-defective mutants, the introduction of transposon insertions is preferable as the number and location of insertions can be determined rapidly. As some, but not necessarily all, determinants that influence the invasion phenotype are likely to be exported products, the transposon Tn*phoA* can be elected for mutagenesis to limit the number of mutants that must be screened to identify invasion-defective mutants. Tn*phoA* carries kanamycin resistance and a portion of the *E. coli* alkaline phosphatase gene (*phoA*).[9] When inserted into the sequence of a gene that produces an exported product, the alkaline phosphatase fusion protein is detected by the appearance of a blue color in the presence of 5-bromo-4-chloro-3-indolyl phosphate. After mutagenesis with Tn*phoA*, blue kan^r colonies are screened for a decreased invasion phenotype by using the tissue culture invasion assay described above. By use of Tn*phoA* mutagenesis in *S. enteritidis*, 13 invasion-defective mutants were identified from approximately 300 blue colonies screened.[3]

Step 2

Construct a library of DNA fragments from the organism of interest. The library should be constructed as described in Method 1 with the following consideration. The library should be constructed in a vector

[8] B. J. Stone and V. L. Miller, unpublished results, 1992.
[9] C. Manoil and J. Beckwith, *Proc. Natl. Acad. Sci. U.S.A.* **82,** 8129 (1985).

that can be introduced efficiently into the invasion-defective mutant(s). Introduction of vector DNA by transformation, electroporation, conjugation, or phage transduction is acceptable. Because *Salmonella* spp. tend to be difficult to transform, a mobilizable cosmid vector (pLAFR2)[10] was chosen for this approach in *S. enteritidis*. Additional transfer functions were supplied *in trans* by a helper plasmid (pRK2013).[11]

Step 3

Introduce library pools into invasion-defective mutants. Delivery of the recombinant plasmids by transformation or electroporation should be feasible in most instances; however, conjugation by a filter method was found to be the most effective for *S. enteritidis*. Overnight cultures (1 ml each) of *E. coli* carrying the recombinant plasmid library, the invasion-defective strain, and the strain harboring the helper plasmid (if necessary) are mixed, passed through a sterile Millipore (Bedford, MA) Swinnex-25 filter apparatus containing a 0.45-μm membrane filter, and collected on the filter. The filter is then removed from the apparatus and incubated, bacteria side up, on a nonselective solid medium (LB agar). After incubation for 4–12 hr at 37° the bacteria are resuspended in LB broth, diluted appropriately, and plated onto medium that will select recipient bacteria (invasion defective mutant) carrying the plasmid library. Using this conjugation protocol we have observed transfer of pLAFR2-based plasmids larger than 30 kb into approximately 0.6% of recipient *S. enteritidis*.

Step 4

To enrich for invasive clones, the exconjugates from step 3 are combined for each library pool. The conjugate pools are cultured overnight with antibiotic selection prior to enrichment for invasive clones using the invasion assay. Approximately 10^7 bacteria from these overnight cultures are added to each well of the invasion assay (one or two wells per "pool"), and the assay is performed as described above. At least two rounds of enrichment are suggested for this approach after which it may become apparent (from increased recovery of bacteria) which library/conjugate pools contain invasion-complementing clones. After identification of

[10] A. M. Friedman, S. R. Long, S. E. Brown, W. J. Buikema, and F. M. Ausubel, *Gene* **18,** 289 (1982).

[11] G. Ditta, S. Stanfield, D. Corbin, and D. R. Helinski, *Proc. Natl. Acad. Sci. U.S.A.* **77,** 7347 (1980).

putative complementing clones it is important to isolate the recombinant plasmid and reintroduce it into the invasion-defective mutant. This is necessary to ensure that the increased invasive activity is due to complementation by the plasmid rather than reversion of the original mutation.

[41] Molecular Cloning and Expression of Internalin in *Listeria*

By JEAN-LOUIS GAILLARD, SHAYNOOR DRAMSI, PATRICK BERCHE, and PASCALE COSSART

Introduction

Listeria monocytogenes is a ubiquitous gram-positive bacterium that causes severe infections in humans and in a variety of warm-blooded animals.[1] This pathogen has emerged as an important agent of foodborne disease in the last decade.[2] The majority of human infections affect pregnant women and result in stillbirths or neonatal sepsis. Meningitis, meningoencephalitis, and bacteremia account for most cases in nonpregnant adults, occurring usually but not exclusively in the setting of immunosuppression or aging.

In natural conditions of infection, *L. monocytogenes* invades the host via the intestine. Experimental data suggest that this pathogen may penetrate into the intestinal epithelial cells, before being ingested by macrophages in the lamina propria.[3] In a cell assay system, *L. monocytogenes* is capable of penetrating the enterocyte-like Caco-2 cells by induced phagocytosis.[4] We identified a surface protein, internalin, which triggers uptake of *Listeria* by cells.[5]

[1] M. L. Gray and A. H. Killinger, *Bacteriol. Rev.* **30,** 309 (1966).

[2] A. Schuchat, B. Swaminathan, and C. V. Broome, *Clin. Microbiol. Rev.* **4,** 169 (1991).

[3] P. Racz, K. Tenner, and E. Mero, *Lab. Invest.* **26,** 694 (1972).

[4] J.-L. Gaillard, P. Berche, J. Mounier, S. Richard, and P. Sansonetti, *Infect. Immun.* **55,** 2822 (1987).

[5] J.-L. Gaillard, P. Berche, C. Fréhel, E. Gouin, and P. Cossart, *Cell (Cambridge, Mass.)* **65,** 1127 (1991).

Isolation of Transposon-Induced Noninvasive Mutants of *Listeria monocytogenes*

Mutagenesis with Conjugative Transposon Tn1545

Mutagenesis with transposon Tn*1545* was used to target the chromosomal region responsible for the invasive phenotype of *L. monocytogenes*. Transposon Tn*1545* is a 26-kb conjugative element that confers resistance to kanamycin, erythromycin, and tetracycline.[6] It was originally identified in *Streptococcus pneumoniae* and was successfully transferred to *L. monocytogenes* strain LO17 (serovar 7) by C. Carlier and P. Courvalin (Institut Pasteur, Paris), giving rise to strain BM4140. Transposon mutagenesis was achieved using classic mating procedures.[7] The donor strain was *L. monocytogenes* BM4140 and the recipient strain was *L. monocytogenes* EGD-SmR, a spontaneous mutant resistant to streptomycin (minimum inhibitory concentration >100 mg/liter) obtained from the virulent strain EGD (serovar 1/2a).[7] The insertion mutants were selected on agar containing streptomycin (100 mg/liter) and tetracycline (10 mg/liter). The frequency of transfer was about 10^{-8} per recipient.

Other transposons can be used in *L. monocytogenes*, for example Tn*917*, which gives single insertions and is more stable than Tn*1545*, or its derivative Tn*917-lac*, which allows to evaluate gene expression because of the presence of *lacZ* gene. Transposons Tn*917* and Tn*917-lac*[8] are carried on thermosensitive plasmids which can be introduced into *Listeria* by electroporation.[9] Transposon insertion events are easily selected after incubation of transformed cells at 42°.

Screening of Insertion Mutants for Loss of Invasiveness

A bank of 2500 Tn*1545* insertion mutants obtained from strain EGD-SmR was screened for the ability of bacteria to enter the human colon carcinoma cell line Caco-2, using a micromethod derived from the gentamicin survival assay (see [28]). The cell line Caco-2, originally obtained from A. Zweibaum and M. Rousset (INSERM U. 178, Villejuif, France), was used between passages 76 and 90. For unknown reasons, older passages are less susceptible to infection by *L. monocytogenes*[10] and should not

[6] F. Caillaud, C. Carlier, and P. Courvalin, *Plasmid* **17,** 58 (1987).

[7] J.-L. Gaillard, P. Berche, and P. Sansonetti, *Infect. Immun.* **52,** 50 (1986).

[8] J. B. Perkins and P. Youngman, *Proc. Natl. Acad. Sci. U.S.A.* **83,** 140 (1986).

[9] E. Michel, K. A. Reich, R. Favier, P. Berche, and P. Cossart, *Mol. Microbiol.* **4,** 2167 (1990).

[10] L. Pine, S. Kathariou, F. Quinn, V. George, J. D. Wenger, and R. E. Weaver, *J. Clin. Microbiol.* **29,** 990 (1991).

be used. Caco-2 cells were grown in Dulbecco's modified Eagle's minimum essential medium (DMEM) (GIBCO, Grand Island, NY), supplemented with 20% fetal calf serum (Gibco) and 1% nonessential amino acids (Flow Laboratories, Inc., MacLean, VA), in an atmosphere of 5% CO_2 in air.

For the screening test (Fig. 1), cells were suspended at a concentration of 5×10^4/ml in culture medium, and 0.1-ml samples were pipetted into 96-well microtiter plates (Costar, Cambridge, MA) with a multipipette. Semiconfluent monolayers were established by 48 to 72 hr of incubation. Caco-2 cells cultured at confluency are not permissive for *L. monocytogenes* and should therefore not be used for invasion assays.[4] The insertion mutants were grown overnight in tryptic soy broth at 37° in 96-well microtiter plates. Caco-2 cells were washed once and the cells in each well were covered with 0.2 ml of nonsupplemented DMEM. Complete culture medium was not used at this step because fetal calf serum inhibits entry of *L. monocytogenes* into cells.[4,10] Ten microliters of bacterial culture from each mutant was added to each well with a multipipette. After 1 hr of incubation at 37° to allow bacterial entry, the cells were washed twice and incubated for 2 hr at 37° in the presence of fresh DMEM containing gentamicin (5 mg/liter) to kill extracellular bacteria. The cells were then washed twice and lysed by adding 0.1 ml of cold distilled water. The microtiter plates were incubated for 15 min at 4° and agitated briefly. Ten microliters of lysate was cultured on tryptic soy agar for 24 hr at 37°.

Isolation of Noninvasive Mutants

The insertion mutants that appeared defective for entry in the screening test, i.e., for which bacterial growth was markedly reduced after culturing cell lysates onto agar plates, were further assayed for invasion by a quantitative macromethod. Bacteria were grown at 37° in tryptic soy broth to an optical density of 0.5 at 600 nm and diluted in DMEM to a concentration of 5×10^7 bacteria/ml, and 1 ml of this bacterial suspension was added to semiconfluent Caco-2 monolayers ($\sim 5 \times 10^5$ cells per plate) grown in 35-mm plates (Corning Glass Works, Corning, NY). After 1 hr of incubation at 37°, the monolayers were washed twice and incubated for 2 hr at 37° with DMEM containing gentamicin (5 mg/liter). The cells were then washed twice and lysed by adding 1 ml of cold distilled water. Viable bacteria released from the cells were titered on agar plates.

Three insertion mutants (BUG5, BUG8, and BUG11) appeared to be defective for entry into Caco-2 cells by the quantitative macromethod and were otherwise indistinguishable from the parental strain. The percentage of inoculated bacteria that survived treatment with gentamicin was 0.02–0.04% for the three mutants versus 2% for the parental strain. Similar

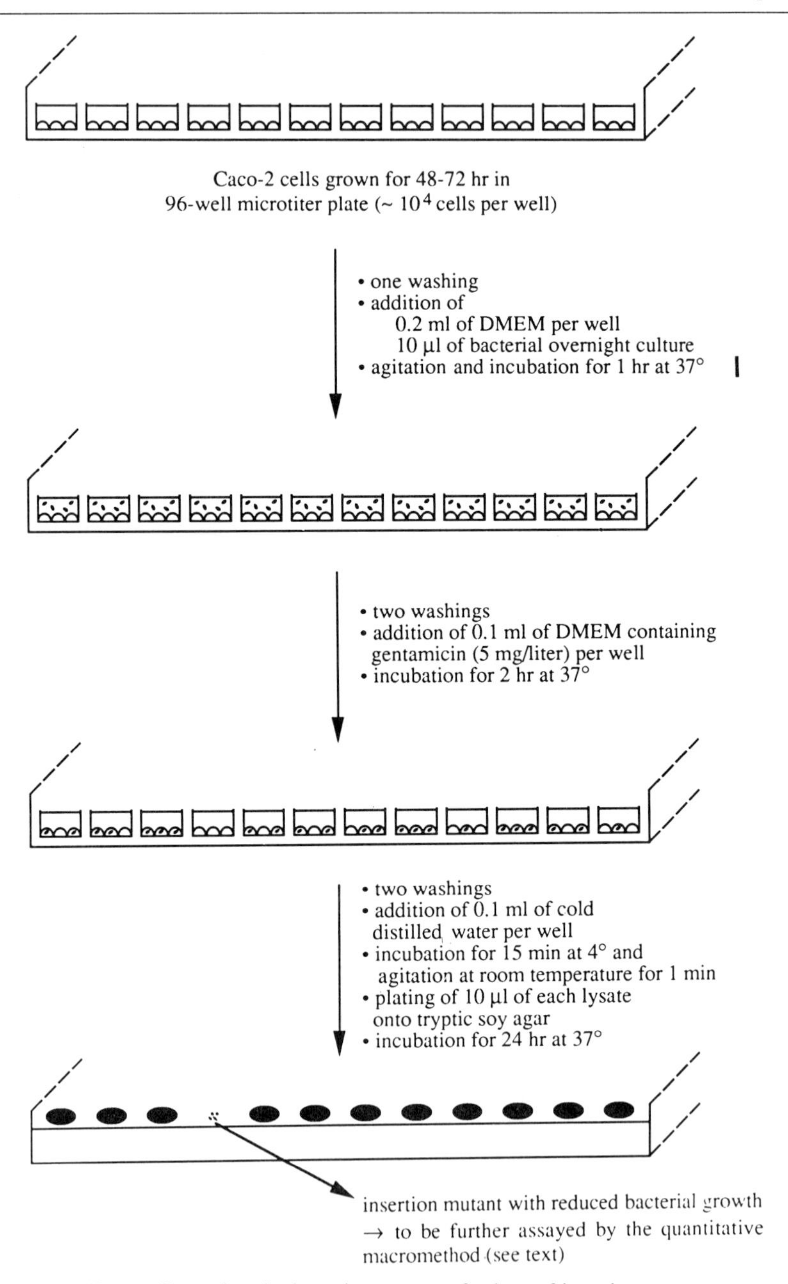

FIG. 1. Screening the insertion mutants for loss of invasiveness.

results were obtained using the epithelial cell lines HT-29 (provided by A. Zweibaum and M. Rousset) and HeLa (purchased from Flow Laboratories). When the Caco-2 cell line is cultured for 48–72 hr, discrete islets are formed that are each composed of 10 to 20 confluent cells. During infection of Caco-2 cells at 37°, wild-type *L. monocytogenes* strains adhere to the cells at the periphery of the islets, penetrate them, and spread from cell to cell toward the center.[11] We observed that the noninvasive mutants, in contrast to the parental strain, were unable to interact significantly with the periphery of the Caco-2 islets (Fig. 2). To determine whether the mutants were defective for adhesion but retained the ability to trigger uptake by cells, we centrifuged each mutant onto Caco-2 cells at 4° for 10 min at 1000 *g*, and invasion was allowed to proceed by raising the temperature to 37°. This procedure did not promote entry of the mutants. These data suggest that the noninvasive mutants have lost the capacity to bind to a mammalian cell receptor that may itself be responsible for uptake of *L. monocytogenes* by cells. The nature of this putative receptor remains unknown.

Identification of *inl* Region

Cloning and Sequencing of Listeria Tn1545 Junction

The endonuclease *Hin*dIII recognizes three sites in Tn*1545*, including one site downstream from the kanamycin resistance gene at the left end of the transposon.[6] We used this site to clone the *Listeria*–Tn*1545* junction from each noninvasive mutant as follows. A library of chromosomal *Hin*dIII DNA fragments cloned in pUC vectors was constructed from each mutant. After transformation of *Escherichia coli* MC1061,[12] clones containing the left part of Tn*1545* and the flanking listerial region were identified by colony hybridization, using a 0.53-kb internal fragment of the kanamycin phosphotransferase gene of Tn*1545* as a probe. The positive clones were all resistant to kanamycin because of the expression of the kanamycin phosphotransferase gene in *L. monocytogenes* [it is in fact possible to select directly the transformants containing the junction between *Listeria* and the left end of Tn*1545* on agar plates containing ampicillin (100 mg/liter) and kanamycin (50 mg/liter)]. The recombinant plasmids present in the positive clones were isolated and digested with *Hin*dIII to determine the size of the listerial fragment (the transposon makes up 6.6 kb of the insert).

[11] J. Mounier, A. Ryter, M. Coquis-Rondon, and P. J. Sansonetti, *Infect. Immun.* **58,** 1048 (1990).

[12] M. J. Casadaban and S. N. Cohen, *J. Mol. Biol.* **138,** 179 (1980).

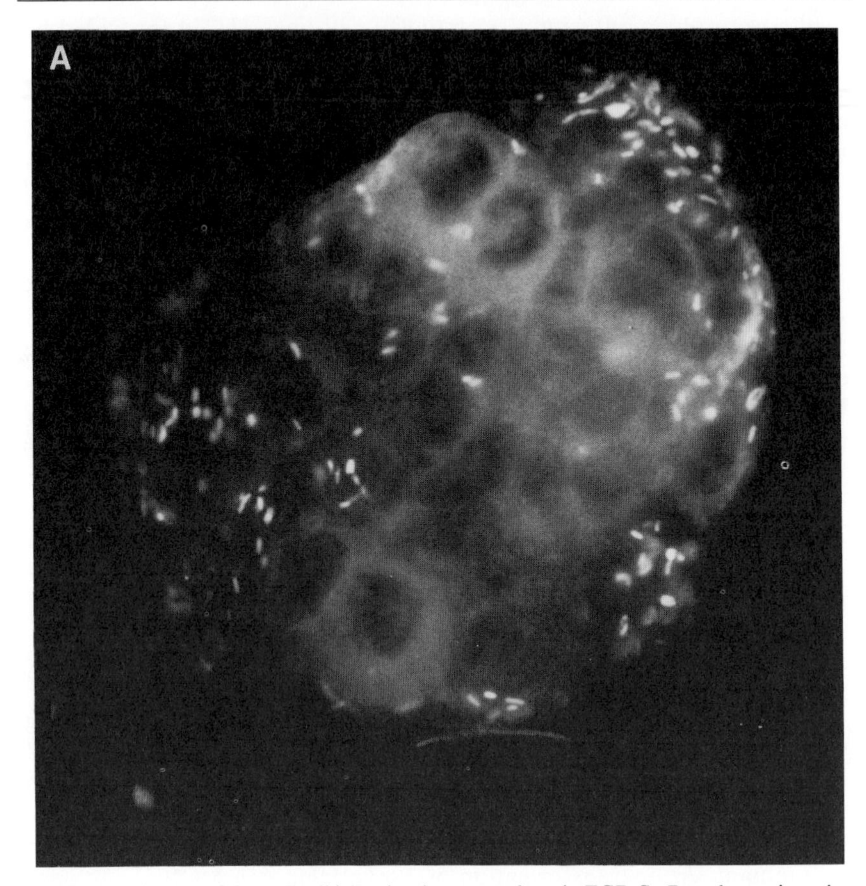

FIG. 2. Infection of Caco-2 cell islets by the parental strain EGD-SmR or the noninvasive mutant BUG5. Caco-2 monolayers were infected with 100 bacteria per cell for 1 hr at 37°, washed twice, and fixed with 2.5% paraformaldehyde. Bacteria were labeled by using a rabbit anti-*Listeria monocytogenes* serum and a goat anti-rabbit fluorescein-conjugated immunoglobulin G. (A) Parental strain EGD-SmR: many bacteria are seen, mostly associated with the cells at the periphery of the islet. (B) Mutant BUG5: few bacteria are seen.

Plasmid DNA was used to sequence directly (Sequenase kit [U.S. Biochemicals, Cleveland, Ohio]) the listerial DNA–Tn*1545* junctions with the primer 5′-CCACTCAATTTACTACTA-3′. The transposon Tn*1545* had inserted in the same chromosomal region in each of BUG5, BUG8, and BUG11 (Fig. 3A,B). This region was termed *inl*, for internalization. In BUG5, between the invariable terminus of the left end of Tn*1545* and the *Listeria* chromosome, we detected a 6-bp stretch that was absent in

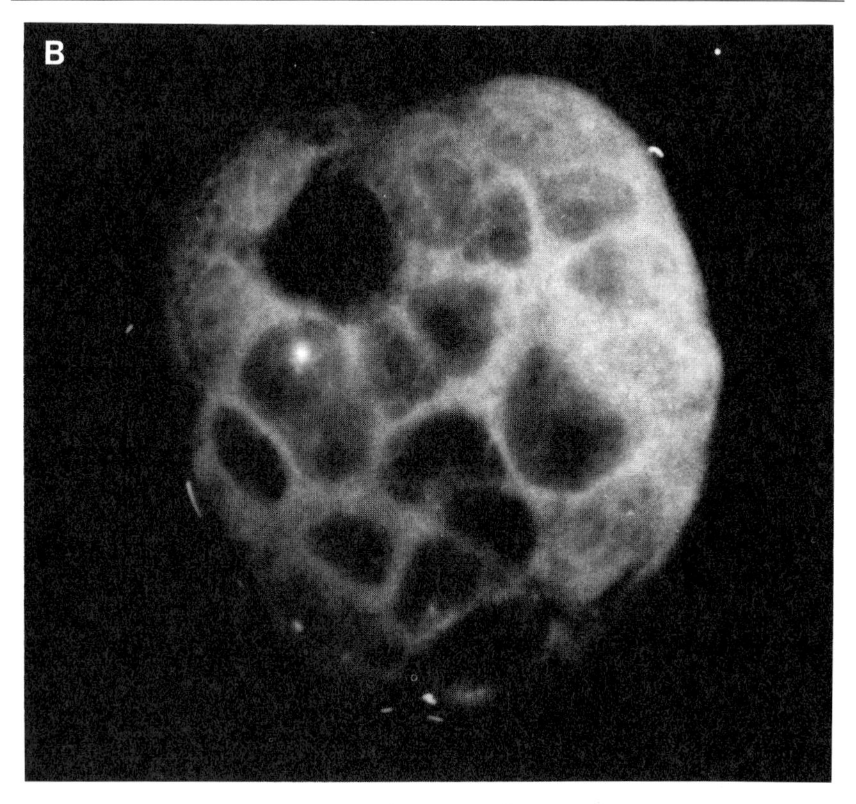

FIG. 2. (*continued*)

BUG8 and BUG11. This motif corresponds to a variable 6-bp sequence known as the core sequence of Tn*1545*. It is present at either end of the transposon and is involved in the excision process.[13]

Cloning and Sequencing of inl Region

The next step was to clone and to sequence the *inl* region from the parental strain EGD-SmR. Libraries of chromosomal *Hin*dIII and *Kpn*I DNA fragments were constructed from this strain. After transformation of *E. coli* MC1061, transformants containing fragments of the *inl* region were identified by colony hybridization, using as a probe the *Hin*dIII–*Acc*I fragment containing the *Listeria*–Tn*1545* junction fragment from the non-

[13] C. Poyart-Salmeron, P. Trieu-Cuot, C. Carlier, and P. Courvalin, *EMBO J.* **8,** 2425 (1989).

A

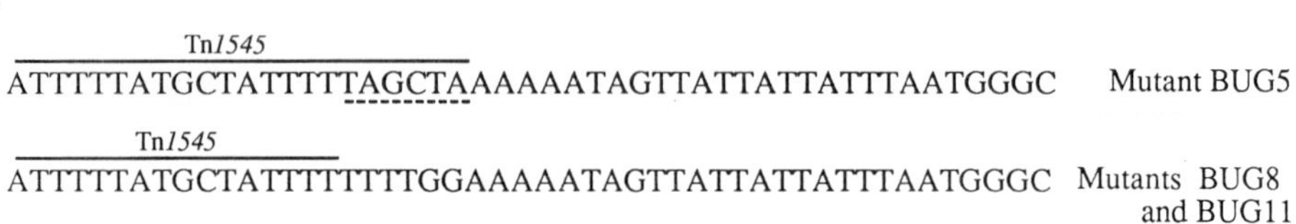

B

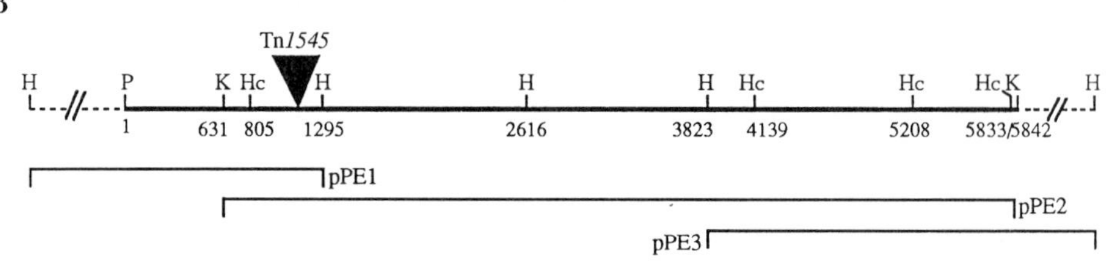

C

FIG. 3. Genetic organization of the *inl* region. (A) The nucleotide sequence of the *Listeria*–Tn*1545* left end junction in the noninvasive mutants. Each of the three mutants is the result of a single insertion in the same chromosomal region. In the mutant BUG5, the transposon Tn*1545* has inserted 6 bp downstream from the insertion site in mutant BUG8 and BUG11, and the invariable terminus of the transposon is prolonged by a 6-bp motif. This 6-bp stretch corresponds to the variable 6-bp sequence known as the Tn*1545* core sequence, which is present at either the left end or the right end of Tn*1545*.[13] Solid line: the invariable terminus of the left end of Tn*1545*; broken line: the core sequence of Tn*1545*. (B) Restriction-enzyme map of the *inl* region. The three inserts cloned into pUC vectors are indicated under the map, i.e., pPE1, pPE2, and pPE3. The determined 5,912-bp sequence (heavy solid line) extends from the *Pst*I site at position 1 to 70 bp downstream from the *Kpn*I site at position 5,842. The transposon insertions are between nucleotides 1109/1110 in mutants BUG8 and BUG11 and nucleotides 1115/1116 in mutant BUG5. Restriction sites: *Hinc*II (Hc), *HindIII (H)*, *Kpn*I (K), *Pst*I (P). (C) Open reading frame map of the *inl* region. Three significant open reading frames are apparent (arrows): ORF A (744 bp), *inlA* (2400), and *inlB* (1890 bp). Ω: putative transcription terminator.

invasive mutant BUG8. Two plasmids containing overlapping *Hin*dIII and *Kpn*I fragments from the *inl* region were obtained and used for DNA sequence determination (pPE1 and pPE2, respectively). A third plasmid (pPE3) was obtained by rescreening the *Hin*dIII library with the distal *Hinc*II fragment of plasmid pPE2.

The *inl* region contains three open reading frames (ORFs): ORFA, *inlA*, and *inlB* (see Fig. 3C). ORFA is 744 bp long and encodes a polypeptide of 248 amino acids that has 30% identity with the ribitol dehydrogenase from *Klebsiella aerogenes*,[14] the protein encoded by the *nodG* gene of *Rhizobium meliloti*,[15] and the *act*III gene from *Streptomyces coelicolor*.[16] 681 bp downstream from the end of ORFA, starts *inlA*, which is 2400 bp long (reexamination[17] of the sequence originally published[5] revealed a one base-pair omission, leading to an erroneous sequence of internalin, i.e. lack of signal sequence; new coordinates and numbers have been introduced in this article). Two-thirds of *inlA* is made up of intragenic repeats that form two regions, region A and region B. It is likely that these regions have been independently formed by internal tandem duplications of ancestral sequences *inlB* starts 85 bp after the end of *inlA* and the two are separated by a putative transcription terminator. *inlB* is 1890 bp long and resembles *inlA* (Fig. 4). There is a sequence similarity, which is strongest in the central region of the genes, suggesting a divergent evolution from a common central block. Like *inlA*, *inlB* contains a region (region A) of repeated units toward its 5′ end. Region A in *inlB* is similar to repeats 2–6 and 10 to 12 of region A in *inlA* but is shorter than that in *inlA* (8 and 15 repeats in *inlB* and *inlA* respectively).

inlA encodes a protein of 800 amino acids with a predicted M_r of 86.4 kd (Fig. 5). We named this molecule internalin because it is necessary for *L. monocytogenes* to invade Caco-2 cells (see below). Internalin is not similar to any known protein; however, it is structurally analogous to certain cell wall proteins, especially the members of the M-protein family from *Streptococcus pyogenes* (Fig. 6)[18–20] Like these proteins, internalin contains: 1) a signal peptide; 2) an N-terminal part made up of extensive internal repeats; 3) a C terminus made up of a proline- and glycine-rich region, followed by a sequence of 20 hydrophobic amino acids and a short charged tail; 4) and a hexapeptide conforming to the consensus sequence LPXTGX, located just before the C-terminal hydrophobic domain. This motif is found in the corresponding position in virtually all sequenced streptococcal cell wall proteins and is thought to be required for anchoring these molecules in the bacterial cell wall.[21] The two specific features of

[14] T. Loviny, P. M. Norton, and B. S. Hartley, *Biochem. J.* **230,** 579 (1985).
[15] F. Debellé and S. B. Sharma, *Nucleic Acids Res.* **14,** 7453 (1986).
[16] S. E. Hallam, F. Malpartida, and D. A. Hopwood, *Gene* **74,** 305 (1988).
[17] S. Dramsi, P. Dehoux, and P. Cossart, *Mol. Microbiol.* **9,** 1119 (1993).
[18] S. K. Hollingshead, V. A. Fischetti, and J. R. Scott, *J. Biol. Chem.* **261,** 1677 (1986).
[19] L. Miller, L. Gray, E. Beachey, and M. Kehoe, *J. Biol. Chem.* **263,** 5568 (1988).
[20] A. Mouw, E. Beachey, and V. Burdett, *J. Bacteriol.* **170,** 676 (1988).
[21] V. A. Fischetti, V. Pancholi, and O. Schneewind, *Mol. Microbiol.* **4,** 1603 (1990).

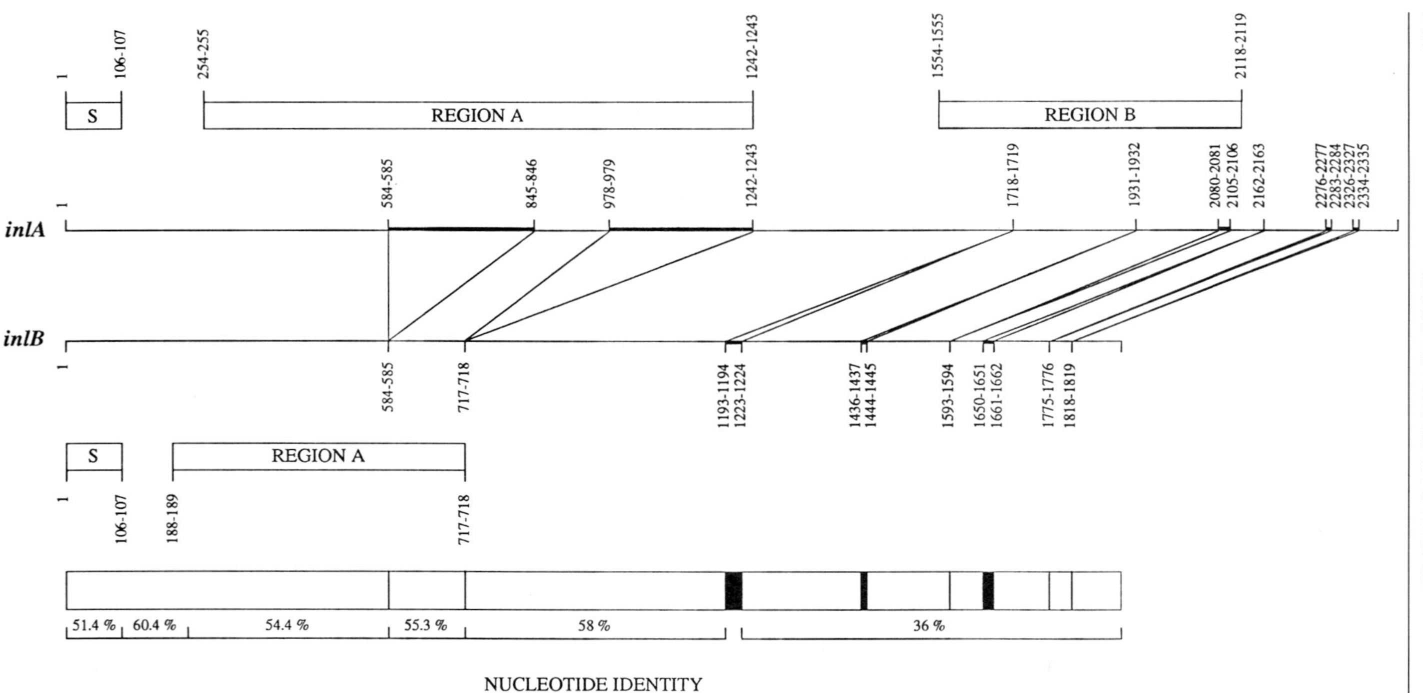

FIG. 4. Comparison of *inlA* and *inlB* sequences. The sequences of *inlA* and *inlB* were aligned to achieve maximal identity with a minimum gap score. The aligned regions are connected by continuous lines and the percentage identity between the sequences is indicated. Gaps are represented by heavy lines.

THE INTERNALIN

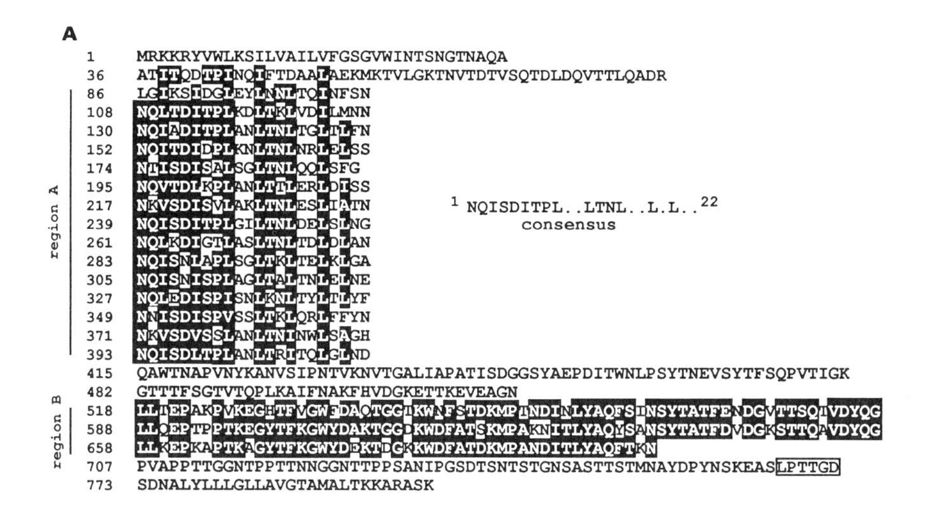

B

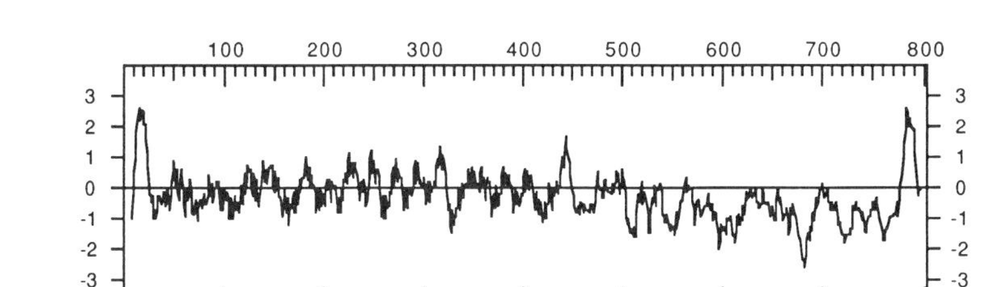

FIG. 5. Structure of internalin. (A) Primary structure. (B) Hydropathy plot. Hydrophilicity is indicated below the abscissa and hydrophobicity above the abcissa.

internalin are that: 1) It has a high content of threonine (13.30%) and serine (7.80%) residues; and 2) the repeat motif (region A) has regularly spaced leucine (or isoleucine) residues that might promote protein-protein interactions[22,23] *inlB* encodes a protein of 630 amino acids, with a predicted M_r of 71 kd. The *inlB* gene product resembles internalin in that it also

[22] A. Nose, V. Mahajan, and C. Goodman, *Cell* **70,** 553 (1992).

[23] D. D. Krantz, R. Zidovetzki, B. L. Kagan, and S. L. Zipursky, *J. Biol. Chem.* **266,** 16801 (1991).

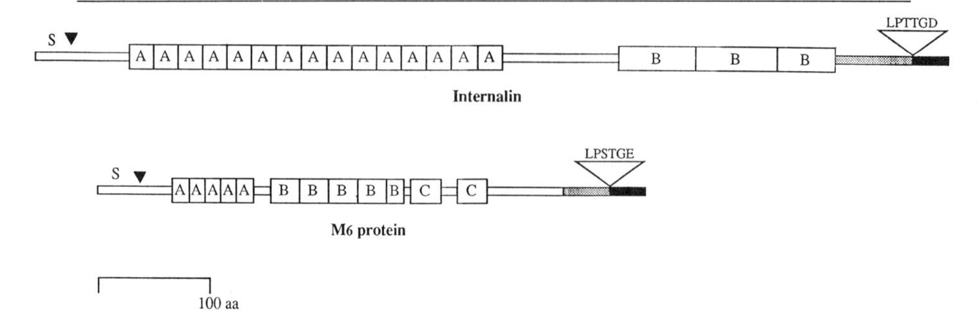

FIG. 6. Structural analogies between internalin and M6 protein. A, B, and C indicate different repeat regions within each molecule. M6 protein is from Hollingshead *et al.*[18] S, Signal peptide; ▼, signal sequence cleavage site; ▨, proline/glycine-rich region; ■, hydrophobic domain and charged tail.

contains a region of leucine-rich repeats in its N-terminal part. Unlike internalin, however, the *inlB* gene product: 1) is rich in lysine (10.30%) and arginine (4.10%) residues, giving a predicted basic *pI* for the molecule (10.11), in contrast to the acidic *pI* (4.55) calculated for internalin; 2) its C-terminal part does not exhibit the features of a cell wall attachment region. The function of the *inlB* gene product remains unknown.

Expression of inlA and inlB in Noninvasive Mutants

In each of the mutants BUG5, BUG8, and BUG11, the Tn*1545* element inserted 422 bp upstream of *inlA*, into a noncoding region. RNA slot-blot analysis was performed in the mutants and the parental strain to determine whether expression of *inlA* and *inlB* had been affected by the transposon insertion. RNAs were extracted after various times of bacterial growth (4 hr, 8 hr, and 29 hr), spotted on Immobilon N membranes (Millipore SA, Molsheim, France) with a slot-blot apparatus, and hybridized using the rapid hybridization system of Amersham (Amersham Corp., Buckinghamshire, England). There was no detectable transcription of either *inlA* and *inlB* in the noninvasive mutants.[5]

Demonstration of Role of *inlA*

The role of *inlA* in the entry process was demonstrated in two ways: A plasmid carrying *inlA* alone was shown to be sufficient (1) to restore the ability of the noninvasive mutants to enter the Caco-2 cell line and (2) to transform the normally noninvasive species *Listeria innocua* into an invasive-competent species.

Various fragments of the *inl* region containing either ORFA (plasmid pGM1); a part of ORFA, *inlA*, and most of *inlB* (plasmid pGM2); a part of ORFA and a part of *inlA* (plasmid pGM3); or *inlA* only (plasmid pGM4) were introduced into *L. monocytogenes* BUG5 and *L. innocua* CLIP 11254 (J. Rocourt, Institut Pasteur, Paris, France), using the *E. coli–Listeria* shuttle vector pAT28.[24] Plasmid pAT28 confers resistance to spectinomycin to both gram-negative and gram-positive bacteria. It is stably maintained in *E. coli* and *Listeria*, owing to the presence of the origins of replication of pUC and of the broad-host-range enterococcal plasmid pAMβ1. It contains the origin of transfer of the IncP plasmid RK2 and can be mobilized by self-transferable IncP plasmids such as pRK212.1. Plasmid pAT28 derivatives were introduced into *E. coli* strain HB101 (pRK212.1)[25] by transformation and transferred from *E. coli* to *Listeria* by conjugation. Transconjugants were selected on tryptic soy agar containing spectinomycin (final concentration 60 mg/liter) and nalidixic acid (final concentration 50 mg/liter).

The introduction of plasmids pGM2 and pGM4 into *L. monocytogenes* BUG5 almost completely restored the ability of this mutant to enter the Caco-2 cell line. Moreover, plasmids pGM2 and pGM4 conferred the invasive phenotype onto *L. innocua*, as evidenced by gentamicin survival assay and electron microscopic observation; however, the *L. innocua* strains harboring these plasmids invaded cultured cells with an efficiency that was slightly lower than that of *L. monocytogenes* (0.5% vs 2% in the gentamicin survival assay).

The *L. monocytogenes* DNA fragment in pGM4, the smallest conferring the invasive phenotype, contains the *inlA* gene and the adjacent upstream 895 bp and downstream 204 bp. Therefore, *inlA* appears to be the gene responsible for the acquired invasive phenotype of *L. monocytogenes* BUG5 and of the recombinant *L. innocua* strains. It can also be concluded that loss of invasive capacity in mutants BUG5, BUG8, and BUG11 is due to loss of expression of *inlA*.

Nonconjugative vectors such as pMK4[26] or pGK12[27] can be used for complementation experiments or expression in *L. innocua*. These plasmids can be introduced by electroporation.[8] pGK12 offers the advantage of a single origin functional in both *E. coli* and *Listeria*.

[24] P. Trieu-Cuot, C. Carlier, C. Poyart-Salmeron, and P. Courvalin, *Nucleic Acids Res.* **18,** 4296 (1990).

[25] D. H. Figurski and D. R. Helinski, *Proc. Natl. Acad. Sci. U.S.A.* **76,** 1648 (1979).

[26] M. Sullivan, R. E. Yasbin, and F. E. Young, *Gene* **29,** 21 (1984).

[27] J. Kok, J. B. M. van der Vossen, and G. Venema, *Appl. Environ. Microbiol.* **48,** 726 (1984).

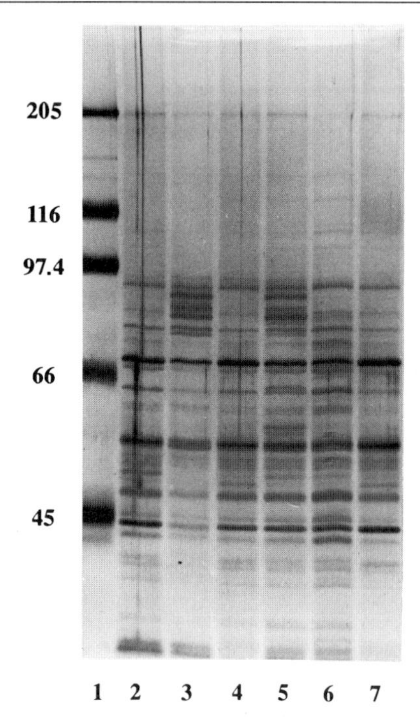

FIG. 7. Surface proteins from *Listeria innocua* strains harboring various fragments of the *inl* region. *L. innocua* surface proteins were prepared by SDS extraction, separated on 7.5% SDS–polyacrylamide gels, and silver-stained. Lane 1: molecular weight standards; lane 2: *L. innocua* CLIP 11254; other lanes: *L. innocua* CLIP 11254 harboring pGM4 (lane 3); pGM3 (lane 4); pGM2 (lane 5); pGM1 (lane 6); and pAT28 (lane 7). The presence of *inlA* results in the appearance of 75–90 kD polypeptides (lane 3 and 5), and of inlB (lane 5) in the appearance of additional 50–60 kD polypeptides.

Expression of Internalin in *Listeria innocua*

Surface proteins from *L. innocua* strains carrying the various pAT28 derivatives (plasmid pGM1, pGM2, pGM3, or pGM4) were prepared by sodium dodecyl sulfate (SDS) extraction, using conditions (1% SDS for 5 min at 22°) under which the bacteria are not lysed.[28] The extracted proteins were then analyzed by SDS–polyacrylamide gel electrophoresis. As shown in Fig. 7, the extracts from strains carrying pGM2 or pGM4 had several additional bands in the range of 75-90 kDa that were absent from strains carrying pGM1 or pGM3. A multiple banding pattern has

[28] C. Kocks, E. Gouin, M. Tabouret, P. Berche, H. Ohayon, and P. Cossart, *Cell* (*Cambridge, Mass.*) **68,** 521 (1992).

previously been observed in the case of proteins containing the consensus LPXTGX sequence, such as the M protein.[29–31] Thus, these multiple bands probably all result from the expression of the internalin gene. This was confirmed by immunoblotting experiments using antisera raised against synthetic peptides from the deduced internalin sequence (not shown). Additional 50- to 60-kD polypeptides were present in extracts from strains carrying pGM2. Whether or not these bands correspond to *inlB* gene product has not been demonstrated.

inlA as Part of Gene Family

The internal 1.2-kb *Hin*dIII fragment of *inlA* was used to probe *Hin*dIII-digested chromosomal DNA from EGD-SmR and from an unrelated *L. monocytogenes* clinical isolate, strain LO28 (serovar 1/2c)[32] (data not shown). This fragment covers most of the 3′ half of the gene. Southern blot hybridization was performed under conditions of high and low stringency. For high-stringency hybridization, prehybridization and hybridization were carried out using the rapid hybridization system of Amersham, and the filters were washed twice for 30 min at 65° in 0.7× SSC, 0.1% SDS. For low-stringency hybridization, prehybridization and hybridization were carried out in 30% formamide at 37°, and the filters were washed twice for 30 min in 0.1× SSC, 0.1% SDS at 42°. Under stringent conditions, a single 1.2-kb *Hin*dIII fragment was recognized by the intragenic probe in strain EGD-SmR. A single band with the same apparent size was also detected in strain LO28. In contrast, under conditions of low stringency, several bands were detected in the two strains tested, with apparent sizes ranging from 1.2 to 5 kb. The patterns were different in the two strains, and in each strain, the various hybridization signals were of differing strength. The results of these Southern blotting experiments suggest that *inlA* belongs to a gene family, probably created by gene duplication and subsequent sequence diversification. Whether the *inlA*-related sequences are expressed is unknown, as are the functions of any possible gene products.

Acknowledgments

We thank Marc Tabouret for help in the surface protein preparation.

[29] V. A. Fischetti, K. F. Jones, B. N. Manjula, and J. R. Scott, *J. Exp. Med.* **159,** 1083 (1984).
[30] V. A. Fischetti, K. F. Jones, and J. R. Scott, *J. Exp. Med.* **161,** 1384 (1985).
[31] E. J. Haanes and P. P. Cleary, *J. Bacteriol.* **171,** 6397 (1989).
[32] M. F. Vicente, F. Baquero, and J. C. Perez-Diaz, *FEMS Microbiol. Lett.* **30,** 77 (1985).

[42] Use of *Staphylococcus aureus* Coated with Invasin Derivatives to Assay Invasin Function

By SUSANNAH RANKIN, GUY TRAN VAN NHIEU, and RALPH R. ISBERG

Introduction

The genetic analysis of outer membrane protein function in gram-negative bacteria may be hampered by the inability to achieve proper localization of potentially informative mutant proteins. The technique described here allows functional analysis of an outer membrane protein with mutant derivatives which cannot be properly expressed on the cell surface. The proteins are purified and subsequently attached to the bacterial cell surface, which allows easy manipulation of the final protein coating concentration. The technique is therefore particularly useful in studying the relative ability of a surface-exposed protein that mediates binding to and uptake by eukaryotic cells.

The *inv* gene of *Yersinia pseudotuberculosis* codes for a 986-amino-acid outer membrane protein called invasin,[1] which confers on *Escherichia coli* K12 the ability to enter mammalian cells.[2] Invasin was subsequently shown to mediate bacterial uptake by binding to a variety of β_1-integrins on the mammalian cell surface.[3] A series of monoclonal antibodies against invasin was used to demonstrate that the C terminus of the protein was surface exposed and involved in integrin binding.[4] To define more precisely the integrin binding domain of invasin, various *malE–inv* gene fusions were constructed in which the *E. coli* gene encoding maltose-binding protein (MBP) is fused at its 3′ end to invasin-derived sequences. The resulting hybrid proteins were purified on amylose resin via their maltose-binding protein moiety, immobilized on plastic dishes, and tested for their ability to mediate mammalian cell binding. The results indicated that the C-terminal 192 amino acids of invasin are necessary and sufficient to mediate the attachment of mammalian cells.[5] Determining whether this cell binding domain was also sufficient to promote bacterial entry into host cells proved problematic. It remained possible that an additional domain of invasin could contain a signal for uptake by mammalian cells,

[1] R. R. Isberg, D. L. Voorhis, and S. Falkow, *Cell* (*Cambridge, Mass.*) **50,** 769 (1987).
[2] R. R. Isberg and S. Falkow, *Nature* (*London*) **317,** 262 (1985).
[3] R. R. Isberg and J. M. Leong, *Cell* (*Cambridge, Mass.*) **60,** 861 (1990).
[4] J. M. Leong, R. S. Fournier, and R. R. Isberg, *Infect. Immun.* **59,** 3424 (1991).
[5] J. M. Leong, R. S. Fournier, and R. R. Isberg, *EMBO J.* **9,** 1979 (1990).

but internal and amino-terminal deletions in the *inv* gene resulted in protein products that were not expressed on the cell surface, making assessment of their ability to promote cellular entry impossible.[5]

The apparent obstacles to this genetic analysis, as well as the availability of the MBP–invasin hybrid proteins generated previously, led to the design of a new protocol for the analysis of invasin function.[6] By attaching MBP–Inv hybrid proteins to the surface of *Staphylococcus aureus,* an organism that is not ordinarily able to enter mammalian cells, it is possible to test various C-terminal fragments of invasin for their ability to mediate bacterial uptake using the previously standardized gentamicin protection assay. This approach could be broadly applicable in the analysis of potential adhesins or invasins from a wide range of unrelated organisms.

Figure 1 illustrates the way *S. aureus* is used to test the ability of various MBP–invasin hybrid proteins to mediate bacterial entry into mammalian cells. Briefly, live *S. aureus,* which expresses protein A on its surface, is incubated with anti-MBP serum. Protein A binds to the Fc region of IgG, and the result is the directional attachment of the antibodies on the bacterial cell surface, with the antigen binding sites facing outward. After a wash step, the bacteria are incubated with the purified MBP–invasin hybrid protein to be tested. The hybrid is attached via the anti-MBP serum, leaving the invasin-derived amino acids decorating the surface of the bacterium (see Fig. 1). The high affinity of these interactions results in a stable construct which can then be used in a standard bacterial internalization assay.

Purification of Maltose-Binding Protein–Invasin Hybrid Proteins

Maltose-binding protein–invasin hybrid proteins are purified in a single step by affinity chromatography using an amylose column. Although the procedure can be rapid and simple for most small hybrids, the purification of certain large hybrids can prove difficult because of contamination with degradation products. In the case of MBP–Inv hybrids, degradation of hybrid proteins can be limited by harvesting the hybrids from a protease-deficient strain. Also, after affinity chromatography, full-length hybrids can be separated from the degradation products by ammonium sulfate precipitation.

Several types of plasmid vectors designed for generating MBP hybrid proteins are commercially available. These vectors contain the *malE* gene, which codes for MBP, followed by a multiple cloning site which allows the introduction of coding sequence at the 3′ end of the *malE* gene. These

[6] S. Rankin, R. R. Isberg, and J. M. Leong, *Infect. Immun.* **60(9),** 3909 (1992).

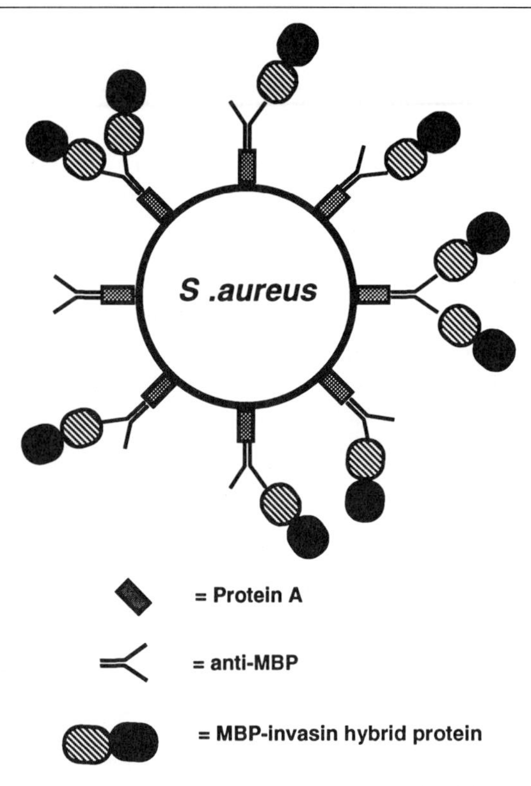

FIG. 1. Use of *Staphylococcus aureus* to test the ability of MBP–invasin hybrid protein to mediate mammalian cell entry. *S. aureus* is coated first with polyclonal anti-MBP antibodies, and then with purified MBP invasin hybrid proteins. The resulting constructs are used in a standard invasion assay.

fusions are generally under P_{lac} control and thus can be induced with isopropyl β-D-thiogalactopyranoside (IPTG). In some vectors, the signal sequence of the *malE* gene has been removed to allow better recovery of the fusion protein.

Preparation of Bacterial Extract

The choice of cell lysis procedure is dependent on the size of the bacterial sample. For up to 500-ml cultures, a sonication procedure, described below, is convenient and can yield as much as 1 to 2 mg of purified protein. For large-scale purification from extracts corresponding to several liters of bacterial culture, lysis by French press is advisable.

Materials and Reagents

The MBP–invasin hybrid proteins used to coat *S. aureus* are produced by cloning fragments of the invasin gene into the plasmid pCGX807 (New England Biolabs, Beverly, MA) in frame with the *malE* gene. This vector still retains the *malE* signal sequence and allows expression of the fused hybrids under the control of the IPTG-inducible *lac* promoter.

For MBP hybrid protein purification, the plasmid constructs are introduced into *E. coli* 71-18 [F′*proAB*+ *lacI*^{q} *lacZ*ΔM15/F$^-$Δ(*lac-proAB*) *thi supE*].[7] For the MBP–Inv proteins that show degradation products, the plasmid constructs are introduced into the protease-deficient *E. coli* SR2 [F′*proAB*+ *lacI*^{q} *lacZ*ΔM15/F$^-$Δ(*lac-proAB*) *lpp5508 degP*::Tn*5* Δ(*phoA*)].[8] A 100 m*M* stock solution of IPTG is made in sterile water. Extraction buffer is 25 m*M* HEPES (pH 7.0) containing 1 m*M* phenylmethylsulfonyl fluoride (PMSF), 1 μg/ml pepstatin, and 2.5 μg/ml leupeptin. The 2× YT medium contains, per liter, 16 g tryptone, 10 g yeast extract, and 5 g NaCl.

Procedure

1. A flask containing 500 ml of 2× YT is inoculated with 5 ml of a fresh overnight culture and grown at 30° until midlogarithmic phase (OD_{550} ~ 0.4).

2. The inducer IPTG is added at a final concentration of 1 m*M*, and the bacteria are grown for another 2 hr at 30° to allow expression of the fusion protein.

3. The flask is chilled on ice for 10 min. The bacteria are collected by centrifugation for 10 min at 5000 g at 4° and resuspended in half the initial culture volume of extraction buffer. All subsequent steps are carried out at 4° unless otherwise stated.

4. The bacteria are again collected by centrifugation at 5000 g for 10 min, resuspended in 10 ml of extraction buffer, and transferred to a 30-ml tube suitable to allow good cooling of the sample during the sonication procedure.

5. The bacterial suspension is subjected to sonication using a cell disruptor (Branson Sonifier 200, Branson Ultrasonics, Danbury, CT) for 5 min at 30% duty cycle, energy output level 7. It is critical that the tube containing the sample be immersed in an ice bath throughout the sonication procedure to avoid overheating of the sample, which can result in loss of protein activity.

[7] C. Yanisch-Perron, J. Viera, and J. Messing, *Gene* **33,** 103 (1985).

[8] K. L. Strauch and J. Beckwith, *Proc. Natl. Acad. Sci. U.S.A.* **85,** 1576 (1988).

6. The extract is cleared by centrifugation for 1 hr at 45,000 rpm in a Beckman 50TI rotor. The cleared extract is processed immediately for affinity chromatography.

Affinity Chromatography

Materials and Reagents. Wash buffer in 25 m*M* HEPES, pH 7.0, containing 1 m*M* PMSF. Elution buffer is 25 m*M* HEPES, pH 7.0, containing 1 m*M* PMSF and 10 m*M* maltose. Amylose resin is purchased from New England Biolabs and prepared according to the manufacturer's instructions.

Procedure

1. Two milliliters of prepared amylose resin is poured in a disposable column (Model 737 1005, Bio-Rad, Richmond, CA). The resin is washed with 10 column volumes of elution buffer, followed by 10 column volumes of wash buffer and 1 column volume of extraction buffer.
2. The cleared extract from step 6 above is loaded on the amylose resin and allowed to pass through the resin over a 2-hr period without allowing the column to run dry.
3. The resin is washed with 10 column volumes of washing buffer.
4. MBP–hybrid proteins are eluted with elution buffer and serial 400-μl fractions are collected. The hybrid protein is generally eluted in the first three fractions.
5. Ten microliters of the eluted fractions is analyzed for purity by sodium dodecyl sulfate-polyacrylamide gel electrophoresis (SDS–PAGE) and Coomassie staining.[9]

Ammonium Sulfate Precipitation

1. Fractions containing the eluted MBP–Inv hybrid proteins are pooled and transferred to a 10-ml beaker containing a magnetic stir bar.
2. Ammonium sulfate is added slowly to the sample, with constant stirring, until the desired final concentration is reached. The ideal final concentration of ammonium sulfate depends on the protein being purified. In the case of MBP–Inv hybrid proteins, a final concentration of 50% saturation (2.9 g ammonium sulfate/10 ml protein solution) is usually sufficient to precipitate the large hybrids. Most of the degradation products that contain intact MBP remain soluble in as much as 75% ammonium sulfate saturation.

[9] U. K. Laemmli, *Nature (London)* **227,** 680 (1970).

3. The precipitates are pelleted by centrifugation at 5000 *g* for 30 min and the supernatants are discarded.

4. The protein pellet is resuspended in at least one-tenth of the original volume of wash buffer, and the suspension is subjected to three rounds of dialysis against wash buffer.

5. Protein preparations not being used immediately are stored at −70°. Some hybrids appear to be more stable to freezing and thawing than others, and in all cases the number of freeze–thaw cycles should be kept to a minimum.

Preparation of *Staphylococcus aureus* Coated with Maltose-Binding Protein–Invasin

Materials and Reagents. The Cowan 1 strain of *S. aureus* is used in these experiments. This strain produces large amounts of protein A and is therefore particularly useful in cases where a high surface concentration of hybrid protein is desired. *S. aureus* is grown in Penassay R broth (Difco). Anti-MBP serum may be purchased from New England Biolabs. Phosphate-buffered saline (PBS) is 20 m*M* KPO_4 (pH 7.4), 150 m*M* NaCl.

Procedure

1. *Staphylococcus aureus* is grown to midlogarithmic phase (OD_{600} = 0.5–0.8) in Penassay R broth. The cultures are cooled on ice and collected by centrifugation for 10 min at 8000 *g* at 4°. The supernatant is discarded and the bacteria are washed with one-half volume of cold phosphate-buffered saline (PBS). After pelleting again, the bacteria are resuspended in PBS at approximately 5×10^{10} cells/ml.

2. Approximately 5×10^9 bacteria are diluted in 1 ml of PBS containing anti-MBP serum. The final surface concentration of hybrid protein is easily manipulated by varying the serum concentration in this first coating step. Samples are prepared using different antiserum concentrations; the highest concentration used is 0.5% antiserum, with serial twofold dilutions down to 0.001%. Anti-MBP serum is mixed with preimmune serum so that the total serum concentration is constant (0.5% total serum). Control samples are incubated with preimmune serum in place of anti-MBP serum. The samples are mixed gently for 1 hr at 4°.

3. To wash away serum, the samples are centrifuged briefly in a microcentrifuge at room temperature (30 sec at 14,000 rpm), the supernatant is discarded, and the bacteria are resuspended in 1 ml PBS by gentle pipetting. This wash step is repeated two more times.

4. The bacteria are resuspended in 0.5-ml final volume of 10–50 μg/ml hybrid protein in PBS and incubated for 1 hr at 4° with gentle

mixing. These concentrations of hybrid protein are usually sufficient to saturate the available antibodies on the cell surface. The bacteria are washed three times with PBS, as above, resuspended in 0.5 ml PBS, transferred to a clean tube, and used in a gentamicin protection assay (see below).

It should be noted that it is possible to perform this assay using a purified protein that is not a hybrid and a monoclonal antibody that does not interfere with the activity of that protein. For example, purified invasin fragments can be attached to the surface of *S. aureus* using a monoclonal antibody that does not block invasin binding to mammalian cells.

Determining Efficiency of Entry Conferred by Hybrid Proteins: Invasion Assay

Because live bacterial cells are used in this procedure, the efficiency of bacterial uptake by mammalian cells conferred by various hybrids can be assessed using the previously standardized technique of gentamicin protection (see [28]).[10] This technique is based on the observation that bacteria that are inside animal cells are protected from the antibiotic gentamicin.

Materials and Reagents. HEp-2 cells (ATCC) are cultured in RPMI 1640 (Irvine Scientific, Santa Ana, CA) supplemented with 5% (v/v) newborn calf serum (Hyclone, Logan, UT). Binding buffer is RPMI 1640 supplemented with 0.4% bovine serum albumin (BSA) and 10 m*M* HEPES, pH 7.0. A 10 mg/ml stock solution of gentamicin sulfate is made in sterile water and diluted prior to use in binding buffer.

Procedure

1. After *S. aureus* cells have been coated with hybrid protein, viable bacteria are titered by plating dilutions on Luria–Bertani (LB) agar. Generally it is assumed that about one-half of the viable counts have been lost during the various wash steps, and therefore 10 μl of a 1 : 10^6 dilution will result in a quantity of colony-forming units (cfu) that can be counted with ease.
2. Approximately 10^7 bacteria coated with hybrid protein are added to 2×10^5 subconfluent HEp-2 cells and incubated for 90 min at 37° in binding buffer in a 5% CO_2 atmosphere.
3. The monolayers are washed three times with PBS and treated with 50 μg/ml gentamicin in binding buffer at 37° for 15 min to kill extracellular bacteria.

[10] T. Vesikari, J. Bromirska, and M. Maki, *Infect. Immun.* **36,** 834 (1982).

4. The monolayers are gently washed five times with PBS, and colony-forming units that remain (following the gentamicin treatment) are determined as follows. The HEp-2 cells are lysed with 0.25 ml sterile H_2O. Lysis is completed by the addition of 0.25 ml 1% Triton X-100. The cells are dispersed by gently pipetting up and down, and 10 μl is immediately plated on LB agar. A 1 : 100 dilution is made in PBS and 10 μl of this is also plated. The efficiency of uptake is defined as the number of colony-forming units surviving gentamicin treatment relative to the number of bacteria originally added to the monolayer, as determined in step 1, above.

Figure 2 shows the results of a typical uptake assay in which a variety of MBP–Inv hybrids were tested for their ability to mediate cellular entry. The uptake efficiency is expressed as percentage survival of input bacteria following gentamicin treatment. The apparent internalization of bacteria, as determined by the gentamicin protection assay, may be confirmed by transmission electron microscopy. The micrograph in Fig. 3 shows the results of incubating HEp-2 cells with *S. aureus* cells that have been coated with MBP–Inv479. The large number of bacteria inside the mammalian cells is consistent with the high efficiency of entry conferred by MBP–Inv479 as indicated by the experiment shown in Fig. 2.

Quantitation of Hybrid Protein on Bacterial Cell Surface

It is possible, using this technique, to estimate the number of invasin molecules that are required to mediate the entry process. This is done by measuring the amount of hybrid protein associated with the bacteria following the coating procedure and normalizing this amount to the number of bacteria in the sample. Quantitation of hybrid protein associated with the bacteria can be achieved by immunoprobe analysis or by fluorometric analysis as described below.

Western Blot Analysis

1. After the bacteria are coated with hybrid protein, approximately one-half of the preparation (about 5×10^8 bacteria) is spun down and resuspended in Laemmli sample buffer.[9]

2. The samples are boiled for 2 min and spun in a table-top microcentrifuge for 15 min at 14,000 rpm at 4° to pellet DNA. The pellet can be removed with a clean toothpick.

3. Proteins are then subjected to sodium dodecyl sulfate–polyacryl-

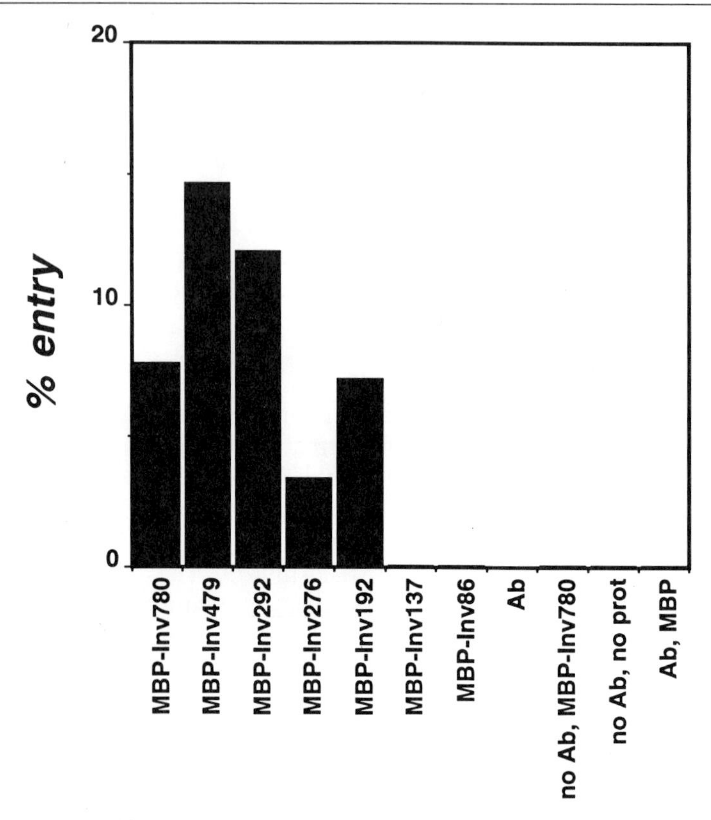

FIG. 2. Entry efficiency conferred on *Staphylococcus aureus* by various invasin derivatives. *S. aureus* cells coated with different MBP–invasin hybrids were tested for their ability to enter animal cells in a standard invasion assay. The number of invasin-derived amino acids in each hybrid is indicated by the name of the protein, shown below each bar (e.g., MBP–Inv292 contains 292 invasin-derived amino acids). Ab: no hybrid protein was used; no Ab, MBP–Inv780: hybrid protein was used, but the bacteria were not first incubated with anti-MBP; MBP: purified MBP was used instead of hybrid protein.

amide gel electrophoresis using standard techniques[9] and transferred to Immobilon filters (polyvinylidene fluoride, Millipore, Bedford, MA).

4. The filter is probed with a monoclonal mouse anti-invasin antibody and a rabbit anti-mouse secondary antibody conjugated to alkaline phosphatase (Bio-Rad, Richmond, CA). Immunoblots are processed as previously described.[5] The amount of hybrid protein is determined by scanning densitometry, comparing sample lanes with lanes containing serial dilutions of known amounts of purified protein.

[11] B. A. Sandow, W. N. B., R. L. Norman, and R. M. Brenner, *Am. J. Anat.* **156,** 15 (1979).

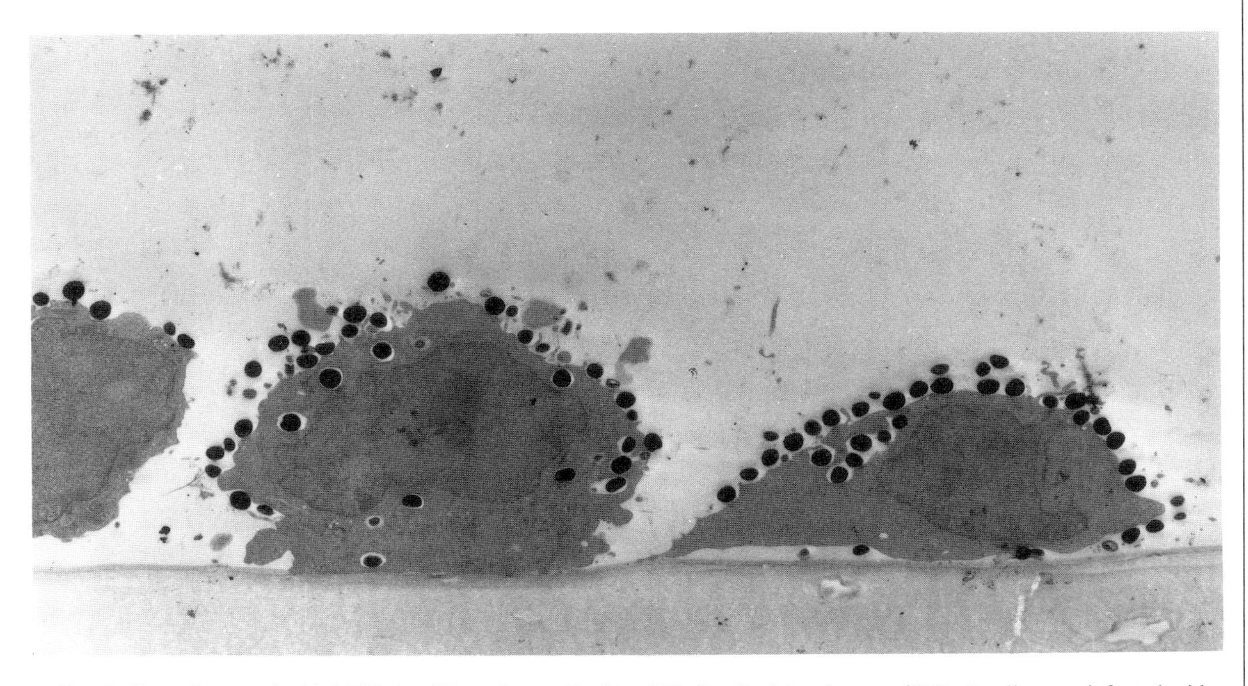

FIG. 3. Bacteria coated with MBP–Inv479 are internalized by HEp-2 cells. Monolayers of HEp-2 cells were infected with *Staphylococcus aureus* cells coated with MBP–Inv479 as described. After being washed several times with PBS, the monolayer was fixed with modified Brenner's broth[11] and embedded in Epon 812 (Polysciences, Gaithersburg, MD). Large numbers of MBP–Inv479-coated bacteria are seen inside the mammalian cells. Bacteria coated with MBP or left uncoated are not seen inside the mammalian cells (not shown).[11]

Fluorometric Analysis

1. The protein to be labeled is dialyzed overnight against PBS at 4° and placed in 1.5-ml plastic centrifuge tube.

2. A 10 mg/ml solution of carboxytetramethylrhodamine succinimidyl ester (CTM-rhodamine) (Molecular Probes, Eugene, OR) is made in dimethylformamide just before use and is kept on ice wrapped in aluminum foil.

3. The CTM-rhodamine is added to the protein solution at a final molar ratio of 10 : 1 (label : protein). The tube is wrapped in aluminum foil and kept on ice for 15 min, during which time the tube is inverted several times by hand.

4. The unreacted ester is inactivated by adding ethanolamine to a final concentration of 50 m*M*.

5. The labeled protein is isolated from free label using a Bio-Gel P-10 column (Bio-Rad). The fluorescence activity of the hybrid is determined using a microfluorometer (Titertek, Flow Labs, Inc.) and the protein concentration is determined by the Bradford assay (Bio-Rad protein assay). The specific activity (fluorescence units per milligram of protein) is calculated.

6. *S. aureus* cells are coated with labeled protein and used in an uptake assay as described above. The amount of labeled hybrid protein per bacterium in a given sample is determined by measuring the fluorescence of a portion of that sample and normalizing it to the number of colony-forming units in that sample.

Conclusion

The technique described here provides a powerful and sensitive means of measuring the ability of various hybrid proteins to mediate bacterial entry into mammalian cells. Because the assay does not rely on cumbersome and time-consuming microscopic analysis, many different hybrids can easily be screened. The ease with which MBP–hybrid proteins can be purified makes the study of a large number of hybrids feasible. For some inherently "sticky" proteins it is apparently even possible to coat *S. aureus* without the use of antibodies.[12]

Using the technique described here, we have been able both to assess the phenotype conferred by various invasin derivatives that are not expressed on the bacterial cell surface and to measure reproducibly the number of invasin molecules required to mediate cellular entry.[6] These

[12] E. Leininger, C. A. Ewanowich, A. Bhargava, M. S. Peppler, J. G. Kenimer, and M. J. Brennan, *Infect. Immun.* **60(6)**, 2380 (1992).

questions were impossible to address using more conventional genetic or biochemical analyses.

Acknowledgments

This work was supported by the Center for Gastroenterology Research on Absorptive and Secretory Processes, USPHS Grant 1-P30 DK-39428 awarded by the National Institute of Diabetes and Digestive and Kidney Diseases, and Grant RO1 AI-32538 from the National Institutes of Health. S.R. is a Mortimer Sackler Scholar.

Author Index

Numbers in parentheses indicate footnote reference numbers and indicate that an author's work is referred to although the name is not cited in the text.

A

B

C

D

E

G

H

K

L

M

N

O

P

Q

R

S

U

V

W

X

Y

Z

Subject Index

A

C

M

N

O

P

Q

R

S

T

U

V

W

Y

ISBN 0-12-182137-4

90038

9 780121 821371